SPRINGER PROCEEDINGS IN PHYSICS 121

SPRINGER PROCEEDINGS IN PHYSICS

106 **Modern Trends in Geomechanics**
Editors: W. Wu and H. S. Yu

107 **Microscopy of Semiconducting Materials**
Proceedings of the 14th Conference, April 11–14, 2005, Oxford, UK
Editors: A.G. Cullis and J.L. Hutchison

108 **Hadron Collider Physics 2005**
Proceedings of the 1st Hadron Collider Physics Symposium, Les Diablerets, Switzerland, July 4–9, 2005
Editors: M. Campanelli, A. Clark, and X. Wu

109 **Progress in Turbulence II**
Proceedings of the iTi Conference in Turbulence 2005
Editors: M. Oberlack, G. Khujadze, S.Guenther, T.Weller,M. Frewer, J. Peinke, S. Barth

110 **Nonequilibrium Carrier Dynamics** in Semiconductors Proceedings of the 14th International Conference, July 25–29, 2005, Chicago, USA
Editors: M. Saraniti, U. Ravaioli

111 **Vibration Problems ICOVP 2005**
Editors: E. Inan, A. Kiris

112 **Experimental Unsaturated SoilMechanics**
Editor: T. Schanz

113 **Theoretical and Numerical Unsaturated Soil Mechanics**
Editor: T. Schanz

114 **Advances inMedical Engineering**
Editor: T.M. Buzug

115 **X-Ray Lasers 2006**
Proceedings of the 10th International Conference, August 20–25, 2006, Berlin, Germany
Editors: P.V. Nickles, K.A. Janulewicz

116 **Lasers in the Conservation of Artworks**
LACONA VI Proceedings, Vienna, Austria, Sept. 21–25, 2005
Editors: J. Nimmrichter,W. Kautek, M. Schreiner

117 **Advances in Turbulence XI**
Proceedings of the 11th EUROMECH European Turbulence Conference, June 25–28, 2007, Porto, Portugal
Editors: J.M.L.M. Palma and A. Silva Lopes

118 **The StandardModel and Beyond**
Proceedings of the 2nd International Summer School in High Energy Physics, Muğla, 25–30 September 2006
Editors: M. SeriT. Aliev, N.K. Pak, and M. Serin

119 **Narrow Gap Semiconductors 2007**
Proceedings of the 13th International Conference, 8-12 July, 2007, Guildford, UK
Editors: B. Murdin, S. Clowes

120 **Microscopy of Semiconducting Materials 2007**
Proceedings of the 15th Conference, 2-5 April 2007, Cambridge, UK
Editors: A.G. Cullis, P.A. Midgley

121 **Time Domain Methods in Electrodynamics**
A Tribute to Wolfgang J. R. Hoefer
Editors: Peter Russer, and Uwe Siart

For other titles published in this series, go to www.springer.com/series/361

Peter Russer · Uwe Siart (Eds.)

Time Domain Methods in Electrodynamics

A Tribute to Wolfgang J. R. Hoefer

With 291 Figures and 21 Tables

Springer

Editors

Prof. Dr. Peter Russer
TU München
Fak. Elektro- und
Informationstechnik
LS Hochfrequenztechnik
Arcisstr. 21
80333 München
Germany
russer@ei.tum.de

Dr. Uwe Siart
TU München
Fak. Elektro- und
Informationstechnik
LS Hochfrequenztechnik
Arcisstr. 21
80333 München
Germany

ISBN: 978-3-540-68766-5 e-ISBN: 978-3-540-68768-9

Library of Congress Control Number: 2008928454

Cover design: WMXDesign GmbH

Printed on acid-free paper

9 8 7 6 5 4 3 2 1

springer.com

All stream that's roaring by
Came out from a needle's eye;
Things unborn, things that are gone,
From needle's eye still goad it on.

William Butler Yeats

Preface

On May 16th 2007 the Faculty of Electrical Engineering and Information Technology of the Technische Universität München bestowed the degree of the doctor honoris causa to Wolfgang J.R. Hoefer for Extraordinary achievements in the theory of electromagnetic fields. On this special occasion a symposium on Time Domain Methods in Modern Engineering Electrodynamics has been held in honor of Professor Wolfgang J.R. Hoefer at the Technische Universität München on May 16 and 17, 2007. The symposium topic was focused on the main area of research of Wolfgang J.R. Hoefer, the time domain methods in computational electromagnetics especially the transmission line matrix method and its applications. The transmission line matrix method has been developed and first published by Johns and Beurle in 1971. In the past 20 years Wolfgang Hoefer has given exemplary contributions to the development of the transmission line method.

Space and time discretizing time domain methods have emerged as key numerical methods in computational electromagnetics. Time domain methods are versatile and can be applied to the solution of wide range of electromagnetic field problems. Computing the response of an electromagnetic structure to an impulsive excitation localized in space and time provides a comprehensive characterization of the electromagnetic properties of the structure in a wide frequency range. The most important methods are the finite difference time domain and the transmission line matrix methods. Whereas finite difference methods are based on the transition from differentials in the Maxwells Equations to finite differences, the transmission line matrix (TLM) method is based on the representation of the discretized electromagnetic by wave pulses propagating in a three-dimensional mesh of transmission lines. The space is discretized by subdivision into cells and the electromagnetic field is modeled by wave pulses propagating between adjacent cells and scattered within the cells. The TLM algorithm is based on the modeling of the propagation of wave pulses through a mesh of transmission lines and the scattering of the wave pulses and in mesh nodes. The simulation of the reaction to a single impulsive electromagnetic excitation yields a large amount information. The versatility of the TLM methods allows straightforward calculation of complex electromagnetic structures. With the computational power of today's computers, this method is a powerful tool for the computer-aided design of complex electromagnetic structures.

The symposium has given the opportunity to colleagues and students of Wolfgang J.R. Hoefer to present their recent advances in the field of time domain electromagnetics and its applications. This book contains extended versions of most of the scientific contributions of the symposium. The arrangement of the contributions in five chapters corresponds to the allocation of the presented material into five sessions.

1. Time-Domain Methods for Electromagnetic Field Modelling
2. The Transmission-Line-Matrix Method
3. Circuit Concepts and Methods
4. Antenna and Ultrawideband System Design
5. Novel Devices and Systems

The first chapter deals with time-domain methods for electromagnetic modeling in general. The second chapter focuses on the TLM method. The third chapter is dedicated to network concepts applied to electromagnetic field modeling. The fourth chapter is dedicated to circuit and system applications, and the fifth chapter dels with broadband devices, systems and measurement techniques.

The honorary doctor degree bestowal ceremony took place in the morning of 16 May 2007. In the beginning of the bestowal ceremony addresses of Professor Ulrich Wagner, the dean on the Faculty of Electrical Engineering and Information Technology of the Technische Universitt Mnchen, and Professor Wolfgang A. Hermann, the president of the Technische Universitt Mnchen were given. After this Professor Peter Russer held the laudatio. This was follwed by the bestowal and the lectio of Wolfgang J.R. Hoefer.

Munich,
December 2007

Peter Russer
Uwe Siart

Laudatio on Professor Wolfgang J.R. Hoefer

Peter Russer

The Faculty of Electrical Engineering and Information Technology of the Technische Universitt München bestows the Honorary Doctor degree on Professor Dr. Wolfgang Hoefer for extraordinary scientific achievements in the theory of electromagnetic fields. Since many years Professor Wolfgang Hoefer is one of the internationally outstanding scientists in the area of numerical methods for electromagnetic field computation. He has given fundamental contributions in the field of technical applications of the electromagnetic theory. In particular he has pioneered the development of the transmission line matrix (TLM) method, an efficient computational tool for the numerical computation of electromagnetic fields by numerous contributions.

Wolfgang Hoefer was born in 1941 in the Rhineland. He received the Dipl.-Ing. degree from the Rheinisch Westflische Technische Hochschule in Aachen and the Docteur Ingenieur degree from the Universit Grenoble. Subsequently he became Research Fellow at the Institut Nationale Polytechnique de Grenoble.

In 1969 Wolfgang Hoefer firstly became Assistant Professor at the University of Ottawa, Canada. In 1975 he became there Associate Professor and in 1980 Full Professor. In 1992 he was appointed to a professorship at the NSERC Industrial Research Chair in RF Engineering at the University of Victoria, B.C. Canada. There he led the Computational Electromagnetics Research Laboratory (CERL) at the Department of Electrical and Computer Engineering until his retirement in July 2006.

In 1996 Professor Hoefer founded the Faustus Scientific Corporation. This spin-off company develops professional electromagnetic field simulation CAD tools for research, development and education in the area of radio frequency analog circuits and high-speed digital circuits.

Professor Hoefer has given pioneering contributions to the development of time domain methods for the numerical computation of electromagnetic fields. He has been among the worldwide first scientists who applied the finite difference time domain method for the numerical modeling of waveguide structures and distributed microwave circuits.

P. Russer
Institute for High Frequency Engineering, Technischen Universität München,
e-mail: russer@tum.de

In these areas Professor Hoefer contributed more than 150 publications in international scientific journals and more than 250 contributions to international conferences. The book Microwave Circuit Modeling Using Electromagnetic Field Simulation, which he published in 2003 together with D.G. Swanson, presents the theoretical fundamentals und methods of the design of microwave circuits by means of electromagnetic CAD tools.

His contributions to the development of the TLM method are seminal [1, 2]. He has been the first who modeled dispersive materials in time domain. Moreover he developed novel methods for the field-based synthesis of optimum microwave structures and methods for the generation of compact circuit models based upon the field modeling of circuits.

In the areas of electromagnetic theory and numerical methods of electromagnetic field modeling Professor Hoefer acted successfully as educator and guided numerous young PhD students to success. In 2006 he received the Distinguished Educator Award of the IEEE MTT Society.

In his research areas Wolfgang Hoefer gained internationally highest recognition. He is Fellow of the IEEE and Fellow of the ASI (Advanced Systems Institute of British Columbia). In 2003 he has been elected Fellow of the Royal Society of Canada. He is member of numerous scientific committees and boards and he is co-founder and editor of the International Journal of Numerical Modelling.

Since 1990 Professor Hoefer is closely related to the Institute for High Frequency Engineering by a continuous scientific cooperation. Within this cooperation numerous joint scientific publications resulted. Professor Hoefer has been Visiting Professor at our institute in 1991 and 1999 and this year he is again Visiting Professor from April to July.

To fully appraise Wolfgang Hoefers contributions to the electromagnetic theory as a scientist and as an educator we first of all take a look at some peculiarities of this area. On the one hand the electromagnetic theory requires mathematical strictness and abstraction, on the other hand creativity in this field is especially promoted by imagery thinking.

Maxwell's theory establishes the analytic and conceptual framework of electromagnetism. As an analytic tool it allows the mathematical modeling of the electromagnetic field in even complex structures. However, prior to this Maxwell's theory allows the construction of concepts and also is the basis of mental imagery. Heinrich Hertz wrote in his treatise on Maxwell's equations of electromagnetism:

> One cannot escape the feeling that these mathematical formulae have an independent existence and an intelligence of their own, that they are wiser than we are, wiser even than their discoverers, that we get more out of them than was originally put into them [3,4].

In scientific research mental imagery plays an essential role. In natural sciences knowledge is constructed on the basis of experimental experience. According to Henri Poincar the mind contains two synthetic a priori intuitions that organize perception into knowledge – the principle of mathematical induction and the intuition of continuous groups that exists in our mind prior to all experience.

Theories are created by invention and not by discovery. Einstein took the view that the axiomatic structure of a theory is built psychologically on the experiences of the world of perceptions. In his inaugural speech at the Royal Prussian Academy of Science on 2nd July 1914 Albert Einstein said:

> Die Methode des Theoretikers bringt es mit sich, dass er als Fundament allgemeine Voraussetzungen, so genannte Prinzipe, benutzt, aus denen er Folgerungen deduzieren kann. Seine Ttigkeit zerfllt also in zwei Teile. Er hat erstens jene Prinzipe aufzusuchen, zweitens die aus den Prinzipen flieenden Folgerungen zu entwickeln. Für die Erfüllung der zweiten der genannten Aufgaben erhlt er auf der Schule ein treffliches Rüstzeug. Wenn also die erste seiner Aufgaben auf einem Gebiete bzw. für einen Komplex von Zusammenhngen bereits gelöst ist, wird ihm bei hinreichendem Flei und Verstand der Erfolg nicht fehlen. Die erste der genannten Aufgaben, nmlich jene, die Prinzipe aufzustellen, welche der Deduktion als Basis dienen sollen, ist von ganz anderer Art. Hier gibt es keine erlernbare, systematisch anwendbare Methode, die zum Ziele führt. Der Forscher muss vielmehr der Natur jene allgemeinen Prinzipe gleichsam ablauschen, indem er an gröeren Komplexen von Erfahrungstatsachen gewisse allgemeine Züge erschaut, die sich scharf formulieren lassen [5].

Fig. 1 The President of the Technische Univesität München Wolfgang A. Herrmann and Wolfgang J.R. Hoefer

> The method of the theoretician involves that he uses general conditions, so-called principles as a basis and to draw the conclusions from there. He first has to find these principles and then to develop the conclusions from there. To fulfill the second task he receives a suitable tool in school. If the first of his tasks already has been solved in an area or for a contextual complex, sufficient diligence and intelligence will yield success. The first mentioned task to establish the principles, which will serve as a basis for deduction is of completely different nature. Here no successful systematic method can be learnt. On the contrary the researcher has to listen to nature's general principles by recognizing general patterns that can be formulated strictly in large complexes of empirical facts. (Translation by the author)

These abilities mentioned by Einstein first to find the principles underlying the phenomena and then to develop the conclusions from there characterize Wolfgang Hoefer impressingly and are the basis not only of his extraordinary scientific creativity but also of his exceptional aptitude as an educator. In his scientific work

Professor Hoefer has shown remarkable creativity up to the present. His scientific achievements are extraordinary by any standard. His name stands for pioneering contributions to the theory of electromagnetic waves and their applications over a period of more than forty years. In the theory of electromagnetic waves, their technological application and academic teaching, he is an outstanding international scientific figure.

For our faculty the bestowal of the honorary doctorate on Professor Wolfgang Hoefer is an event of great significance and pleasure.

References

1. W. Hoefer, "The transmission line matrix method-theory and applications," *IEEE Trans. Microw. Theory Techn.*, vol. 33, pp. 882–893, Oct. 1985.
2. W. Hoefer, "The transmission line matrix (TLM) method," in *Numerical Techniques for Microwave and Millimeter Wave Passive Structures* (T. Itoh, ed.), pp. 496–591, New York: J. Wiley., 1989.
3. H. Hertz, *Gesammelte Werke, Bd. 2, Untersuchungen über die Ausbreitung der elektrischen Kraft*. Johann Ambrosius Barth Leipzig 1894.
4. F. Wilczeck, "A piece of magic – the Dirac equation," in *It must be Beautiful – Great Equations of Modern Science* (G. Farmelo, ed.), London, New York: Granta, 2002.
5. A. Einstein, "Antrittsrede des Herrn Einstein," *Sitzungsberichte der königlich preußischen Akademie der Wissenschaften*, vol. SB II, pp. 739–742, July 1914.

Contents

List of Contributors

Ajibola Ajayi
The George Green Institute for Electromagnetics Research, University Park, Nottingham, NG7 2RD, UK, e-mail: eezks@gwmail.nottingham.ac.uk

Andrey Baev
Moscow Aviation Institute (State Technical University) Volokolamskoe sh. 4, A-80, GSP-3, Moscow, 125993, Russia, e-mail: baev@mai-trt.ru

John W. Bandler
McMaster University, Hamilton, ON, Canada L8S 4K1, e-mail: bandler@mcmaster.ca

Andreas Barchanski
Institut für Theorie Elektromagnetischer Felder, Technische Universität Darmstadt, Schlossgartenstr. 8, 64289 Darmstadt, Germany, e-mail: barchanski@temf.de

John C. Batchelor
Department of Electronics, The University of Kent, Canterbury, Kent, UK, CT2 7NT, e-mail: j.c.batchelor@kent.ac.uk

Bruno Biscontini
Munch University of Technology, Institute for High Frequency Engineering, Arcisstr. 21 80333, Munich Germany, e-mail: bicontini@mytum.de

Geovany A. Borges
Dep. de Eng. Elétrica - Universidade de Brasília, CEP 70910-919 – Brasília – DF, Brazil

Jens Bornemann
Department of Electrical and Computer Engineering, University of Victoria, P.O. Box 3055 STN CSC, Victoria, B.C., V8W 3P6, Canada

Stephan Braun
Institute for High-Frequency Engineering, Technische Universität München, Germany, e-mail: stephan.braun@tum.de

Joachim Brose
Institute for High Frequency Technology, Faculty for Electronics and Information Theory, Bundeswehr University, Munich, 85577 Neubiberg, Germany

C. Caloz
Ecole Polytechnique, Montréal, Canada, e-mail: christophe.caloz@polymtl.ca

Andreas C. Cangellaris
Department of Electrical and Computer Engineering, University of Illinois at Urbana-Champaign, USA

Andreas C. Cangellaris
Department of Electrical and Computer Engineering, University of Illinois at Urbana-Champaign, USA

Yi Cao
Dept of Electronics, Carleton University, 1125 Colonel By Dr., Ottawa, Canada K1S 5B6

Zhizhang (David) Chen
Department of Electrical and Computer Engineering, Dalhousie University, Halifax, Canada, e-mail: z.chen@dal.ca

Christos Christopoulos
The George Green Institute for Electromagnetics Research, School of Electrical and Electronic Engineering, University of Nottingham, University Park, Nottingham, NG7 2RD, UK, e-mail: eezks@gwmail.nottingham.ac.uk

G. Csaba
Institute for Nanoelectronics Technical University of Munich, D-80333 Munich, Germany

Leonardo R.A.X de Menezes
Dep. de Eng. Elétrica – Universidade de Brasília, CEP 70910-919 – Brasília - DF, e-mail: Brazil,leonardo@ene.unb.br

Huilian Du
Department of Electrical and Computer Engineering, University of Victoria, P.O. Box 3055 STN CSC, Victoria, B.C., V8W 3P6, Canada

Nikolaus Fichtner
Institute for High-Frequency Engineering, Technische Universität München, Arcisstr. 21, 80333 München, Germany, e-mail: fichtner@tum.de

Arnd Frech
Institute for High-Frequency Engineering, Technische Universität München, Germany, e-mail: frech@tum.de

Erion Gjonaj
Institut für Theorie Elektromagnetischer Felder, Technische Universität Darmstadt, Schlossgartenstr. 8, 64289 Darmstadt, Germany, e-mail: gjonaj@temf.de

Wolfgang J. R. Hoefer
Department of Electrical and Computer Engineering, University of Victoria, Victoria, BC V8W 3P6 Canada, e-mail: whoefer@ece.uvic.ca

Fengxi Huang
Formerly with the Department of Electronics, The University of Kent, Canterbury, Kent, UK, CT2 7NT

Dr. Savvas Kosmopoulos
Space Engineering S.p.A., via dei Berio No.91, 00155, Rome, Italy

Slawomir Koziel
McMaster University, Hamilton, ON, Canada L8S 4K1,
e-mail: koziels@mcmaster.ca
Now with Reykjavik University, Reykjavik, Iceland, e-mail: koziel@ru.is

Yury Kuznetsov
Moscow Aviation Institute (State Technical University) Volokolamskoe sh. 4, A-80, GSP-3, Moscow, 125993, Russia, e-mail: kuznetsov@mai-trt.ru

Hung-Jui Lam
Department of Electrical and Computer Engineering, University of Victoria, P.O. Box 3055 STN CSC, Victoria, B.C., V8W 3P6, Canada

A. Lauer
IMST GmbH, D-47475 Kamp-Lintfort, Germany

Stefan Lindenmeier
Institute for High Frequency Technology, Faculty for Electronics and Information Theory, Bundeswehr University, Munich, 85577 Neubiberg, Germany

Petr Lorenz
Rohde&Schwarz GmbH & Co. KG, Mühldorfstr. 15, 81671 München, Germany, e-mail: petr.lorenz@ieee.org

Yinying Lu
Department of Electrical and Computer Engineering, University of Victoria, P.O. Box 3055 STN CSC, Victoria, B.C., V8W 3P6, Canada

P. Lugli
Institute for Nanoelectronics Technical University of Munich, D-80333 Munich, Germany

Davide Mencarelli
Dipartimento di Elettromagnetismo e Bioingegneria - Universitá Politecnica delle Marche, Ancona 60100, Italy, e-mail: d.mencarelli@univpm.it

Michel M. Ney
Telecom Breta Institute, Brest, France, e-mail: michel.ney@telecom-bretagne.eu

John Paul
George Green Institute for Electromagnetics Research, School of Electrical and Electronic Engineering, University of Nottingham, Nottingham, NG7 2RD, UK, e-mail: john.paul@nottingham.ac.uk

F. Peretti
Institute for Nanoelectronics Technical University of Munich, D-80333 Munich, Germany

Luca Pierantoni
Dipartimento di Elettromagnetismo e Bioingegneria – Universitá Politecnica delle Marche, Ancona 60100, Italy, e-mail: l.pierantoni@univpm.it

A. Rennings
IMST GmbH, D-47475 Kamp-Lintfort, Germany, e-mail: A.Rennings@ieee.org

Amin Rida
Georgia Electronic Design Center, School of Electrical and Computer Engineering, Georgia Institute of Technology, Atlanta, GA 30332-0250, USA

Tullio Rozzi
Dipartimento di Elettromagnetismo e Bioingegneria - Universitá Politecnica delle Marche, Ancona 60100, Italy, e-mail: t.rozzi@univpm.it

Johannes A. Russer
Department of Electrical and Computer Engineering, University of Illinois at Urbana-Champaign, USA

Peter Russer
Munich University of Technology, Institute for High Frequency Engineering, Technische Universität München, Arcisstr. 21, 80333 Munich, Germany, e-mail: russer@tum.de

Benito Sanz-Izquierdo
Department of Electronics, The University of Kent, Canterbury, Kent, UK, CT2 7NT, e-mail: b.sanz@kent.ac.uk

Phillip Sewell
The George Green Institute for Electromagnetics Research, University Park, Nottingham, NG7 2RD, UK, e-mail: eezks@gwmail.nottingham.ac.uk

Uwe Siart
Munch University of Technology, Institute for High Frequency Engineering, Technische Universität München, Arcisstr. 21 80333, Munich Germany, e-mail: uwe.siart@tum.de

Nikolaos Sidiropoulos
Space Engineering S.p.A., via dei Berio No.91, 00155, Rome, Italy

Poman P.M. So
Department of Electrical Engineering, Department of Electrical and Computer Engineering, University of Victoria, P.O. Box 3055 STN CSC, Victoria, BC, Canada, V8W 3P6, e-mail: Poman.So@ECE.UVic.CA

Mohammed I. Sobhy
Department of Electronics, The University of Kent, Canterbury, Kent, UK, CT2 7NT

Prasad S. Sumant
Department of Electrical and Computer Engineering, University of Illinois at Urbana-Champaign, USA

Manos M. Tentzeris
Georgia Electronic Design Center, School of Electrical and Computer Engineering, Georgia Institute of Technology, Atlanta, GA 30332-0250, USA

Peter Thoma
Institut für Theorie Elektromagnetischer Felder, Technische Universität Darmstadt, Schlossgartenstr. 8, 64289 Darmstadt, Germany; Computer Simulation Technology GmbH, Bad Nauheimerstr. 19, 64289 Darmstadt, Germany,
e-mail: peter.thoma@cst.com

David W. P. Thomas
George Green Institute for Electromagnetics Research, School of Electrical and Electronic Engineering, University of Nottingham, Nottingham, NG7 2RD, UK

Anya Traille
Georgia Electronic Design Center, School of Electrical and Computer Engineering, Georgia Institute of Technology, Atlanta, GA 30332-0250, USA

Robert Weigel
Institute for Electronics Engineering, University of Erlangen-Nuremberg, Cauerstrasse 9, 91058 Erlangen, Germany, e-mail: weigel@lfte.de

Thomas Weiland
Institut für Theorie Elektromagnetischer Felder, Technische Universität Darmstadt, Schlossgartenstr. 8, 64289 Darmstadt, Germany, e-mail: thomas.weiland@temf.de

I. Wolff
IMST GmbH, D-47475 Kamp-Lintfort, Germany

Li Yang
Georgia Electronic Design Center, School of Electrical and Computer Engineering, Georgia Institute of Technology, Atlanta, GA 30332-0250, USA

Hristomir Yordanov
Munich University of Technology, Institute for High Frequency Engineering, Arcisstr. 21, 80333 Munich, Germany, e-mail: yordanov@tum.de

M. Zedler
Technische Universität München, Germany, e-mail: michael.zedler@tum.de

Qi-Jun Zhang
Department of Electronics, Carleton University, 1125 Colonel By Dr., Ottawa, Canada K1S 5B6

Martin L. Zitzmann
BMW Group, Development Ignition Systems, Hufelandstrasse 4, 80788, Munich, Germany, e-mail: Martin-Ludwig.Zitzmann@bmw.de

In Search of the Intangible – 43 Years of Research in Electromagnetics

Wolfgang J.R. Hoefer

Magnifizenz, Spektabilität, Professor Russer, Honored Guests, Ladies and Gentlemen:

This is a day of great honor and deep emotions. Receiving this honorary doctorate from the prestigious Technische Universität München is not only a splendid professional accolade, but also a deeply gratifying and humbling distinction, bestowed on me by one of the leading academic institutions of my native country.

1 Words of Appreciation and Gratitude

I wish to open this lecture with heartfelt words of appreciation and gratitude. First and foremost, I thank the academic leaders of this university, President Herrmann, Dean Wagner, and the distinguished members of Faculty Council, for honoring me beyond all expectations. My sincere gratitude and admiration go to my colleague and friend, Professor Peter Russer, not only for initiating and promoting the bestowal process, but also for being my trusted colleague, inspiring fellow researcher, loyal friend and gracious host at his Institute of High Frequency Engineering for over twenty years. Speaking of divine justice – I was delighted to learn that he himself received a well-deserved honorary doctorate from the renowned Moscow University of Aerospace Technologies (MAI) less than a month ago.

This is the right time to pay tribute to all those who, through their love, support, friendship and collaboration have sustained and enriched my life and career. They deserve a significant share of the honor and recognition I receive today. My mother, in spite of her 91 years, did not hesitate to travel all the way from Koblenz to München so she could celebrate this event with us. After my father perished in the war in 1944, she single-mindedly and at considerable personal sacrifice, ensured that my sister and I received the best possible education. Without her unfailing love

Wolfgang J.R. Hoefer
Department of Electrical and Computer Engineering, University of Victoria, Victoria, BC V8W 3P6 Canada, e-mail: whoefer@ece.uvic.ca

P. Russer, U. Siart (eds.), *Time Domain Methods in Electrodynamics*,

and determination this bestowal would not have taken place. I am also indebted to my sister Marlies and her family for their unconditional support. My dear wife Diana has redefined my life in more than one way and deserves a special doctorate of her own – amoris causa! My children Christian and Elise who cannot be with us today, have always been loving and supportive, even though I had to be away on many of their birthdays or special school events. So many colleagues, associates, students and friends have shared their talents and insights with me over all these years that it is impossible to mention them all, but I would like to single out Professor Poman So who has been the keystone of my research team for more than twenty years. His exceptional talent for bringing electromagnetic fields alive on a computer has significantly shaped and enabled the evolution of our research and was critical to the development of the commercial electromagnetic simulator MEFiSTo. To Poman and to all my former students and associates, several of them present among us today, go my sincere feelings of gratitude and appreciation. I am delighted to share this honor with all of you!

2 The Intangible

Let me preface the account of my search for the Intangible with these mystical words:

> Durch alle Töne tönet
> Im bunten Erdentraum
> Ein leiser Ton gezogen
> Für den, der heimlich lauschet.
>
> Beneath the thousand sounds
> Of Earth's colorful dream
> There rings a constant gentle tone
> For all who secretly listen.

This motto, taken from a poem by the romantic philosopher and poet Friedrich von Schlegel (1722–1829), is usually cited in connection with one of Robert Schumann's most ambitious piano works, the Fantasy in C Major, Op. 17. Like many works of poetry, these lines transcend their immediate context and convey a much deeper insight. Clearly, they capture the essence of romanticism and appear to relate more to the esoteric than to the scientific. However, I could not think of a more fitting way to describe the secret attraction that entices not only the artist but also the scientific researcher to reach for the Intangible. Research is not merely a professional occupation but a life-long passion and commitment. Its evolution is therefore closely intertwined with the phases of the human existence. Again I have chosen a rather romantic aphorism – not entirely without tongue-in-cheek – namely a series of four paintings by Thomas Cole (1801–1848) titled "The Voyage of Life" to paraphrase this evolution. I leave it to you to make your own connections between Thomas Cole's paintings and my journey as we go along.

The Voyage of Life series is an allegory of the four stages of life: childhood, youth, adulthood, and old age. In each painting, accompanied by a guardian angel, the voyager rides in a boat on the River of Life. The landscapes, depicting the seasons of the year, play a major role in telling the story. In childhood, the infant glides from a dark cave into a rich, green pasture full of promise (Fig. 1). As a youth (Fig. 2), the voyager takes control of the boat and aims for a shining castle in the sky, a vivid symbol of the Intangible that exerts an irresistible attraction upon the voyager and propels him forward. The attentive viewer will note that the river soon takes an ominous turn towards distant cliffs and treacherous rapids. In adulthood, (Fig. 3) the voyager relies on prayer, faith and steadfastness to sustain him through rough waters and a threatening landscape. As the voyager reaches old age, the angel guides him to heaven across the waters of eternity (Fig. 4). He is still reaching for something intangible, which is now more meta-physical in nature than the shining castle of youth.

Fig. 1 The Voyage of Life – Childhood

Fig. 2 The Voyage of Life – Youth

Fig. 3 The Voyage of Life – Adulthood

Fig. 4 The Voyage of Life – Old Age

In spite of their religious and moral intent, these allegoric paintings strongly resonate with anyone who contemplates the human condition. On a lighter note, they remind us gently but firmly that we are inevitably approaching that fourth phase in our lives by the time we receive an honorary doctorate.

3 First Steps

My own "Voyage of Life" began in 1941 in Urmitz, a small village on the western border of the river Rhine. I still remember posing for the picture Fig. 5 at age seven with my sister Marlies. We were given something to do, so the picture would look more natural. While my sister was occupied with weaving two-dimensional patterns,

Fig. 5 First attempts at 3D modeling

I am shown here making my first attempts at three-dimensional modeling using uniform finite building blocks. Obviously, this picture turned out to be somewhat prophetic.

4 Studies at the RWTH Aachen

In October 1959 I began my studies of Electrical and Communications Engineering in Aachen, the former Imperial City of Charlemagne. From early on I felt very much attracted by the intangible nature of electric and magnetic fields and their ability to transmit information and energy. I wrote my engineering diploma thesis in Professor Döring's Institut für Hochfrequenztechnik, on a project to fabricate rotational ferrite ellipsoids of various aspect ratios, and to predict and measure their interaction with electromagnetic fields in rectangular waveguide components.

The bottles on the table in Fig. 6a suggest a chemistry lab rather than a microwave lab, but since my lab coat is not white but grey, I cannot be a chemist. The liquids are simply fluids for cleaning the tiny ferrite samples I am making, and the vertical tubes at the back of the table form part of a home-made differential control system to stabilize the air pressure for the grinding mill shown in Fig. 7. The drawing on the wall, reproduced again in Fig. 6b, shows the magnetic coupling between a standing TE_{10} wave in the horizontal guide and the two vertical guides through coupling slots via the magnetized ferrite sample. This was the arrangement to be calculated, realized and measured.

The mill for grinding the ferrite samples was a cylindrical pillbox covered inside with fine sandpaper (Fig. 7). Compressed air enters tangentially through a narrow hole and exits through the perforated side walls of the mill. A small cylindrical ferrite sample would normally be ground down to a sphere in about an hour, but in order to obtain an oblong ellipsoid, I placed the mill between the poles of a strong electro-magnet. Careful control of air pressure, field strength and grinding time yielded an ellipsoid of desired aspect ratio.

(a) (b)

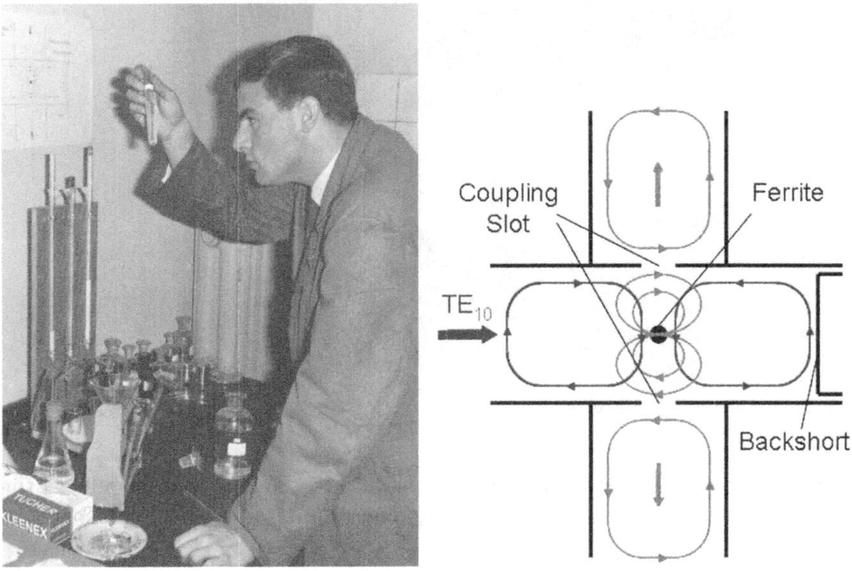

Fig. 6 (**a**) Making gyromagnetic ferrite ellipsoids; (**b**) Waveguide cross-coupler with ferrite ellipsoid

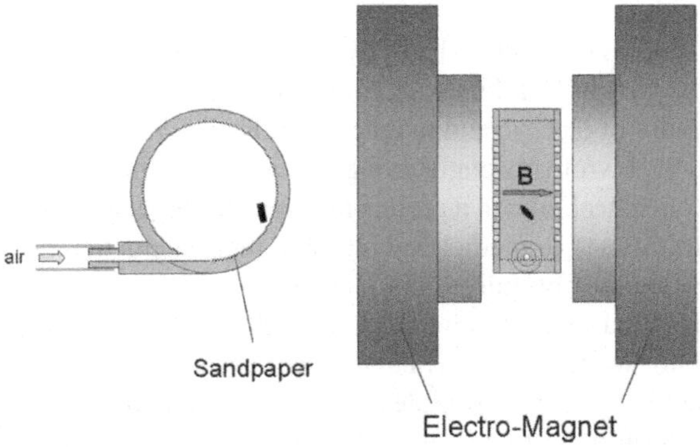

Fig. 7 A mill for grinding perfect ellipsoids. A DC magnetic field determines the ellipticity

In the early 1960's most of the electronic and mechanical components of an experimental arrangement had to be hand-made. For this project I needed to control the pressure of the air at the ferrite mill to ensure consistent results. In Fig. 8 I am building my own air pressure control unit. Some interesting examples of electronic and microwave equipment available at the time can be seen here as well. In the end, everything worked perfectly, and I got my engineering diploma. But what should I do next?

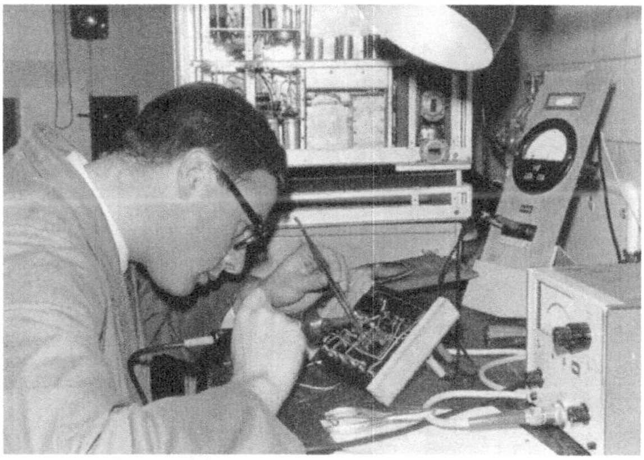

Fig. 8 Building the air pressure control unit

5 Doctoral Studies at the Université de Grenoble

I was not yet ready for a nine-to-five job. My idea was to continue the free life of a student, but with a different dimension. So I explored the possibility of a graduate scholarship in France, and ended up in Grenoble on February 9, 1966, just three days after turning 25. I really liked my new environment and had a great time in France, but I still needed to decide on a topic of study. I first looked into computer engineering and informatics, a field that was in full expansion at that time, but I finally gravitated towards the "Laboratoire d'Electromagnétisme" which was linked to the reputed "Institut Joseph Fourier", where I could conduct research on my favorite topic of electromagnetics and microwaves.

Researchers at the French National Scientific Research Center (CNRS) had developed a technology for fabricating Yttrium-Iron Garnet (YIG) spheres of exceptional purity and surface quality. The resonant bandwidth of these spheres was so narrow that it could not be measured with available instrumentation. My task was to create a measurement system that was sensitive enough to measure resonant bandwidths as narrow as several tenths of an Oersted. I first needed to build a very stable X-band source, shown in Fig. 9 against a blackboard with a drawing of the RF part of the stabilizer, that included a high-Q cavity discriminator driving a differential control amplifier. All components, from the cavity to the amplifier, were to be designed and made in-house (Fig. 10).

The complete measurement system required a cart with three shelves, and it worked perfectly: it down-converted the swept microwave response of a cavity containing the YIG sphere to 30 MHz and compared it to the response of a calibrated resonant circuit of adjustable Q-factor. The experimental technique was based on the analysis of coupling between an electromagnetic cavity and a YIG sphere of very narrow bandwidth, described in my first paper titled *Couplage d'une cavité electromagnétique avec un échantillon de grenat à raie très étroite* (Coupling between an electromagnetic cavity and a garnet sample of very narrow resonant bandwidth). It

Fig. 9 A stabilized klystron X-band source

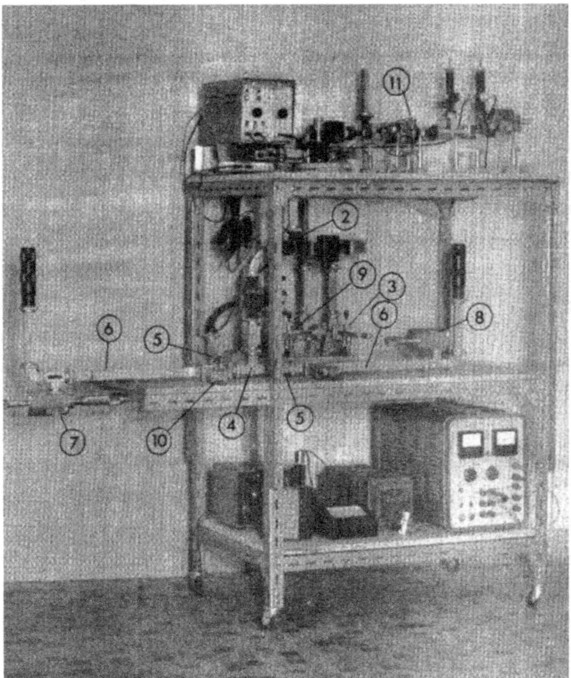

Fig. 10 Complete Q-factor measurement system

was published in the Transactions of the French Academy of Sciences. Each transaction paper must be sponsored and transmitted by one of the members of the French Academy. My sponsor was Professor Louis Néel who, three years later, received the Nobel Prize in Physics for his work on the nature and properties of ferrites. Naturally, I was very proud of this paper and the prestigious circumstances of its publication (Fig. 11).

The design of the cavities and the analysis of the field interaction between the cavities and the YIG samples required some advanced electromagnetic modeling, but the available computational tools of the day were just as elementary as the laboratory equipment and included the slide rule, tables of logarithms, trigonometric and higher mathematical functions, and collections of formulae and integrals. A digital computer could only be found in the computing center. A FORTRAN program was punched into a stack of perforated cards and fed to the compiler. The results came back a day or two later, printed on yards and yards of paper, that had a tendency to accumulate into large piles in students' offices. I remember submitting a program for

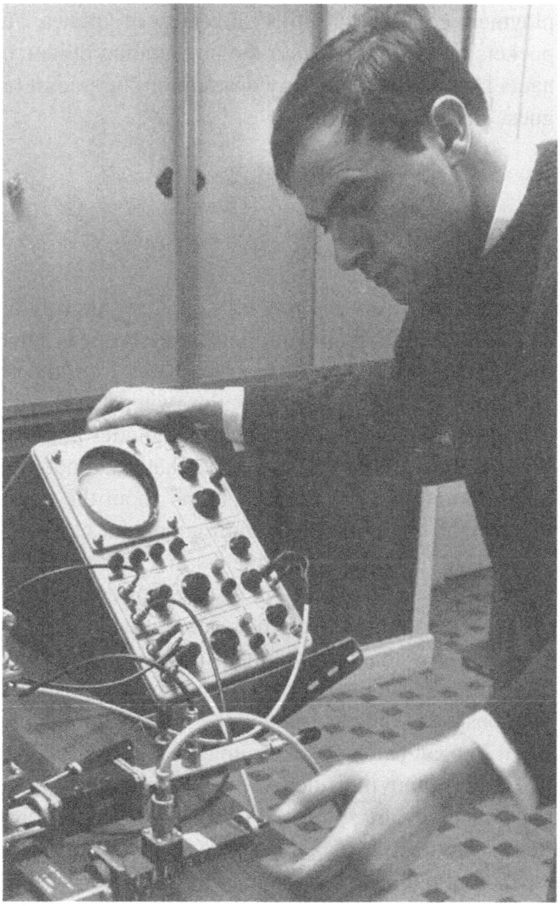

Fig. 11 Measuring the Q-factor of a high quality YIG sphere

finding the first root of a transcendental equation that involved Bessel and Neumann functions. When I returned the next day I found my pile of perforated cards with a note that my job had been rejected because it required more than 12 K of RAM. I needed to get a special permission from the director of the computing center to run such a computationally intensive job.

6 Teaching and Research at the University of Ottawa

When I completed and defended my doctoral thesis in June 1968, France was in upheaval. Student protests and general strikes paralyzed the country. After one year of postdoctoral research and teaching in Grenoble I was definitely ready to move on. At that time, Canada had entered a phase of intense economic and social development which made it highly attractive to start an academic career in that country. So I decided to embark upon the big journey across the Atlantic, bought a one-way airplane ticket, and on August 4, 1969, landed in Ottawa with two suitcases, an employment contract with the University of Ottawa, and four hundred dollars in my pocket, just two weeks after the first landing of man on the moon. Unlike the Astronauts I wanted to stay at my destination for at least one year. As we know now, that guess was very wrong.

7 New Challenges

The transition from the role of student to the role of teacher and researcher, from the well-established institutionalized European environment to the evolving and competitive North-American system, and from the old to the new world, presented numerous challenges. In particular, when faced with the mandate to develop new courses in microwaves and electromagnetics and to build an infrastructure for cutting edge research, questions arose that I had not given much thought before. They ranged from philosophical questions about the physical nature of electromagnetic fields to the computational requirements for real-world problem solving. My ambition was to teach electromagnetics in a way that would engage not only the analytical faculties of the brain but also its powerful processing abilities associated with visual perception and intuitive integration of physical and mathematical relationships, abilities that are essential for innovation and the creative process.

One fundamental aspect that I had never considered seriously was the nature and relationship of the electric and the magnetic fields. Yet, this became a key question when I began teaching electromagnetics. Consider for example two identical positive charges moving side by side at a constant velocity v, as shown in Fig. 12. We know that they are subject to a repulsive electric force F_e. However, they also experience an attractive magnetic force F_m since moving charges represent an equivalent local current i, and parallel currents attract each other. Which force is greater?

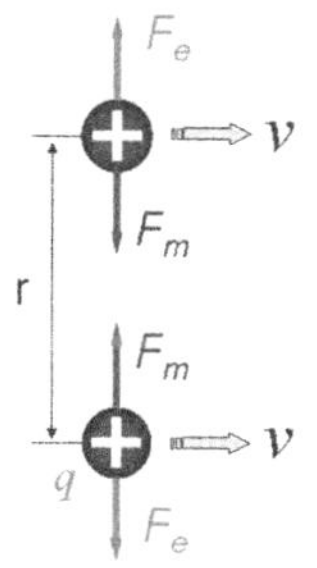

Fig. 12 Forces between two moving charges

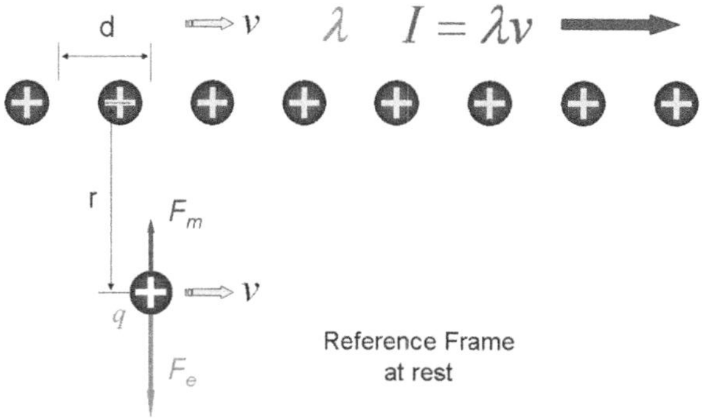

Fig. 13 Classical electromagnetic analysis of moving charges

We can, of course, answer this question correctly with classical field theory. Let us use a slightly different configuration, shown in Fig. 13, that leads us to the answer more easily. We, as observers in the reference frame at rest, look at a line of equidistant charges that move at constant velocity v. A single charge q moves along with the line of charges at the distance r. We can define a linear charge density λ (the number of positive charges per unit length) and an equivalent current $I = \lambda v$. The electric force F_e acting upon the single charge will then be, according to Coulombs law for line charges:

$$F_e = qE = q\frac{\lambda}{2\pi\varepsilon r} \tag{1}$$

For the magnetic force F_m we find

$$F_m = q\left(\vec{v}\times\vec{B}\right) = -qv\frac{\mu I}{2\pi r} = -qv\frac{\mu\varepsilon\lambda v}{2\pi\varepsilon r} = -q\frac{\lambda}{2\pi\varepsilon r}\frac{v^2}{c^2} \tag{2}$$

where we have introduced the velocity of light as $c = (\mu\varepsilon)^{-0.5}$. We see immediately that these two opposing forces are proportional to each other, and that the magnetic force is always much smaller than the electric force for all realistic velocities v. The net force

$$F_t = F_e + F_m = q\frac{\lambda}{2\pi\varepsilon r}\left(1 - \frac{v^2}{c^2}\right) \tag{3}$$

is clearly dominated by the electric term. While this "classical" exercise gives us the correct answer, it does not tell us much about the underlying reason. Of course, we need special relativity to better understand this relationship. Let us now jump on a reference frame that moves along with the charges such that they all appear at rest (Fig. 14). The first fact we notice is that the distance between the charges d' has been slightly increased by a factor $1/\sqrt{1-v^2/c^2}$, while the distance r has not changed. Since the line of charges is no longer moving, the equivalent current I is zero, and the force $F_e{}'$ on the charge q is purely static. There is thus no magnetic field and hence, no magnetic force. According to Coulomb's law, the force on q is:

$$F_e' = q\frac{\lambda'}{2\pi\varepsilon r} = q\frac{\lambda}{2\pi\varepsilon r}\sqrt{1-\frac{v^2}{c^2}} \tag{4}$$

If we now transform this force back into the previous reference frame using the Lorentz transform of special relativity, we find exactly the result obtained with classical electromagnetic theory, namely:

$$F_t = F_e'\sqrt{1-\frac{v^2}{c^2}} = q\frac{\lambda}{2\pi\varepsilon r}\left(1-\frac{v^2}{c^2}\right) \tag{5}$$

This *Gedankenexperiment* suggests that the magnetic field is an entirely fictitious concept that has been invented simply to account for the relativistic distortion of electric forces between moving charges.

Electromagnetic theory has indeed accounted for relativistic effects long before the theory of relativity was formulated. Because there exist both positive and negative electric charges of exactly equal and opposite value, the net electric forces between moving charges can cancel each other completely. The much smaller magnetic forces can then be clearly perceived. By contrast, no such effect has been observed in classical mechanics, because it knows only positive mass. Our simple example reveals that the magnetic field is not only intangible, but it is also completely fictitious!

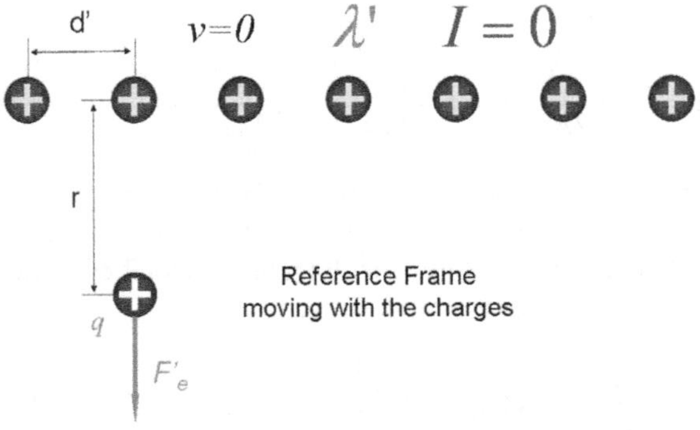

Fig. 14 Relativistic electromagnetic analysis of moving charges

A new exciting development in the early 1970s was the appearance of personal programmable calculators. My first acquisition was the impressive HP-65 hand-held programmable calculator that accepted up to 100 program lines which could be stored on a small magnetic strip. It now became possible to evaluate large sums of trigonometric functions and to solve transcendental equation without the need to go to the computing center. My first conference paper (which I never presented due to lack of travel funds) took advantage of this new powerful capability. It was titled *Computerized evaluation of scattering on thin transverse obstacles in microstrip.* The "computerized" part of the paper was the term-by-term evaluation with an electronic calculator of the following sum:

$$Z_{11} + Z_{12} = 2Z = -\sum_{m=2}^{\text{even m}} \left(\frac{2}{m\pi}\right)^2 \left(\frac{A}{d}\right)^2 \frac{k}{\kappa_m} \sin^2\left(\frac{m\pi}{2}\frac{d}{A}\right) \tag{6}$$

It would have been impractical – to say the least – to evaluate such an expression with a slide rule or a table of trigonometric functions, particularly since such series are not converging very quickly. However, the real breakthrough in computational electromagnetics began when the first personal computers appeared on the market in the early 1980s. By 1989, Poman So who then was my graduate student, had completed the first 2D TLM simulator running on IBM PCs under the DOS operating system, the precursor of MEFiSTo-2D Classic, which incorporated all the advanced features my research team had developed since 1980.

8 A New Electromagnetic Solution – First Encounter with Peter B. Johns

My interest in explicit time domain modeling for solving electromagnetic problems on a computer was awakened in the summer of 1975 when I first met Peter B. Johns after he presented one of his early papers on Transmission Line Matrix (TLM) modeling at the 1975 IEEE MTT Microwave Symposium in Palo Alto, titled *Three-dimensional numerical analysis of microwave cavities using the TLM method.*

Johns' new transmission-line based model of electromagnetic wave propagation was an incredibly simple concept that promised to enable us to solve the most complex electromagnetic problems. That was exactly what I set out to do in the coming years, and many of my former graduate students and research associates have made their seminal contributions to TLM theory and applications during these pioneering years. Peter B. Johns continued to develop the TLM method in Nottingham and proposed the famous Symmetrical Condensed Node (SCN) in 1986. Peter and I had just founded together the *International Journal of Numerical Modelling* at John Wiley & Sons in 1988 and had initiated a joint project on parallel computing in TLM, when Peter suddenly passed away.

9 From Ottawa to Victoria

At the beginning of the 1990s it was again time for me to move on. During a one-year sabbatical leave in Rome, Nice and Munich I was presented with an Industrial Research Chair at the University of Victoria, an offer I could not refuse. My state of mind at the time is captured in the following short verses that I wrote in the hexameter metric for the official inauguration ceremony of the Chair. Before reading or trying to recite it, a brief explanation of the hexameter is in order. The hexameter is an archaic form of poetry, best known from the great epic works of Homer, the Iliad and the Odyssey.

Each line of a hexameter contains six metrons or feet, as shown in Fig. 15. Each metron consists of one long and two short syllables; it is also referred to as a dactyl or finger, that has one long and two short sections, called the thesis and the arsis. The arsis is sometimes replaced by one long syllable, particularly in the last section of a line (spondee). A regular hexameter sounds like this: "Dum-diddi Dum-diddi Dum-diddi Dum-diddi Dum-diddi Dum-dum", with some variations to accommodate the flow of the language, particularly at the end of a line, where the English language is more difficult to bend to the metric form than Greek or Latin. Here is the Hexameter from the MPR/NSERC Research chair:

Analog microwave structures and high-speed digital circuits
both require extensive electromagnetic conception.
With traditional means and tedious trial-and-error
nobody can design a superior product selection.

In the hectic fray of the modern competitive market
Short development time is a major asset indeed.
MPR Teltech and NSERC, united in generous vision,
thus on Vancouver Island to found a Chair agreed.

"This is a worthwhile endeavor", the B.C. Government stated.
Wisdom and clever perception to join the team it persuaded.
Thus in Victoria in a Research Chair I am sitting,
Trying to comprehend how in all this I will be fitting.

MPR Teltech is the company that sponsored the Research Chair, *NSERC* is the "Natural Science and Engineering Research Council" of Canada that provided

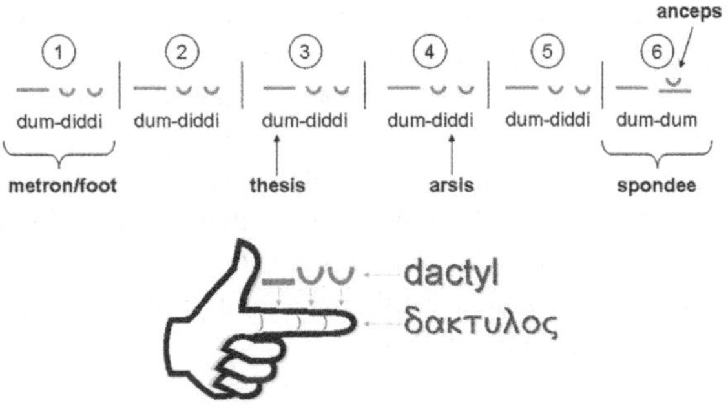

Fig. 15 The structure of the dactylic hexameter

federal funding for the Chair, and B.C. is the Province of British Columbia. The NSERC/MPR Teltech Industrial Research Chair has been an extraordinary opportunity for me to translate academic research into useful software tools and to transfer technology to industrial partners. I was very fortunate to have had outstanding students and researchers in my research team in Victoria.

Let me give you a flavor of some exciting new research that we pursued during the past few years. A class of artificial electromagnetic materials with exotic properties, known as metamaterials, has recently been the subject of intense interest in both the physics and the engineering communities. We have successfully created and implemented numerical models that simulate the extremely complex behavior of electromagnetic fields in such materials, and that make it possible to design novel microwave and lightwave components with revolutionary properties. We focused our effort on so-called Negative Refractive Index Metamaterials. They have both negative permeability and negative permittivity and a negative refractive index, resulting in hitherto unknown wave behavior that has not been observed in nature.

Two unusual properties characterize such metamaterials (Fig. 16). The first is their negative phase velocity that is opposed to the group velocity and the flow of energy. The second is the negative refraction angle that occurs when a wave penetrates from a regular medium into a metamaterial which gives rise to a strange focusing effect.

The phase velocity and propagation vectors form a left-handed triplet with the E and H vectors, hence the term "left-handed materials". The negative refraction is a direct consequence of the negative phase velocity. The refracted angle θ_2 must be negative in order to preserve the continuity of phase across the boundary between air and metamaterial.

When we simulate the propagation of a monochromatic electromagnetic wave across a section of metamaterial with refractive index $n = -2$ $(\varepsilon_r = \mu_r = -2)$ that is sandwiched between two air sections, it will be quite different from the propagation of the same wave through a regular dielectric with $n = +2$ $(\varepsilon_r = \mu_r = +2)$. Since the impedances of all sections are matched to each other, no scattering takes place at the interfaces, and all energy is entirely transmitted. While the wave pattern moves forward in the regular dielectric, it moves backward in the metamaterial section.

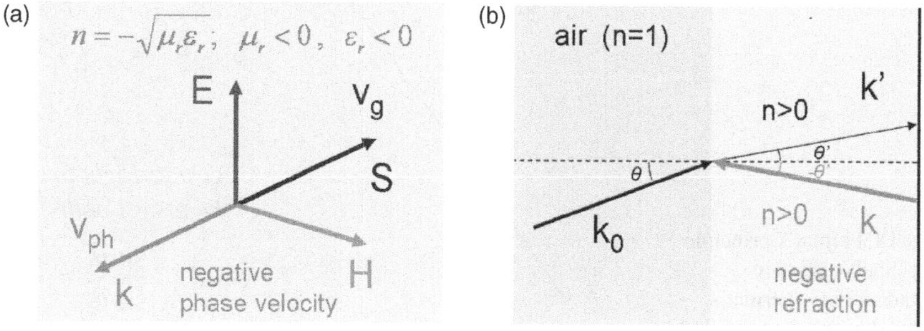

Fig. 16 Unusual properties of metamaterials: (**a**) negative phase velocity, (**b**) negative refraction

This wave behavior is counterintuitive, but the simulations correctly predict what happens in reality.

Refraction is governed by Snell's law which is a manifestation of Fermat's Principle. Let us first consider the transition of light from air to a regular dielectric material of refractive index $n_2 = 2$ as shown in Fig. 17. Fermat noted that light traveling from point A to point B always takes a path of least time. If we check out some possible light rays by moving along them in equal time steps, we see that the fastest path is path 1 due to its optimal compromise between distance and speed. Note that the speed in the material is half the speed in air. The corresponding incident and refracted angles θ_1 and θ_2 are related to the refractive indices n_1 and n_2 by Snell's law. However, in a metamaterial with negative refractive index and negative phase velocity (Fig. 18), Fermat's principle is no longer valid in its original formulation. While Snell's law still applies for negative refractive indices, Fermat's principle must be restated to say that the optical path length between two points must be an extremum,

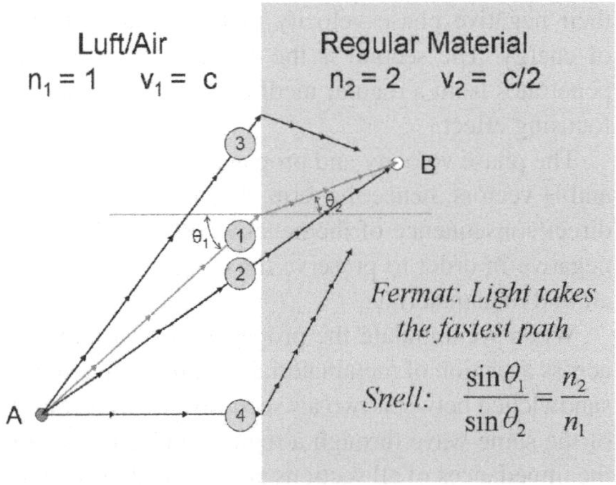

Fig. 17 Fermat's principle and Snell's law in regular materials

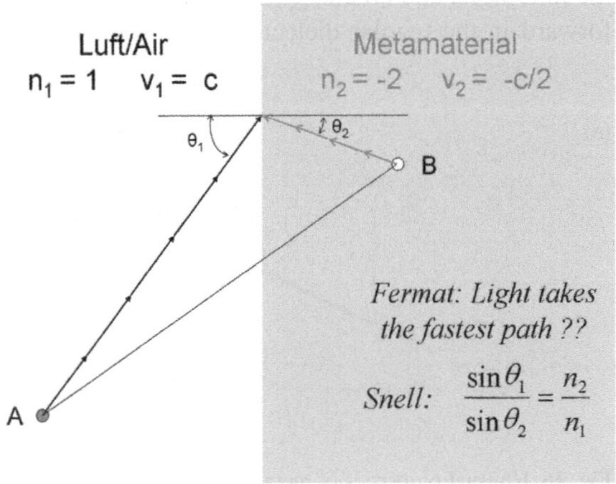

Fig. 18 Fermat's principle and Snell's law in metamaterials. Fermat's principle must be modified for metamaterials

i.e. it can be a minimum, a maximum, or even zero, depending on the geometry and the values of the refractive indices in a particular situation.

An even stranger behavior can be observed when a point source is placed near a metamaterial slab with negative refractive index (Fig. 19). Even though the slab is not curved at all, it acts just like a traditional convex lens. This is due to the negative refraction of the rays at the two faces of the slab. However, the electromagnetic behavior of this flat lens is much more complicated than the simple ray diagram suggests, and it can be comprehended in its full complexity only by an animated numerical simulation (Fig. 20).

Only poetry can provide a more compact account of what happens in such complex circumstances, and I have thus attempted to summarize the essence of our observations in the form of a limerick called "New Physics". A limerick (named after the Irish town of Limerick whose inhabitants were famous for talking a lot of nonsense) is a short poem of five lines that usually has a kind of twist to it. If you recite a limerick and nobody laughs, people are either disgusted, or they didn't get it.

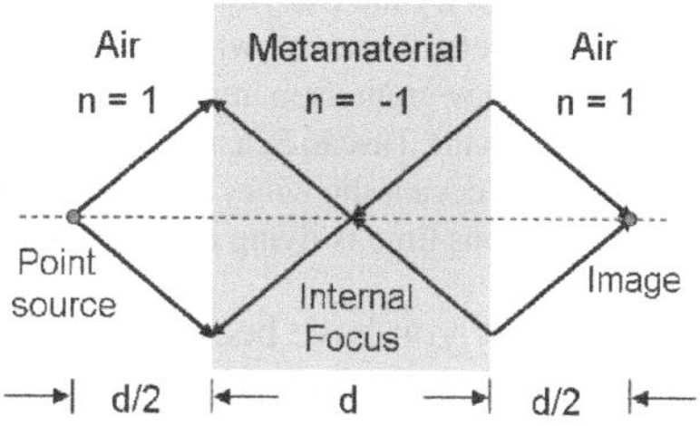

Fig. 19 Subwavelength focusing by a planar metamaterial slab (Pendry Lens)

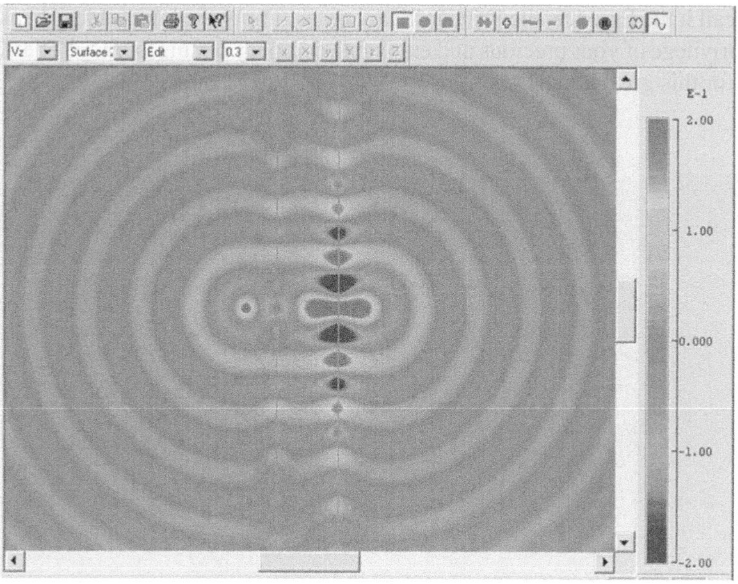

Fig. 20 Full electromagnetic simulation of the planar metamaterial lens

9.1 New Physics

A traveling photon named Beta
encountered Material Meta.
Upon penetration
it had a sensation
that something went wrong with its theta.

10 Conclusion and Outlook

Let me conclude the account of my personal journey with some advice that the Roman poet Horace gave to his lover Leuconoë in 23 BC, advice that seems to make more and more sense as we enter the final stretch of our journey.

Ask not – we shall never know – what end the gods have set for me, for you;
and do not try the Babylonian horoscopes, Leuconoë.
It is far better to accept whatever life has in store for us.
Jupiter may grant us many more winters, or this may be our last
that hurls the Tuscan Sea against the cliffs!
Be wise, decant the wine; and since life is brief, trim back far-reaching plans!
The jealous time is flying as we speak: seize the day, and place little trust in tomorrow!

Many of you have been my travel companions, some for a short time, others for a long stretch, and we have, in the course of time, listened together for the constant gentle tone that rings through the thousand sounds of Earths colorful dream, the tone that entices us to wonder and to search for whatever lies beyond the horizon of perception, call it the Essence, the Deeper Meaning, or simply the Intangible. Thank you for the privilege of your precious and enjoyable company on this journey! Thank you, TUM, for this great honor!

Full-Wave Simulation of Integrated Circuit Packages on a Parallel Architecture

Erion Gjonaj, Andreas Barchanski, Peter Thoma and Thomas Weiland

Abstract The time domain electromagnetic field simulation based on massive parallelization is presented as a tool for the signal integrity analysis of complex IC packages. The simulations are based on a specialized domain partitioning method which allows for highly balanced parallel computations. As a real-world example the analysis of a large computer chip spreader is performed. Numerical results including signal wave forms and delay times are given.

1 Introduction

With increasing demands on speed and package density in Integrated Circuit (IC) technology, the trend continues to move toward more complex designs and higher operation frequency. In view of the fast development in semiconductor technology, with chip devices of clock speeds close to 4 GHz, the industry is working on interconnects and packages capable of supporting high frequency signals without loss of signal integrity. The traditional design process is based on low frequency models with lumped elements, used within the framework of circuit simulators. However, as the operation frequency is increased and the package dimensions are reduced, field interference effects within the circuitry become important. Typical EMC concerns

Erion Gjonaj
Institut für Theorie Elektromagnetischer Felder, Technische Universität Darmstadt, Schlossgartenstr. 8, 64289 Darmstadt, Germany, e-mail: gjonaj@temf.de

Andreas Barchanski
Institut für Theorie Elektromagnetischer Felder, Technische Universität Darmstadt, Schlossgartenstr. 8, 64289 Darmstadt, Germany, e-mail: barchanski@temf.de

Peter Thoma
Computer Simulation Technology GmbH, Bad Nauheimerstr. 19, 64289 Darmstadt, Germany, e-mail: peter.thoma@cst.com

Thomas Weiland
Institut für Theorie Elektromagnetischer Felder, Technische Universität Darmstadt, Schlossgartenstr. 8, 64289 Darmstadt, Germany, e-mail: thomas.weiland@temf.de

P. Russer, U. Siart (eds.), *Time Domain Methods in Electrodynamics*,

at high frequencies, such as radiation, larger delays and mutual coupling should be taken into account in the design of modern ICs.

Attempts have been made to account for high frequency effects in the simulation of IC packages. As an example, in [1] a macromodel approach is proposed, in which first the high frequency characteristics (S-parameters) of IC-subnetworks are extracted and then a reduced order equivalent circuit model for the overall system is derived. The validity of such approximations is, however, difficult to be verified over the whole range of device parameters. Additionally, most of the simulation approaches are applied to small and/or simplified parts of IC devices. Therefore, 3D electromagnetic simulation becomes more and more necessary, as package complexity and operating frequencies increase [2]. Because of the geometrical complexity of IC packages containing thousands of through-hole vias and planar signal traces mounted on several layers, it is a widespread belief that a full-wave simulation taking into account all effects and geometric details is impossible. However, recent algorithmic progress makes such simulations possible, and fully reliable. The key concept is the massive parallelization of field solvers. The use of distributed computing provides sufficient computational resources for the simulation of complete ICs in full 3D geometry within short to moderate simulation times.

In this work, a time domain, full-wave simulation procedure and the results obtained for a very complex computer chip spreader are presented. The work is organized as follows: In Sect. 2, the discretization technique used in the simulations, the Finite Integration Technique, is described. In Sect. 3, the parallelization strategy based on a balanced domain partitioning technique is introduced. In Sect. 3.1 the numerical performance of the parallelization algorithm is investigated using a geometrically simple example, while details of the simulated structure are presented in Sect. 4. Finally, in Sect. 5, a signal integrity study, including computed delay time and transmission/crosstalk waveforms is presented.

2 Discrete Maxwell Equations

The framework for the time domain simulations presented here is provided by the Finite Integration Technique (FIT) [3]. The spatial discretization of FIT utilizes a pair of dual-orthogonal staggered grids $(G,\widetilde{G})$. Denoting by $(\widehat{\mathbf{e}}, \widehat{\mathbf{h}})$ the electromagnetic degrees of freedom and by $\widehat{\widehat{\mathbf{j}}}$ the source currents, the time discrete update equations of FIT read

$$\begin{pmatrix} \widehat{\mathbf{e}}^{n+1} \\ \widehat{\mathbf{h}}^{n+\frac{1}{2}} \end{pmatrix} = \begin{pmatrix} 1 - \Delta t^2 \mathbf{M}_\varepsilon^{-1} \widetilde{\mathbf{C}} \mathbf{M}_\mu^{-1} \mathbf{C} & \Delta t \mathbf{M}_\varepsilon^{-1} \widetilde{\mathbf{C}} \\ \Delta t \mathbf{M}_\mu^{-1} \mathbf{C} & 1 \end{pmatrix} \begin{pmatrix} \widehat{\mathbf{e}}^{n} \\ \widehat{\mathbf{h}}^{n-\frac{1}{2}} \end{pmatrix} - \begin{pmatrix} \Delta t \mathbf{M}_\varepsilon^{-1} \widehat{\widehat{\mathbf{j}}}^{n+\frac{1}{2}} \\ 0 \end{pmatrix} \quad (1)$$

where the unknowns $(\widehat{\mathbf{e}}, \widehat{\mathbf{h}})$ are the integrals of the electric and magnetic fields along the edges of G and $\widetilde{G}$, respectively. The curl-operators $(\mathbf{C}, \widetilde{\mathbf{C}})$ are of pure topological nature, whereas the metric information obtained by discretization is contained in the symmetric positive definite material matrices, $(\mathbf{M}_\varepsilon, \mathbf{M}_\mu)$. A detailed description of the FIT method applied in time and frequency domain simulations is found, e.g., in [3].

The compact writing in terms of matrix operators reveals that the only critical computation to be performed is a sparse matrix-vector multiplication. This can be very efficiently implemented, e.g., by employing specialized algorithms which make best use of local memory operations (cf. [4]).

3 Parallization Strategy

The huge amount of degrees of freedom needed for the 3D discretization of the full IC structure can only be handled in a parallel computing environment.

The parallelization model used in this work is based on the distribution of computational tasks and data among a number of memory independent processors (multiple-instruction-multiple-data). In this model, there are three main factors which do affect the efficiency of the parallel computation.

- **Interprocessor communication**: this is the additional overhead in parallel computations related to the necessity of exchanging field data between the processors involved in the simulation.
- **Workload per processor**: ideally, the computational time needed to perform its own task by an individual processor.
- **Workload balancing**: the workloads assigned to each processor need to be well balanced, in order to avoid idle times of the less loaded processors, waiting for the ones with heavier workloads to finish their task.

We perform a geometric partitioning of the computational domain into subdomains, such that the above criteria are optimally taken into account. Each subdomain is assigned to a single processor (node in the cluster) which is responsible for the solution of (1) within the subdomain (cf. [5]).

The idea of the parallel partitioning approach is shown schematically for a three-node cluster of computers in Fig. 1. Starting with the whole computational domain, an orthogonal recursive domain bisection is applied. The procedure results in a binary tree structure, whose internal nodes are *intermediate* subdomains, whereas the leaf nodes correspond to the *active* subdomains which are used in the computation. Each processor is responsible for two types of operations:

- performing the electromagnetic update equations (1) for the grid cells contained within the own subdomain, and
- exchanging field data with processors assigned to neighboring subdomains.

Because of the local nature of the discrete operators in (1), only field degrees of freedom residing at the boundaries of active subdomains need to be exchanged between the processors. In particular, for regular hexahedral meshes, the orthogonal decomposition approach yields the smallest number of such boundary cells and, therefore, it minimizes the communication overhead in the simulation. The recursive bisection procedure is performed on the basis of *computational weights*, W_i, which are

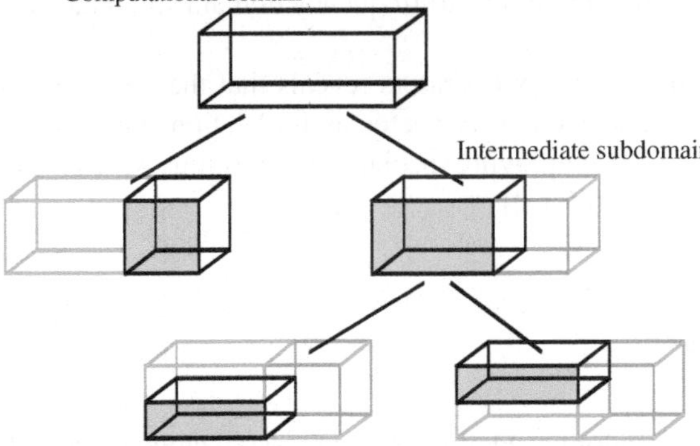

Fig. 1 Example of the orthogonal domain bisection procedure on a three-node cluster

assigned to each grid cell and represent the total number of floating point operations needed for a time-update of the cell's unknowns. The total computational load associated with an intermediate subdomain is, thus, $W = \sum W_i$, where the summation includes only grid cells contained within the subdomain. If the subdomain has to be distributed among N processors, the bisection boundary is chosen such that

$$\frac{W_{\text{left}}}{W_{\text{right}}} = \frac{N_{\text{left}}}{N_{\text{right}}}, \tag{2}$$

with

$$\begin{aligned} N_{\text{left}} &= \frac{N}{2} \\ N_{\text{right}} &= \frac{N}{2} \\ N &= N_{\text{left}} + N_{\text{right}}. \end{aligned} \tag{3}$$

where $(W_{\text{left}}, W_{\text{right}})$ and $(N_{\text{left}}, N_{\text{right}})$ are the computational weights and the number of processors, respectively, associated with the two subdomains created by subdivision.

The above algorithm allows for an almost ideally balanced distribution of computational workloads. In addition, it can be applied to simulations involving an arbitrary number of processors. Furthermore, by selecting appropriate cell weights in (2), well balanced parallel computations can be performed on *arbitrary, heterogeneous clusters*.

3.1 Test Example

In order to assess the performance of the algorithm, a simple example is considered. It consists of a rectangular cavity homogeneously filled with air and discretized with

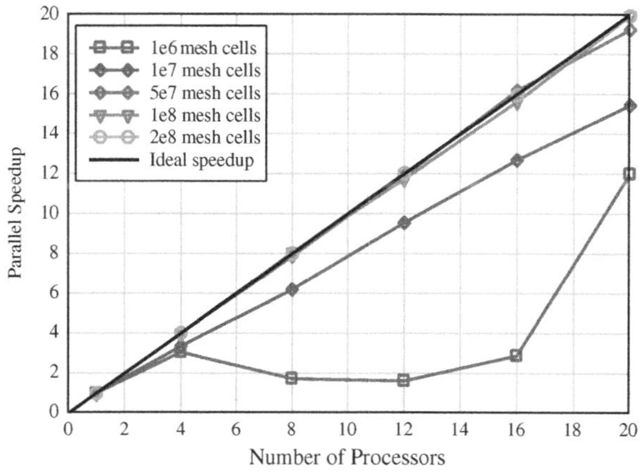

Fig. 2 Parallel speedup of the partitioning algorithm for different grid sizes and number of processors

a regular hexahedral mesh. Different discrete models, leading to meshes containing between 1 million and 200 million mesh cells, were used in the simulation tests.

The algorithm's performance in terms of parallel speedup, for the different grid sizes, is demonstrated in Fig. 2. The speedup curves show a perfect behavior, lying almost on the theoretical limit. Exceptions are the "tiny" discrete models of 1 million and 10 million grid cells, respectively. As expected, in these cases the interprocessor communication dominates the computational workload of the individual processors. Since the interprocessor communication increases with increasing number of processors, the performance speedup for the small models deteriorates faster than for the larger ones. The simulations shown in Fig. 2 were performed on a 20-node cluster of 3.4 GHz Intel processors and a conventional 1 Gbit/s Ethernet network, using the MPI standard. Note that, in this simple performance test neither absorbing boundaries nor conductivity currents or excitation signals were considered. Such additional computations may slightly affect the parallel performance in real-world simulations.

4 The IC Package

The package considered in the following is a multilayer computer chip spreader. It connects the semiconductor die with its larger support which in turn is soldered into the printed circuit board (PCB). A schematic view of the stack-up is shown in Fig. 3. The structure contains 8 metallization layers embedded into a dielectric substrate. The metallization is copper and the dielectrics used for the substrate are ABF-GX13 ($\varepsilon_r = 3.2$) and BT-679FG ($\varepsilon_r = 4.2$). Most of the signal traces propagate on the layer FC3. The signal is connected on the top layer to these traces by via and then routed almost directly to the solder points on the PCB side of the package.

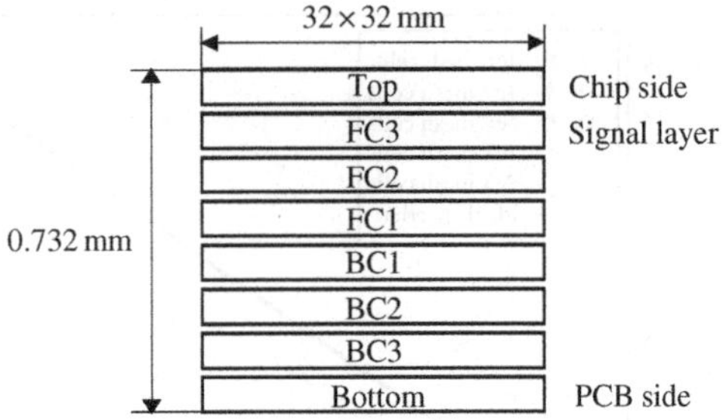

Fig. 3 Schematic view of the 8 metallization layers and the total dimensions of the package

The geometry of the package was provided to the authors by the IBM Thomas J. Watson Research Center [6] in form of a Cadence Allegro design layout. The data consisted of more than 40,000 2D-polygonal elements (pins, vias, traces, etc.). The CAD model of the fully 3D geometry was extracted from these data using the CST STUDIO SUITE software [7]. This step turned out to be the most challenging in the simulation procedure. Besides the huge size of the data to be processed, the main difficulty in extracting the geometry results from the errors contained in the input data. Typically, Allegro design layouts contain a small but non-negligible number of self-intersecting polygons. These errors had to be fixed manually, with considerable effort, in order to generate a consistent 3D-geometry description. Figure 4 shows the complete view of the 3D package geometry.

The geometrical complexity of the device is demonstrated in the detailed view of Fig. 5. In order to give an idea about the proportions: the approximate area of the zoomed-in section is 2×3.5 mm, representing only about 0.7% of the total device area. In the Figure, the 6 signal traces used in the simulations (see Sect. 5) are shown in blue/dark color. The width of these traces is 25 μm with a minimum trace-to-trace

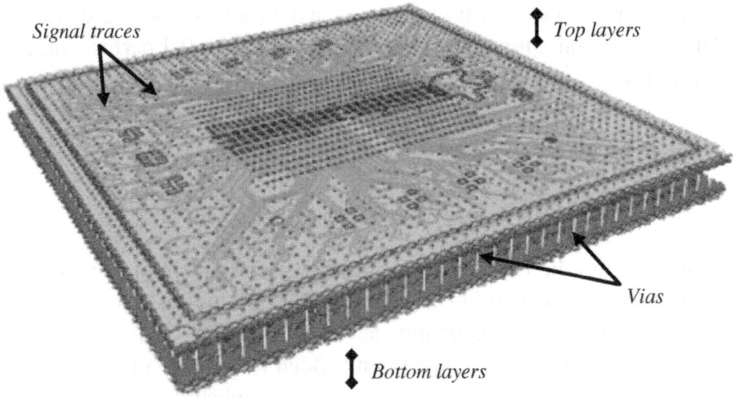

Fig. 4 Complete view of the package including the 8 layers, signal traces and vias (the vertical size is not to scale as the structure is extremely flat)

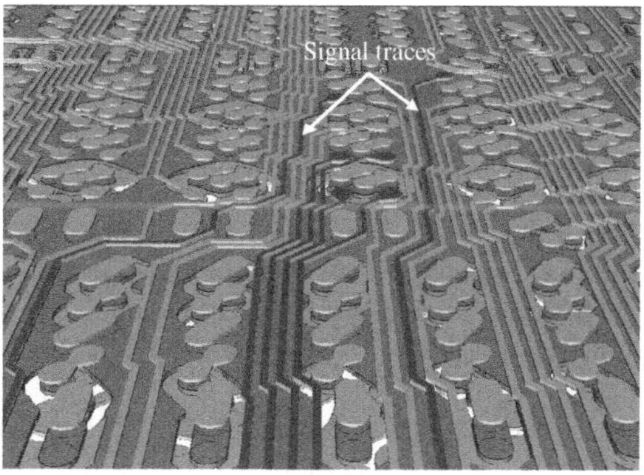

Fig. 5 Chip side view of a small section of the package. Six selected signal traces, used in the simulations are shown in blue/dark color

spacing of only 20 μm. The thickness of the single layers varies between 13 μm and 21 μm. The overall aspect ratio of largest-to-smallest geometrical dimension is, thus, higher than 1000:1. As shown in Sect. 5, in order to resolve all geometrical details of the structure, a huge computational mesh of several hundred millions of cells is needed. Simulations of this size can hardly be handled on a sequential computing platform.

The third step in the geometric modeling phase includes meshing, i.e., the recognition of the electromagnetic material distribution by the discretization grid and the generation of the material operators in (1). Because a huge number (of several hundred millions) of mesh cells are needed in the simulation for resolving the geometrical details of the structure, this step can be hardly performed on a sequential computing platform.

On the other hand, since meshing is a purely local operation, it can be very effectively implemented in a parallel scheme. Using the domain partitioning approach of Sect. 3, only the fractions of geometry data and mesh cells relevant to each partition are actually used in local computations. Thus, the more processors are included in the parallel simulation, i.e., the smaller are the partition domains the less critical becomes the meshing procedure (cf. also Table 1).

Table 1 Main parameters of the full-wave simulation

Parameter	Value
Signal rise time	39.5 ps
Signal duration	600 ps
Total number of cells	594,000,000
Total number of unknowns	3,564,000,000
Number of processors	20
Peak memory/processor	7.5 GB
CPU time for meshing	$\approx$ 3 hrs
Simulation time/signal	$\approx$ 68 hrs

5 Simulation and Results

The full IC package was discretized using a regular Cartesian grid. In order to resolve the smallest geometrical details of the structure, a minimum grid step of 8 μm in the vertical plane was used. In the horizontal plane a uniform grid of 10 μm step was applied. This choice assured a discretization with 2–3 grid cells within the cross-section of the signal traces (25 μm in width), which are the most critical geometrical components in the simulation. The whole discretization procedure resulted in a grid of ca. 600 million cells distributed mainly along the metallization planes.

An exemplary domain partitioning of the model on a 15-node cluster is shown in Fig. 6. The distribution of active partitions is seemingly irregular. However, all partitions do contain (approximately) the same number of mesh cells, as required for parallel coherence.

Table 1 gives a summary of the main parameters in the actual simulations. Note that only 3 hrs were needed in the meshing phase as compared to 38 hrs needed in the transient simulation. This figure scales down linearly with the number of processors used in the simulation.

Figure 7 shows the normalized absolute value of the electric field on the signal layer (FC3) shortly after the excitation is applied on line 1. Apart from the signal propagation along the trace, the near-end crosstalk to one of the neighboring pins is clearly seen.

Figure 8 shows the recorded wave forms obtained when exciting trace 2. The displayed signals are the input signal, the transmitted signal, as well as the near-end (NE) and far-end (FE) cross-talks with traces 3 and 6.

Figure 9 shows similar voltage curves when the input line is trace 6. The signal delay time seen at the half-widths of the signal rise times for the input and transmitted curves, respectively, amounts to 119.9 ps, which is in good agreement with

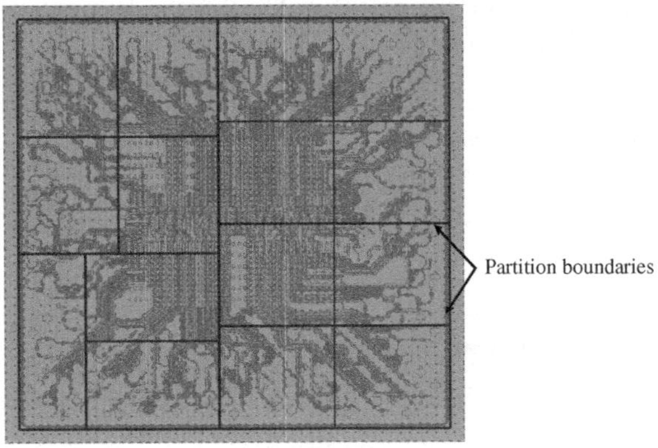

Fig. 6 Exemplary domain partitioning of the model on a 15-node cluster. The partition boundaries are shown as black lines

Fig. 7 Electric field strength (a.u.) on the FC3 layer, 30 ps after the start of the pulsed excitation at trace 1

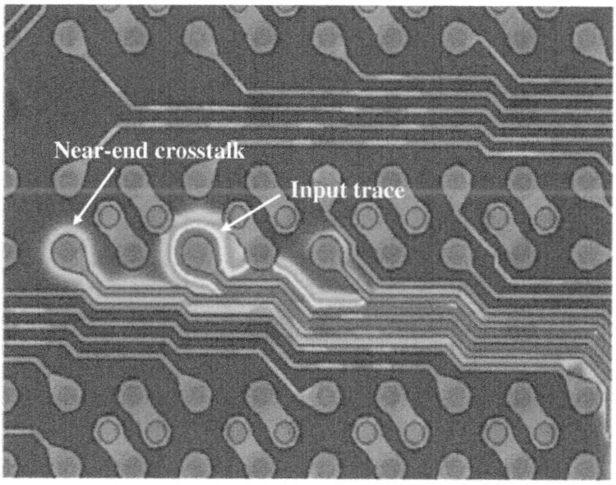

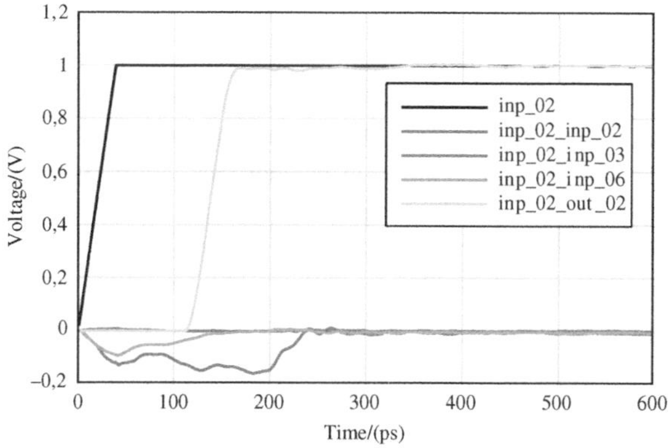

Fig. 8 Some of the voltage profiles at the input (labeled "inp 02 inp 0x") and output ("inp 02 out 0x") ports when the excited line is trace 2

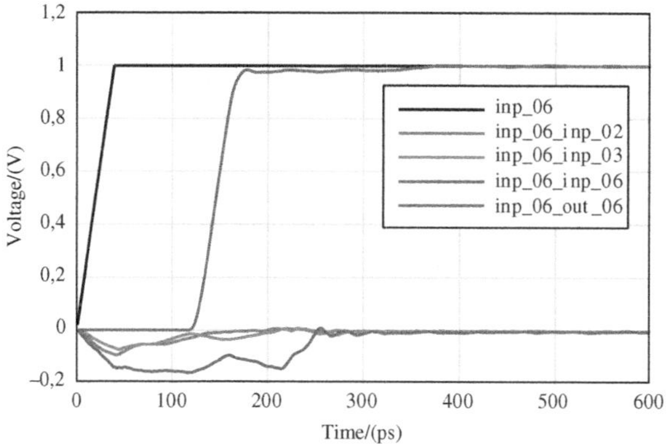

Fig. 9 Some of the voltage profiles monitored at the input and output ports when the excited line is trace 6

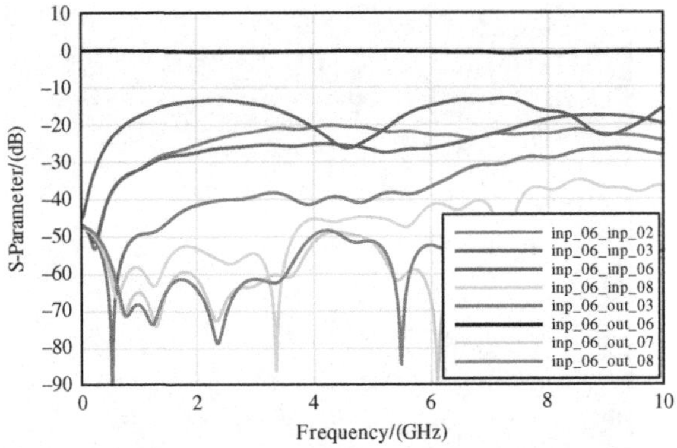

Fig. 10 S-parameters extracted from a time domain simulation with a pulsed excitation using trace 6 as the excitation line

the measured value of 119.3 ps [6]. Additionally, note the reciprocity of the voltage profiles obtained, e.g., for the near-end crosstalk between traces 2 and 6.

Figure 10 shows the S-parameters of the device in the range from 0 to 10 GHz using trace 6 as the excitation line. The far-end coupling is also here (except for trace 3) negligible. The strength of the near-end coupling with traces 2 and 3, however, is close to 20 dB. This effect becomes slightly stronger with increasing frequency.

6 Conclusion

A large scale, full-wave simulation of a complete IC package is presented. The use of distributed computing with a specialized domain partitioning approach provides sufficient computational resources for overcoming the simulation burdens related to the geometrical modeling and to the huge number of field degrees of freedom involved. In view of the ever-increasing need for high speed/high density packaging, the method has the potential of becoming a standard tool in the design of modern ICs.

Acknowledgements The authors wish to thank A. Deutsch at the IBM Thomas J. Watson Research Center for providing the geometry of the package and the measurement data.

References

1. E.-P. Li, E.-X. Liu, and L.-W. Li (2004) "A Coupled Efficient and Systematic Full-Wave Time-Domain Macromodeling and Circuit Simulation Method for Signal Integrity Analysis of High-Speed Interconnects", IEEE Trans. on Adv. Packaging, Vol. 27, No. 1, Feb. 2004, pp. 213–223
2. Rd Goering (2004) "When Bad Packages Kill Good pc Boards", EE Times Online, 2004

3. T. Weiland (1996) "Time Domain Electromagnetic Field Computation with Finite Difference Methods", Int. J. Numer. Model., Vol. 9, 1996, pp. 295–319
4. S. Toledo (1997), "Improving Memory-System Performance of Sparse Matrix-Vector Multiplication", IBM J. Res. Develop., Vol. 41, No. 6, 1997, pp. 711–725
5. F. Wolfheimer, E. Gjonaj, and T. Weiland (2005), "A Parallel 3D Particle-In-Cell Code with Dynamic Load Balancing", Nucl. Inst. Meth. Vol. A 558, 2005, pp. 202–204
6. IBM, Thomas J. Watson Research Center, Yorktown Heights, NY, private communications
7. Computer Simulation Technology (CST) GmbH, www.cst.com

Recent Progress in Unifying the Time- and Frequency-Domain Methods

Zhizhang (David) Chen and Michel M. Ney

Abstract Along with other numerical methods such as finite-difference methods, Method of Moments (MoM) or Method of Weighted Residual (MWR) has traditionally been applied in the frequency domain and has been shown to be very effective and efficient in computing open structure problems. It is generally considered to be a numerical method that is different from other numerical methods. In this paper, we summarize our recent progress in using the MoM as a general framework to unify most of numerical methods developed so far, either in frequency-domain or in time-domain. As the result, numerical issues can now be understood and derived with a common procedure. The significance of such unification is that new grid-based numerical methods, particularly effective and efficient for specific structures, can now be developed with the MoM procedure using new expansion and testing functions.

Keywords Method of moments (MoM) · or method of weighted residuals (MWR) · frequency-domain method · time-domain methods · mode matching and finite-different time-domain method

1 Introduction

Many numerical methods have been developed for simulation of electromagnetic structures. There are mainly two types of numerical methods: frequency-domain and time-domain methods. The frequency-domain methods solve Maxwell's equations in the temporal spectral domain while the time-domain methods in the original time domain in which Maxwell's equations were originally formulated. The frequency-domain methods [1] include finite-difference methods, finite-element

Zhizhang (David) Chen
Department of Electrical and Computer Engineering, Dalhousie University, Halifax, Canada,
e-mail: z.chen@dal.ca

Michel M. Ney
TELECOM Bretagne Institute, Brest, France,
e-mail: michel.ney@telecom-bretagne.eu

P. Russer, U. Siart (eds.), *Time Domain Methods in Electrodynamics*,

methods, method of lines and mode matching. The time-domain methods include the finite-difference time-domain methods [2], transmission-line-matrix methods [3], time-domain finite-element methods [4], and time-domain integral equation methods [5]. These methods have been applied widely in solving electromagnetic structure problems. And they have been shown to be derived and developed independently based on different mathematical bases.

In this paper, we present our recent progress in unifying these methods with the method of moments (MoM) [6], or method of weighted residuals (MWR). We will show that the frequency-domain and time-domain methods can be derived with the Method of Moment (MoM). More specifically, solutions of numerical methods can be obtained by expanding solutions in terms of sets of basis functions and by minimizing the errors with sets of testing functions. Differences between different methods are simply the uses of different expansion and testing (or weighting) functions.

The work presented in this paper is the summary of our recent work. The intention is not to reduce the significance of other ways of deriving numerical methods but to provide an alternative means to understand the existing methods and to possibly create new methods, particularly effective to certain classes of electromagnetic problems.

2 The Method of Moments

Although Maxwell's equations for electromagnetic fields involve vectorized electric and magnetic field quantities **E** and **H**, they can be de-vectorized and expressed in a system of scalar equations such as:

$$L\phi - f = 0 \tag{1}$$

where L is a mathematical operator that can be either differential, integral or mixed differential and integral operators. ϕ is a component of **E** or **H**, or a one-column vector containing all the field components. f is the known source function.

With the Method of Moments (MoM) [6], (1) was solved with two computing steps: solution expansion and error minimization.

In the solution expansion step, a pre-selected set of known basis functions in both space and time is first chosen and then used to expand ϕ. Suppose that the basis functions in time and space for ϕ are $\Phi_m(\mathbf{r})$ and $T_n(t)$, respectively. n and m are the indices for each basis function which is independent of each other. Then,

$$\phi = \sum_{m=1}^{\infty}\sum_{n=1}^{\infty} A_{mn}\Phi_m(\mathbf{r})\,T_n(t) \tag{2}$$

Here A_{mn} are the expansion coefficients that are to be found. $T_n(t)$ is expansion basis function in time. For the frequency-domain methods, it does not exist as the frequency-domain methods deal with a temporal frequency rather than the time factor.

In the error minimization step, the residual error of placing (2) in (1) is:

$$R = L\left[\sum_{m=1}^{\infty}\sum_{n=1}^{\infty} A_{mn}\Phi_m(\mathbf{r})\,T_n(t)\right] - f \tag{3}$$

Making R to be completely zero in the whole solution domain is difficult to achieve. By the way of MoM, a preselected known testing or weighting functions, denoted as $W_k(\mathbf{r})$ and $P_l(t)$, is chosen. It is then used to make R null in terms of its inner product with the testing functions:

$$< R, W_k(\mathbf{r})\,P_l(t) >= 0 \tag{4}$$

More specifically, substitution of (3) into (4) leads to a system of equations that contain the expansion coefficient A_{mn}:

$$\sum_{m=1}^{\infty}\sum_{n=1}^{\infty} A_{mn} < L\Phi_m(\mathbf{r})T_n(t), W_k(\mathbf{r})P_l(t)] >=< f, W_k(\mathbf{r})P_l(t) > \tag{5}$$

The above equation is normally solvable, in some cases in a recursive fashion. As a result, the approximated solutions are obtained with (2).

It should be noted that the expansion basis function and testing functions have to be chosen carefully to avoid solution divergence. Readers are referred to [7, 8] for more details.

3 Derivations of Frequency-Domain Numerical Methods

Many frequency numerical methods have been developed for solving frequency-domain Maxwell's equations [1]. Due to limitation of space, we will consider two of them: the spectral domain and mode matching methods.

3.1 Spectral Domain Method

The spectral domain method was developed specifically for planar structures types, such as microstrip lines ([1] and references therein). Based on the frequency-domain Maxwell's equations, an integral equation is first developed where current densities or charges on the metal strips are the unknown functions to be solved for. Then the method of moment is applied and a system of linear equations is obtained for the expansion coefficients. The key is that the elements of the coefficient matrix of the system of linear equations are efficiently found through the use of the Green's functions in the spectral (or spatial frequency) domain rather than directly in the spatial domain [9]. Therefore, the spectral domain method falls within the framework of MoM.

3.2 Mode Matching Method

Mode matching is one of the most frequently used methods for solving boundary-value problems of waveguide structures ([1] and references therein).

The first step of the mode matching is to expand the unknown fields in the individual regions in terms of their respective modes. The expanded field components are then matched at the interfaces of the two adjacent regions. By utilizing the orthogonality property of the mode functions, a set of linear simultaneous equations can be established for the unknown modal expansion coefficients. Such a process can be considered exactly the same as that of the MoM. More specifically, it can be described as follows.

Suppose that the problem to be solved is:

$$\begin{array}{ll} \text{Region \#1} & \text{Region \#2} \\ L\phi_1 - f_1 = 0 & L\phi_2 - f_2 = 0 \end{array} \tag{6}$$

$$B_1(\phi_1) - B_2(\phi_2) = 0 \quad \text{at interface} \quad \mathbf{r} = \mathbf{r}_{interface} \tag{7}$$

where B_1 and B_2 are the linear operators for the interface conditions.

Then, ϕ_1 and ϕ_2 are expanded in terms of the known mode functions ϕ_{1j} and ϕ_{2j} that satisfy (6) and the boundary conditions in region #1 and region #2, respectively:

$$\phi_1 = \sum_j a_j \phi_{1j} \tag{8}$$

$$\phi_2 = \sum_j b_j \phi_{2j} \tag{9}$$

Substitution of them into the interface conditions (7) leads to the residual:

$$R = \sum_j a_j B_1(\phi_{1j}) - \sum_j b_j B_2(\phi_{2j}) \tag{10}$$

Minimization of R by integration of (10) with the mode functions reads:

$$\begin{aligned} \sum_j a_j \int B_1(\phi_{1j})\,\phi_{1i} ds &= \sum_j b_j \int B_2(\phi_{2j})\,\phi_{1i} ds \\ \sum_j a_j \int B_1(\phi_{1j})\,\phi_{2i} ds &= \sum_j b_j \int B_2(\phi_{2j})\,\phi_{2i} ds \end{aligned} \tag{11}$$

The above equations allow the solutions of the expansion coefficients a_j and b_j.

3.3 Other Frequency-Domain Methods

Other frequency-domain numerical methods can be derived in a similar way. Readers are referred to [10].

4 Derivations of Time-Domain Numerical Methods

Like its frequency-domain counterparts, many time-domain methods have been developed. In the following paragraphs, we present our recent results in deriving these time-domain methods with MoM.

4.1 Finite-Difference Based Time-Domain (FDTD) Methods

The finite-differenced based time domain methods widely used for solving electromagnetic structure problems. They include finite-difference time-domain (FDTD) method, transmission-line-matrix (TLM) method, multi-resolution time-domain (MRTD) method, pseudo-spectral time-domain (PSTD) method, Crank-Nicolson FDTD, alternating-direction-implicit (ADI) FDTD and unconditionally stable FDTD using weighted Laguerre polynomials. They all can be derived from MoM. However, because of space limitations here, only the conventional FDTD of Yee's scheme and the PSTD are derived with MoM. For details on the derivations of other finite-difference based methods, readers are referred to [11].

Before the derivation, the following rooftop function T is introduced:

$$T(\xi,\xi_0,\Delta\xi) = \begin{cases} 1 - \frac{|\xi-\xi_0|}{\Delta\xi} & \text{when } |\xi-\xi_0| \le \Delta\xi \\ 0 & \text{otherwise} \end{cases} \tag{12}$$

Its graphical presentation is shown in Fig. 1:

In the finite-difference time-domain method, the equations to be solved are time-domain Maxwell's equations with six field components in the x, y, and z, directions, respectively. For instance, one of the equations can be expressed as:

$$\varepsilon\frac{\partial E_x}{\partial t} = \frac{\partial H_z}{\partial y} - \frac{\partial H_y}{\partial z} \tag{13}$$

By the way of MoM, the field components are expanded as follows:

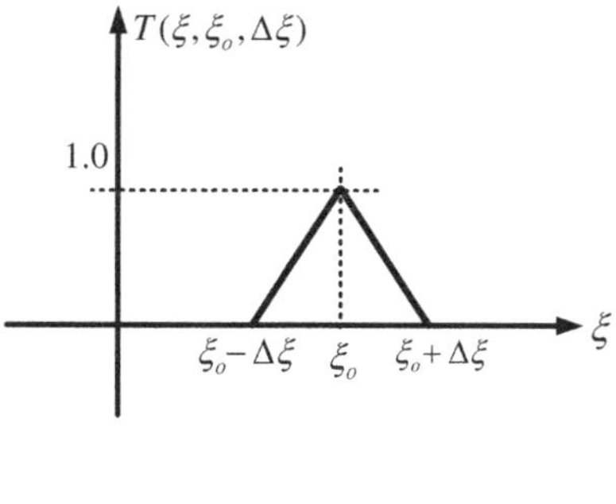

Fig. 1 Illustration of the triangle function T

$$
\begin{aligned}
E_x &= \sum_{i_x,i_y,i_z,n} E_x\Big|^{n}_{i_x+\frac{1}{2},i_y,i_z} T\left[x,\left(i_x+\frac{1}{2}\right)\Delta x,\Delta x\right] \\
&\quad T\left(y,i_y\Delta y,\Delta y\right)T\left(z,i_z\Delta z,\Delta z\right)T\left(t,n\Delta t,\Delta t\right) \\
H_y &= \sum_{i_x,i_y,i_z,n} H_y\Big|^{n+\frac{1}{2}}_{i_x+\frac{1}{2},i_y,i_z+\frac{1}{2}} T\left[x,\left(i_x+\frac{1}{2}\right)\Delta x,\Delta x\right] \\
&\quad T\left(y,i_y\Delta y,\Delta y\right)T\left[z,\left(i_z+\frac{1}{2}\right)\Delta z,\Delta z\right]T\left[t,\left(n+\frac{1}{2}\right)\Delta t,\Delta t\right] \\
H_z &= \sum_{i_x,i_y,i_z,n} H_z\Big|^{n+\frac{1}{2}}_{i_x+\frac{1}{2},i_y+\frac{1}{2},i_z} T\left[x,\left(i_x+\frac{1}{2}\right)\Delta x,\Delta x\right] \\
&\quad T\left[y,\left(i_y+\frac{1}{2}\right)\Delta y,\Delta y\right]T\left(z,i_z\Delta z,\Delta z\right)T\left[t,\left(n+\frac{1}{2}\right)\Delta t,\Delta t\right]
\end{aligned} \tag{14}
$$

where Δx, Δy and Δz are the spatial steps and Δt is the time step. i_x, i_y, i_z and n are the spatial and temporal indices. $E_x|^{n}_{i_x+\frac{1}{2},i_y,i_z}$, $H_y|^{n+\frac{1}{2}}_{i_x+\frac{1}{2},i_y,i_z+\frac{1}{2}}$ and $H_z|^{n+\frac{1}{2}}_{i_x+\frac{1}{2},i_y+\frac{1}{2},i_z}$ are expansion coefficients. Note that because of the use of the rooftop function, expansion coefficients happen to be the field values at the grid points. For instance, $E_x|^{n}_{i_x+\frac{1}{2},i_y,i_z} = E_x(t=n\Delta t, x=(i_x+\frac{1}{2})\Delta x, y=i_y\Delta y, z=i_z\Delta z)$

Substitution of (14) into (13) leads to

$$
\begin{aligned}
&\varepsilon\frac{E_x|^{n+1}_{i_x+\frac{1}{2},i_y,i_z} - E_x|^{n}_{i_x+\frac{1}{2},i_y,i_z}}{\Delta t} = \frac{H_z|^{n+\frac{1}{2}}_{i_x+\frac{1}{2},i_y+\frac{1}{2},i_z} - H_z|^{n+\frac{1}{2}}_{i_x+\frac{1}{2},i_y-\frac{1}{2},i_z}}{\Delta y} \\
&\quad - \frac{H_y|^{n+\frac{1}{2}}_{i_x+\frac{1}{2},i_y,i_z+\frac{1}{2}} - H_{yz}|^{n+\frac{1}{2}}_{i_x+\frac{1}{2},i_y,i_z-\frac{1}{2}}}{\Delta z} = 0
\end{aligned} \tag{15}
$$

This is exactly the same as the FDTD equation derived by replacing differentials with their finite-difference counterparts [2].

For derivations of other finite-difference based time-domain methods with Method of Moments (MoM) including the derivations of PSTD and Crank Nicolson methods, readers are referred to [11].

4.2 Derivations of the Time-Domain Finite-Element Methods

The time-domain finite-element methods have gained much attention recently due to demands for wideband and transient simulations of modern communication devices and components. In the derivations of the time-domain FEM, MoM procedure was applied for expansion and error testing in the spatial domain [4, 12]. In the time domain, the finite-differences were used to replace the temporal derivatives [12]. Since a finite-difference approach can be derived from the MoM as shown above for the derivation of the FDTD method, the time-domain FEM is then derivable from the MoM. More details can be found in [12].

4.3 Derivations of the Time-Domain Integral Equation (TDIE) Methods

TDIE methods are another type of numerical methods where solutions can be obtained from field integration over interfaces and boundaries with appropriate Green's functions [13]. They have the advantage of reducing problem complexity by one dimension, i.e. three-dimensional structures are solved with two-dimensional equations and two-dimensional structures with one-dimensional equation. However, computational inefficiency and late-time instability have prevented them from becoming effective simulation tools. Much of the recent research efforts have been focused on resolving the instability issue.

There are many different integral equations derived from Maxwell's equations such as the electric field integral equation (EFIE). They have been solved by following the solution steps of MoM, expansion of field quantities and minimization of residual errors [12].

5 Numerical Instability

In a physically realizable problem, field quantities are of finite values. Therefore, expansion (2) should be finite at any spatial location and at any time. In a normal situation, expansion basis functions $\Phi_m(\mathbf{r})$ and $T_n(t)$ are selected to be bounded. Therefore, to ensure that (2) is finite, the expansion coefficient A_{mn} should be finite in its value in particular when $n \to \infty$ and $m \to \infty$. This has translated into the well-known CFL stability condition in the FDTD method; it ensures that the expansion coefficients do not grow with time. More information for the FDTD case can be found in [11]. For other methods, appropriate conditions can also be developed based on the MoM perspective.

6 Numerical Dispersion

Since numerical methods are shown to provide expanded approximate solutions, it is very desirable to assess the errors of such approximations and impacts of discretization parameters Δx, Δy, Δz and Δt on these errors.

Direct evaluation of residual R represented by (3) is normally difficult. An alternative way is to examine the approximate solution (2) in the spectral domain. In other words, because the solutions are approximate, relationship between the spatial frequencies and temporal frequency of the approximate solutions will be different from the one for the exact solutions. This leads to the so-called numerical dispersion errors. In the FDTD case, such a numerical dispersion can be found in a simple analytical form as described in [11]. In other cases, they can be found only numerically.

7 Discussions and Conclusions

In this paper, we summarize our recent results in unifying time- and frequency-domain numerical methods for computing electromagnetic structures with a common MoM framework. It has been shown that numerical methods are derivable with the MoM procedure. The differences among the different methods are the use of different expansion and testing functions. Therefore, not only numerical methods are unified under the framework of MoM, but also new methods, particularly effective and efficient for specific structures, can now be developed with a common procedure. In addition, numerical instability and dispersions can be explained relatively easily within the framework.

It should be mentioned that the work presented is not intended to exclude existing or other ways of developing numerical methods. It is meant to provide another perspective of understanding numerical methods and their associate physical interpretations. It is hoped that it may give another dimension in advance of computational electromagnetics.

References

1. T. Itoh, Ed., *Numerical Techniques for Microwave and Millimeter-wave Passive Structures*, John Wiley & Sons, 1989
2. A. Taflove and S. Hagness, *Computational Electrodynamics*: The *Finite-difference Time-domain Method,* 3rd ed. Artech House, 2005
3. W. J. R. Hoefer, "The transmission-line matrix method-theory and applications," *IEEE Trans. Microwave Theo.* Tech., Vol. 33, No. 10, Oct., 1985, pp. 882–893
4. J. Lee, R. Lee and A. Cangellaris, "Time-domain finite-element methods," *IEEE Trans. Antennas Propagat.,* No. 3, Mar. 1997, pp. 430–442
5. D. Weile, G. Pisharody, N.-W. Chen, B. Shanker and E. Michielssen, "A novel scheme for the solution of the time-domain integral equations of electromagnetics," *IEEE Trans. Antennas Propagat.,* No. 1, Jan. 2004, pp. 283–285
6. R. Harrington, *Field Computation by Moment Methods,* IEEE Press, 1993
7. T. K. Sarkar, A. R. Djordjevic and E. Arvas, "On the choice of expansion and weighting functions in the numerical solution of operator equations," *IEEE Trans. Antennas Propagat.,* Vol. AP-33, No. 9, Sept. 1985, pp. 988–996
8. T. K. Sarkar, "A note on the choice weighting functions in the method of moments," *IEEE Trans. Antennas Propagat.*, Vol. 33, No. 4, April 1985, pp. 436–41

9. D. Davidson and J. Aberle, "An introduction to spectral domain method-of-moments formulations," *IEEE Antennas and Propagation Magazine,* Vol. 46, No. 3, June 2004, pp. 11–19
10. Z. Chen and M. Ney, "The method of weighted residuals: a general approach to deriving time- and frequency-domain numerical methods," Submitted for publication in *IEEE Antennas and Propagation Magazine*
11. Z. Chen and S. Luo, "Generalization of the finite-difference-based time-domain methods using the method moments," *IEEE Trans. Antennas Propagat.,* Vol. 54, No. 9, Sept. 2006, pp. 2515–2524
12. Z. Lou and J. Jin, "A new explicit time-domain finite-element method based on element-level decomposition," *IEEE Trans. Antennas Propaga.,* Vol. 54, No. 10, Oct. 2006, pp. 2990–2999
13. G. Pisharody and D. Weile, "Robust solution of time-domain integral equations using loop-tree decomposition and band limited extrapolation," *IEEE Trans. Antennas Propagat.,* Vol. 53, No. 6, June 2005, pp. 2089–2098

Time-Domain Neural Network Approaches to EM Modeling of Microwave Components

Qi-Jun Zhang and Yi Cao

Abstract Time-domain modeling of EM behaviors with geometrical parameters as variables is addressed in this paper. Two approaches, a combined-state-space equation/neural network technique (SSE-NN), and a recurrent neural network (RNN) technique are described. In the SSE-NN approach, the model is a hierarchical structure with two levels. In the lower level, a neural network maps the geometrical/physical parameters of the passive component into coefficient matrices of state equations. In the higher level, the coefficient matrices are transformed into the state space equation to compute the EM response in time domain circuit design. The RNN approach provides a direct time-domain model trained from data generated by time-domain EM simulation. A time-domain neural network structure called RNN is exploited to model the transient EM responses for varying material and geometrical parameters. An automated RNN modeling technique is introduced to efficiently determine the training waveform distribution and internal RNN structure during the offline training process. Through two examples it's demonstrated that the trained time domain neural network model provides fast EM solutions with variable values of the geometrical parameter in the model.

1 Introduction

Increasing complexity, stringent specifications and the need for shorter-design cycles for today's microwave circuits, demand fast and accurate tools. EM-based modeling and design is essential. During EM based design, geometrical parameters of the EM structure are variables, leading to repetitive EM simulations with changing values of geometry. This expensive computational challenge is an active topic of microwave CAD. Neural Networks and Space Mapping are recognized recent developments addressing such challenges [1]. Neural networks can be trained to learn

Qi-Jun Zhang
Department of Electronics, Carleton University, 1125 Colonel By Dr., Ottawa, Canada K1S 5B6

Yi Cao
Department of Electronics, Carleton University, 1125 Colonel By Dr., Ottawa, Canada K1S 5B6

P. Russer, U. Siart (eds.), *Time Domain Methods in Electrodynamics*,

EM data with varying geometrical parameters, and trained neural networks become fast models for circuit design with geometrical optimization. Recent research has led to a variety of applications such as modeling and design of microstrip lines, vias, spiral inductors, transistors, VLSI interconnects, coplanar waveguide discontinuities, printed antennas, and embedded passives, circuit synthesis, optimization, and yield analyses [2, 3, 4].

This paper describes time-domain neural network approaches of EM modeling. In the past, neural network techniques have been used to model frequency domain behavior of EM structures, such as embedded passives [5]. However, such neural network models, trained to learn S-parameters data, cannot be used directly in time-domain circuit simulation and optimization. Aimed at developing passive neural network based models from EM data that can be used directly in both time and frequency domain circuit design, a hybrid approach based on the state-space equation and neural network (SSE-NN) [6] has been presented. The SSE-NN model is hierarchical with two levels. At the higher level, a set of linear state equations are used to provide the EM response in time domain. At the lower level, a neural network maps the geometrical parameters of the EM structure into the coefficients of the state equations. The final combined model is converted into a set of mathematical expressions that can be directly used to carry out high-level circuit design in time-domain simulators. By allowing geometrical/physical parameters as design variables, the SSE-NN model helps to improve circuit performance and reduce design/manufacture cost.

Recently, with the maturation of time-domain EM solvers based on algorithms such as transmission line matrix (TLM) [7, 8], direct time-domain EM modeling has become an important subject. Time-domain EM solvers are efficient in obtaining wideband information of EM structures in a single transient simulation. Using the fact that the transient EM responses to an applied input are described by Maxwell's partial differential equations (PDE), through discretization the time sequence of the input-output can be converted into a finite difference relationship involving multiple time points (history). As the precise finite difference equation representing the time-domain behavior is unknown, a time-domain neural network structure called recurrent neural network (RNN) [9, 10] has been applied to learn and model the port dynamics of arbitrary structures by training with the discrete-time EM responses. As geometrical parameters vary, the RNN formulation avoids the tracking of pole-zero movements [11] as required by indirect state-space [12, 13, 14] or pole-zero [15, 16, 17] based approaches for time-domain EM modeling. To develop the RNN model with minimum human efforts, an automatic RNN modeling technique is also introduced to automatically drive the EM solver to generate the appropriate training data and adjust the RNN structure during training [18].

This paper is organized as follows. Firstly we provide a brief overview of neural network modeling for microwave problems. Then we describe the formulation and training process of the combined SSE-NN modeling approach. The RNN macromodel and its application to transient EM modeling are given in the next section. Two numerical examples are presented to highlight the accuracy and efficiency of time-domain EM modeling techniques.

2 Overview of Neural Network for RF and Microwave Modeling

In the past few years, both frequency-domain and time-domain EM modeling have been addressed by neural network approaches. Between the time-domain and frequency domain, neural network model was first developed for frequency-domain case because of the simplicity of the frequency-domain formulation. In this section, we introduce the key concepts of neural network based EM modeling. For simplicity of the presentation, the frequency domain illustrations are used. The basic neural network concepts introduced here will be used to further develop advanced time domain formulations in the subsequent sections.

Let $\boldsymbol{x}$ represent an N_x-vector containing physical/geometrical parameters of a microwave component, e.g., length of an embedded resistor. Let $\boldsymbol{y}$ represent an N_y-vector containing the responses of the component under consideration, e.g., S-parameter. The physics/EM relationship between $\boldsymbol{x}$ and $\boldsymbol{y}$ can be represented as $\boldsymbol{y} = \boldsymbol{y(x)}$. The theoretical model for $\boldsymbol{y(x)}$ may be computationally too expensive for online microwave design and repetitive optimization (e.g., 3D full-wave EM analysis inside a Monte Carlo statistical design loop). The objective now is to develop a fast model that will accurately represent the original $\boldsymbol{x}$-$\boldsymbol{y}$ relationship. Since neural networks are efficient computational tools that can easily handle nonlinear and multidimensional problems [1], they become good candidates to replace the original EM models during microwave design.

A neural model is developed by teaching/training a neural network to learn the microwave problem through a set of measured/simulated sample pairs called training data $\{(\boldsymbol{x}_p, \boldsymbol{d}_p), \ p \in T_R\}$, where $\boldsymbol{d}_p$ represents the measured/simulated output $\boldsymbol{y}$ for the input $\boldsymbol{x}_p$, and T_R is the index set of training data. The neural model is defined as [1]

$$\tilde{\boldsymbol{y}} = \tilde{\boldsymbol{y}}\left(\boldsymbol{x},\boldsymbol{w}\right), \tag{1}$$

where $\boldsymbol{w}$ is the weight vector containing N_w trainable parameters of the neural network. For the purpose of training, we define error function $E(\boldsymbol{w})$ as

$$E\left(\boldsymbol{w}\right) = \sum_{p \in T_R} \sum_{k=1}^{N_y} \left(\tilde{y}_k(\boldsymbol{x}_p,\boldsymbol{w}) - d_{pk}\right)^2, \tag{2}$$

where d_{pk} is the kth element of $\boldsymbol{d}_p$ and $\tilde{y}_k(\boldsymbol{x}_p,\boldsymbol{w})$ is the kth output of the neural network for input sample $\boldsymbol{x}_p$. The objective of neural network training is to find $\boldsymbol{w}$ such that $E(\boldsymbol{w})$ is minimized. In general, the structure of the neural network is specifically determined by the definition of $\boldsymbol{w}$, and the approach by which $\tilde{\boldsymbol{y}}$ is computed from $\boldsymbol{x}$ and $\boldsymbol{w}$.

The most widely used neural network structure is the multi-layer perceptrons (MLP) neural network [1], which is part of the feed-forward neural network (FFNN) family. A typical 3-layer MLP neural network is shown in Fig. 1. For a general MLP, $\tilde{y}$ is computed starting from the input layer $z_i^0 = x_i$, and then proceeding through the hidden layers as

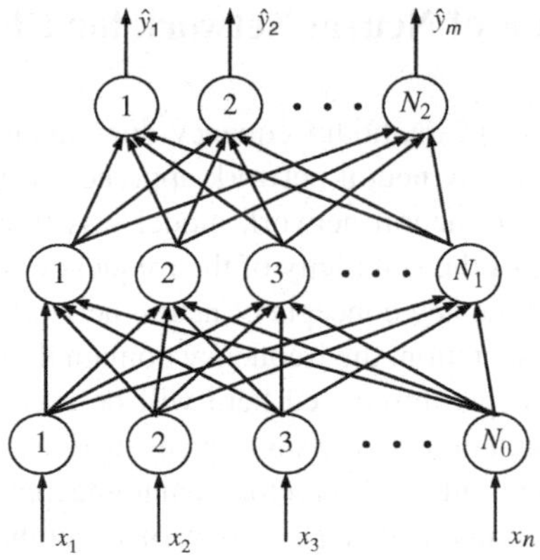

Fig. 1 Illustration of the feedforward multilayer perceptrons (MLP) structure

$$z_i^l = \sigma\left(\sum_{j=1}^{N_{l-1}} w_{ij}^l z_j^{l-1} + w_{i0}^l\right),$$

$$i = 1,\ldots,N_l, \quad l = 1,\ldots,L+1 \tag{3.1}$$

$$\tilde{y}_k = z_k^{L+1} \quad k = 1,\ldots,N_y \tag{3.2}$$

where x_i is the ith input to the neural network, N_l is the number of neurons in layer l, z_i^l is the output of ith neuron of lth layer, w_{ij}^l represents weight of the link between jth neuron of $l-1th$ layer and ith neuron of lth layer, w_{i0}^l is the bias parameter of ith neuron of lth layer, and L is the total number of layers. The input and output layers are denoted as layer *0* and layer $L+1$, respectively. In Eq. (3.1), $\sigma(\cdot)$ reprensets the neuron activation function, which is usually a sigmoid function for the hidden layers and a linear function for the output layer.

Training is an essential step in neural network model development. An appropriate structure cannot guarantee to achieve a good model, unless trained by a suitable training algorithm. Due to the complexity of training error $E(w)$, iterative algorithms are often used to explore the weight space. Popularly used training algorithms include conjugate gradient and quasi-Newton methods [1]. After training, the neural model should be tested with a separate set of data (from the same problem) which has not been used during training. If the neural model can match the test data accurately, then the resulting model is ready to be used in microwave design.

Because the FFNN structures described above do not provide any time-domain dynamic feature, they cannot be used directly in time-domain circuit simulation and optimization. In next two sections, we will describe two advanced neural network formulations for time-domain modeling where the feedforword MLP structure shown in Fig. 1 and described in this section is used as a portion of the overall model structure. To facilitate the training of time-domain neural models, existing training algorithms for FFNNs also need to be modified to accommodate different forms of training data and different types of structures.

3 Combined SSE-NN Modeling Approach

3.1 Frequency-Domain Formulation

Through full wave EM simulation/measurement, EM data of a passive component with respect to different geometrical/physical parameters can be obtained. For a given frequency range, we can use transfer functions (polynomial rational functions) to represent the electrical behavior (e.g., admittance **Y** matrix) of a passive component. For any two-port passive component, the following three transfer functions are adequate to represent Y_{11}, Y_{21}, and Y_{22}, respectively,

$$H_1(s) = \frac{b_0 + b_1 s + \cdots + b_{n-1} s^{n-1} + b_n s^n}{a_0 + a_1 s + \cdots + a_{n-1} s^{n-1} + s^n} \tag{4.1}$$

$$H_2(s) = \frac{d_0 + d_1 s + \cdots + d_{n-1} s^{n-1} + d_n s^n}{a_0 + a_1 s + \cdots + a_{n-1} s^{n-1} + s^n} \tag{4.2}$$

$$H_3(s) = \frac{c_0 + c_1 s + \cdots + c_{n-1} s^{n-1} + c_n s^n}{a_0 + a_1 s + \cdots + a_{n-1} s^{n-1} + s^n} \tag{4.3}$$

where $s = j\omega$, and n is the number of effective order of the passive. Let us define a real coefficient vector, as $\boldsymbol{g}_v = \{a_0,\ a_{1,\dots}\ a_{n-1};\ b_0,\ b_1, \dots b_n;\ c_0,\ c_1, \dots c_n;\ d_0,\ d_1, \dots d_n\}$. Using space-mapping concept [19], a relationship exists between the coefficients and geometrical/physical parameters. However, the relationship would be highly nonlinear and too complicated. Therefore, neural networks are exploited to learn the highly nonlinear relationship between the coefficients and geometrical/physical parameters.

3.2 State Space Equation for Time-Domain Simulation

Using the coefficients $\boldsymbol{g}_v$ in (4), four matrices are defined

$$\boldsymbol{A} = \begin{bmatrix} 0 & 1 & 0 & \cdots & 0 & 0 & 0 & \cdots & 0 & 0 \\ 0 & 0 & 1 & \cdots & 0 & 0 & 0 & \cdots & 0 & 0 \\ \vdots & \vdots & \vdots & \ddots & 0 & \vdots & \vdots & \ddots & \vdots & \vdots \\ -a_0 & -a_1 & -a_2 & \cdots & -a_{n-1} & 0 & 0 & \cdots & 0 & 0 \\ 0 & 0 & \cdots & 0 & 0 & 0 & 1 & 0 & \cdots & 0 \\ 0 & 0 & \cdots & 0 & 0 & 0 & 0 & 1 & \cdots & 0 \\ \vdots & \vdots & \ddots & \vdots & \vdots & \vdots & \vdots & \vdots & \ddots & \vdots \\ 0 & 0 & \cdots & 0 & 0 & -a_0 & -a_1 & -a_2 & \cdots & -a_{n-1} \end{bmatrix}_{2n \times 2n} \tag{5}$$

$$\boldsymbol{B} = \begin{bmatrix} 0 & 0 & \cdots & 1 & 0 & 0 & \cdots & 0 \\ 0 & 0 & \cdots & 0 & 0 & 0 & \cdots & 1 \end{bmatrix}^T_{2n\times 2} \qquad \boldsymbol{D} = \begin{bmatrix} b_n & d_n \\ d_n & c_n \end{bmatrix}_{2\times 2}$$

$$\boldsymbol{C} = \begin{bmatrix} b_0 - a_0 b_n & \cdots & b_{n-1} - a_{n-1} b_n & d_0 - a_0 d_n & \cdots & d_{n-1} - a_{n-1} d_n \\ d_0 - a_0 d_n & \cdots & d_{n-1} - a_{n-1} d_n & c_0 - a_0 c_n & \cdots & c_{n-1} - a_{n-1} c_n \end{bmatrix}_{2\times 2n}$$

to form the state space equation,

$$\begin{cases} \dot{\boldsymbol{x}}(t) = \boldsymbol{A}\boldsymbol{x}(t) + \boldsymbol{B}\boldsymbol{u}(t) \\ \boldsymbol{y}(t) = \boldsymbol{C}\boldsymbol{x}(t) + \boldsymbol{D}\boldsymbol{u}(t) \end{cases} \tag{6}$$

where $\boldsymbol{x}(t)$ is a vector of internal states, $\boldsymbol{u}(t)$ and $\boldsymbol{y}(t)$ are vectors of the input and output signals, e.g., input voltages and output currents of the passive components respectively. Following the state space equation in Eq. (6) or the transfer functions in Eq. (4), the combined model can be implemented into a time domain circuit simulator or a frequency domain circuit simulator, respectively.

3.3 Structure of the Combined SSE-NN Model

There are two levels in the hierarchical structure of the combined SSE-NN model. At the lower level, a neural network maps the geometrical/physical parameters into $\boldsymbol{g}_v$ vector. At the higher level, we insert the coefficient vectors into the state equations to compute the EM response in frequency or time domain simulation. Figure 2 shows the structure of the combined SSE-NN model. For circuit CAD tools in time domain, we export our SSE-NN into SPICE sub-circuit format using Eqs. (5) and (6). The lower-level neural network will be described by a set of mathematical equations, which calculate the coefficient values based on different geometrical/physical parameters and pass them into the higher level state space equations.

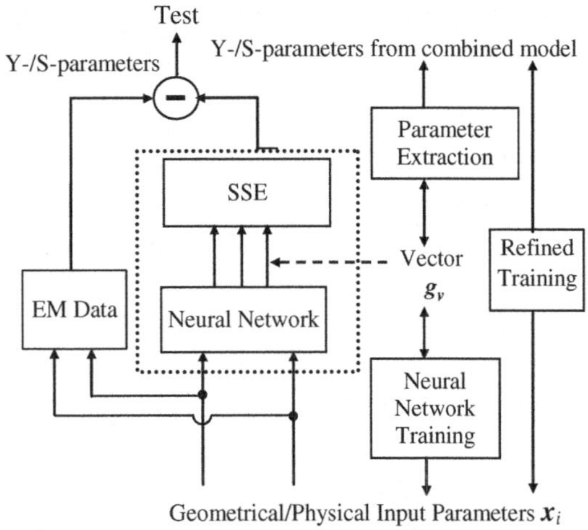

Fig. 2 Structure of the combined SSE-NN models illustrating the model development process and the testing phase

3.4 Combined SSE-NN Model Training

Firstly, the EM data are generated from simulation/measurement with component's geometrical/physical parameters and frequency as inputs and S-parameters as outputs. The next phase is parameter extraction, which is carried out for each geometry over the entire frequency range. The objective here is to determine the coefficient values that best fit the original EM data. Different geometrical parameter values and their corresponding coefficient values are then re-arranged into neural network training data. A 3-layer MLP neural network is trained using quasi-Newton algorithm in *NeuroModeler* [20], to map the geometrical dimensions of the component to the elements of vector $\boldsymbol{g}_v$. We combine the state equation with the neural model using our hierarchical setup to obtain the overall combined model. The inputs to the combined model are the geometrical dimensions of the embedded component. The final outputs of the combined model are component's EM behavior, e.g., S-parameters. In the test phase, an independent set of test data containing S-parameters versus new geometrical parameter values (i.e., never seen during training) is used to test the accuracy of the combined model. In the final phase, the combined model is transformed into a set of mathematical expressions such that the model can be directly used in time-domain simulators for high-level circuit design.

4 RNN-Based Modeling Approach

4.1 RNN Macromodel

RNN is a suitable choice for time-domain modeling due to the presence of feedback (recurrency) and memory (history) in the structure. Previously, the RNN structure has been used to model the dynamic behavior of nonlinear circuits [10]. The RNN macromodel is illustrated in Fig. 3 for a time-varying input signal $\boldsymbol{u}(t)$ and output signal $\boldsymbol{y}(t)$. Mathematically let g_{RNN} represent the RNN as

$$\begin{aligned} y(kT-\tau) &= g_{\mathrm{RNN}}(u(kT),\boldsymbol{w},\boldsymbol{p}) \\ &= f_{\mathrm{FFNN}}(y((k-1)T-\tau),\ldots,y((k-M_y)T-\tau), \\ &\quad u(kT),u((k-1)T),\ldots,u((k-M_u)T),\boldsymbol{w},\boldsymbol{p}) \end{aligned} \tag{7}$$

where k is the index for time step, T is the time step size, M_y is the feedback order, M_u is the input history, f_{FFNN} represent the input-output relationship of a FFNN, $\boldsymbol{w}$ is a vector containing the internal weights of the FFNN, and $\boldsymbol{p}$ is the vector of time-independent parameters. The number of delayed time steps M_y and M_u, of $\boldsymbol{y}$ and $\boldsymbol{u}$ respectively, represent effective order of the original circuit as seen from input-output data.

To train/teach a FFNN to learn the training data, nonlinear gradient-based optimization techniques such as conjugate gradient and quasi-Newton are commonly

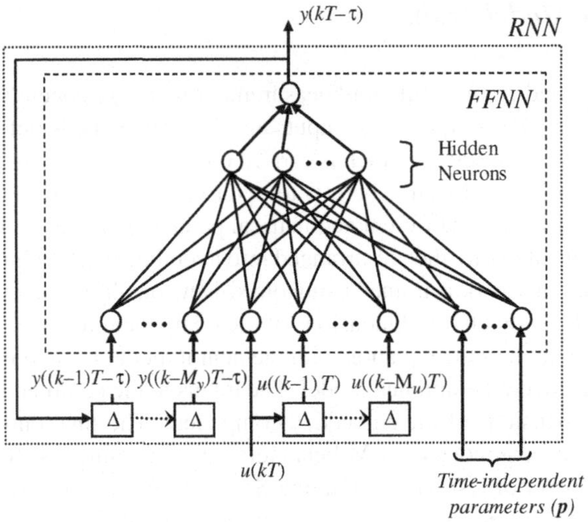

Fig. 3 RNN structure with output feedback (M_y). The RNN is a discrete time structure trained with sampled input-output data

used to find the optimal internal weights (***w***) [2]. For the RNN, the presence of feedback adds an additional complexity since the output is not only a function of the present inputs but also the previous outputs. To address this issue, the back-propagation through time (BPTT) concept [21] is applied to calculate gradients that include the recurrent nature of the RNN. Using the BPTT gradients, the RNN can then be trained with time-domain input/output sequence data. However to achieve good training for a given set of training waveforms, the RNN must have sufficient feedback order and hidden neurons. For a given RNN modeling problem, the precise number of hidden neurons and order required is not known *a priori*. As a result, developing a good RNN model may require substantial manual efforts leading to the inefficient model development process. Recently, an automatic model generation (AMG) for RNN is introduced by integrating the various RNN development subtasks, such as data generation of training waveforms, choosing the number of hidden neurons and order, training, and validation, into one single automated process [18].

4.2 RNN Training with Transient EM Data

Figure 4 illustrates the EM simulation setup for the purpose of RNN data generation. Only three sets of port responses are required for RNN training of 2-port passive structures. The input excitation $u_{inc}(t)$ applied to Port 1 should be capable of establishing a dominant mode of propagation within the structure and can represent either

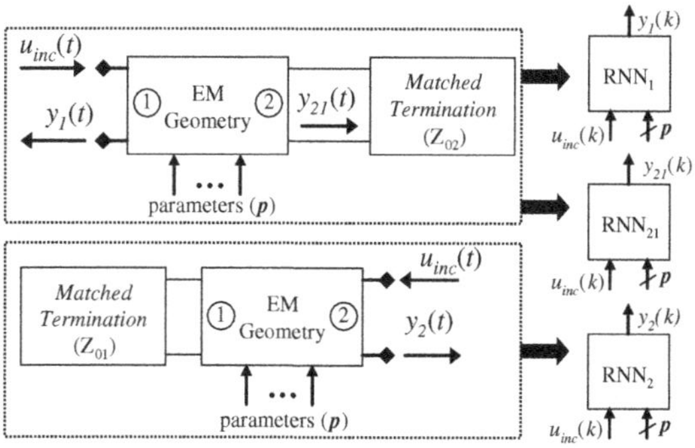

Fig. 4 EM simulation setup for EM data generation

voltage, E-field, or H-field as allowed by the time-domain EM solver. A variety of waveforms can be used as the input excitation $u_{inc}(t)$ for generating RNN training data. A suitable excitation waveform should have sufficient bandwidth to produce port responses to train the model. In this technique, the excitation used is a Gaussian pulse signal [18]. The resulting port responses $y_1(t)$ and $y_{21}(t)$ will be used to train RNN_1 and RNN_{21} respectively. The same excitation is then applied to Port 2 and the port response $y_2(t)$ is used for RNN_2 training. The time-domain simulations are then repeated for different combinations of geometrical and material parameters (i.e. time-independent RNN inputs $\boldsymbol{p}$) to build the entire data sets for RNN training.

Usually, a relatively long initial output delay (τ) may exist between the input excitation and the EM transient response y_{21}. Direct training of RNN with such transient data may suffer from slow convergence and poor accuracy. To avoid that, the initial delay (τ) can be removed during training by setting $y_{21}(kT) = y_{21}(kT + \tau)$. The removed delay (τ) is later reintroduced when the final RNN model is implemented into the circuit simulator.

The time-domain EM simulation should run until all the transient responses to the excitation input decay to zero. As a result, the port responses can be quite long with many samples, which will cause an additional slow down of RNN training. A simple heuristic is to reduce the length of the training sequences by considering fewer time samples in accordance with the Nyquist interval. The RNN sampling interval T in Eq. (7) can be set according to

$$T_{EM} < T < Nyquist\ Interval = \left(\frac{1}{2(BW)}\right). \tag{8}$$

where BW is the bandwidth of the input excitation signal, and T_{EM} is the actual simulation time step. Re-sampling the EM data using the interval T in Eq. (8) prevents aliasing problems while reducing the length of the training sequences. The final choice of the sampling interval should be the maximum value allowed by Eq. (8) before significant sampling distortion of the EM responses occurs.

5 Numerical Examples

5.1 SSE-NN Modeling of an Embedded Resistor

This example demonstrates the SSE-NN modeling of an embedded resistor as shown in Fig. 5 [6], which represents an emerging technology in high-speed multilayer print circuit board (PCB) design. The EM data of the embedded resistor is automatically generated from EM simulation of *Sonnet* [22]. Length (L) and width (W) are used as inputs. The outputs are real and imaginary parts of S_{11} and S_{21} in the EM data.

We have trained multiple SSE-NN models based on various orders of state equations in SSE-NN modeling development and the best results are obtained with the 3rd order SSE-NN model where the testing error is 1.12% [6]. The agreement between 3rd order SSE-NN model and EM data is achieved even though the independent testing data was never seen in training.

Using the trained SSE-NN models, we performed Monte-Carlo analysis of a three-coupled transmission line circuit with geometrical parameters as statistical design variables [6]. The output of Monte-Carlo analysis is measured at the buffer input port. Signal integrity curves generated from the Monte-Carlo analysis are shown in Fig. 6. The total simulation time for 500 output curves based on the geometry tolerance around the nominal design center is 8.24 min using SSE-NN models by

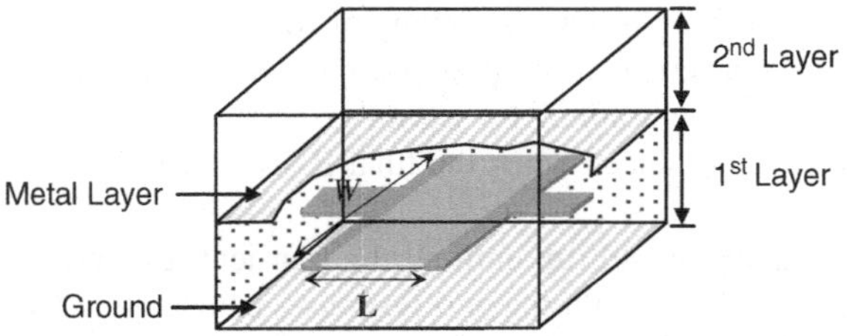

Fig. 5 3-D physical structure of embedded resistor

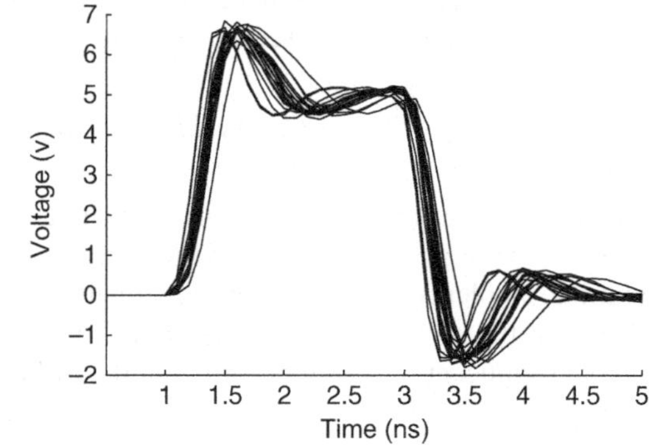

Fig. 6 Output of Monte-Carlo analysis of the 3-coupled transmission line circuit using the SSE-NN models of embedded passives. Here 20 randomly chosen curves are shown out of 500 simulations of the circuit. The 20 waveforms are obtained using the SSE-NN models with 20 random perturbations of the geometrical parameters of the embedded passives

Hspice [23]. However, the required time of *Ansoft-HFSS* [24] for 500 different geometry is more than 8 h. The combined SSE-NN models retain the advantages of neural network learning, speed, and accuracy, and provide EM effects in high-level circuit design.

5.2 RNN Time Domain Modeling of a Microstrip Filter

In this example, a RNN macromodel is developed for a microstrip filter [9] using EM data generated from a time-domain EM solver *MEFiSTo* [25]. The top view of the 2-port filter structure is shown in Fig. 7 with a user-defined dimension, L.

It is desired to model the 2-port behavior of the filter over a bandwidth of 4.5 Ghz for L between 5 and 19 mm. An input TEM Gaussian pulse with an approximate bandwidth of 4.5 Ghz ($\sigma \approx 58.9$ ps) is launched for a simulation of 4097 time steps ($T_{EM} = 1.66782$ ps/step) until all the port responses decay to zero. The resulting E-field in the z-direction (E_z) at the ports is used for RNN training. An output delay of 293 samples ($\tau = 293T_{EM}$) in $f_{21}(t)$ is removed before RNN training. The Nyquist interval is calculated as 111.1 ps, so selecting a sampling interval of $T = 25T_{EM} = 41.7$ ps leads to port responses that are shortened without adding a significant amount of sampling distortion.

The automated RNN technique is used to train all three RNN to represent the port dynamics of the filter. Table 1 shows the final RNN structures with good training results. The transient port responses for geometries not used in training are shown in Fig. 8 to demonstrate the accurate generalization capability.

The transient port responses of 15 filter geometries using TLM require approximately 39 s while the RNN macromodel takes about 10 s. For complex EM structures, the speed up will become even more pronounced.

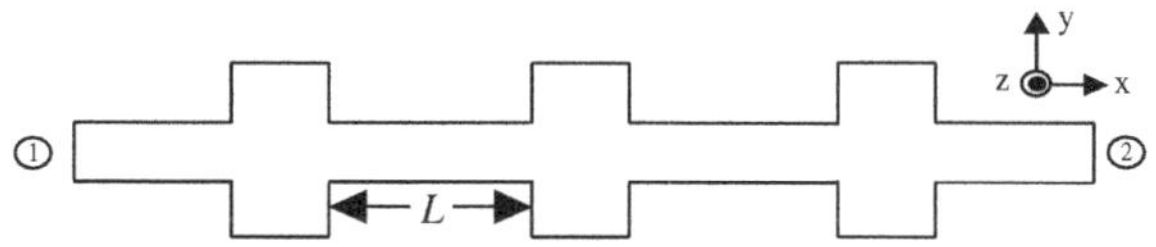

Fig. 7 Microstrip filter with dimension L

Table 1 RNN training results for microstrip filter example

RNN	# of Hidden neurons	Final RNN order (My=Mu)	Average L2 error (%) (15 geometries)
RNN_1	20	17	0.272
RNN_{21}	20	17	0.440
RNN_2	20	17	0.281

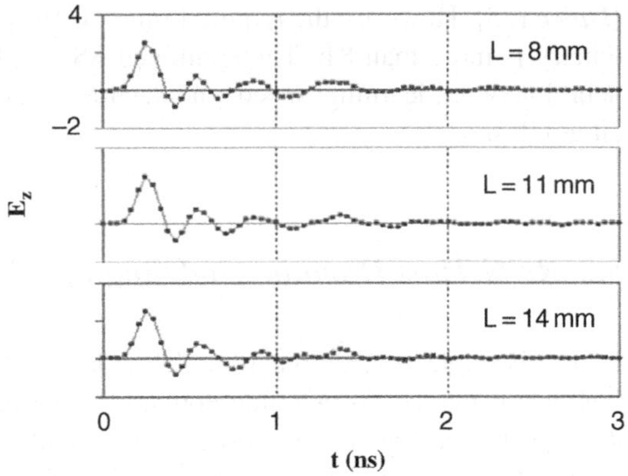

Fig. 8 Comparison between microstrip RNN_1 responses (-) and TLM responses (■) for *L* not used in training. This demonstrates the accurate generalization capability of the RNN

6 Conclusion

This paper has reviewed two types of neural network approaches for time-domain EM modeling of passive microwave components. As an indirect method, the combined SSE-NN model is a two-level hierarchical structure that is developed from the frequency-domain EM data. The RNN modeling approach, on the other hand, provides a direct means to model the EM behaviors from the data generated by time-domain EM simulation. As demonstrated by the examples, these neural network approaches provide fast and accurate models with EM effects and allow geometrical/physical parameters to become design variables in circuit simulation.

References

1. Q.J. Zhang, and K.C. Gupta, *Neural Networks for RF and Microwave Design*, Boston: Artech House, 2000.
2. Q.J. Zhang, K.C. Gupta, and V.K. Devabhaktuni, "Artificial neural networks for RF and microwave design—from theory to practice," *IEEE Trans. Microw. Theory Tech.*, vol. 51, no. 4, pp. 1339–1350, April 2003.
3. P. Burrascano, S. Fiori, and M. Mongiardo, "A review of artificial neural networks applications in microwave computer-aided design," *Int. J. RF and Microw. CAE*, vol. 9, pp. 158–174, 1999.
4. J.E. Rayas-Sanchez, "EM-based optimization of microwave circuits using ANN: the state-of-the-art," *IEEE Trans. Microw. Theory Tech.*, vol. 52, pp. 420–435, Jan. 2004.
5. Q.J. Zhang, M.C.E. Yagoub, X. Ding, D. Goulette, R. Sheffield, and H. Feyzbakhsh, "Fast and accurate modeling of embedded passives in multi-layer printed circuits using neural network approach," *Elect. Compon. & Tech. Conf.*, San Diego, CA, May 2002, pp. 700–703.
6. X. Ding, J.J. Xu, M.C.E. Yagoub, and Q.J. Zhang, "A combined state space formulation/equivalent Circuit and neural network technique for modeling of embedded passives in multilayer printed circuits," ACES Journal, vol. 18, pp. 158–174, 2003.
7. M.H. Bakr, P.P.M. So, and W.J.R. Hoefer, "The generation of optimal microwave topologies using time-domain field synthesis," *IEEE Trans. Microw. Theory Tech.*, vol. 50, no. 11, pp. 2537–2544, November 2002.

8. D.G. Swanson Jr., and W.J.R. Hoefer, *Microwave Circuit Modeling Using Electromagnetic Field Simulator*, Boston: Artech House, 2003.
9. H. Sharma and Q.J. Zhang, "Transient electromagnetic modeling using recurrent neural networks," *2005 IEEE MTT-S Int. Microw. Symp. Dig.*, Long Beach, CA, June 2005.
10. Y. Fang, M.C.E. Yagoub, F. Wang, and Q.J. Zhang, "A new macromodeling approach for nonlinear microwave circuits based on recurrent neural networks," *IEEE Trans. Microw. Theory Tech.*, vol. 48, no. 12, pp. 2335–2344, December 2000.
11. B.C. Kuo, *Automatic Control Systems*, 7th Edition, New York: John Wiley & Sons, 1995.
12. X. Ding, V.K. Devabhaktuni, B. Chattaraj, M.C.E. Yagoub, M. Deo, Jianjun Xu, and Q.J. Zhang, "Neural-network approaches to electromagnetic-based modeling of passive components and their applications to high-frequency and high-speed nonlinear circuit optimization," *IEEE Trans. Microw. Theory Tech.*, vol. 52, no. 1, pp. 436–449, January 2004.
13. S. Grivet-Talocia, "Package macromodeling via time-domain vector fitting," *IEEE Microw. Wirel. Compon. Lett.*, vol. 13, no. 11, pp. 472–474, November 2003.
14. B. Gustavsen and A. Semlyen, "A robust approach for system identification in the frequency domain," *IEEE Trans. Power Deliver.*, vol. 19, no. 3, pp. 1167–1173, July 2004.
15. R. Achar and M.S. Nakhla, "Simulation of high-speed interconnects," *Proc. of the IEEE*, vol. 89, no. 5, pp. 693–728, May 2001.
16. B. Gustavsen and A. Semlyen, "Rational approximation of frequency domain responses by vector fitting," *IEEE Trans. Power Deliver.*, vol. 14, no. 3, pp. 1052–1061, July 1999.
17. V. Antonini, "SPICE equivalent circuits of frequency-domain responses," *IEEE Trans. Electromagnetic Compatibility*, vol. 45, no. 3, pp. 502–512, August 2003.
18. H. Sharma and Q.J. Zhang, "Automated time domain modeling of linear and nonlinear microwave circuits using recurrent neural networks," *Int. J. RF and Microw. CAE*, (accepted).
19. J. Bandler, M. Ismail, J. Rayas-Sanchez, and Q. Zhang, "New directions in model development for RF/microwave components utilizing artificial neural networks and space mapping," *IEEE APS Int. Symp. digest*, Orlando, FL, July 1999, pp. 2572–2575.
20. *NeuroModeler v.* 1.3, Q.J. Zhang, Dept. of Electronics, Carleton University, Ottawa, Canada.
21. P.J. Werbos, "Backpropagation through time: what it does and how to do it," *Proc. of the IEEE*, vol. 78, no. 10, pp. 1550–1560, October 1990.
22. *Sonnet v* 7.0, Sonnet Software, Liverpool, NY, USA.
23. *Hspice v.*2001.2, Avant! Corp., Fremont, CA, USA.
24. *Ansoft HFSS v.*8.0, Ansoft Corp., Pittsburg, PA, USA.
25. *MEFiSTo 3D Pro*, Faustus Scientific Corporation, Victoria, BC, Canada, 2005.

Modeling of Curved Boundaries in the Finite-Difference Time-Domain Method using a Lagrangian Approach

Johannes A. Russer, Prasad S. Sumant and Andreas C. Cangellaris

Abstract We present a Lagrangian approach for increasing the accuracy of the Finite-Difference Time-Domain method in modeling electromagnetic wave interactions in geometries involving curved boundaries. The methodology provides for a mapping of the physical computational domain onto a reference domain with all its boundaries parallel to a Cartesian coordinate system. The Lagrangian formulation offers a convenient way to define a modified set of Maxwell's equations on the reference domain. This modified set of equations is then discretized using the staggered Cartesian grids of the Yee's lattice. The attributes of the method are demonstrated through its application to the calculation of the eigenfrequencies of a two-dimensional metallic resonator and the transmission properties of a two-dimensional periodic structure.

1 Introduction

The Finite-Difference Time-Domain (FDTD) method with its spatial and temporal discretization of Maxwell's curl equations allows for an efficient numerical solution to electromagnetic wave phenomena in complex structures [15]. The success of the FDTD method and its popularity are in major part due to the simplicity of the so-called Yee's scheme for the spatial discretization of the curl operators on two staggered Cartesian grids [16]. It is because of its simplicity that the utilization of a Cartesian grid remains the most popular in the application of the method to the modeling of wave interactions in domains involving arbitrarily-shaped, non-Cartesian

Johannes A. Russer
Department of Electrical and Computer Engineering, University of Illinois at Urbana-Champaign

Prasad S. Sumant
Department of Electrical and Computer Engineering, University of Illinois at Urbana-Champaign

Andreas C. Cangellaris
Department of Electrical and Computer Engineering, University of Illinois at Urbana-Champaign

P. Russer, U. Siart (eds.), *Time Domain Methods in Electrodynamics*,

material boundaries. Despite significant innovative advances in avoiding the modeling error resulting from the staircase approximation of a curved material boundary [2, 7, 10, 12, 13, 17], the simplicity of the Yee's lattice leads many electromagnetic practitioners to its adoption at the cost of utilizing a finer grid in order to minimize the modeling error [3].

The aforementioned advanced techniques for improving modeling fidelity without increasing the spatial sampling of the fields come at the cost of increased complexity in both mesh generation and in the development of the discrete forms of the curl operators. In this paper we present an interesting alternative to these techniques, which is motivated by ideas borrowed from the application of Lagrangian techniques for solving problems in mechanics and coupled electro-mechanical problems [4, 11]. The basic idea of the method is to map the physical domain onto a reference domain with all its boundaries parallel to the planes of the reference Cartesian coordinate system. Once such a domain can be defined, the mapped version of Maxwell's curl equations are cast in a form that lends itself to their direct discretization using Yee's staggered Cartesian grids on the reference domain. The attributes of such an approach were first demonstrated in [14].

In this paper we present in detail the Lagrangian formalism and its use for the systematic development of the modified system of Maxwell's equations on the reference Cartesian grid. This is followed by the validation of the method through its application to the numerical calculation of the eigenfrequencies of a two-dimensional cavity with perfectly conducting walls and the prediction of the pass-band and stop-band characteristics of a two-dimensional, parallel-plate waveguide with periodical, non-rectangular perforations of one of its perfectly conducting walls.

2 The Lagrangian Approach

To introduce the concept of the Lagrangian approach it is useful to consider problems in mechanics where the Lagrangian concept is frequently used.

Figure 1 depicts the reference configuration B and the deformed configuration b of one body. We find the particle $P(\alpha,\gamma)$ displaced to $p(x,z)$ where $\boldsymbol{x} = \boldsymbol{\alpha} + \boldsymbol{u}$ and $\boldsymbol{u}$ is the displacement vector. There are two different approaches that are commonly

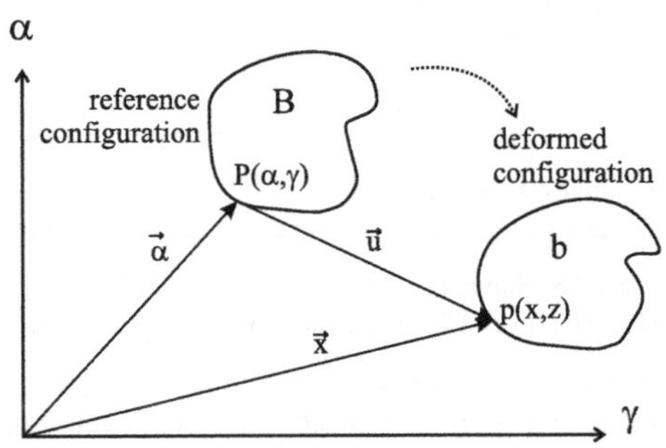

Fig. 1 Reference and deformed configuration

taken in mechanics to formulate a mathematical model describing moving particles [4]. The first approach describes a particle position as a function of its position at an initial reference configuration and its time evolution. The second approach focuses on a fixed position in space and the evolution of physical quantities at this point. The latter suggests a Lagrangian approach, as we find it, for example, in fluid dynamics, where the knowledge of the pressure at some point tends to be more valuable than the actual position of each of the fluid's particles. The Lagrangian approach allows for a mapping such that we solve for the quantities of interest at fixed locations in the reference configuration.

2.1 Application of Lagrangian Formulation in FDTD

Since we are interested in utilizing the standard Yee's lattice for the discretization of the curl operators in Maxwell's equations the preferred choice for the reference domain is one with its boundaries parallel to the planes of the reference Cartesian coordinate system. Let us assume that a reference, Cartesian domain has been established and a uniform, Cartesian grid is used to provide for its discretization. With reference to Fig. 2, in the spirit of the Lagrangian approach, in order to simplify the modeling of electromagnetic wave interactions inside the domain with the curved top boundary on the right we displace, in analogy to the moving particle problem in mechanics, the grid points on the boundaries of the uniform grid in the reference domain until they coincide with the curved boundary on the domain on the right. Thinking of the reference domain as an elastic membrane, the displacement of the boundary grid points will result in displacement of the remaining grid points of the uniform grid. Thus, a deformed grid results over the deformed domain. The definition of a mapping between the deformed grid and the reference grid allows us, through the use of the Lagrangian formulation, to apply the spatial discretization of the curl operators in Maxwell's equations over the Cartesian grid of the reference domain. The key difference is that the discretization is applied to a modified version of Maxwell's equations, the form of which is governed by the mapping between the

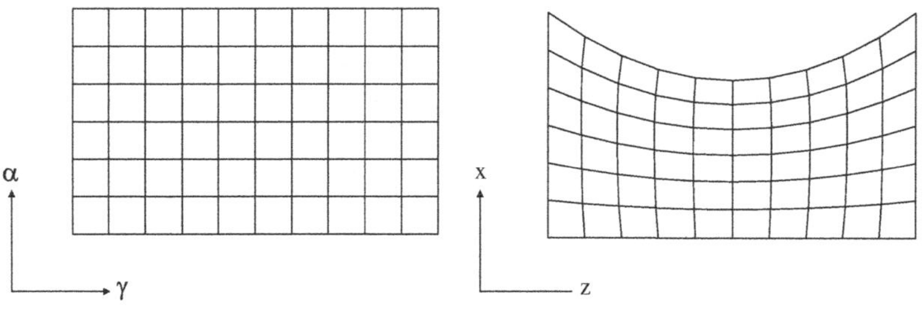

(a) Reference configuration. (b) Deformed configuration.

Fig. 2 Uniform (Cartesian) and non-uniform (deformed) grids for *reference and deformed configuration*, respectively

two domains. Furthermore, all boundary conditions at the material boundaries on the physical (deformed) domain can be enforced in a much simpler manner on the corresponding maps of the curved boundaries on the reference domain.

We proceed, next, to develop the general form of the modified system of Maxwell's curl equations on the reference, Cartesian domain. Without loss of generality, we assume that the media are linear, isotropic and time-invariant. Thus, Maxwell's curl equations in the deformed (physical) domain are

$$\nabla \times \mathbf{E} = -\mu \frac{\partial \mathbf{H}}{\partial t} \tag{1}$$

$$\nabla \times \mathbf{H} = \varepsilon \frac{\partial \mathbf{E}}{\partial t} + \mathbf{J}. \tag{2}$$

We will take advantage of the two-dimensional geometries depicted in Fig. 2 to facilitate our discussion. While Fig. 2(a) depicts a regular grid in two dimensions, let us assume that the actual grid is over the domain of a rectangular parallelepiped, with $x-z$ cross-sectional geometry as depicted in the figure. The discretization of (2) is straightforward if the field quantities are defined on a regular grid like the one shown. Figure 2(b) depicts the $x-z$ cross section of the deformed grid over the physical domain, the geometry of which differs from that of the reference, rectangular parallelepiped only in terms of the indentation of the top surface (which, without loss of generality, is assumed to be independent of y).

Let $(\hat{\alpha}, \hat{\beta}, \hat{\gamma})$ denote the unit vectors along the three axes of the Cartesian coordinate system for the reference configuration. Let $(\hat{x}, \hat{y}, \hat{z})$ denote the unit vectors in the reference Cartesian coordinate system for the physical (deformed) configuration. For the deformation of the grid from its Cartesian shape in the reference domain to the deformed shape in the deformed domain use is made of a quasi elastic finite element (FEM) solver [6, 14]. Each grid point $p(x,y,z)$ in the deformed configuration corresponds to a point $P(\alpha, \beta, \gamma)$ in the reference configuration (see Fig. 1). This relationship is cast in the form

$$d \begin{pmatrix} x \\ y \\ z \end{pmatrix} = \mathscr{F} d \begin{pmatrix} \alpha \\ \beta \\ \gamma \end{pmatrix}, \tag{3}$$

with the deformation gradient, $\mathscr{F}$, given by

$$\mathscr{F} = \begin{bmatrix} \frac{\partial x}{\partial \alpha} & \frac{\partial x}{\partial \beta} & \frac{\partial x}{\partial \gamma} \\ \frac{\partial y}{\partial \alpha} & \frac{\partial y}{\partial \beta} & \frac{\partial y}{\partial \gamma} \\ \frac{\partial z}{\partial \alpha} & \frac{\partial z}{\partial \beta} & \frac{\partial z}{\partial \gamma} \end{bmatrix}. \tag{4}$$

Furthermore, the map between vector quantities in the two domains is given in terms of the following relation

$$\begin{pmatrix} A_x \\ A_y \\ A_z \end{pmatrix} = \mathscr{G} \begin{pmatrix} A_\alpha \\ A_\beta \\ A_\gamma \end{pmatrix}, \tag{5}$$

where the matrix $\mathscr{G}$ is given by

$$\mathscr{G} = \begin{bmatrix} \frac{\partial \alpha}{\partial x} & \frac{\partial \beta}{\partial x} & \frac{\partial \gamma}{\partial x} \\ \frac{\partial \alpha}{\partial y} & \frac{\partial \beta}{\partial y} & \frac{\partial \gamma}{\partial y} \\ \frac{\partial \alpha}{\partial z} & \frac{\partial \beta}{\partial z} & \frac{\partial \gamma}{\partial z} \end{bmatrix}. \tag{6}$$

The above relationship is utilized for the mapping of the electric and the magnetic field vectors between the two domains. In addition, we need to derive the pertinent equations for the mapping of the first-order spatial derivatives present in the curl operator. Towards this, it is useful to define the following Jacobian matrix,

$$D_{x_1 x_2}^{\alpha_1 \alpha_2} := \frac{\partial(\alpha_1, \alpha_2)}{\partial(x_1, x_2)} := \det \begin{bmatrix} \frac{\partial \alpha_1}{\partial x_1} & \frac{\partial \alpha_1}{\partial x_2} \\ \frac{\partial \alpha_1}{\partial x_1} & \frac{\partial \alpha_2}{\partial x_2} \end{bmatrix}, \tag{7}$$

where α_1, α_2 assume values from the set $\{\alpha, \beta, \gamma\}$, while x_1, x_2 assume values from the set $\{x, y, z\}$. Using the above definitions, it is straightforward to show after some manipulation that the mapped form of Faradays's law on the deformed (physical) configuration

$$\frac{\partial}{\partial t}\mathbf{E} = \frac{1}{\varepsilon}\nabla \times \mathbf{H} \tag{8}$$

onto the reference, Cartesian domain is given by

$$\frac{\partial}{\partial t}\tilde{\mathbf{E}} = \frac{1}{\varepsilon}\mathscr{G}^{-1}\tilde{\mathscr{D}}\nabla \times \tilde{\mathbf{H}}. \tag{9}$$

where the ˜ is used to denote the vector in the reference domain. In a similar fashion, we have for Ampère's law in a source-free region (2)

$$\frac{\partial}{\partial t}\tilde{\mathbf{H}} = -\frac{1}{\mu}\mathscr{G}^{-1}\tilde{\mathscr{D}}\nabla \times \tilde{\mathbf{E}} \tag{10}$$

In the above equations, the matrix $\tilde{\mathscr{D}}$ is given by

$$\tilde{\mathscr{D}} = \begin{bmatrix} D_{yz}^{\beta\gamma} & -D_{yz}^{\alpha\gamma} & D_{yz}^{\alpha\beta} \\ -D_{xz}^{\beta\gamma} & D_{xz}^{\alpha\gamma} & -D_{xz}^{\alpha\beta} \\ D_{xy}^{\beta\gamma} & -D_{xy}^{\alpha\gamma} & D_{xy}^{\alpha\beta} \end{bmatrix}. \tag{11}$$

We observe that the mapping results in a modified system of Maxwell's equations over the reference domain that can be interpreted in terms of an anisotropic medium with the electric permittivity and magnetic permeability tensors defined by

$$\bar{\varepsilon} = \tilde{\mathscr{D}}^{-1}\mathscr{G}\varepsilon \tag{12}$$

$$\bar{\mu} = \tilde{\mathscr{D}}^{-1}\mathscr{G}\mu. \tag{13}$$

Thus, we conclude that use of the Lagrangian approach for Maxwell's curl equations on a domain with curved boundaries amounts to re-casting the electromagnetic boundary value problem in terms of one involving a general anisotropic medium over a domain with boundaries parallel to the planes of a reference Cartesian coordinate system. Application of the FDTD method to the simulation of electromagnetic wave interactions in the transformed domain is readily and accurately carried out using a Cartesian Yee's lattice.

2.2 Discretization of the Lagrangian Form of Maxwell's Equations: Two-Dimensional Case

To demonstrate the discrete form of the Lagrangian form of Maxwell's equations in the context of the FDTD method, we consider the two-dimensional case where the geometry and the field excitation are invariant in the $\hat{y}$ direction; hence, $\hat{\beta} \equiv \hat{y}$. This results in two possible field polarizations, namely, the transverse electric (TE) and the transverse magnetic (TM). For the TM polarization it is $E_y = 0$, $H_x = 0$ and $H_z = 0$. Furthermore, because of the two-dimensional nature of the problem, it is $\frac{\partial\alpha}{\partial y} = \frac{\partial\gamma}{\partial y} = 0$ and $\frac{\partial\beta}{\partial y} = 1$. Making use of these relations in (6) and (11) results in the following, reduced forms of (9) and (10)

$$\frac{\partial H_\beta}{\partial t} = -\frac{1}{\mu}\det\begin{bmatrix}\frac{\partial\alpha}{\partial x} & \frac{\partial\gamma}{\partial x}\\ \frac{\partial\alpha}{\partial z} & \frac{\partial\gamma}{\partial z}\end{bmatrix}\left(\frac{\partial E_\alpha}{\partial\gamma} - \frac{\partial E_\gamma}{\partial\alpha}\right) := \frac{1}{\mu}\mathscr{H}\left(\frac{\partial E_\alpha}{\partial\gamma} - \frac{\partial E_\gamma}{\partial\alpha}\right) \tag{14}$$

$$\frac{\partial}{\partial t}\begin{pmatrix}E_\alpha\\ E_\gamma\end{pmatrix} = \frac{1}{\varepsilon}\begin{bmatrix}\frac{\partial\alpha}{\partial x} & \frac{\partial\gamma}{\partial x}\\ \frac{\partial\alpha}{\partial z} & \frac{\partial\gamma}{\partial z}\end{bmatrix}^{-1}\begin{bmatrix}\frac{\partial\gamma}{\partial z} & -\frac{\partial\alpha}{\partial z}\\ -\frac{\partial\gamma}{\partial x} & \frac{\partial\alpha}{\partial x}\end{bmatrix}\begin{pmatrix}-\frac{\partial H_\beta}{\partial\gamma}\\ \frac{\partial H_\beta}{\alpha}\end{pmatrix} := \frac{1}{\varepsilon}\mathscr{I}\begin{pmatrix}-\frac{\partial H_\beta}{\partial\gamma}\\ \frac{\partial H_\beta}{\alpha}\end{pmatrix}. \tag{15}$$

The above equations are discretized on Yee's staggered grid as shown in Fig. 3 where the electric field quantities E_α and E_γ are shown as arrows and the H_β field, which is protruding out of the plane, is depicted by dotted circles. More specifically, we denote $E_\alpha|_{i,k}^n = E_\alpha\left(\left(i-\frac{1}{2}\right)\Delta\alpha,\ (k-1)\Delta\gamma, n\Delta t\right)$, $E_\gamma|_{i,k}^n = E_\alpha((i-1)\Delta\alpha,$ $\left(k-\frac{1}{2}\right)\Delta\gamma, n\Delta t)$ and $H_\beta|_{i,k}^n = E_\alpha\left(\left(i-\frac{1}{2}\right)\Delta\alpha, (k-\frac{1}{2})\Delta\gamma, n\Delta t\right)$. $\Delta\alpha$ and $\Delta\gamma$ represent the grid spacing, while Δt represents the discrete time step. Furthermore, $\mathscr{H}|_{i,k}$

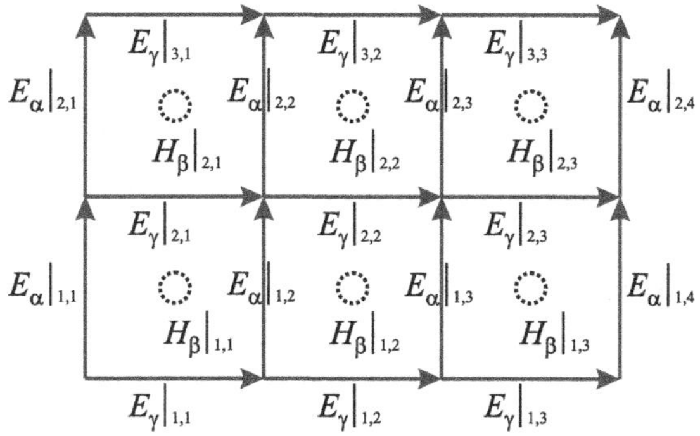

Fig. 3 Yee's staggered grid on the Cartesian reference domain

denotes the determinant of (14) evaluated at the position of the E field component associated with it in the same equation. In a similar manner, $\mathscr{I}|_{i,k}$ denotes the matrix of (15) evaluated at the position of $H_\beta|_{i,k}$, with $\mathscr{I}|_{i,k}(q,r)$ specifying the (q,r) element of the matrix. This yields the following discrete form of the Lagrangian formulation of Maxwell's curl equations

$$H_\beta|_{i,k}^{n+1/2} = H_\beta|_{i,k}^{n-1/2} + \frac{\Delta t}{\mu}\mathscr{H}|_{i,k}\left(\frac{E_\gamma|_{i+1,k}^n - E_\gamma|_{i,k}^n}{\Delta\alpha} - \frac{E_\alpha|_{i,k+1}^n - E_\alpha|_{i,k}^n}{\Delta\gamma}\right) \tag{16}$$

$$E_\alpha|_{i,k}^{n+1} = E_\alpha|_{i,k}^{n} + \frac{\Delta t}{\varepsilon}\left(-\mathscr{I}|_{i,k}(1,1)\frac{H_\beta|_{i,k}^{n+1/2} - H_\beta|_{i,k-1}^{n+1/2}}{\Delta\gamma} + \mathscr{I}|_{i,k}(1,2)\right.$$
$$\left.\times\frac{H_\beta|_{i+1,k-1}^{n+1/2} - H_\beta|_{i-1,k-1}^{n+1/2} + H_\beta|_{i+1,k}^{n+1/2} - H_\beta|_{i-1,k}^{n+1/2}}{4\Delta\alpha}\right) \tag{17}$$

$$E_\gamma|_{i,k}^{n+1} = E_\gamma|_{i,k}^{n} + \frac{\Delta t}{\varepsilon}\left(-\mathscr{I}|_{i,k}(2,1)\left(\frac{H_\beta|_{i-1,k+1}^{n+1/2} - H_\beta|_{i-1,k-1}^{n+1/2}}{4\Delta\gamma}\right.\right.$$
$$\left.\left.+\frac{H_\beta|_{i,k+1}^{n+1/2} - H_\beta|_{i,k-1}^{n+1/2}}{4\Delta\gamma}\right) + \mathscr{I}|_{i,k}(2,2)\frac{H_\beta|_{i,k}^{n+1/2} - H_\beta|_{i-1,k}^{n+1/2}}{\Delta\alpha}\right) \tag{18}$$

These equations are used for the discretization of Maxwell's equations in the validation studies presented in the next section.

3 Validation Studies

In order to validate the proposed method we have chosen to apply it to the electromagnetic modeling of two structures, a two-dimensional rectangular cavity with an indentation in one of its four metallic walls, and a parallel-plate waveguide with a non-rectangular, periodic corrugation in one of its two walls.

3.1 Eigen-Analysis of a Deformed Rectangular Metallic Resonator

The resonator we consider has the dimensions $a = 3\,\text{cm}$ by $b = 4\,\text{cm}$ and is filled with air. The walls are perfect electric conductors (PEC) and we are interested in calculating the eigenfrequencies of the TM modes of the resonator in the presence of an indentation in the top wall. In the absence of any indentation the resonant frequencies of the modes are given by

$$f_{m,n} = \frac{1}{2\pi\sqrt{\mu\varepsilon}}\sqrt{\left(\frac{m\pi}{a}\right)^2 + \left(\frac{n\pi}{b}\right)^2} \tag{19}$$

where the indices m, n define the mode number. The deformed resonator is shown in Fig. 4. The indentation of the top plate is defined mathematically in terms of the function

$$\mathrm{a}'(z) = \begin{cases} \mathrm{a} & |z - \mathrm{b}/2| \geq \mathrm{b}/4 \\ \left(1 - 0.1 * \left(\frac{4z}{\mathrm{b}} - 1\right)\right)\mathrm{a} & \mathrm{b}/4 < z \leq \mathrm{b}/2 \\ \left(1 - 0.1 * \left(3 - \frac{4z}{\mathrm{b}}\right)\right)\mathrm{a} & \mathrm{b}/2 < z < 3/4\mathrm{b} \end{cases} \tag{20}$$

with $0 \leq z \leq \mathrm{b}$, yielding a triangular shaped dent. For an analytic means of approximating the resonant frequencies of the deformed resonator we use a perturbational technique [8]. The shift $\Delta\omega_r$ in the resonant frequency ω_r is calculated through the expression

$$\frac{\Delta\omega_r}{\omega_r} \approx \frac{\iint_{\Delta S}(\mu|H_0|^2 - \varepsilon|E_0|^2)ds}{\iint_S(\mu|H_0|^2 + \varepsilon|E_0|^2)ds}. \tag{21}$$

Referring to Fig. 4(b), S is the area of the non-deformed rectangular box, whereas ΔS denotes the area of the indentation. H_0 and E_0 are the fields of the mode for

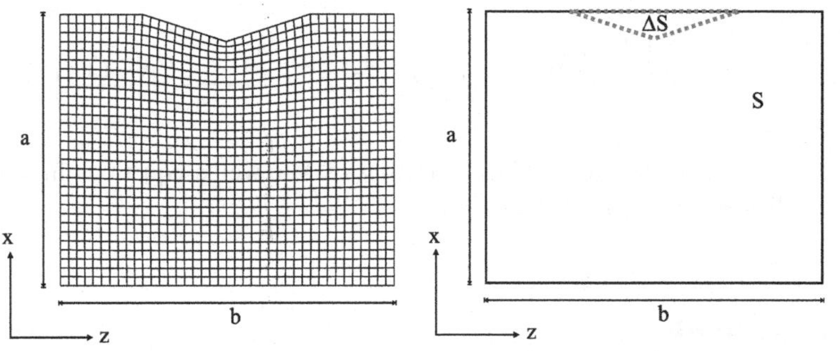

(a) Deformed resonator mesh.

(b) Areas used for the calculation of the eigenfrequencies using perturbational techniques.

Fig. 4 Geometry and gridding of the deformed resonator

the unperturbed resonator. We expect the analytic result obtained using the above equation to serve as a good indicator of the frequency shift caused to the resonant frequencies of the unperturbed cavity caused by the indentation.

Our calculation of the resonant frequencies of the perturbed resonator relies on its excitation by a pulse and the subsequent Fourier transform of the transient response obtained from the Lagrangian FDTD simulation. In addition, we have computed the resonance frequencies using the Transmission Line Method (TLM) field solver MEFiSTo [1, 9], which utilizes a rectangular grid.

The results obtained from the three approaches are listed in Table 1, along with the values of the resonant frequencies of the unperturbed resonator. The results of the Lagrangian FDTD and the TLM solution follow closely the trend predicted by (21) and are in very good agreement with each other when the grid size in the reference, Cartesian grid is 1 mm. As the grid becomes coarser, with its grid size assuming a value of 2 mm, the values for the resonant frequencies obtained by the Lagrangian FDTD method change only very little. Since the dent extends into the resonator by 3 mm, a grid resolution of 2 mm in a staircase approximation of the indentation will be too coarse to model the shape of the metallic wall accurately. However, the Lagrangian FDTD method enables us to obtain accurate results even when using a coarser grid.

Table 1 Resonant frequencies of the metallic resonator depicted in Fig. 4

	Mode order				
m	0	1	1	0	1
n	1	0	1	2	2
	Resonant frequency (GHz)				
Unperturbed resonator	3.74	4.99	6.24	7.49	9.00
Equation (21) perturbed resonator	3.59	5.23	6.17	7.64	9.22
Lagrangian FDTD (1 mm grid spacing)	3.67	5.11	6.18	7.53	9.15
Lagrangian FDTD (2 mm grid spacing)	3.67	5.11	6.18	7.51	9.14
TLM (1 mm grid spacing)	3.62	5.11	6.14	7.56	9.16

3.2 Analysis of a Corrugated Parallel-Plate Waveguide

For this study the objective is the calculation of the transmission properties of a corrugated, air-filled, parallel–plate waveguide under TM excitation. The PEC walls have a maximum plate separation of $d = 1\,\text{cm}$. The periodic corrugation is present in the top wall only, and has a period of $a = 4\,\text{cm}$. We have cascaded 15 sections to form the periodic section of the waveguide that will be used as a filter. Figure 5 depicts the deformed grid of the parallel–plate waveguide for two cascaded sections.

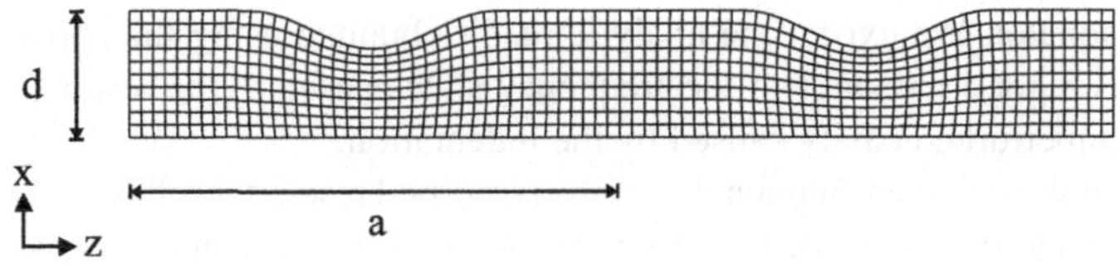

(a) Deformed grid for Lagrangian FDTD Approach.

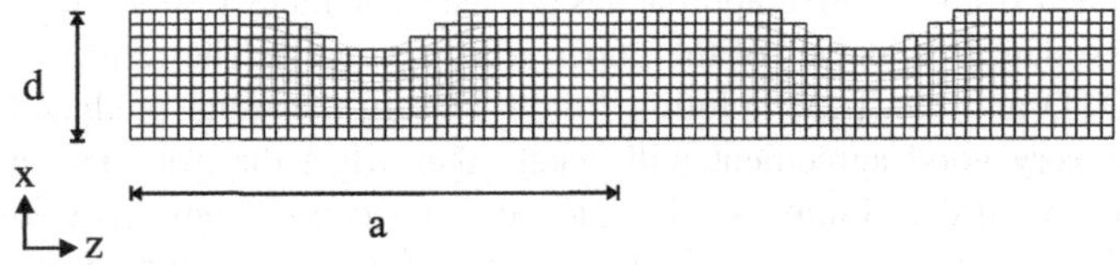

(b) Stair case approximation for standard FDTD (1 mm grid spacing).

Fig. 5 Two sections of the corrugated parallel–plate waveguide

At its narrowest, the separation of the plates is reduced to 70% of the original separation. The distance of the top wall from the bottom wall versus z is described by

$$d'(z) = \begin{cases} d & |z - a/2| \geq a/4 \\ (1 - 0.3 * \cos^2([2z/a - 1]\pi))d & a/4 < z < 3a/4 \end{cases} \tag{22}$$

with $0 \leq z \leq a$, as shown in Fig. 5. The waveguide is excited in the fundamental mode with a modulated Gaussian pulse with its center frequency at 4.7 GHz and a 3 dB bandwidth of 2.5 GHz. Because the indentations serve as a periodic capacitive loading of the guide, we expect a stopband at around 3.75 and 7.5 GHz [5]. The magnitude of the transmission coefficient, S_{21}, is plotted in Fig. 6 with the solid line showing the results obtained from the Lagrangian FDTD method and the dashed lines those obtained using a standard FDTD method with staircase approximation of the corrugated boundary. Figures 6(a) and 6(c) show $|S_{21}|$ obtained with grid spacings of 0.5, 1 and 2 mm. The vertical lines in the figures mark the frequencies where the value of $|S_{21}|$ meets the -3 dB level. Numerical values for the -3 dB frequencies are given in Table 2.

As it is evident from the plots and the data, the Lagrangian FDTD method sustains better accuracy as the grid becomes coarser. For the case of the finest grid resolution, with a grid size of 0.5 mm, the results obtained from Lagrangian FDTD and the standard FDTD method are in very good agreement. For coarser grids the accuracy of the standard FDTD method degrades, especially at higher frequencies, as expected. The staircase approximation of the standard FDTD in a grid with a spacing of 2 mm is too coarse to resolve accurately the 3 mm deep corrugation. However, the Lagrangian FDTD method is capable of predicting the -3 dB frequencies for the falling and rising edges of the upper stopband with sustained accuracy even at a grid resolution of 2 mm.

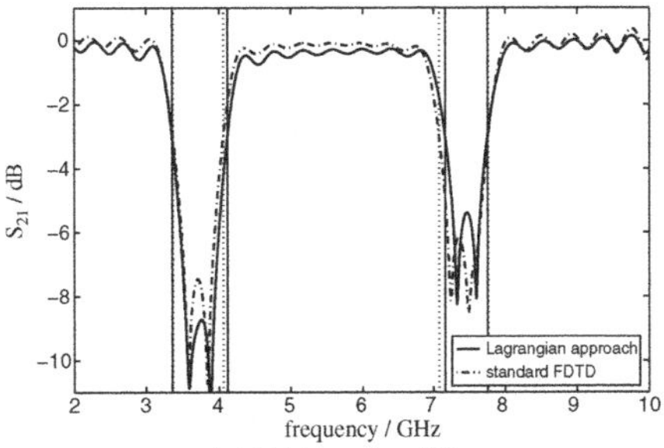

(a) Grid spacing 0.5 mm.

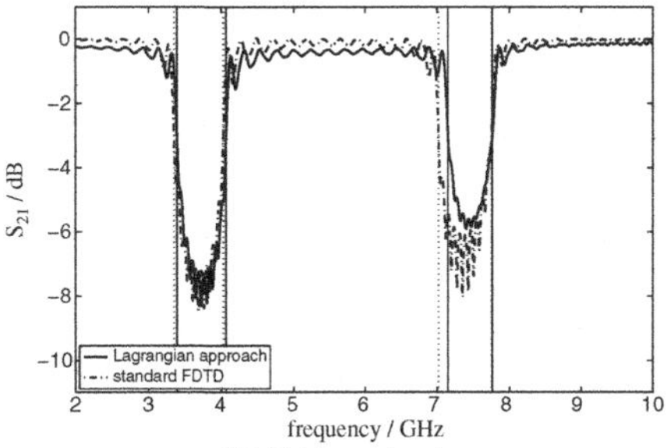

(b) Grid spacing 1 mm.

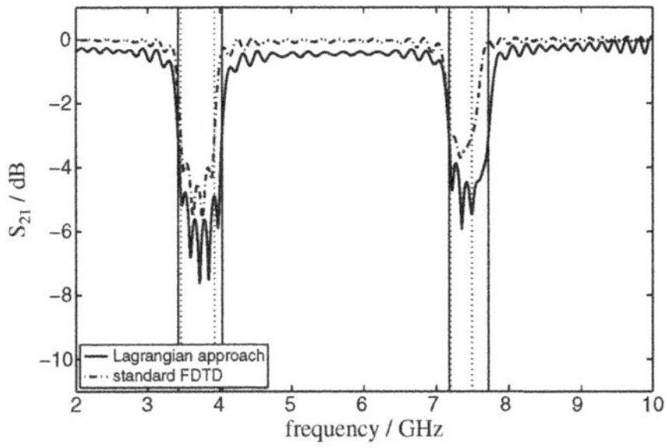

(c) Grid spacing 2 mm.

Fig. 6 Magnitude of the transmission coefficient, $|S_{21}|$, of the corrugated waveguide for different grid resolutions. Vertical lines mark -3 dB frequency points for Lagrangian (*solid line*) and standard FDTD (*dotted line*) method

Table 2 -3 dB frequency points of the falling (F) and rising (R) edges of the $|S_{21}|$ graph

Method	Grid spacing/mm	−3 dB frequency/GHz			
		Lower		Upper stopband	
		F	R	F	R
Lagrangian FDTD	0.5	3.35	4.11	7.16	7.74
	1.0	3.38	4.07	7.15	7.76
	2.0	3.42	4.03	7.17	7.71
Standard FDTD	0.5	3.36	4.05	7.08	7.76
	1.0	3.34	4.02	7.02	7.75
	2.0	3.46	3.92	7.19	7.49

4 Conclusion

We have presented a Lagrangian approach for the FDTD modeling of electromagnetic phenomena in domains with curved boundaries with good accuracy and improved modeling and computational efficiency. We have presented examples in two dimensions to validate the method. Through the definition of a mapping of the deformed (physical) domain onto a reference domain with Cartesian boundaries, the map of Maxwell's equations on the reference domain is interpreted in terms of a Maxwellian system describing wave propagation in a general anisotropic medium inside a domain with boundaries parallel to the planes of the Cartesian reference coordinate system. Thus, the standard FDTD scheme on a Cartesian Yee's lattice can be employed for the spatial discretization of the curl operators. The staircase approximation of the curved boundaries is avoided and boundary conditions are applied accurately, therefore resulting in enhanced accuracy in the numerical solution without the need for finer grid resolution. Numerical studies in two dimensions were used to demonstrate these attributes and the accuracy of the proposed Lagrangian FDTD method.

Acknowledgements This material is based upon work supported in part by the U.S. Army Research Office as a Multi-disciplinary University Research Initiative on Standoff Inverse Analysis and Manipulation of Electronic Systems under grant number W911NF-05-1-0337.

References

1. *MEFiSTo (Multi-purpose Electromagnetic Field Simulation Tool)*. Faustus Scientific Corporation, Victoria, B.C., Canada, 2007. http://www.faustcorp.com.
2. A. C. Cangellaris, C.-C. Lin, and K. K. Mei. Point-matched time domain finite element methods for electromagnetic radiation and scattering. *IEEE Trans. Antenn. Propag.*, 35:1160–1173, 1987.
3. A. C. Cangellaris and D. B. Wright. Analysis of the numerical error caused by the stair-stepped approximation of a conducting boundary in FDTD simulations of electromagnetic phenomena. *IEEE Trans. Antenn. Propag.*, 39:1518–1525, Oct. 1991.
4. D. S. Chandrasekharaiah, and L. Debnath. *Continuum Mechanics*. Academic Press, San Diego, 1994.
5. R. E. Collin. *Foundations for Microwave Engineering*. McGraw-Hill, New York, 1966.
6. R. D. Cook, D. S. Malkus, M. E. Plesha, and R. J. Witt. *Concepts and Applications of Finite Element Analysis*. John Wiley, New York, 2001.
7. C. Fumeaux, D. Baumann, and R. Vahldieck. Finite-volume time-domain analysis of a cavity-backed archimedean spiral antenna. *IEEE Trans. Antenn. Propag.*, 54:844–851, Mar. 2006.
8. R. F. Harrington. *Time Harmonic Electromagnetic Fields*. John Wiley, New York, 2001.
9. W. J. R. Hoefer. The transmission line matrix method – theory and applications. *IEEE Trans. Microw. Theory Tech.*, 33:882–893, 1985.
10. R. Holland. Finite difference solutions of Maxwell's equations in generalized nonorthogonal coordinates. *IEEE Trans. Nucl. Sci.*, 30:4589–4591, 1983.
11. G. Li and N. R. Aluru. A Lagrangian approach for electrostatic analysis of deformable conductors. *J. Microelectromech. Syst.*, 11:245–254, 2002.
12. R. M. Mäkinen, H. De Gersem, T. Weiland, and M. A. Kivikoski. Modeling of lossy curved surfaces in 3-D FIT/FDTD techniques. *IEEE Trans. Antenn. Propag.*, 54:3490–3498, Nov. 2006.
13. K. K. Mei, A. Cangellaris, and D. J. Angelakos. Conformal time domain finite difference method. *Radio Sci.*, 19:1145–1147, 1984.
14. J. A. Russer, P. S. Sumant, and A. C. Cangellaris. A Lagrangian approach for the handling of curved boundaries in the finite-difference time-domain method. *IEEE MMT-S International Microwave Symposium*, pages 717–720, June 2007.

15. A. Taflove and S. Hagness. *Computational Electrodynamics: The Finite-Difference Time-Domain Method*. Artech House, Boston, 2005.
16. K. S. Yee. Numerical solution of initial boundary value problems involving Maxwell's equations in isotropic media. *IEEE Trans. Antenn. Propag.*, 14:302–307, May 1966.
17. W. Yu, Y. Liu, T. Su, N.-T. Huang, and R. Mittra. A robust parallel conformal finite-difference time-domain processing package using the MPI library. *IEEE Antenn. Propag. Mag.*, 47:39–59, June 2005.

Computing the Transmission Line Parameters of an On-chip Multiconductor Digital Bus

Hristomir Yordanov and Peter Russer

The task execution on modern digital integrated circuits requires internal data rates in the range of several gigabit per second. On the other hand circuit miniaturization leads to greater resistance of the digital interconnection bus conductors. Therefore the auto-interference effects like pulse distortion along the line and crosstalk increase and become a limiting factor for the bus performance. In order to optimize the data throughput along such an interconnection line coding techniques can be implemented [1]. The design of these methods requires an estimation of the auto-interference effects.

In this work a quasi-analytical model of an on-chip digital bus with equidistant conductors of equal cross section is presented. The quasi-static parameters are computed under the assumption for symmetry using even-odd mode analysis [2] and Schwarz-Cristoffel transformation [3, 4]. The obtained results are used to solve the multiconductor transmission line equations in frequency domain, thus obtaining the frequency response of the digital interconnect [5]. Then the pulse distortion is computed using Fourier transformation. The obtained data have been verified using full-wave analysis and SPICE simulation of the lumped-element equivalent circuit of the interconnection bus.

The problem of estimating the electrostatic parameters of an interconnection bus reduces to computation of rectangular coupled lines between parallel grounded plates. The first attempts for analytical modelling of such structures were based on the assumption of zero conductor thicknesses and adding a correction term for the finite thickness case. Cohn, for example, has calculated the transmission line parameters for the zero-thickness case using conformal mapping methods [6]. Getsinger has extended Cohn's work to computing the fringing field capacitances at the conductor edges, but he has imposed restrictions on the conductor dimensions [7].

Hristomir Yordanov
Munich University of Technology, Institute for High Frequency Engineering, Arcisstr. 21, 80333 Munich, Germany, e-mail: yordanov@tum.de

Peter Russer
Munich University of Technology, Institute for High Frequency Engineering, Arcisstr. 21, 80333 Munich, Germany, e-mail: russer@tum.de

P. Russer, U. Siart (eds.), *Time Domain Methods in Electrodynamics*,

In this work a numerical inversion of the Schwarz-Christoffel conformal mapping [8] is used to produce exact results for the even and odd mode capacitances. The only assumption that is made is for symmetry. No restrictions on the width to height ratio of the conductor cross section are made. The multiconductor transmission line equations are also solved numerically.

1 Theory

In this section we present in short the analytical solution of the multiconductor transmission line equations, which has been used to compute the frequency response of the interconnection bus. Some notes on the Shwarz-Christoffel mapping are also given here.

1.1 Multiconductor Transmission Lines

In this section the TEM modes of a multiconductor transmission line filled with homogenous isotropic dielectric material used for digital interconnect bus are treated. A cross section of such line with three signal conductors between two ground planes is presented in Fig. 1. In order to fully characterize a line, consisting of n conductors and reference ground plane, the voltage $v_k(z,t)$ and the current $i_k(z,t)$ of each line are needed. These variables can be summarized in the following vectors

$$\mathbf{v}(z,t) = \left[v_1(z,t), v_2(z,t), \ldots, v_n(z,t)\right]^T, \tag{1}$$

$$\mathbf{i}(z,t) = \left[i_1(z,t), i_2(z,t), \ldots, i_n(z,t)\right]^T. \tag{2}$$

The relation between these two vectors can be expressed via the frequency-domain telegrapher's equation in matrix form

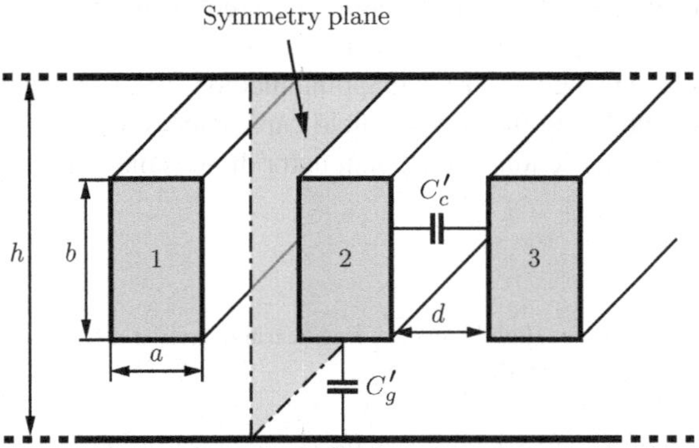

Fig. 1 A cross section of three-wire digital interconnect embedded between ground plates with a coupling and a ground capacitance

$$\frac{\mathrm{d}^2\mathbf{V}(z)}{\mathrm{d}z^2} = \mathbf{Z}'\mathbf{Y}'\mathbf{V}(z), \tag{3}$$

$$\frac{\mathrm{d}^2\mathbf{I}(z)}{\mathrm{d}z^2} = \mathbf{Y}'\mathbf{Z}'\mathbf{I}(z), \tag{4}$$

where

$$\mathbf{Z}' = \mathbf{R}' + j\omega\mathbf{L}', \tag{5}$$

$$\mathbf{Y}' = \mathbf{G}' + j\omega\mathbf{C}'. \tag{6}$$

are respectively the complex impedance and complex admittance per unit length. $\mathbf{R}'$ is the resistance per unit length, which models the ohmic losses in the conductors, $\mathbf{G}'$ is the conductivity per unit length, which accounts for the dielectric losses in the material, in which the line is embedded, and $\mathbf{C}'$ and $\mathbf{L}'$ are respectively the capacitance and inductance per unit length. In the case of quasi-TEM modes the following relation between $\mathbf{C}'$ and $\mathbf{L}'$ holds

$$\mathbf{L}'\mathbf{C}' = \mathbf{C}'\mathbf{L}' = \frac{1}{c^2}\mathbf{1}, \tag{7}$$

where c is the phase velocity of the TEM mode and $\mathbf{1}$ is the unity matrix.

In order to find a close form solution of the transmission line equations the uncoupled modal voltages and currents are introduced

$$\tilde{\mathbf{V}} = \mathbf{M}_V\mathbf{V}, \tag{8}$$

$$\tilde{\mathbf{I}} = \mathbf{M}_I\mathbf{I}. \tag{9}$$

The transformation matrices $\mathbf{M}_V$ and $\mathbf{M}_I$ can be found by diagonalizing the product $\mathbf{Z}'\mathbf{Y}'$ as follows

$$\tilde{\gamma}^2 = \tilde{\mathbf{Z}}'\tilde{\mathbf{Y}}' = \mathbf{M}^{-1}\mathbf{Z}'\mathbf{Y}'\mathbf{M}, \tag{10}$$

where $\tilde{\gamma}$ is the modal propagation coefficient diagonal matrix, and $\tilde{\mathbf{Z}}'$ and $\tilde{\mathbf{Y}}'$ are the diagonalized $\mathbf{Z}'$ and $\mathbf{Y}'$ matrices

$$\tilde{\mathbf{Z}}' = \mathbf{M}_V^{-1}\mathbf{Z}'\mathbf{M}_I = \mathrm{diag}[\tilde{Z}_1, \tilde{Z}_2, \ldots, \tilde{Z}_n], \tag{11}$$

$$\tilde{\mathbf{Y}}' = \mathbf{M}_I^{-1}\mathbf{Y}'\mathbf{M}_V = \mathrm{diag}[\tilde{Y}_1, \tilde{Y}_2, \ldots, \tilde{Y}_n]. \tag{12}$$

The relation between $\mathbf{M}_V$, $\mathbf{M}_I$, and $\mathbf{M}$ is given by

$$\mathbf{M} = \mathbf{M}_V = (\mathbf{M}_I^T)^{-1}. \tag{13}$$

The characteristic impedance matrix is defined as

$$\mathbf{Z}_0 = \mathbf{M}\tilde{\gamma}\mathbf{M}^{-1}\tilde{\mathbf{Z}}. \tag{14}$$

The boundary conditions for solving equations (3) and (4) are applied by introducing matrices $\mathbf{Z}_S$, accounting for the source impedances, and $\mathbf{Z}_L$, accounting for the load impedances, as well as vectors $\mathbf{V}_S$ and $\mathbf{V}_L$, accounting for the impressed source and load voltages.

We need to solve for the modal currents $\tilde{\mathbf{I}}$ the following matrix equation

$$\begin{bmatrix} (\mathbf{Z}_0+\mathbf{Z}_S)\mathbf{M} & (\mathbf{Z}_0-\mathbf{Z}_S)\mathbf{M} \\ (\mathbf{Z}_0-\mathbf{Z}_L)\mathbf{M}\mathrm{e}^{-\tilde{\gamma}l} & (\mathbf{Z}_0-\mathbf{Z}_L)\mathbf{M}\mathrm{e}^{\tilde{\gamma}l} \end{bmatrix} \begin{bmatrix} \tilde{\mathbf{I}}^{(+)} \\ \tilde{\mathbf{I}}^{(-)} \end{bmatrix} = \begin{bmatrix} \mathbf{V}_S \\ \mathbf{V}_L \end{bmatrix}, \tag{15}$$

where $\tilde{\mathbf{I}}^{(+)}$ and $\tilde{\mathbf{I}}^{(-)}$ are the amplitudes of the solution of the diagonalized equation (4)

$$\tilde{\mathbf{I}}(z) = \mathrm{e}^{-\tilde{\gamma}z}\tilde{\mathbf{I}}^{(+)} + \mathrm{e}^{\tilde{\gamma}z}\tilde{\mathbf{I}}^{(-)}. \tag{16}$$

Finally we can compute the transmission line voltages $\mathbf{V}$ in frequency domain as a function of the distance along the line z using the following relation

$$\mathbf{V} = \mathbf{Z}_0\mathbf{M}\left(\mathrm{e}^{-\tilde{\gamma}z}\tilde{\mathbf{I}}^{(+)} + \mathrm{e}^{\tilde{\gamma}z}\tilde{\mathbf{I}}^{(-)}\right). \tag{17}$$

1.2 Schwarz-Christoffel Mapping of Simply Connected Polygonal Domains

The Schwarz-Christoffel mapping is a conformal mapping technique, which maps the upper half of the complex plane Ω_z into a polygonal domain by preserving the local angles [3, 4]. This transformation is described by the Schwarz-Christoffel formula

$$f(z) = A + C\int^{z} \prod_{k=1}^{n-1} (\zeta - z_k)^{\alpha_k - 1}\mathrm{d}\zeta. \tag{18}$$

In this equation z_k are the points on the real axis of the Ω_z domain, which are mapped into the polygon vertices, α_k are the internal angles of the polygon in counterclockwise direction, normalized to π, and A and C are complex constants. Under this transformation a capacitance, which characterizes the original structure characterizes the image too.

A simply connected planar domain can be defined as the interior of a planar closed line, that does not contain any holes. Using the inverse of the SC transformation any simply connected polygon can be mapped into the upper half of the image plane. Using one more SC transformation we can map the upper half-plane into a canonical shape like a rectangle (see Fig. 2). This is the mapping that can be used for symmetric multiconductor transmission lines, as shown in the next section.

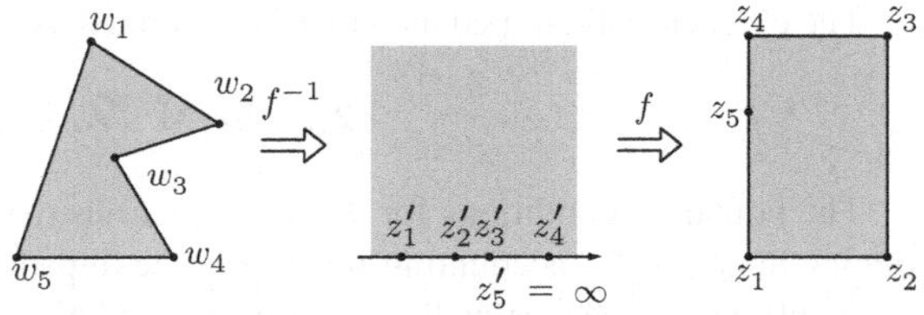

Fig. 2 Mapping an arbitrary simple bounded polygon to a rectangle

2 Application

In order to compute the response of the digital interconnection bus to a trapezoidal pulse the following methodology was used. First the capacitance per unit length $\mathbf{C}'$ has been computed from the bus geometry using Schwarz-Christoffel mapping. Then the inductance and resistance per unit length matrices are calculated. The obtained results along with load and source data have been used to solve (15) in frequency domain, thus obtaining the frequency response of the bus. The obtained data has been multiplied with the spectrum of an excitation trapezoidal pulse and inverse Fourier transform has been applied to the result to obtain the output pulse shape.

2.1 Computation of the Capacitance Matrix

In order to compute the static capacitance matrix of the transmission line even-odd mode analysis is utilized. If an electric wall inserted at the plane of symmetry between the signal conductors (see Fig. 1) only one signal conductor can be considered. The capacitance between this conductor and ground is the odd-mode capacitance C_o'. If a magnetic wall is inserted at the plane of symmetry the capacitance of the resulting structure is the even-mode capacitance C_e' (see Fig. 3). The ground and the coupling capacitances are connected with the even and odd mode capacitance with the following equations

$$C_g' = C_o', \tag{19}$$

$$C_c' = \frac{1}{4}(C_o' - C_e'). \tag{20}$$

The even and odd mode capacitances have been computed by making use of the symmetry of the structure, as depicted in Fig. 3. Schwarz-Christoffel mapping is applied to the quoter of the structure, shown in bold lines. The capacitance per unit

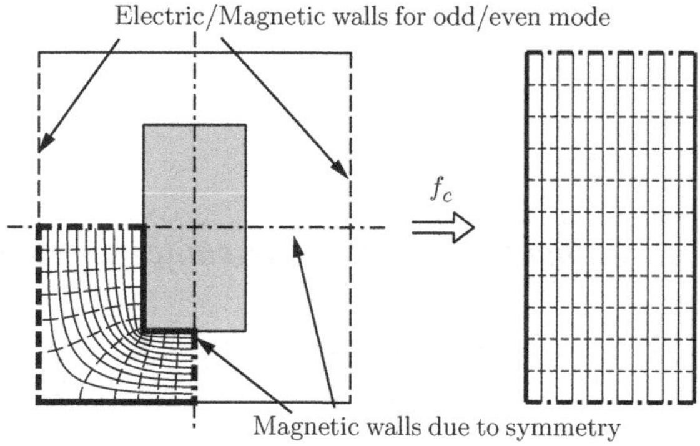

Fig. 3 Mapping a quoter of an even/odd mode equivalent structure to a rectangle

of length of the resulting rectangle is computed as the ratio of the electric wall length to the magnetic wall length times the dielectric permitivity of the medium. The Schwaz-Christoffel mapping is performed numerically using an integrated MATLAB toolbox [8].

Up to now it was considered that the ground capacitances of all signal conductors are equal. This does not hold for the end bus lines in the case of small distance between conductors d and long extended ground planes. In order to find the ground capacitance of the end conductors we consider the SC mapping, shown in Fig. 4, where the ground planes have been extended to infinity. Here we can use only one symmetry plane. The ground capacitance of the end conductors is equal to twice the capacitance of the equivalent rectangle.

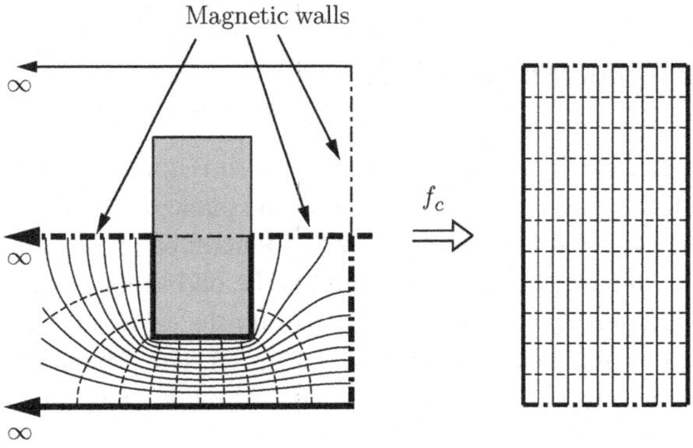

Fig. 4 Mapping half of an end-conductor equivalent structure to a rectangle

Finally the capacitance per unit of length matrix is constructed. Its diagonal elements are the sum of all capacitances, connected to the corresponding conductor, and its off-diagonal elements are the coupling capacitances between the corresponding conductors with a negative sign, as shown below

$$\mathbf{C}' = \begin{bmatrix} C'_g + C'_c & -C'_c & \dots & 0 \\ -C'_c & C'_g + 2C'_c & \dots & 0 \\ \vdots & \vdots & \ddots & \vdots \\ 0 & 0 & \dots & C'_g + C'_c \end{bmatrix}. \tag{21}$$

2.2 Computation of the Inductance and Resistance Matrices

The inductance per unit of length matrix $\mathbf{L}'$ is computed from the capacitance per unit of length matrix $\mathbf{C}'$ from the following equation

$$\mathbf{L}' = \frac{1}{c^2}\mathbf{C}'^{-1}. \tag{22}$$

This equation is only approximative for lossy lines, but due to the low resistivity of the used conductor this approximation can be used.

The clock frequency of the digital bus under consideration is several hundred megahertz and the bus dimensions are in the range of several hundred nanometers, therefore we can assume an uniform current distribution in the conductors. Therefore the resistance per unit length matrix elements can be computed from the metal conductivity σ and the bus dimensions as follows

$$R'_{ij} = \begin{cases} \dfrac{1}{\sigma ab} & \text{for } i = j, \\ 0 & \text{otherwise.} \end{cases} \tag{23}$$

3 Results

The above described computational procedure was implemented on a MATLAB code. The results for the capacitances per unit length for a transmission line, filled in with silicon, are presented in Figs. 5 and 6. Since the capacitance is only a function of the ratio of the line dimensions, all geometry is normalized to the distance between the ground planes h.

As a real case study a 4-line digital bus with the dimensions, shown in Table 1 is investigated. In order to solve (15) the following values were chosen. The load and source impedance matrices $\mathbf{Z}_L$ and $\mathbf{Z}_S$ were chosen as diagonal matrices with a value of 50 Ω for every diagonal element. This was done in order to compare the results

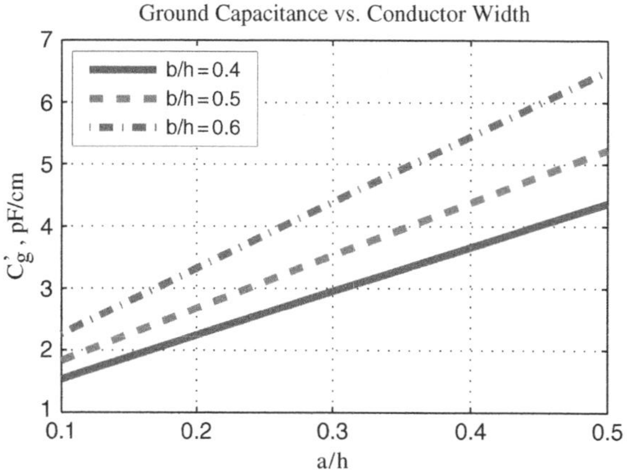

Fig. 5 Ground capacitance vs. geometry for digital transmission line, filled in with silicon. Here $d/h = 0.125$

Fig. 6 Coupling capacitance vs. geometry for digital transmission line, filled in with silicon. Here $a/h = 0.25$

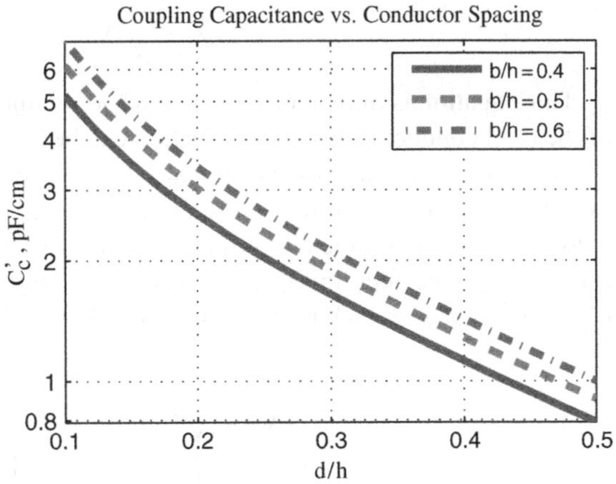

Table 1 Real digital bus dimensions

Description	Notation	Dimension
Conductor width	a	200 nm
Conductor height	b	400 nm
Distance between conductors	d	100 nm
Distance between ground plates	h	800 nm
Line length	l	1 mm

with the data, obtained from the commercial MoM-based simulator MOMENTUM from Advanced Design Systems. The load voltages vector $\mathbf{V}_L$ was set to zero, and the source voltage vector $\mathbf{V}_S$ was set to zero except for one element, which was set to one, providing an excitation, as follows

$$\mathbf{V}_S = [1 \quad 0 \quad 0 \quad 0]^T. \tag{24}$$

The frequency response for the crosstalk voltage at the far end of the line is presented in Fig. 7. The simulation time required by MOMENTUM on a Pentium 4 based computer with 2.4 GHz clock frequency and 1 GB of RAM was in the order of 60 min, while the equation-solving routine requires less than a second on the same machine.

In order to compute the pulse distortion more realistic values for the load and source impedance matrices $\mathbf{Z}_L$ and $\mathbf{Z}_S$ were chosen. The source impedance matrix $\mathbf{Z}_S$ has been set to zero, and the diagonal elements of the load impedance matrix $\mathbf{Z}_L$ were set to 10 MΩ. After performing a Fourier transform the shape of the distorted pulse at the far end of the line, shown in Fig. 8, was obtained. The results are compared with the data, obtained from SPICE simulation of the equivalent lumped-element circuit of the bus, using the previously computed capacitance and resistance values.

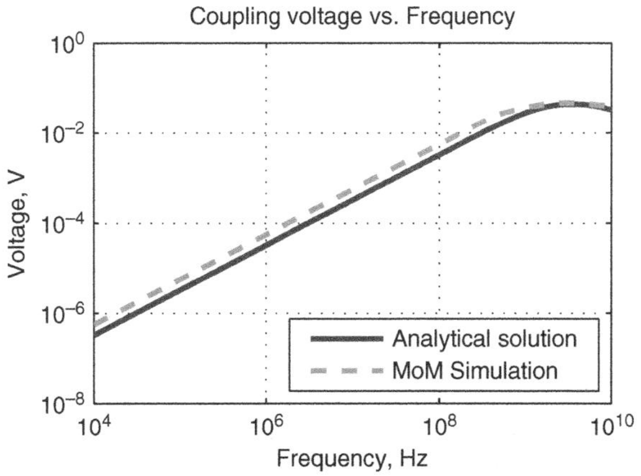

Fig. 7 Frequency response of the crosstalk voltage at the far end of the line. Comparison of the results, obtained by solving the transmission line equations, and the data, obtained by full-wave analysis

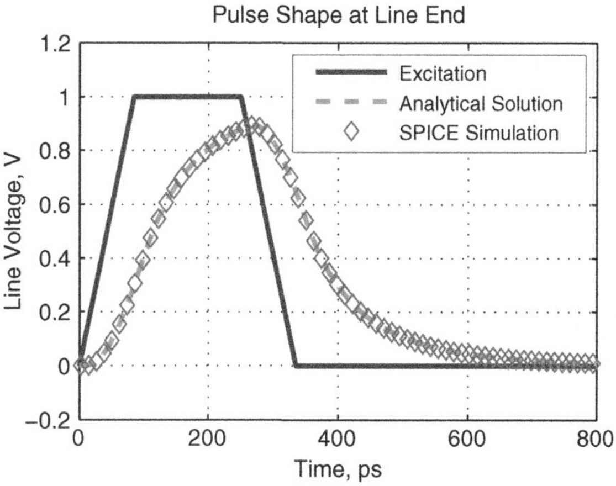

Fig. 8 Pulse distortion at the far end of the line. Comparison of the data, obtained with the proposed method, and the results from a SPICE simulation of the equivalent lumped-element circuit of the bus

4 Conclusions

A precise electromagnetic model of a multiconductor transmission line, used as an on-chip digital bus was presented. The structure was investigated under static conditions using numerical inversion of the Schwarz-Christoffel transformation. Using the even-odd mode analysis techniques the capacitance, inductance and resistance per unit length matrices were calculated. The dependance of the electrostatic parameters on the transmission line dimensions was explored. Using the obtained results the frequency response and the pulse distortion of a real-case on-chip digital bus was obtained. The advantages of the proposed method are its accuracy, the lack of geometrical limitations and the algorithm efficiency.

Acknowledgements This work has been supported by the Deutsche Forschungsgemeinschaft.

References

1. M. T. Ivrlac and J. A. Nossek, "Chalanges in coding for quantized MIMO systems," in *Proc. IEEE International Symposium on Information Theory*, Jul. 2006, pp. 2114–2118.
2. D. M. Pozar, *Microwave Engineering*, 3rd ed. New York, NY: John Wiley & Sons, 2005.
3. P. Russer, *Electromagnetics, Microwave Circuit and Antenna Design for Communications Engineering*, 2nd ed. Nordwood, MA: Artec House, 2006.
4. T. A. Driscoll and L. N. Trefethen, *Schwarz-Christoffel Mapping*, ser. Cambridge Monographs on Applied Computational Mathematics. Cambridge, UK: Cambridge University Press, 2002.
5. C. R. Paul, *Analysis of Multiconductor Transmission Lines*. New York, NY: Wiley, 1994.
6. S. B. Cohn, "Shielded coupled-strip transmission line," *IEEE Trans. Microw. Theory Tech.*, vol. 8, pp. 29–38, Oct. 1955.
7. W. J. Getsinger, "Coupled rectangular bars between parallel plates," *IEEE Trans. Microw. Theory Tech.*, vol. 10, pp. 65–72, Jan. 1962.
8. T. A. Driscoll, "Algorithm 756: A MATLAB toolbox for schwarz-christoffel mapping," *AMC Trans. Math. Softw.*, vol. 22, no. 2, pp. 168–186, Jun. 1996.

Two Decades of SCN Modelling and Beyond

Dr. Poman So, P. Eng.

Abstract This chapter gives a brief overview of the history of SCN development in the last twenty years. Some of the important features for developing TLM based simulation tools are addressed. A number of simulators developed by the author are used to illustrate the importance of applying object-oriented software design practice to computational electromagnetics. The conclusion ends with an invitation to the TLM community to create an open source TLM simulation package.

1 Introduction — A Snapshot of SCN Development

Two decades ago, the SCN acronym was brought into the computational electromagnetic world by P. B. Johns [1]; his innovative symmetrical condensed node eliminated the field space staggering problem in the Yee's FDTD [2] and the expanded TLM node modelling procedures [3, 4]. The sudden passing of P. B. Johns was a great loss to the TLM computational electromagnetic community. Hoefer et al. named the removed-branch TLM impulse response stream the Johns Matrix in 1989 [5] to honour P.B. Johns' many contributions in computational electromagnetics.

While most TLM research and development effort were on the application side, there were persistent activities to improve the SCN node; HSCN (hybrid SCN), SSCN (Super SCN) and GSCN (Generalized SCN) nodes were published in the later years [6, 7, 8, 9, 10, 11, 12, 13]. Much of the work was done at the George Green Institute of Electromagnetics Research at Nottingham by V. Trenkic et al. Jin and Vahldieck, on the other hand, developed a frequency domain scattering procedure for the SCN method — FDTLM [14]. New scattering procedures with minimum number of floating point operations per scattering were developed by Tong et al. [15] and Trenkic et al. [13]. Russer et al. exploited the mutually exclusive property of the neighbouring cell during odd and even time steps; without loss of

Dr. Poman So, P. Eng.
Department of Electrical Engineering, University of Victoria, Victoria, BC, Canada,
e-mail: Poman.So@ECE.UVic.CA

P. Russer, U. Siart (eds.), *Time Domain Methods in Electrodynamics*,

accuracy, their ATLM (Alternating TLM) procedure reduces the memory and CPU time requirement by 50 percent [16, 17]. Concurrently with the above development, there were progress in other areas of SCN research; parallel and distribution computing [18, 19], PML [20, 21, 22], nonlinear device [23, 24, 25, 26], wire models [27, 28, 29], graded mesh [30, 31, 32], multiple grids [33, 34, 35], unstructured mesh [36, 37], and meta-materials [38, 39] are just some of them.

In these two decades of SCN research and development, millions of lines of code were written by SCN researchers. Some of these codes were put together as simulation tools [5, 40, and 41]. Are these code modules easily accessible by the EM community at large? No, not yet; because virtually all computational electromagnetics research activities were in algorithm developments with no foresight for module integration. This was naturally so because software development did not belong to the traditional engineering electromagnetics discipline; hence the "dirty" module integration task was left behind to the software industry. Can we afford to be complacent about this situation? Shall we develop an open-source SCN simulator framework to benefit the EM community at large? Before addressing the above questions, let's look at what has been done in the TLM simulator area.

2 Available TLM Simulators

This session summarizes the last twenty years of research and development work in TLM simulators by Hoefer et al. Fig. 1 depicts a frequency halver modelled using two varactor diodes [5] with a 2D-TLM simulator [40], the time and frequency

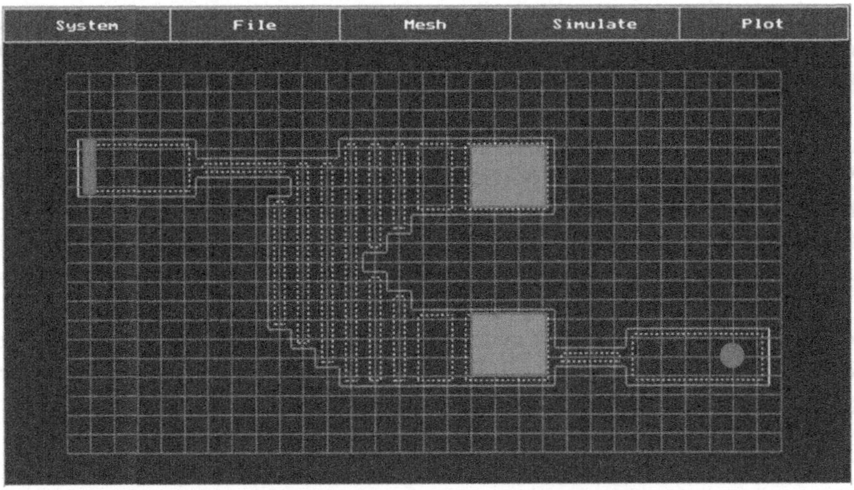

Fig. 1 A frequency halver modelled with two varactor diodes (identified with two *solid squares*). The *solid rectangle* at the left and the *solid circle* at the right are the input and output regions, respectively

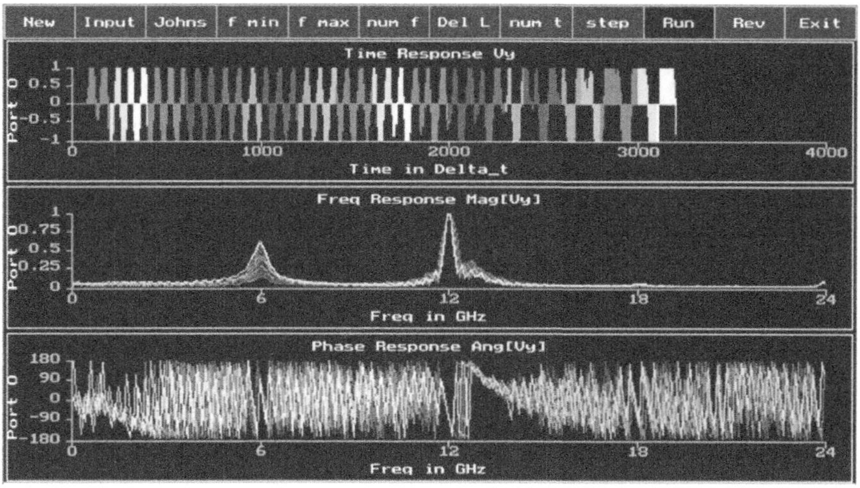

Fig. 2 The time and frequency domain outputs of the frequency halver depicted in Fig. 1 when the input frequency was 12 GHz

domain responses of the device are shown in Fig. 2. This simulator was the first TLM simulator equipped with a full-fledged graphical user interface. Multiple output regions and three-dimensional field animation were built-in features of the simulator, see the screen images in Figs. 3 and 4.

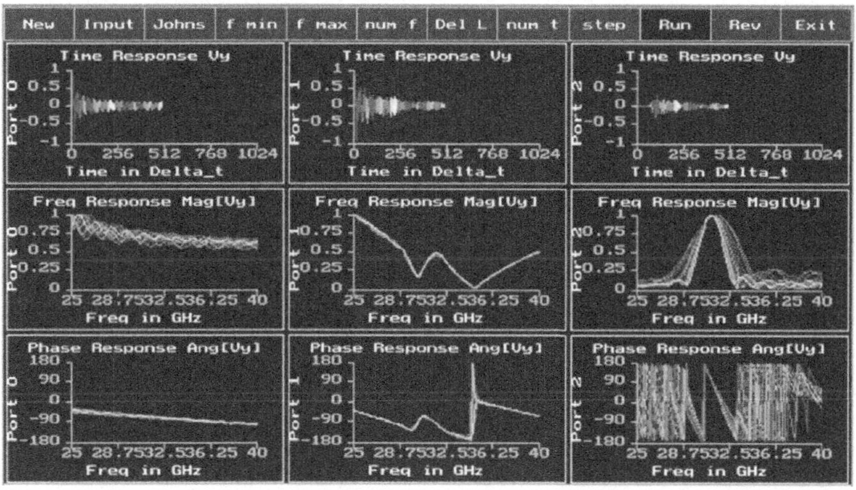

Fig. 3 The first 2D-TLM simulator, [40], was equipped with multiple output region capability. The screen image shows the time and frequency responses of a filter at its input and output ports (*port 1 and 2*) as well as the signal at the reference port (*port 0*)

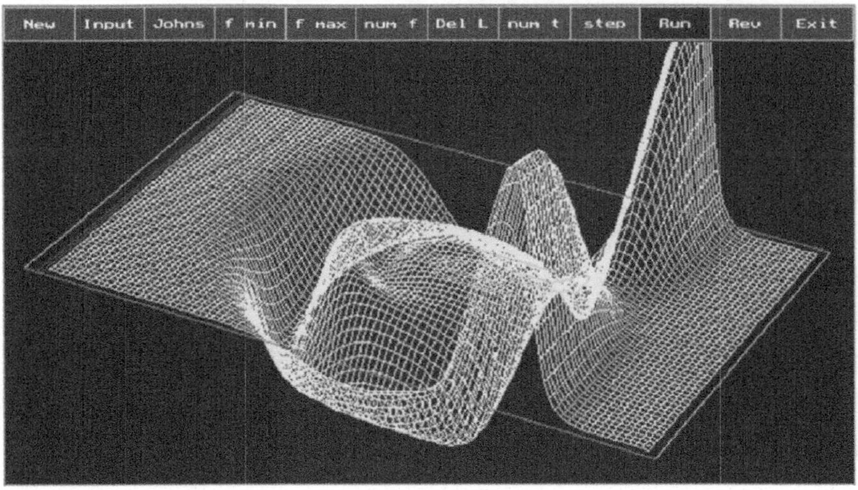

Fig. 4 The field animation capability of the 2D-TLM simulator published by John Wiley and Sons, [40]

The next TLM simulator from Hoefer et al. was a full-fledged three-dimensional field simulator based on the SCN algorithm, [41]. The initial version, known only to the group, was for MS-DOS, Fig. 5; the released version was for X-Window, Fig. 6.

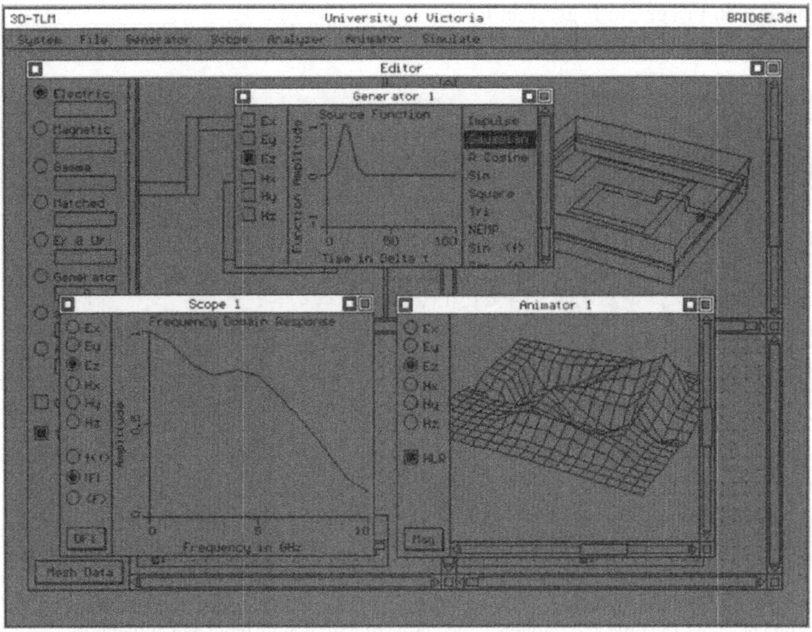

Fig. 5 The first full-fledged 3D-TLM simulator created by Hoefer et al.; this was the prototype for the X-Window version shown in Fig. 6

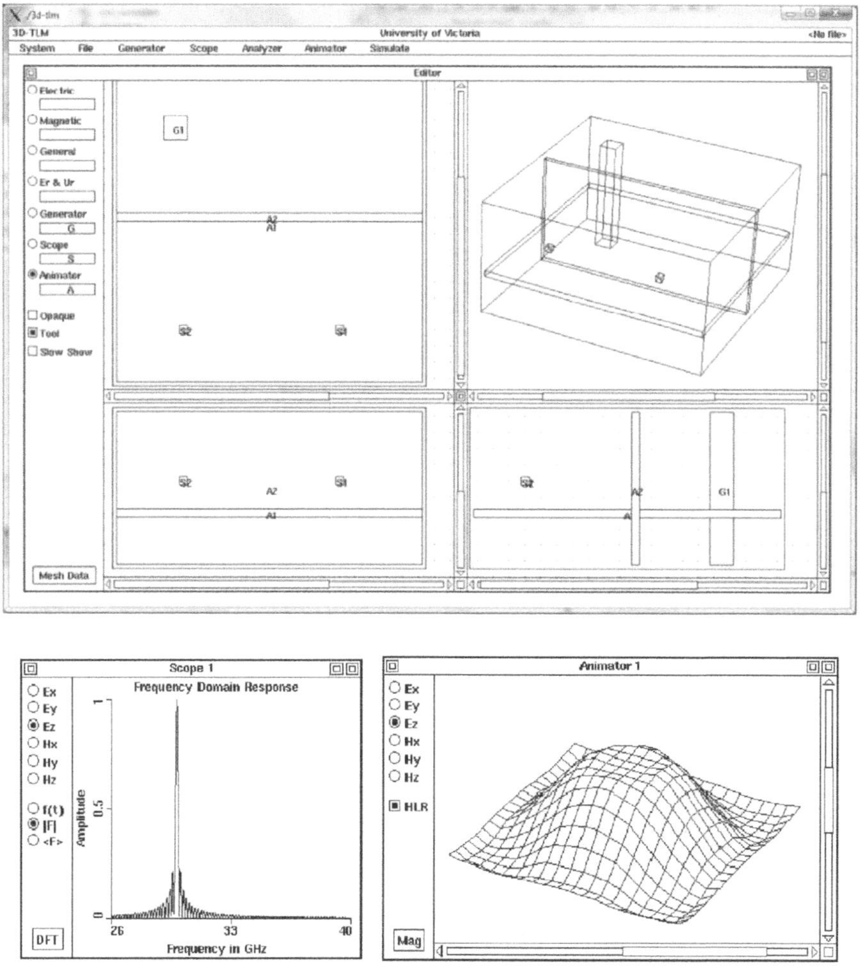

Fig. 6 The first full-fledged X-Windows 3D-TLM simulator created by Hoefer et al., [41]. The image above shows an air filled cavity and some computational elements — G1 is the source region, S1 is the sampling location, and A1 and A2 are two output regions for field animation

The work that was invested in developing the software tool would not benefit the computational community if the simulators were only accessible to a small group of people. Hence Windows versions of the simulators were developed and released as freeware. Figures 7 and 8 show the screen images of the two simulators. The author purposely used the examples shown in Figs. 1 and 4 in the last two figures to emphasize the relationship between the old and new versions. The 2D-Shunt and 3D-SCN algorithms are used in both the old and new versions; the user interfaces have similar design; most importantly, which is not obviously to the end users, is that the software are implemented in an object-oriented paradigm so that new idea can be incorporated into the framework with minimum impact to the existing code.

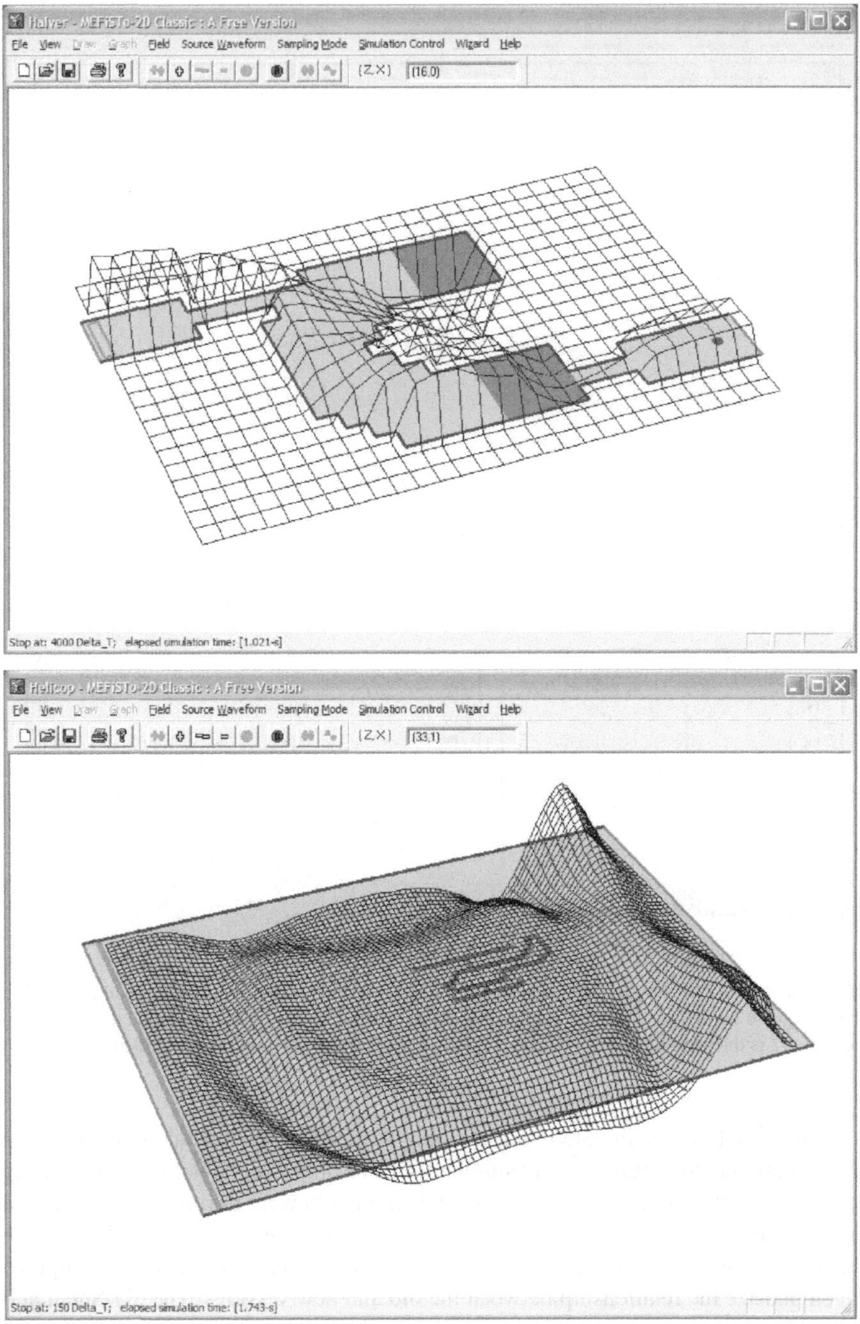

Fig. 7 MEFiSTo-2D Classic for Windows; a free 2D-TLM simulator by Faustus Scientific Corporation

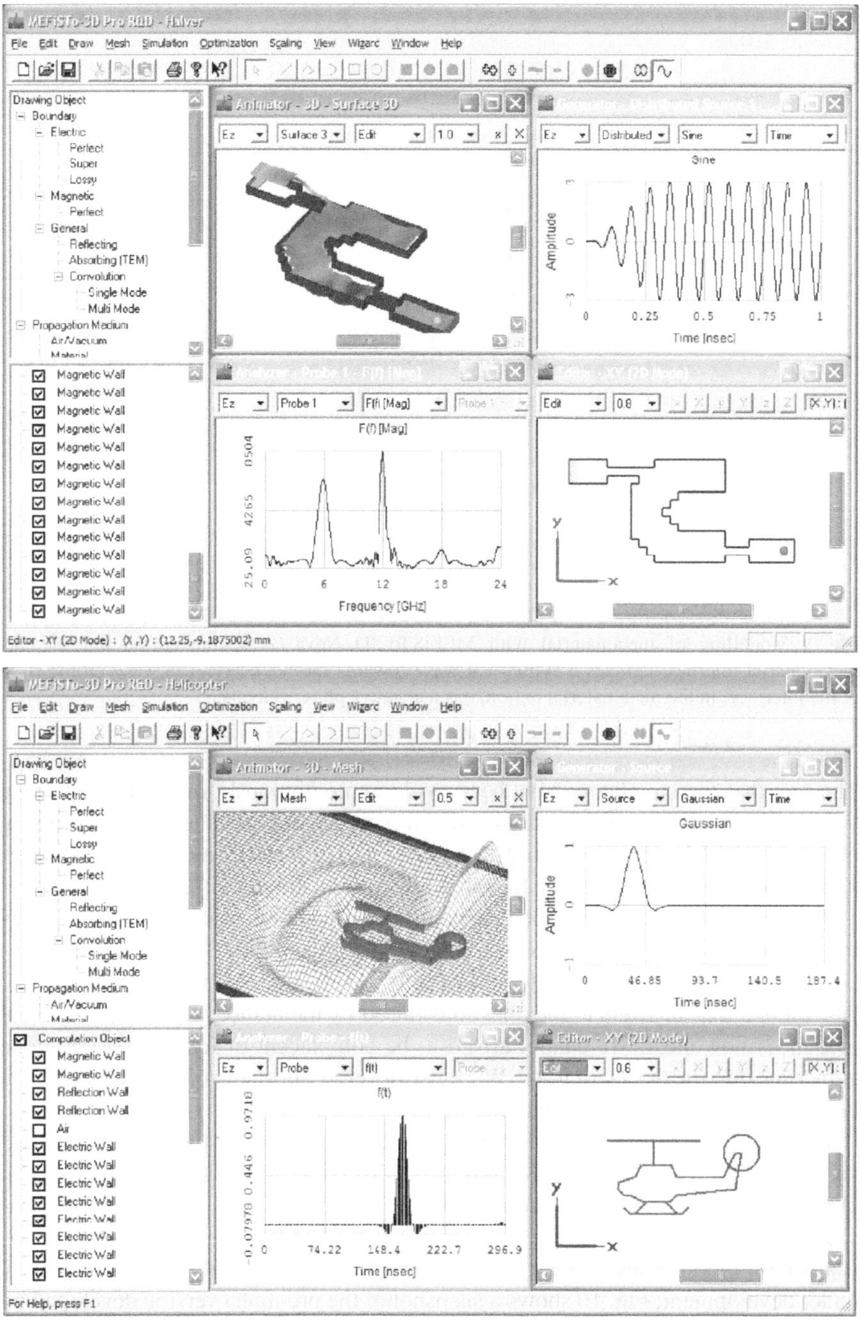

Fig. 8 MEFiSTo-3D Nova for Windows; a free 3D-TLM simulator by Faustus Scientific Corporation

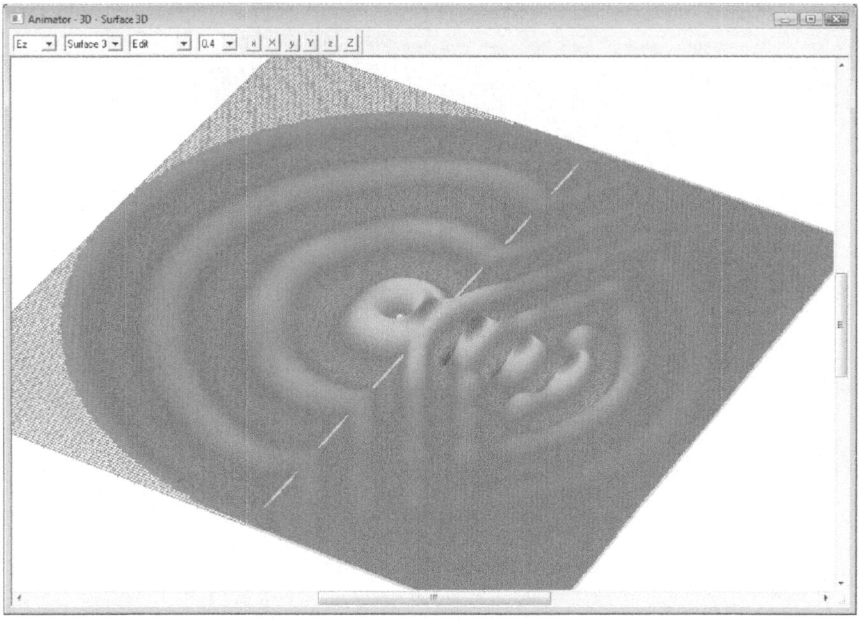

Fig. 9 Modelling of metamaterial with MEFiSTo-3D Nova. A source at the left the air-metamaterial interface radiates a cylindrical wave towards the boundary. The wave on the right of the interface, in the metamaterial region, has a negative phase velocity

The fact that the simulators have been evolving continuously over the last twenty years with virtually complete backward compatibility is a good testimony for using object-oriented design. MEFiSTo-3D Nova is free but it has many advanced features known today; Fig. 9 depicts the focusing effect of metamaterial [38].

What the author has described is the effort of just one research group in making research tool available as freeware. How much would the TLM community accomplish if there was coordinated effort in releasing source code and software for general use. The answers to the questions raised by the author before this historical recount is now clear. We cannot afford to be complacent of letting research code gathering dust in the laboratory. We must develop an open-source framework to benefit the EM community at large.

YATPAC (Yet Another TLM Package) is a free and powerful SCN based field simulator developed at the Technische Universität München, Germany [42]. Different from MEFiSTo-2D Classic and MEFiSTo-3D NOVA, YATPAC is an open-source product. The SCN engine in this package is written in FORTRAN. The visualization modules are written in *C*. A graphical user interface front is currently under development; Fig. 10 shows a snapshot of the pre-alpha version developed by the YATPAC team.

FORTRAN and *C* are procedural languages that do not support object-oriented programming. Implementing electromagnetics algorithms and graphical user

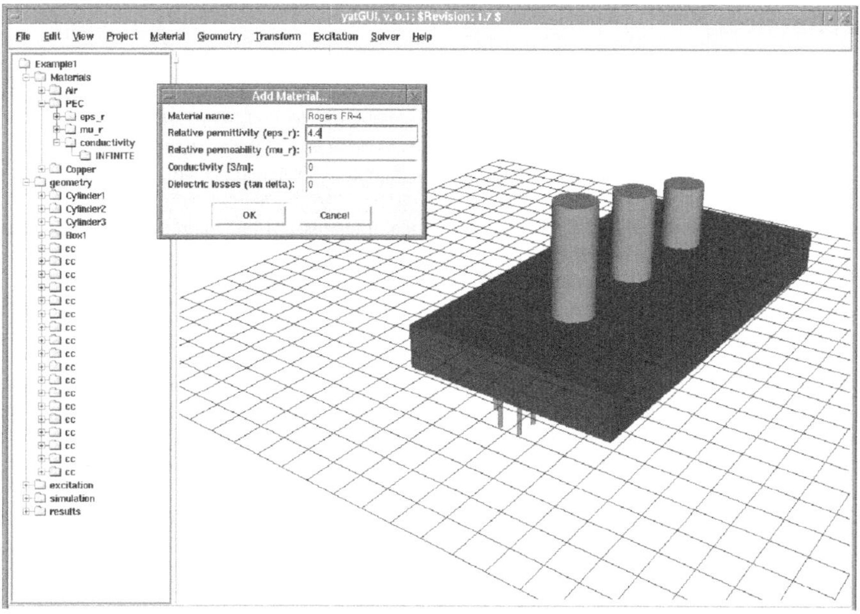

Fig. 10 YatGUI, a graphical user interface for Yatpac

interface in these languages is like doing research in electromagnetics without vector calculus, or geometric calculus for those who are already in the new frontier.

3 Conclusion

Open source practice is not new to engineering. One of the very first project of this kind would be the Berkeley SPICE project. Open source is more common in computer science due to the nature of their work in algorithms and software development; some of the famous open source systems and projects are the X Window System, LINUX, Cygwin, eclipse, NetBeans, JAVA, MONO, and last but not least, all the GNU compilers and software. The list is far from complete. For electromagnetics, NEC2 would be among the first well known open source package; two recent additions to the family are MIT MEEP and TUM YATPAC.

Much about SCN has been discovered and implemented in the last twenty years. The birth places of TLM and YATPC, the University of Nottingham and the Technische Universität München, respectively, are two of the "goldmines" of TLM codes for the TLM community. Since much of the legacy TLM codes were developed in the procedural programming era, it is time to re-package those codes in the object-oriented programming paradigm. YATPAC, an already open source package, would be the package to begin with.

References

1. P.B. Johns, "A symmetrical condensed node for the TLM method", IEEE Transactions Microwave Theory and Techniques, vol. 35, issue 4, pp. 370–377, April 1987.
2. K.S. Yee, "Numerical solution of initial boundary value problems involving Maxwell's equations in isotropic media", IEEE Transactions Antennas Propagation, vol. 14, pp. 302–307, May 1966.
3. S. Akhtarzad and P.B. Johns, "Three-dimensional transmission-line matrix computer analysis of microstrip resonators", IEEE Transactions on Microwave Theory and Techniques, vol. 23, issue 12, pp. 990–997, December 1975.
4. S. Akhtarzad and P.B. Johns, "Solution of 6-component electromagnetic fields in three space dimensions and time by the TLM method", Electronics Letters, vol. 10, issue 25, pp. 535–536, December 1974.
5. P.P.M. So, C. Eswarappa and W.J.R. Hoefer, "A two-dimensional transmission line matrix microwave field simulator using new concepts and procedures", IEEE Transactions on Microwave Theory and Techniques, vol. 37, issue 12, pp. 1877–1898, December 1989.
6. R. Scaramuzza and A.J. Lowery, "Hybrid symmetrical condensed node for the TLM method", Electronics Letters, vol. 8, issue 23, pp. 1947–1949, November 1994.
7. V. Trenkic, C. Christopoulos and T.M. Benson, "Generally graded TLM mesh using the symmetrical supercondensed node", Electronics Letters, vol. 30, issue 10, pp. 795–797, May 1994.
8. P. Berini and K. Wu, "A Pair of hybrid symmetrical condensed TLM nodes", IEEE Microwave and Guided Wave Letters, vol. 4, issue 7, pp. 244–246, July 1994.
9. V. Trenkic, C. Christopoulos and T.M. Benson, "Theory of the symmetrical super-condensed node for the TLM method", IEEE Transactions on Microwave Theory and Techniques, vol. 43, issue 6, pp. 1342–1348, June 1995.
10. V. Trenkic, C. Christopoulos and T.M. Benson, "On the time step in hybrid symmetrical condensed TLM node", IEEE Transactions on Microwave Theory and Techniques, vol. 41, issue 9, pp. 2172–2174, September 1995.
11. V. Trenkic, C. Christopoulos and T.M. Benson, "Development of a general symmetrical condensed node for the TLM method", IEEE Transactions on Microwave Theory and Techniques, vol. 44, issue 12, pp. 2129–2135, December 1996.
12. V. Trenkic, C. Christopoulos and T.M. Benson, "Advanced node formulations in TLM — the adaptable symmetrical condensed node", IEEE Transactions on Microwave Theory and Techniques, vol. 44, issue 12, pp. 2473–2478, December 1996.
13. V. Trenkic, C. Christopoulos, and T.M. Benson, "Optimization of TLM schemes based on the general symmetrical condensed node", IEEE Transactions on Microwave Theory and Techniques, vol. 45, issue 3, pp. 457–465, March 1997.
14. H. Jin and R. Vahldieck, "The frequency-domain transmission line matrix method-A new concept", IEEE Transactions on Microwave Theory and Techniques, vol. 40, issue 12, pp. 2207–2218, December 1992.
15. C.E. Tong and Y. Fujino, "An efficient algorithm for transmission line mairix analysis of electromagnetic problems using the symmetrical Condensed node", IEEE Transactions on Microwave Theory and Techniques, vol. 39, issue 8, pp. 1420–1424, August 1991.
16. P. Russer and B. Bader, "The alternating transmission line matrix (ATLM) scheme", IEEE MTT-S Digest, vol. 1, pp. 19–22, May 16–20 1995.
17. S. Lindenmeier and P. Russer, "The alternating rotated TLM scheme (ARTLM) for fast simulations in time domain", 26th European Microwave Conference, vol. 1, pp. 493–496, October 1996.
18. C. Eswarappa, P.P.M. So and W.J.R. Hoefer, "Efficient field-based CAD of microwave circuits on massively parallel processor computer using TLM and Prony's methods", IEEE MTT-S Symposium Digest, vol. 3, pp. 1531–1534, May 1994.
19. P.J. Parsons, S.R. Jaques, S.H. Pulko and F.A. Rabhi, "TLM modeling using distributed computing", IEEE Microwave and Guided Wave Letters, vol. 6, issue 3, pp. 141–142, March 1996.

20. J. Xu, Z. Chen and J. Chuang, "Numerical implementations of PML boundary conditions in the TLM-based SCN FDTD grid", IEEE Transactions on Microwave Theory and Techniques, vol. 45, issue 8, Part 1, pp. 1263–1267, August 1997.
21. N. Pena and M.M. Ney, "Absorbing-boundary conditions using perfectly matched-layer (PML) technique for three-dimensional TLM simulations", IEEE Transactions on Microwave Theory and Techniques, vol. 45, issue 10, Part 1, pp. 1749-1755, October 1997.
22. J-L. Dubard and D. Pompei, "Optimization of the PML efficiency in 3-D TLM method", IEEE Transactions on Microwave Theory and Techniques, vol. 48, issue 7, Part 1, pp. 1081–1088, July 2000.
23. P. Russer, P.P.M. So and W.J.R. Hoefer, "Modeling of nonlinear active regions in TLM", IEEE Microwave and Guided Wave Letters, vol. 1, issue 1, pp. 10–13, January 1991.
24. C.J. Smartt and C. Christopoulos, "Modelling nonlinear and dispersive propagation problems by using the TLM method" IEE Proceedings on Microwaves, Antennas and Propagation, vol. 145, issue 3, pp. 193–200, June 1998.
25. J. Paul, C. Christopoulos and D.W.P. Thomas, "Generalized material models in TLM – Part 3: materials with nonlinear properties", IEEE Antennas and Propagation, vol. 50, issue 7, pp. 997–1004, July 2002.
26. V. Janyani, J.D. Paul, A. Vukovic, T.M. Benson and P. Sewell, P., "TLM modelling of non-linear optical effects in fibre Bragg gratings", IEE Proceedings on Optoelectronics, vol. 151, issue 4, pp. 185–192, August 2004.
27. A.P. Duffy, T.M. Benson, C. Christopoulos and J.L. Herring, "New methods for accurate modelling of wires using TLM", Electronics Letters, vol. 29, issu 2, pp. 224–226, January 1993.
28. P. Sewell, Y.K. Choong and C. Christopoulos, "An accurate thin-wire model for 3-D TLM simulations", IEEE Transactions on Electromagnetic Compatibility, vol. 45, issue 2, pp. 207–217, May 2003.
29. J. Paul, J.C. Christopoulos, D.W.P. Thomas and X. Liu, "Time-domain modeling of electromagnetic wave interaction with thin-wires using TLM", IEEE Transactions on Electromagnetic Compatibility, vol. 47, issue 3, pp. 447–455, August 2005.
30. V. Trenkic, C. Christopoulos and T.M. Benson, "Generally graded TLM mesh using the symmetrical supercondensed node", Electronics Letters, vol. 30, issue 10, pp. 795–797, May 1994.
31. V. Trenkic, C. Christopoulos and T.M. Benson, "A graded symmetrical super-condensed node for the TLM method", IEEE AP-S Symposium Digest, vol. 2, pp. 1106–1109, June 1994.
32. J. Hesselbarth and R. Vahldieck, "Mesh grading and cutoff frequencies in the frequency-domain TLM method", IEEE MTT-S Symposium Digest, vol. 3, pp. 1551–1554, June 1998.
33. M.I. Sobhy, M.H. Abd El-Azeem and K.W. Royer, "A new multi-grid 3-D TLM algorithm for simulation of microwave FSS", IEEE MTT-S Symposium Digest, vol. 2, pp. 439–42, June 1996.
34. P. Sewell, J. Wykes, A. Vukovic, D.W.P. Thomas, T.M. Benson and C. Christopoulos, "Multi-grid interface in computational electromagnetics", Electronics Letters, vol. 40, issue 3, pp. 162–163, February 2004.
35. L. Pierantoni and T. Rozzi, "A general multigrid-subgridding formulation for the transmission line matrix method", IEEE Transactions on Microwave Theory and Techniques, vol. 55, issue 8, pp. 1709–1716, August 2007.
36. P. Sewell, J.G. Wykes, T.M. Benson, C. Christopoulos, D.W.P. Thomas and A. Vukovic, "Transmission-line modeling using unstructured triangular meshes", IEEE Transactions on Microwave Theory and Techniques, vol. 52, issue 5, pp. 1490–1497, May 2004.
37. P. Sewell, T.M. Benson, C. Christopoulos, D.W.P. Thomas, A. Vukovic and J.G. Wykes, "Transmission-line modeling (TLM) based upon unstructured tetrahedral meshes", IEEE Transactions on Microwave Theory and Techniques, vol. 53, issue 6, Part 1, pp. 1919–1928, June 2005.
38. P.P.M. So, H. Du and W.J.R. Hoefer, "Modeling of metamaterials with negative refractive index using 2-D shunt and 3-D SCN TLM networks", IEEE Transactions on Microwave Theory and Techniques, vol. 53, issue 4, Part 2, pp. 1496–1505, April 2005.

39. M. Zedler, C. Caloz and P. Russer, "A 3-D isotropic left-handed metamaterial based on the rotated transmission-line matrix (TLM) scheme", IEEE Transactions on Microwave Theory and Techniques, vol. 55, issue 12, Part 2, pp. 2930–2941, December 2007.
40. W.J.R. Hoefer and P.P.M. So, "The electromagnetic wave simulator — a dynamic visual electromagnetics laboratory based on the two-dimensional TLM method", John Wiley and Sons Inc., New York 1991.
41. P.P.M. So and W.J.R. Hoefer, "3-D TLM time domain electromagnetic wave simulator for microwave circuit modeling", IEEE MTT-S Symposium Digest, vol. 2, pp. 631–634, June 1991.
42. Yatpac — http://www.yatpac.org.

Calculation of Instantaneous Power and Energy Quantities in TLM Simulations

John Paul, Christos Christopoulos, and David W. P. Thomas

A method for the calculation of instantaneous power and energy quantities in three-dimensional TLM simulations based on the symmetrical condensed node is presented. The approach involves augmenting the TLM algorithm with extra equations to calculate the curl quantities. These auxiliary variables are used to evaluate the divergence of the Poynting vector and the energy density. The formulation of the technique is described and the numerical algorithm is cast into a digital filter system. Results are shown for electromagnetic power divergence and energy deposition in a block of Debye material under steady-state and pulse illumination.

1 Introduction

In the simulation of electromagnetic wave propagation in the time-domain, it is often of interest to calculate the divergence of the Poynting vector and the energy density. Knowledge of these quantities are essential in the validation of numerical codes and may be of interest in applications involving nonlinear materials, high power pulse processing of materials, coupled thermal-electromagnetic problems and others. Direct simulations of the time evolution of the fields also give invaluable insight into electromagnetic physics and offer the possibility of employing software

John Paul
George Green Institute for Electromagnetics Research, School of Electrical and Electronic Engineering, University of Nottingham, Nottingham, NG7 2RD, U.K.,
e-mail: john.paul@nottingham.ac.uk

Christos Christopoulos
George Green Institute for Electromagnetics Research, School of Electrical and Electronic Engineering, University of Nottingham, Nottingham, NG7 2RD, U.K.,
e-mail: eezks@nottingham.ac.uk

David W. P. Thomas
George Green Institute for Electromagnetics Research, School of Electrical and Electronic Engineering, University of Nottingham, Nottingham, NG7 2RD, U.K.

P. Russer, U. Siart (eds.), *Time Domain Methods in Electrodynamics*,

validation, investigation and optimization of electromagnetic devices and processes to complement theoretical predictions and practical experimentation.

In previous work, i.e., [1, 2, 3], we have developed procedures for the time-domain simulation of electromagnetic wave interaction with various kinds of materials displaying complex material properties, e.g., Debye, Drude, Lorentz, non-linear, chiral and others. The TLM scattering rules developed to account for these materials were based on using $\mathcal{Z}$-transforms to design digital filter systems to describe the frequency-dependence and nonlinearity. The idea of using $\mathcal{Z}$-transforms to deal with complex material properties in TLM was introduced by de Menezes and Hoefer [4]. In this contribution, the digital filter approach is employed to design algorithms which are coupled to the TLM scattering scheme to output power and energy quantities.

The main difference between TLM based on the symmetrically condensed node (SCN) and the other major time-domain approach, the finite-difference time-domain method (FDTD) [5], is that in FDTD the sample points of the fields $\underline{E}$ and $\underline{H}$ are not located at the same space and time points. This may be illustrated by the term "leap-frog method" which is often used to describe the kind of discretization used in FDTD. Because calculation of the Poynting vector requires both $\underline{E}$ and $\underline{H}$, to evaluate it in FDTD would require both time averaging and spatial interpolation of the fields. In addition, to obtain the divergence of the Poynting vector requires integration over a closed surface, so an FDTD solution would require fields belonging to adjacent cells and while feasible, such a process would not be easy to implement in a general way. More seriously, such averaging would introduce spatial and temporal filtering, thus the quantities obtained from such a process would have to be treated with caution.

Since in TLM based on the SCN, the field samples are available colocated in space and time, as detailed below, it is straightforward to develop procedures to evaluate the instantaneous divergence of the Poynting vector and the electromagnetic energy density at the node centres on a cell-by-cell basis.

Although it is easy to calculate the Poynting vector at the centres of TLM cells from $\underline{E} \times \underline{H}$, this does not yield the divergence of the Poynting vector nor the energy density. In addition, it may be argued that the divergence of the Poynting vector $\nabla \cdot \underline{S}$ may be evaluated on the faces of the cells from the average of the powers carried by the incident and reflected waves employed in the transmission-line model. While this a perfectly valid approach, here $\nabla \cdot \underline{S}$ is obtained from the physical fields observed at the centre of the cell. Finally, as shown below, the calculation of the energy density at source cells requires the solution of $\underline{E} \cdot \underline{J}_{ef}$ and if nonphysical magnetic sources are employed in a simulation, then $\underline{H} \cdot \underline{J}_{mf}$ must be obtained. Here, $\underline{J}_{ef}$ and $\underline{J}_{mf}$ are the free electric (magnetic) current densities. Because $\{\underline{E}, \underline{H}, \underline{J}_{ef}, \underline{J}_{mf}\}$ are physical quantities and their sample points are located at the centres of the TLM cells, it appears more natural to use these to obtain the power divergence and energy densities rather than the nonphysical incident and reflected waves of the transmission-line model.

The rest of this contribution is organized as follows, in Sect. 2 the details of the approach are given: In Sect. 2.1 Poynting's theorem is reviewed and is cast into a normalized format which is compatible with the TLM formulation of reference [1].

In Sect. 2.2, the development of the TLM scattering procedure for regular Cartesian meshes is reviewed and the calculation of the curl quantities by augmenting of the scattering process with a simple algorithm is described. This leads to Sect. 2.3 where a scheme for the calculation of the divergence of the Poynting vector is given. In Sect. 2.4, the evaluation of the instantaneous energy density is shown. In Sect. 3 the approach is validated by investigation of electromagnetic wave interaction with a block of Debye material in the steady-state (i.e. a sinusoidal excitation). By employing simple postprocessing to perform time-averaging, it is shown that the time-domain method yields results in agreement with elementary frequency-domain analysis. Finally, results are shown for the interaction of a pulsed electromagnetic wave with the material block. The time-domain results for the divergence of the Poynting vector and the energy density shown here give a deeper insight into electromagnetic field-material interaction in the case of lossy dielectric materials. The conclusions and ideas for further investigation are in Sect. 4.

2 Formulation

2.1 Poynting's Theorem

Maxwell's curl equations are

$$\begin{bmatrix} \nabla \times \underline{H} \\ -\nabla \times \underline{E} \end{bmatrix} - \frac{\partial}{\partial t}\begin{bmatrix} \underline{D} \\ \underline{B} \end{bmatrix} = \begin{bmatrix} \underline{J}_{ef} \\ \underline{J}_{mf} \end{bmatrix} \tag{1}$$

In (1), all symbols have their usual meaning except here the quantity $\partial \underline{D}/\partial t$ includes the electric conduction current density. By applying $\underline{E}\cdot$ to the top line of (1) and $\underline{H}\cdot$ to the bottom line of (1) and adding the results gives Poynting's theorem [6], i.e.,

$$-\underline{E}\cdot\nabla\times\underline{H}+\underline{H}\cdot\nabla\times\underline{E}+\underline{E}\cdot\frac{\partial \underline{D}}{\partial t}+\underline{H}\cdot\frac{\partial \underline{B}}{\partial t}=-\underline{E}\cdot\underline{J}_{ef}-\underline{H}\cdot\underline{J}_{mf} \tag{2}$$

The first two terms in this equation may be combined to give

$$-\underline{E}\cdot\nabla\times\underline{H}+\underline{H}\cdot\nabla\times\underline{E}=\nabla\cdot(\underline{E}\times\underline{H})=\nabla\cdot\underline{S} \tag{3}$$

where $\underline{S}$ is the Poynting vector (i.e. power flow density). The next two terms in (2) are written as

$$\underline{E}\cdot\frac{\partial \underline{D}}{\partial t}+\underline{H}\cdot\frac{\partial \underline{B}}{\partial t}=\frac{\partial W_e}{\partial t}+\frac{\partial W_m}{\partial t}=\frac{\partial W}{\partial t} \tag{4}$$

where W_e and W_m are the stored electric and magnetic energy densities and W is the total stored energy density. Note that because $\partial \underline{D}/\partial t$ includes the conduction current density, W_e and W account for the energy lost due to conduction currents. The final two terms of (2) are

$$-\underline{E}\cdot\underline{J}_{ef}-\underline{H}\cdot\underline{J}_{mf}=-p_e-p_m=-p \tag{5}$$

where p_e and p_m are the electric and magnetic source power densities and p is the total source power density. Using (3)–(5) in (2) gives

$$\nabla \cdot \underline{S} + \frac{\partial W}{\partial t} = -p \tag{6}$$

which is a statement of the conservation of energy. In the next sections, it is shown how the instantaneous values of $\nabla \cdot \underline{S}$, W_e and W_m may be calculated directly in a TLM simulation.

2.2 Electromagnetic Wave Simulation using Regular TLM

For simplicity, assume regular Cartesian mesh TLM based on the symmetrical condensed node (SCN) as shown in Fig. 1 [7, 8]. In this cell, the space-steps are all equal, i.e. $\Delta x = \Delta y = \Delta z = \Delta \ell$. More details of the development of the scattering algorithm using the notation adopted here are available in [1]. The quantities in (1), $\{\underline{E}, \underline{H}, \underline{J}_{ef}, \underline{J}_{mf}, \underline{D}, \underline{B}\}$ and the space and time derivative operators are normalized using the relationships

$$\begin{array}{llll} \underline{E} & = -\underline{V}/\Delta\ell & , \underline{H} & = -\underline{i}/(\Delta\ell\eta_0) \\ \underline{J}_{ef} & = -\underline{i}_{ef}/(\Delta\ell^2\eta_0) & , \underline{J}_{mf} & = -\underline{V}_{mf}/(\Delta\ell^2) \\ \underline{D} & = -\underline{d}\,\Delta t/\Delta\ell^2 & , \underline{B} & = -\underline{b}\,\Delta t/(\Delta\ell\eta_0) \\ (\nabla\times\cdot) & \rightarrow (\bar{\nabla}\times\cdot)/\Delta\ell & , \partial/\partial t & \rightarrow (1/\Delta t)\partial/\partial T \end{array} \tag{7}$$

where η_0 is the intrinsic impedance of free-space, Δt is the time-step, $\partial/\partial T$ is the normalized time derivative operator and $\bar{\nabla}$ is the normalized space derivative operator. In (7), the normalized quantities $\{\underline{V}, \underline{i}, \underline{i}_{ef}, \underline{V}_{mf}, \underline{d}, \underline{b}\}$ all have dimensions of voltage.

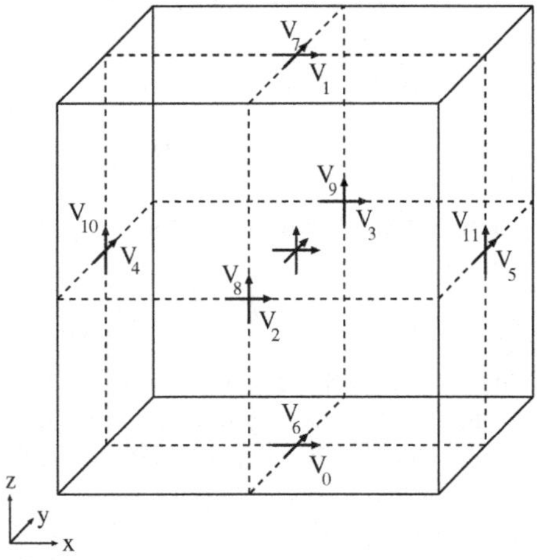

Fig. 1 Symmetrical condensed node (SCN)

Replacing the physical quantities in Maxwell's curl equations (1) with the normalized quantities defined in (7) gives

$$\begin{bmatrix} \bar{\nabla}\times \underline{i} \\ -\bar{\nabla}\times \underline{V} \end{bmatrix} - \frac{\partial}{\partial T}\begin{bmatrix} \underline{d} \\ \underline{b} \end{bmatrix} = \begin{bmatrix} \underline{i}_{ef} \\ \underline{V}_{mf} \end{bmatrix} \tag{8}$$

In succinct form, (8) may be written as

$$\underline{C} - \frac{\partial \underline{\phi}}{\partial T} = \underline{F}_f \tag{9}$$

where $\underline{C}$ is the vector of normalized curls, $\underline{\phi}$ is the vector of normalized fluxes and $\underline{F}_f$ is the vector of normalized source currents. For a three-dimensional TLM mesh having regular space steps, the maximum time-step is given by $\Delta t = \Delta\ell/(2c)$ [8] and the conversion from the mixed derivatives to the wave format is

$$\begin{bmatrix} (\bar{\nabla}\times \underline{i})_x \\ (\bar{\nabla}\times \underline{i})_y \\ (\bar{\nabla}\times \underline{i})_z \\ -(\bar{\nabla}\times \underline{V})_x \\ -(\bar{\nabla}\times \underline{V})_y \\ -(\bar{\nabla}\times \underline{V})_z \end{bmatrix} - 2\frac{\partial}{\partial T}\begin{bmatrix} V_x \\ V_y \\ V_z \\ i_x \\ i_y \\ i_z \end{bmatrix} = 2\begin{bmatrix} (V_0^i + V_1^i + V_2^i + V_3^i) \\ (V_4^i + V_5^i + V_6^i + V_7^i) \\ (V_8^i + V_9^i + V_{10}^i + V_{11}^i) \\ -(V_6^i - V_7^i - V_8^i + V_9^i) \\ -(V_{10}^i - V_{11}^i - V_0^i + V_1^i) \\ -(V_2^i - V_3^i - V_4^i + V_5^i) \end{bmatrix} - 4\begin{bmatrix} V_x \\ V_y \\ V_z \\ i_x \\ i_y \\ i_z \end{bmatrix} \tag{10}$$

where for example, V_j^i is the incident voltage on port j of the SCN illustrated in Fig. 1. Equation (10) may be expressed as

$$\begin{bmatrix} C_{ix} \\ C_{iy} \\ C_{iz} \\ -C_{Vx} \\ -C_{Vy} \\ -C_{Vz} \end{bmatrix} - 2\frac{\partial}{\partial T}\begin{bmatrix} V_x \\ V_y \\ V_z \\ i_x \\ i_y \\ i_z \end{bmatrix} = 2\begin{bmatrix} V_x \\ V_y \\ V_z \\ -i_x \\ -i_y \\ -i_z \end{bmatrix}^r - 4\begin{bmatrix} V_x \\ V_y \\ V_z \\ i_x \\ i_y \\ i_z \end{bmatrix} \tag{11}$$

In succinct form, (11) is

$$\underline{C} - 2\frac{\partial \underline{F}}{\partial T} = 2\underline{F}^r - 4\underline{F} \tag{12}$$

2.3 Instantaneous Power Divergence Density

By applying the normalizations (7) to (3), the normalized power divergence density is

$$-\underline{V}\cdot\bar{\nabla}\times\underline{i} + \underline{i}\cdot\bar{\nabla}\times\underline{V} = \bar{\nabla}\cdot(\underline{V}\times\underline{i}) = \bar{\nabla}\cdot\underline{s} \tag{13}$$

where $\underline{s} = \underline{S}\Delta\ell^2\eta_0$ is the normalized Poynting vector. As indicated in (13), to calculate $\bar{\nabla}\cdot\underline{s}$, the vector of curls $\underline{C}$ needs be evaluated. This is achieved by applying the bilinear $\mathcal{Z}$-transform

$$\frac{\partial}{\partial T} \rightarrow 2\left(\frac{1-z^{-1}}{1+z^{-1}}\right) \tag{14}$$

to (12) to give

$$\underline{C} = 2\underline{F}^r + z^{-1}\underline{S}_c \tag{15}$$

$$\underline{S}_c = -8\underline{F} + 2\underline{F}^r - \underline{C} \tag{16}$$

where $\underline{S}_c$ is the curl accumulator vector. The digital filter implementation of (15) and (16) is shown in Fig. 2. The block $\underline{\underline{t}}(z)$ in this figure is the material function of the TLM scattering process introduced in [1].

$$\underline{F} = \underline{\underline{t}}(z) \cdot \underline{F}^r \tag{17}$$

For free-space, the material function matrix is simply

$$\underline{\underline{t_0}}(z) = (0.5)\underline{\underline{I_6}} \tag{18}$$

where $\underline{\underline{I_6}}$ is the 6-by-6 unit matrix. For the simulation of general dielectric and magnetic materials, $\underline{\underline{t}}(z)$ describes a digital filter system.

The divergence of the normalized power density in (13) naturally splits into two parts,

$$\bar{\nabla} \cdot \underline{s}_e + \bar{\nabla} \cdot \underline{s}_m = \bar{\nabla} \cdot \underline{s} \tag{19}$$

where $\underline{s}_e$ and $\underline{s}_m$ may be interpreted as the normalized electric and magnetic power flow densities. In a Cartesian co-ordinate system, the divergences of these are obtained using:

$$\bar{\nabla} \cdot \underline{s}_e = -\underline{V} \cdot \bar{\nabla} \times \underline{i} = -(V_x\, C_{ix} + V_y\, C_{iy} + V_z\, C_{iz}) \tag{20}$$

$$\bar{\nabla} \cdot \underline{s}_m = \underline{i} \cdot \bar{\nabla} \times \underline{V} = i_x\, C_{Vx} + i_y\, C_{Vy} + i_z\, C_{Vz} \tag{21}$$

Writing (19), (20) and (21) in vector form gives

$$\bar{\nabla} \cdot \underline{s} = -\underline{F} \cdot \underline{C} \tag{22}$$

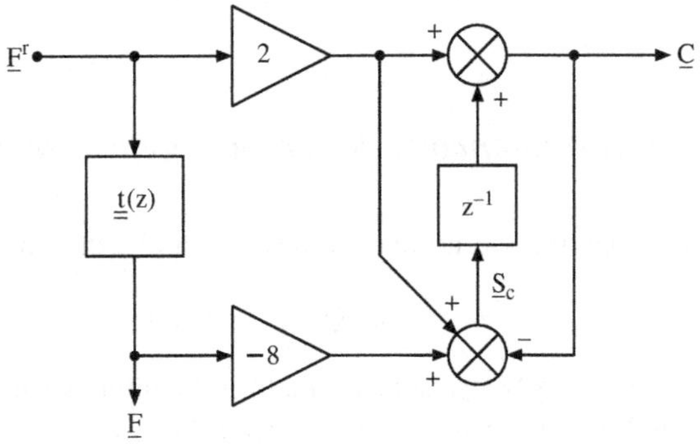

Fig. 2 Digital filter used to evaluate the curl vector $\underline{C}$

Thus to calculate the divergence of the power density requires the implementation of (15), (16) and (22) in software running alongside the TLM scattering system matrix $\underline{\underline{t}}(z)$. The algorithm requires the storage of the six real numbers in $\underline{S}_c$

2.4 Instantaneous Energy Density

Defining the normalized energy densities

$$w_e = W_e \Delta\ell^3 \eta_0/\Delta t, \quad w_m = W_m \Delta\ell^3 \eta_0/\Delta t, \quad w = W \Delta\ell^3 \eta_0/\Delta t \tag{23}$$

and applying (7) to (4), the normalized time-derivative of the energy density is

$$\underline{V} \cdot \frac{\partial \underline{d}}{\partial T} + \underline{i} \cdot \frac{\partial \underline{b}}{\partial T} = \frac{\partial w_e}{\partial T} + \frac{\partial w_m}{\partial T} = \frac{\partial w}{\partial T} \tag{24}$$

Using the normalized Maxwell curl equations (9) in (24) gives

$$\frac{\partial}{\partial T}\begin{bmatrix} w_e \\ w_m \end{bmatrix} = \begin{bmatrix} \underline{V} \cdot \ (\underline{C}_i \ - \underline{i}_{ef}) \\ \underline{i} \cdot (-\underline{C}_V - \underline{V}_{mf}) \end{bmatrix} \tag{25}$$

where

$$\underline{C}_i = \begin{bmatrix} C_{ix} \\ C_{iy} \\ C_{iz} \end{bmatrix}, \ \underline{C}_V = \begin{bmatrix} C_{Vx} \\ C_{Vy} \\ C_{Vz} \end{bmatrix} \tag{26}$$

From (20) and (21), (25) is

$$\frac{\partial}{\partial T}\begin{bmatrix} w_e \\ w_m \end{bmatrix} = -\begin{bmatrix} \bar{\nabla} \cdot \underline{s}_e \\ \bar{\nabla} \cdot \underline{s}_m \end{bmatrix} - \begin{bmatrix} \underline{V} \cdot \underline{i}_{ef} \\ \underline{i} \cdot \underline{V}_{mf} \end{bmatrix} \tag{27}$$

Applying the bilinear transform to (27) gives

$$\begin{bmatrix} w_e \\ w_m \end{bmatrix} = \frac{1}{2}\left\{ -\begin{bmatrix} \bar{\nabla} \cdot \underline{s}_e \\ \bar{\nabla} \cdot \underline{s}_m \end{bmatrix} - \begin{bmatrix} \underline{V} \cdot \underline{i}_{ef} \\ \underline{i} \cdot \underline{V}_{mf} \end{bmatrix} + z^{-1}\begin{bmatrix} S_{we} \\ S_{wm} \end{bmatrix} \right\} \tag{28}$$

where the energy accumulators are

$$\begin{bmatrix} S_{we} \\ S_{wm} \end{bmatrix} = 2\begin{bmatrix} w_e \\ w_m \end{bmatrix} - \begin{bmatrix} \bar{\nabla} \cdot \underline{s}_e \\ \bar{\nabla} \cdot \underline{s}_m \end{bmatrix} - \begin{bmatrix} \underline{V} \cdot \underline{i}_{ef} \\ \underline{i} \cdot \underline{V}_{mf} \end{bmatrix} \tag{29}$$

To calculate the energy density, it is necessary to solve for $\bar{\nabla} \cdot \underline{s}_e$ and $\bar{\nabla} \cdot \underline{s}_m$ using the scheme described in Sect. 2.3, along with (28) and (29). Implementation requires the storage of the quantities $\{\underline{S}_c, \ S_{we}, \ S_{wm}\}$, i.e. eight real numbers.

3 Results: Electromagnetic Interaction with a Block of Debye Material

In this section some results obtained using the method are reported. The simple example of plane wave interaction with a block of Debye material was studied. The problem space used is shown in Fig. 3 and consisted of a line of $20+60+20=100$ cells in the x-direction with $\Delta\ell=0.1$ m. Periodic boundary conditions were used in y and z and were arranged to support plane wave propagation. The 20 cells at each end of the space had the properties of free-space and the 60 cells in the centre simulated a material having a relative permittivity $\varepsilon_r(\omega)$. In this case a Debye material was assumed with

$$\varepsilon_r(\omega)=\varepsilon_\infty+\frac{\varepsilon_s-\varepsilon_\infty}{1+j\omega\tau} \tag{30}$$

where ε_∞ (ε_s) are the high (low) frequency relative permittivities and τ is the relaxation time. The material properties were taken as

$$\varepsilon_\infty=2.0,\ \varepsilon_s=15.0,\ \tau=16.7\text{ns}$$

Details of the construction of the material matrix $\underline{\underline{t}}(z)$ for this type of material is available in [1].

For both simulations reported in the next subsections, the electric field E_y, the divergences of the power flows $\{\nabla\cdot\underline{S},\ \nabla\cdot\underline{S}_e,\ \nabla\cdot\underline{S}_m\}$ and the energy densities $\{W,\ W_e,\ W_m\}$ were observed at point 0 in Fig. 3 located at the centre of the block. Both simulations were iterated for $600\Delta t$.

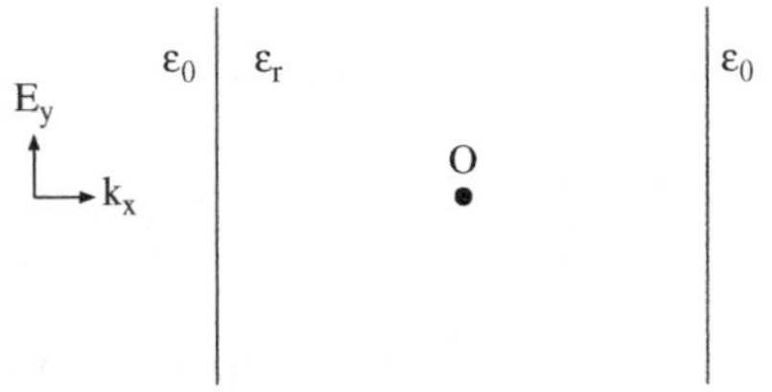

Fig. 3 Problem space

3.1 Sinusoidal Excitation

In this simulation, a y-polarized monochromatic plane wave was launched from the free-space region on the left hand side of the material block. The source had a frequency of 150 MHz and was switched on gradually over a few cycles.

Figure 4 shows the electric field E_y, the divergence of the Poynting vector $\nabla\cdot\underline{S}$ and the energy density W at point 0. The figures show that $\nabla\cdot\underline{S}$ pulsates at twice the frequency of E_y and has a negative offset because of the loss in the Debye

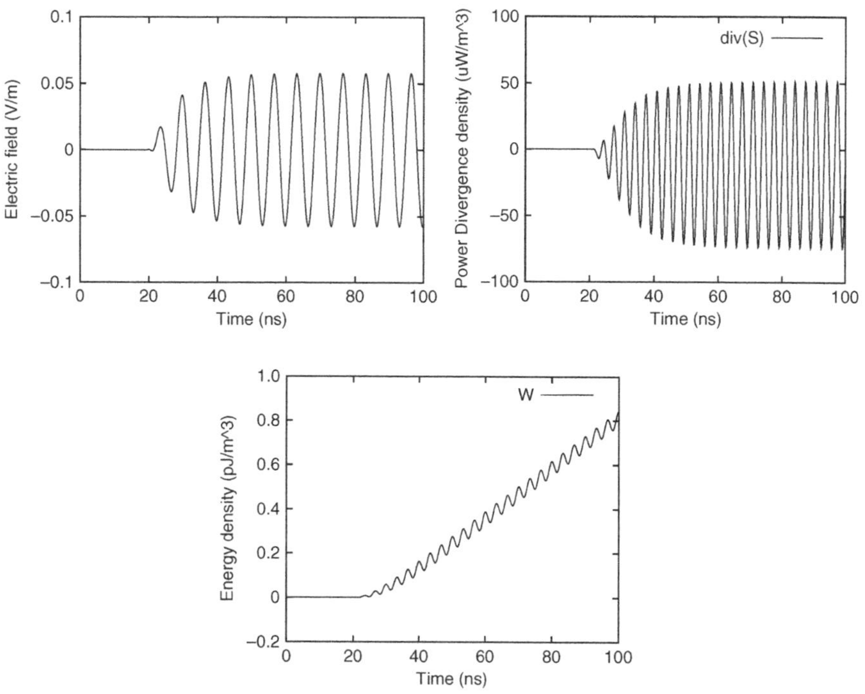

Fig. 4 Sinusoidal excitation: electric field, power divergence and energy densities at 0

material. The graph of W shows the energy density in the observation cell is increasing demonstrating loss. In addition, there is a small oscillation in W at twice the source frequency as reactive energy is stored and released.

In Fig. 5, the divergences of the electric and magnetic power flow densities are shown. The top figure shows that $\nabla \cdot \underline{S}_e$ has a negative offset due to the power lost in the material and oscillation due to reactive power. The middle figure shows that $\nabla \cdot \underline{S}_m$ has no offset as the material has no magnetic losses. The bottom figure shows the difference, $\nabla \cdot \underline{S}_e - \nabla \cdot \underline{S}_m$ which from equipartition of reactive electric and magnetic powers gives the power loss with no reactive component.

Figure 6 shows the energy densities recorded at 0. The electric energy density W_e is in the top figure. As expected, W_e is increasing as energy is deposited in the material. Again there is a superimposed oscillation in W_e due to reactive energy. The middle figure shows that W_m is purely reactive and the bottom figure shows that the energy difference is smoothly increasing because the reactive energy is suppressed by taking $W_e - W_m$.

3.1.1 Sinusoidal Excitation, Numerical Calculation

To estimate the divergence of the Poynting vector and energy density in this simulation, the classic frequency-domain theory of the complex permittivity is used,

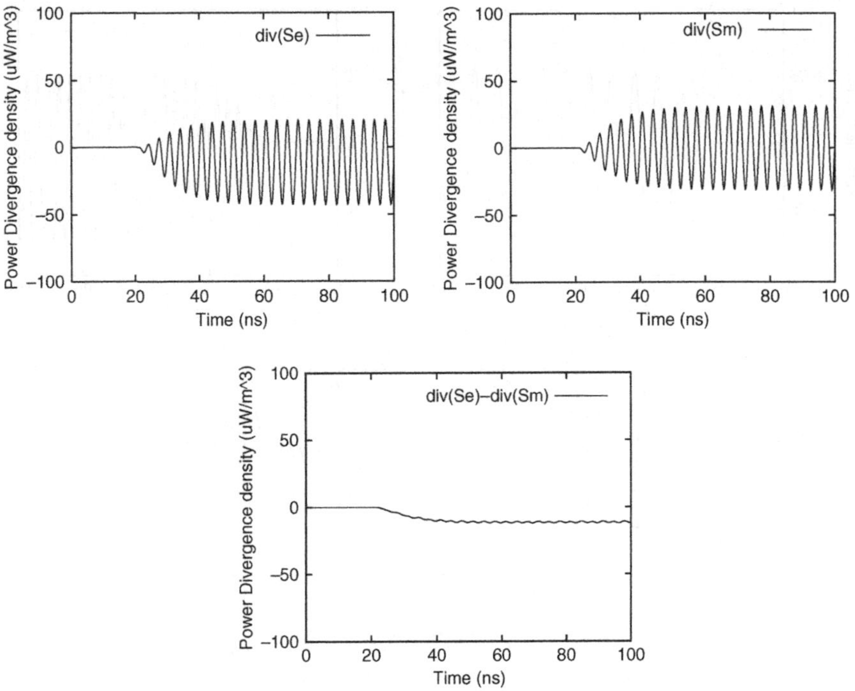

Fig. 5 Sinusoidal excitation: electric and magnetic power divergence densities and their difference at 0

e.g. [9]. Firstly, the relative permittivity of (30) is split into its real and imaginary parts, i.e.

$$\varepsilon_r(\omega) = \varepsilon_r'(\omega) - j\varepsilon_r''(\omega) \tag{31}$$

where the imaginary part is

$$\varepsilon_r''(\omega) = \frac{(\varepsilon_s - \varepsilon_\infty)\omega\tau}{1+\omega^2\tau^2} \tag{32}$$

The effective conductivity is

$$\sigma_e^{eff}(\omega) = \varepsilon_0 \omega \varepsilon_r''(\omega) \tag{33}$$

For a sinusoidal electic field having a peak value of E_y^{pk} in a material having an effective conductivity of σ_e^{eff}, the time average of the divergence of the Poynting vector is

$$< \nabla \cdot \underline{S}(\omega) >= |E_y^{pk}|^2 \sigma_e^{eff}(\omega)/2 \tag{34}$$

For $\omega = 2\pi \times 150 \times 10^6$, $\sigma_e^{eff} \sim 6.903\,\mathrm{mS/m}$. The magnitude of the peak value of the electric field observed at 0 in the simulation was

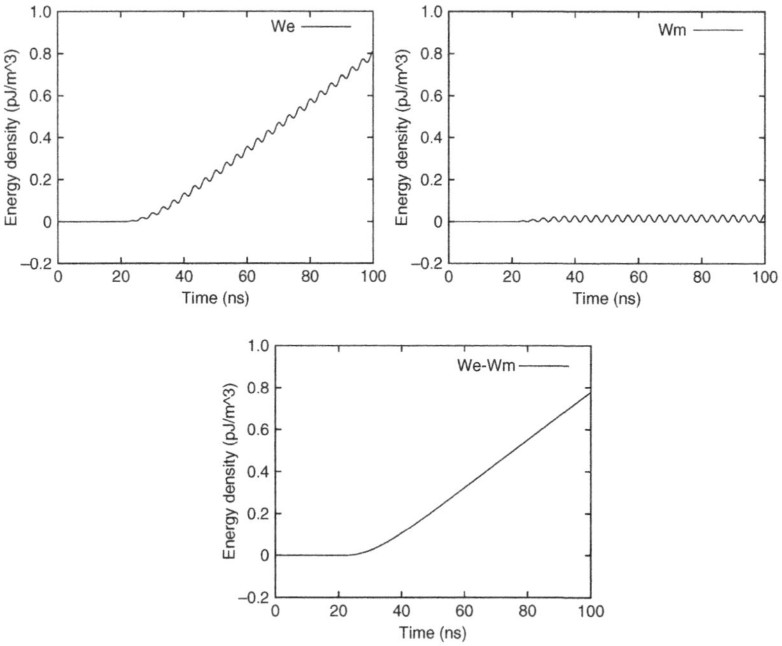

Fig. 6 Sinusoidal excitation: electric and magnetic energy densities and their difference at 0

$$|E_y^{pk}| \sim 0.0577\text{V/m}$$

Substituting these values into (34) gives

$$< \nabla \cdot \underline{S} > \sim 11.5\mu\text{W/m}^3$$

This results is in agreement with those shown in Figs. 4 and 5. The rate of change of the energy density difference is

$$\frac{d(W_e - W_m)}{dt} \sim 11.5\mu\text{W/m}^3$$

which is in accordance with the result shown in Fig. 6.

3.2 Pulse excitation

In this simulation, a y-polarized pulse plane wave was launched from the free-space region on the left hand side of the material block. The source was a sinusoid of 150 MHz modulated by a Gaussian function. The pulse consisted of about five cycles of the 150 MHz carrier.

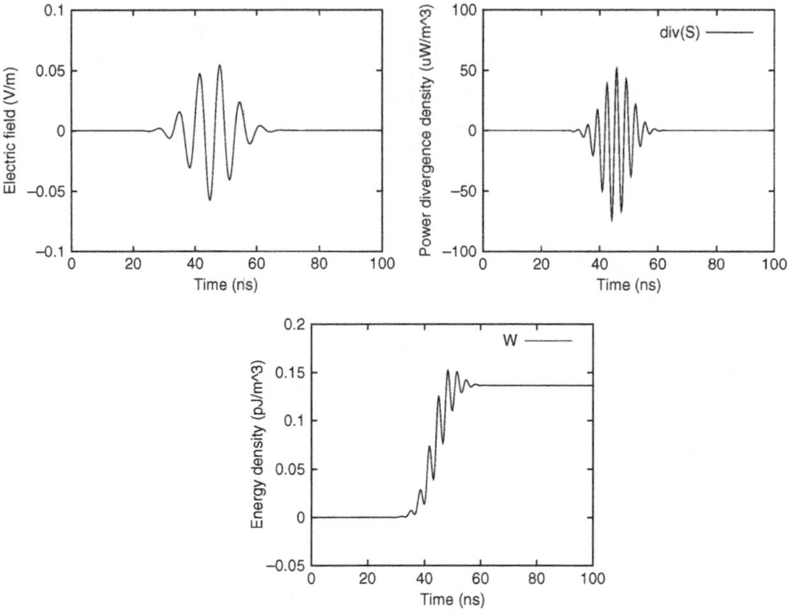

Fig. 7 Pulse excitation: electric field, power divergence density and energy density at 0

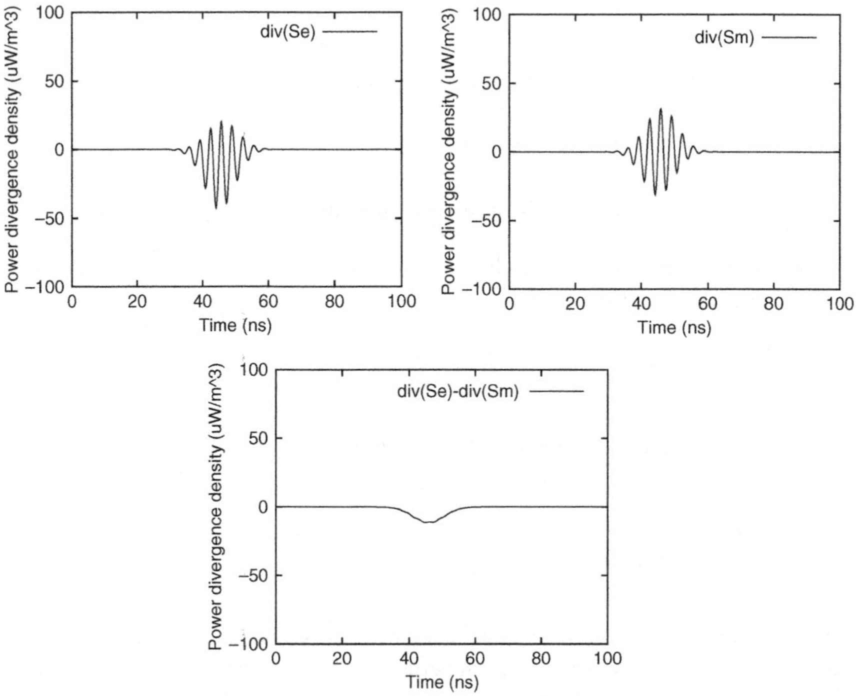

Fig. 8 Pulse excitation: electric and magnetic power divergence densities at 0

Figure 7 shows the electric field E_y, the divergence of the Poynting vector $\nabla \cdot \underline{S}$ and the energy density W at point 0. As in the previous simulation, the figures show that $\nabla \cdot \underline{S}$ pulsates at twice the frequency of E_y and has a negative offset because of loss. The energy density W shows the energy density increases during the pulse demonstrating loss. The energy density after the pulse has subsided is constant giving the total energy density delivered by this pulse.

The divergences of the electric and magnetic power flow densities are in Fig. 8. As in the sinusoidal case, the top figure shows that $\nabla \cdot \underline{S}_e$ has oscillation due to reactive power and a negative offset due to the power lost in the material. The middle figure shows no offset in $\nabla \cdot \underline{S}_m$ due to the absence of magnetic loss. The difference $\nabla \cdot \underline{S}_e - \nabla \cdot \underline{S}_m$ is shown in the bottom figure. Because reactive power is cancelled by taking the difference, this curve clearly demonstrates the power lost during the pulse.

Finally, Fig. 9 shows the energy densities observed at 0. The electric energy density W_e is in the top figure. As expected, W_e is increasing during the pulse as energy is deposited and there is a superimposed oscillation due to reactive energy. The middle figure shows that W_m is purely reactive. The bottom figure shows that the energy difference $W_e - W_m$ is smoothly increasing during the pulse.

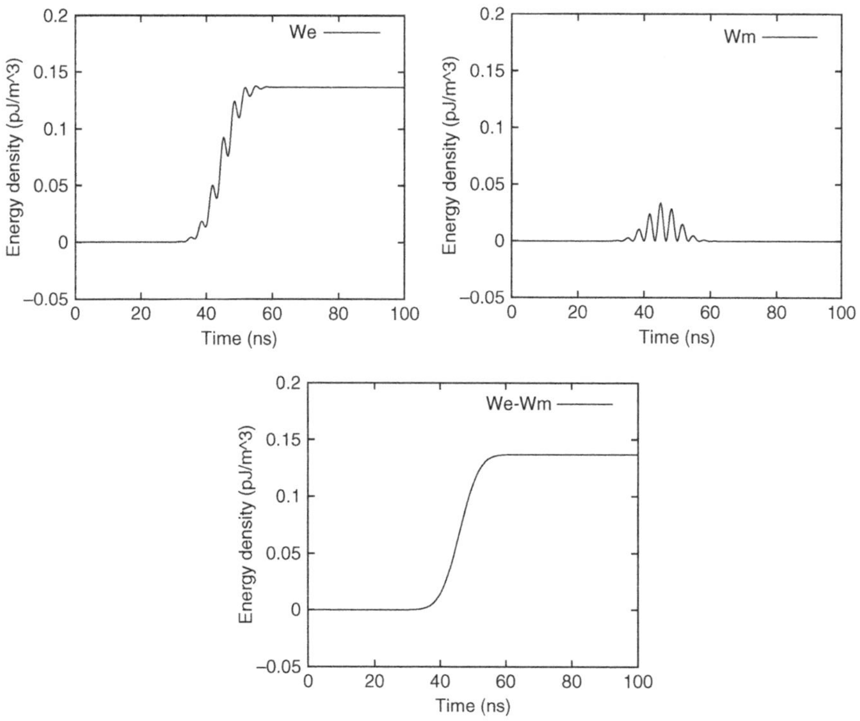

Fig. 9 Pulse excitation: electric and energy densities and their difference at 0

4 Conclusion

This contribution has described a method for calculation of power and energy quantities directly in the time-domain TLM. The technique was based on the $\mathcal{Z}$-transform approach and involved using the physical fields.

The approach is useful for validation of computational electromagnetic codes and simulations involving nonlinear materials, pulse excitations, and/or coupled thermal-electromagnetic problems where instantaneous values are required.

In further work, the approach may be extended to calculate the mechanical forces exerted by electromagnetic fields by evaluation of the Maxwell Stress Tensor.

References

1. J. Paul, C. Christopoulos and D. W. P. Thomas. Generalized Material Models in TLM—Part 1: Materials with Frequency-Dependent Properties. *IEEE Transactions on Antennas and Propagation*, 47(10):1528–1534, October 1999.
2. J. Paul, C. Christopoulos and D. W. P. Thomas. Generalized Material Models in TLM—Part 2: Materials with Anisotropic Properties. *IEEE Transactions on Antennas and Propagation*, 47(10):1535–1542, October 1999.
3. J. Paul, C. Christopoulos and D. W. P. Thomas. Generalized Material Models in TLM—Part 3: Materials with Nonlinear Properties. *IEEE Transactions on Antennas and Propagation*, 50(7):997–1004, July 2002.
4. L. de Menezes and W. J. R. Hoefer. Modeling of General Constitutive Relationships using SCN TLM. *IEEE Transactions on Microwave Theory and Techniques*, 44(6):854–861, June 1996.
5. A. Taflove. *Computational Electrodynamics: The Finite-Difference Time-Domain Method*. Artech House, Norwood, MA, 1995.
6. J. A. Kong. *Electromagnetic Wave Theory*. Wiley, New York, 1986.
7. P. B. Johns. A Symmetrical Condensed Node for the TLM method. *IEEE Transactions on Microwave Theory and Techniques*, 35(4):370–377, April 1987.
8. C. Christopoulos. *The Transmission-Line Modeling Method: TLM*. IEEE Press, Piscataway, NJ, 1995.
9. A. von Hippel. *Dielectrics and Waves*. The M.I.T. Press, Cambridge, MA, 1954.

The Combined Schrödinger-Maxwell Problem in the Electronic/Electromagnetic Characterization of Nanodevices

Luca Pierantoni, Davide Mencarelli and Tullio Rozzi

1 Introduction

After their discovery in 1991 in the soot produced by discharges between carbon electrodes, carbon nanotubes have been the object of an increasing amount of research. Their basic properties have been investigated in depth as they show promise to overcome the fundamental physical barriers to size reduction of transistors made by conventional silicon technology. High carrier mobility, lightness, absence of lattice defects, almost ideal mono-dimensionality of carrier transport, are just a few of the unique properties of carbon nanotubes. Potential applications range over a wide area, including not only transistors [1, 2], but also transmission lines, interconnects [3], nanotweezers [4], field emission devices [5], antennas [6], tunneling structures, [7], nanowires [8, 9, 10, 11, 12, 13]. Moreover, other interesting nanodevices, such as quantum wires, quantum dots, 2DEG confined in heterostructures, electron waveguides, etc, are suitable to be analyzed by the approaches discussed in the following.

Despite the enormous possibilities offered by carbon nanotubes for practical applications, owing to their surprising properties as mono-dimensional channels and although the above applications have been, at least partially, realized, the fabrication processes, i.e. nanotube growing and nano-manipulating, are not mature yet and do not allow massive production of devices with uniform and stable characteristics. A carbon nanotube consists of graphene sheet (i.e. a mono-atomic layer of graphite) rolled up into a cylinder, typically a few nanometers radius and length up to centimeters. Thanks to their small dimensions, nanotubes are typically defect-free.

Luca Pierantoni
Dipartimento di Elettromagnetismo e Bioingegneria – Universitá Politecnica delle Marche, Ancona 60100, Italy, e-mail: l.pierantoni@univpm.it

Davide Mencarelli
Dipartimento di Elettromagnetismo e Bioingegneria - Universitá Politecnica delle Marche, Ancona 60100, Italy, e-mail: d.mencarelli@univpm.it

Tullio Rozzi
Dipartimento di Elettromagnetismo e Bioingegneria - Universitá Politecnica delle Marche, Ancona 60100, Italy, e-mail: t.rozzi@univpm.it

Moreover, in nanotubes shorter than 70–80 nm transport is nearly ballistic, since the carriers do not couple with phonons and retain their coherence as waves. For all these reasons, the demand for the accurate characterization of the transport and the electromagnetic properties of carbon nanotubes under operational conditions is a real challenge. The difficulty arises in modeling quantum phenomena coupled to Maxwell equations.

In the time-domain, the Schrödinger (SE) equation is an intrinsically complex diffusion equation.

In the frequency-domain, it reduces to a wave equation, lending itself to be treated by the transmission line approach. In particular, the propagation of electrons and holes along a monodimensional system can be seen as a traveling guided-wave solution of a wave equation. It has been highlighted that interference effects may take place between co-propagating electron-waves, similarly to what happens for waveguide modes. Examples of these mono-dimensional systems are provided, besides nanotubes, by electron wave-guides and nanowires [9]. Typically, the analysis of such devices is carried out directly in the phasor domain [10]. The effects of an external electric field onto a nanodevice are computed by considering the internal dynamics of carrier transport as instantaneous with respect to the time constant of the electromagnetic transient. For example, the ac operation of nano-transistor devices has been predicted by means of a small-signal equivalent circuit, that makes use of the electrostatic limit, i.e. the solution of the Poisson-Schrödinger system. The challenge of the present work is to overcome this kind of simplified approach in order to develop a full-wave analysis of carrier transport, including the time-transient. Space-discreting methods, like finite-difference time-domain (FDTD) and transmission line matrix (TLM), are well known techniques that allow the EM full-wave modeling of $3D$ structures with nearly arbitrary geometry for a wide range of applications, from electromagnetic compatibility to optics [14, 15, 16, 17, 18, 19, 20]. As they are derived by direct discretization of Maxwell equations, they are intrinsically oriented towards solving deterministic field problems. In the literature, only very few papers have addressed the question of employing these techniques for quantum-mechanical problems. In [21] the time-dependent Schrödinger equation is solved in one space dimension by TLM. In [22], an extension of the method is applied to solve the Schrödinger equation in $3D$-domains by FDTD. Both the aforementioned contributions do not consider the interaction between the quantum device and the electromagnetic field in which the structure is embedded. The goal of the present contribution is to introduce two novel techniques, acting in the time- and in the frequency-domain, dealing with the Maxwell-Schrödinger system and the Poisson-Schrödinger system, respectively.

In Chap. 2, we present a full-wave technique in which, for the first time, the self-consistent problem of the simultaneous solution of Maxwell and Schrödinger equations in the time-domain is investigated. The total electromagnetic (EM) field includes possibly impressed and/or external field and the self-generated EM field, arising from the macroscopic current density, which, in turn, is computed from the solution of the Schrödinger equation. The proposed technique develops as follows: (i) the $3D$-domain isdiscretized by TLM using the symmetrical condensed node

(SCN) approach. (ii) Quantum phenomena are introduced in a subregion of said domain, e.g. in a one-dimensional chain of nodes (see Fig. 1), from one electrode ($z = z_1$) to another ($z = z_2$), thus simulating the presence of a nanotube. (iii) At an arbitrary time-step t, the EM field provides appropriate boundary conditions for the nanotube subregion, by means of the vector and scalar potentials $\boldsymbol{A}(\boldsymbol{r},t)$, $\phi(\boldsymbol{r},t)$, directly calculated by the same EM field [22]. (iv) $\boldsymbol{A}(\boldsymbol{r},t)$, $\boldsymbol{\Phi}(\boldsymbol{r},t)$ together with a $V_p(\boldsymbol{r})$ potential profile, that depends on the quantum properties of the nanotube and materials bounding the domain, constitute the additional terms of the Schrödinger equation [22], which is solved by means of a finite difference scheme for that time-step t. (iv) The wave solution $\psi(\boldsymbol{r},t)$ generates, along the domain of the nanotube, the quantum-mechanical current $\boldsymbol{J}(\boldsymbol{r},t)$ [23], which, in turn, represents an equivalent distribution of current sources for the TLM algorithm. These sources, located along the chain of nodes from ($z = z_1$) to ($z = z_2$), as shown in Fig. 1, affect the EM-field for the next time-step ($t = t + 1$). At that time-step ($t = t + 1$), the TLM scheme generates a new distribution of potentials $\boldsymbol{J}(\boldsymbol{r},t)$ for the Schrödinger equation, and so on. This scheme is the basis of an iterative process by which the time-dependent Maxwell equations and the Schrödinger equation are solved simultaneously. In the results section we present a first investigation of the dynamic behaviour of an electron wavepacket injected in a nanotube between two metallic electrodes. The electromagnetic near field is also shown.

In Chap. 3, we present two independent approaches to the analysis of CNT's, operating in the frequency domain.

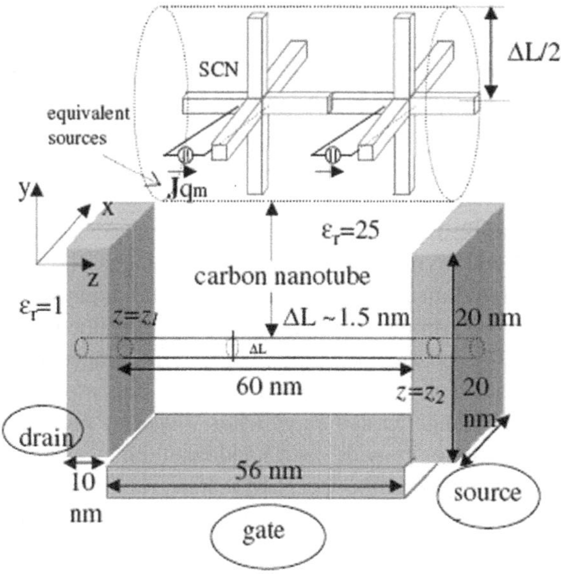

Fig. 1 The analyzed structure: two thick metallic plates (dimensions: $20 \times 20 \times 10$ nm) are separated by a distance of 60 nm. A carbon nanotube is placed between the source-drain electrodes. A gate contact is inserted below

We first discuss a generalization of the self-consistent analysis of CNT-transistors to the case of multi-wall/multi-band coherent carrier transport. The contribution to charge diffusion, due to different walls and sub-bands of a multi-wall mw-CNT, is shown to be non-negligible, especially for high applied external voltages and "large" diameters.

In the second part of this chapter we exploit the formal analogy between Maxwell and Schrödinger equations, holding in some important cases such as low-dimensional quantum problems, in order to make use of electromagnetic simulators for the analysis of carrier transport. In particular, we apply for the first time standard e.m. solvers to CNT's, with a view to derive their main properties, such as dispersion curves and effective masses.

2 Full-Wave Modelling Technique in the Time-Domain

2.1 Maxwell Equations Modeled by TLM

The Transmission Line Matrix (TLM) method introduced by Johns [15] is a space- and time-discretizing method of electromagnetic field computation in which the continuous space is segmented into cells by defining intersecting planes. Ports are defined at the tangential planes between two neighbouring cells and a scattering center is defined at the centre of each cell with the ports, [15, 16, 17, 18]. This physical model is called node and comprises the scattering centre which is connected via transmission lines to the ports at the tangential planes between neighbouring cells. Pulses are scattered at the nodes and propagate along these transmission lines to the neighbouring nodes where they are scattered again. The propagation and the scattering of the wave amplitudes are expressed by operator equations [15, 16, 17, 18, 19]; TLM is considered as the implementation of the Huygens principle [19].

The $3D$-domain of the present problem (see Fig. 1) including the physical objects is modeled by means of the TLM with Symmetrical Condensed Node (SCN), as in [15, 16, 17, 18, 19]. Inside the three-dimensional TLM region, a one-dimensional chain of nodes is considered as the one-dimensional domain of the nanotube where the quantum dynamics is described by means of the time-domain Schrödinger equation. At each time step, the EM field is sampled along this domain, thus providing additional source terms for the Schrödinger equation, which is solved by means of a finite difference scheme [14]. Its solution provide distributed quantum mechanical current sources, which, in turn, inject energy into the TLM nodes. The $3D$-domain of the present problem (see Figs. 1 and 2) including the physical objects is modeled by means of the SCN-TLM scheme as in [18, 19]. Inside the three-dimensional TLM region, a one-dimensional chain of nodes is considered as the one-dimensional domain of the nanotube where quantum dynamics is described by means of the

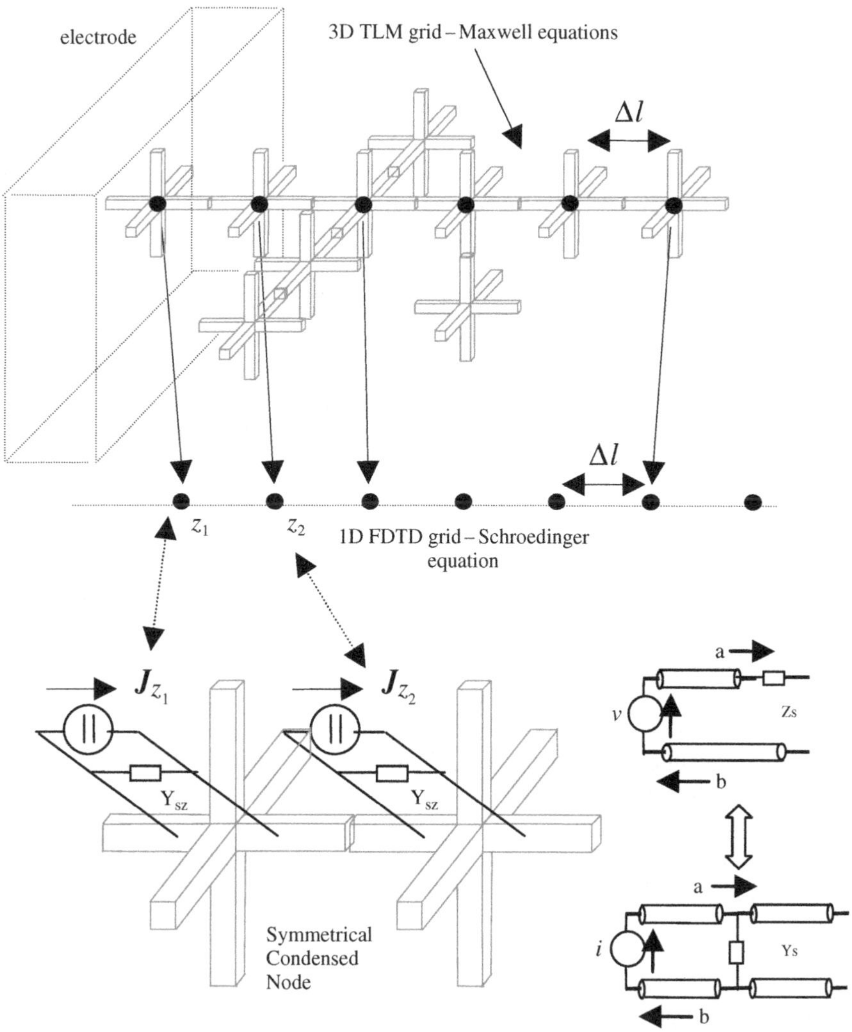

Fig. 2 The modeling of the global problem: three-dimensional Transmission Line Matrix discretization and one-dimensional Finite-Difference Time-Domain discretization. Equivalent source distribution and their connection to the nodes. Stub admittances of the active voltage/current sources

time-domain Schrödinger equation. At each time step, the EM field is sampled along this domain, thus providing additional source terms for the Schrödinger equation, which is solved by means of a finite difference scheme [14], that is best suited for diffusion-like equations. Its solution provide distributed quantum mechanical current sources, which, in turns, inject energy into the TLM nodes.

2.2 The Schrödinger Equation

The quantum mechanical behaviour of a charge carrier is described by the time-dependent Schrödinger equation:

$$i\hbar\frac{\partial\psi(\mathbf{r},t)}{\partial t}=\left(-\frac{\hbar^2}{2m}\nabla^2+V_p(\mathbf{r})\right)\psi(\mathbf{r},t)\quad \hbar=\frac{h}{2\pi}\tag{1}$$

The wave function $\psi(r,t)$ has no direct physical meaning, but all relevant physical parameters can be determined from it. It is complex, in fact, even though it is in the time domain. Meaningful physical quantities are determined once the complex conjugate $\psi^*(r,t)$ is employed too. The basic requirement for the solutions is the following normalization condition, stating that the probability of the particle being somewhere is one:

$$\int_{-\infty}^{+\infty}|\psi(\mathbf{r},t)|^2\,d\mathbf{r}=1\tag{2}$$

The real function $V_p(r)$ is the potential profile, having the units of energy $[eV]$, $h=6.26*10^{-34}$ $[J*sec]$ is the Plank's constant, m is the mass of the particle being represented by the Schrödinger equation, usually an electron, but it might be the "effective mass" of that particle in a particular semiconductor. The vector $\boldsymbol{r}=(x,y,z)$ indicates the position in a rectangular coordinates system, t is the time-variable, i is the imaginary unit $(i=-j)$. In the presence of an electromagnetic field, Eq. (1) is rewritten as follows [23]:

$$\left\{\frac{1}{2m}[\hat{\mathbf{p}}-q\mathbf{A}(\mathbf{r},t)]^2+q\phi(\mathbf{r},t)+V_p(\mathbf{r})\right\}\psi(\mathbf{r},t)=i\hbar\frac{\partial\psi(\mathbf{r},t)}{\partial t}\tag{3}$$

where $\boldsymbol{A}(\boldsymbol{r},t)$, $\phi(\boldsymbol{r},t)$ are vector and scalar potentials, directly related to the EM field through the appropriate gauge, e.g. the "temporal" gauge, $q=-e=1.602*10^{-19}[C]$ is the electron charge.

In Eq. (3) appears the canonical momentum, $\hat{\boldsymbol{p}}$, whereas the kinematic momentum, $\hat{\boldsymbol{k}}$, provides the electromagnetic field contribution to the kinetic energy:

$$\hat{\mathbf{p}}=-i\hbar\nabla\qquad \hat{\mathbf{k}}=\hat{\mathbf{p}}-q\mathbf{A}(\mathbf{r},t)\tag{4}$$

By referring to Fig. 1, the EM field acts over a $3D$-domain, whereas the Schrödinger equation describes quantum dynamics along the $1D$-domain $(z_1<z<z_2)$.

Now, by taking into account the algebraic properties of operators $\hat{\boldsymbol{p}}$ and $\hat{\boldsymbol{k}}$ as, e.g., in [23], Eq. (3) becomes:

$$
\begin{aligned}
i\hbar\frac{\partial\psi(\mathbf{r},t)}{\partial t} &= \frac{1}{2m}\left(\widehat{\mathbf{p}}^2 + e\widehat{\mathbf{p}}\cdot\mathbf{A} + e\mathbf{A}\cdot\widehat{\mathbf{p}} + e^2\mathbf{A}^2\right)\psi(\mathbf{r},t) + (\phi + V(\mathbf{r}))\psi \\
&= \frac{1}{2m}\Big\{-\hbar^2\nabla^2 - ei\hbar\left[(\nabla\cdot\mathbf{A})\psi + \mathbf{A}\cdot(\nabla\psi)\right] \\
&\quad - ei\hbar\mathbf{A}\cdot(\nabla\psi) + e^2\mathbf{A}^2\Big\} + (\phi + V(\mathbf{r}))\psi(\mathbf{r},t)
\end{aligned}
\tag{5}
$$

that is:

$$
\begin{aligned}
&i\hbar\frac{\partial\psi(\mathbf{r},t)}{\partial t}\frac{1}{2m}\left\{-\hbar^2\nabla^2 - ei\hbar(\nabla\cdot\mathbf{A})\psi - 2ei\hbar\mathbf{A}\cdot(\nabla\psi) + e^2\mathbf{A}^2\right\} \\
&+ (\phi + V(\mathbf{r}))\psi
\end{aligned}
\tag{6}
$$

Now, by projecting Eq. (6) along the z-direction, we obtain:

$$
\begin{aligned}
&\frac{1}{2m}\Big\{-\hbar^2\frac{\partial^2\psi(z,t)}{\partial z^2} - 2ei\hbar A_z(z,t)\frac{\partial\psi(z,t)}{\partial z} \\
&\quad - ei\hbar\psi(z,t)\frac{\partial A_z(z,t)}{\partial z}\Big\} + (V_p(z) - e\Phi(z,t))\psi(z,t) \\
&= i\hbar\frac{\partial\psi(z,t)}{\partial t} + \frac{e^2}{2m}|A_z(z,t)|^2\psi(z,t)
\end{aligned}
\tag{7}
$$

By observing Eq. (7) we note that, with respect to the z-direction: (i) $A_z(\boldsymbol{r},t)$ is the component of the $\boldsymbol{A}(\boldsymbol{r},t)$ potential vector, (ii) $\phi(\boldsymbol{r},t)$ is the scalar potential. The $\boldsymbol{r} = z$ vector describes the axial locations of the nanotube, the $V_p(\boldsymbol{r})$ potential profile is also sampled along the z-direction as $V_p(z)$. In order to avoid complex quantities, we split Eq. (7) into two parts, by separating the real and imaginary components of $\psi(z,t) = \psi_R(z,t) + i\psi_I(z,t)$ in the z-direction. This leads to the two following coupled equations:

$$
\frac{\partial\psi_R}{\partial t} = -\frac{\hbar}{2m}\frac{\partial^2\psi_I}{\partial z^2} + \frac{1}{\hbar}\left\{\frac{e^2}{2m}|A_z|^2 - e\Phi + V_p\right\}\psi_I - \frac{e}{2m}\left\{2A_z\frac{\partial\psi_R}{\partial z} + \psi_R\frac{\partial A_z}{\partial z}\right\}
\tag{8}
$$

$$
\frac{\partial\psi_I}{\partial t} = \frac{\hbar}{2m}\frac{\partial^2\psi_R}{\partial z^2} - \frac{1}{\hbar}\left\{\frac{e^2}{2m}|A_z|^2 - e\Phi + V_p\right\}\psi_R - \frac{e}{2m}\left\{2A_z\frac{\partial\psi_I}{\partial z} + \psi_I\frac{\partial A_z}{\partial z}\right\}
\tag{9}
$$

In a one-dimensional space, we approximate $\psi(z,t)$ by $\psi^n(k) = \psi(k\triangle, n\triangle t)$ where k and n are indices and $\triangle z = \triangle t$ and t are the spatial and temporal steps, respectively. In view of the complex, diffusion-line nature of Eqs. (8) and (9), the latter are then solved by a finite difference-time domain scheme, as in [22], giving

$$\psi_R^{n+1}(k) = \psi_R^n(k) - \frac{\hbar}{2m}\frac{\Delta t}{(\Delta z^2)}\left\{\psi_I^{n+1/2}(k+1) - 2\psi_I^{n+1/2}(k) + \psi_I^{n+1/2}(k-1)\right\}$$
$$\frac{\Delta t}{\hbar}\left\{\frac{e^2}{2m}\left|A_z^{n+1/2}(k)\right|^2 - e\phi_z^{n+1/2}(k) + V_p(k)\right\}\psi_I^{n+1/2}(k)$$
$$-\frac{e\Delta t}{2m}\left\{2A_z^{n+1/2}(k)\left[\frac{\psi_R^n(k+1) - \psi_R^n(k-1)}{2\Delta z}\right]\right.$$
$$\left. + \psi_R^n(k)\left[\frac{A_z^{n+1/2}(k+1) - A_z^{n+1/2}(k-1)}{2\Delta z}\right]\right\} \tag{10}$$

and:

$$\psi_i^{n+1}(k) = \psi_i^n(k) + \frac{\hbar}{2m}\frac{\Delta t}{(\Delta z^2)}\left\{\psi_R^{n+1/2}(k+1) - 2\psi_R^{n+1/2}(k) + \psi_R^{n+1/2}(k-1)\right\}$$
$$-\frac{\Delta t}{\hbar}\left\{\frac{e^2}{2m}\left|A_z^{n+1/2}(k)\right|^2 - e\phi_z^{n+1/2}(k) + V_p(k)\right\}\psi_R^{n+1/2}(k)$$
$$-\frac{e\Delta t}{2m}\left\{2A_z^{n+1/2}(k)\left[\frac{\psi_I^n(k+1) - \psi_I^n(k-1)}{2\Delta z}\right]\right.$$
$$\left. + \psi_I^n(k)\left[\frac{A_z^{n+1/2}(k+1) - A_z^{n+1/2}(k-1)}{2\Delta z}\right]\right\} \tag{11}$$

Finally, the wave function $\psi^n(k)$ of Eqs. (10) and (11) is derived once an initial $\psi(z, t = 0)$ distribution is defined and appropriate boundary conditions are imposed for $z = z_1$, $z = z_2$.

2.3 Stability

The choices of the time-step Δt and of the space-step $\Delta l = \Delta z$ are crucial. We first choose the space-step Δl as the same for both Transmission Line Matrix and Finite-Difference Time-Domain, $\Delta l_{TLM} = \Delta z_{FDTD}$. Then, we have to select Δt as the smaller value between: (i) Δt_{FDTD} that ensures stability of the one-dimensional Finite-Difference Time-Domain algorithm of the Schrödinger equations (7), (8), (9), (10) and (11), and (ii) $\Delta t_{TLM} = \Delta l/(Hc)$, that ensures the stability of the Transmission Line Matrix algorithm, being c the free space light velocity, H the stabilization factor [15, 16, 17, 18, 19, 20, 21]. Concerning (i), as reported in [22], in the case of $\boldsymbol{A}(\boldsymbol{r},t) = 0$, $\phi(\boldsymbol{r},t) = 0$, Δt has to be selected so that $\Delta t \leq \Delta t_{critical}$, being:

$$\Delta t_{critical} = \frac{\hbar}{\frac{\hbar^2}{m\Delta z^2} + |V_P|_{\max}} \tag{12}$$

where $|V_p|$ is the maximum absolute value of the potential $V_p(z)$ along the nanotube. In the case of $\boldsymbol{A}(\boldsymbol{r},t) \neq 0$, $\phi(\boldsymbol{r},t) \neq 0$, the spatial derivatives of $\boldsymbol{A}(\boldsymbol{r},t)$ and the terms proportional to $|\boldsymbol{A}(\boldsymbol{r},t)|$ give contributions that should be added to the denominator of Eq. (12). As a matter of fact, these contributions have been tested to provide a value of t critical an order of magnitude smaller than that of Eq. (12). In order to ensure stability, we choose $\triangle t_{FDTD}$ that is two or three orders of magnitude smaller than that of Eq. (12): $\triangle t_{FDTD} \leq \triangle t_{critical} * 10^{-2}$. This choice is also a constraint, since we observe that we always have $\triangle t_{TLM} < \triangle t_{FDTD}$ by some orders of magnitude, thus leading to the position: $\triangle t = \triangle t_{TLM} = \triangle t_{FDTD}$.

2.4 The Quantum-Mechanical Current

Once Eqs. (8) and (9) are solved for a time-step $t = n\triangle t$, the quantum-mechanical current density $[\mathrm{A}/m^2]$ is given by [23]:

$$\mathbf{J}(\mathbf{r},t) = q\left\{\frac{\hbar}{2im}\left(\psi^*\nabla\psi - \psi\nabla\psi^*\right) - \frac{q}{m}|\psi|^2\mathbf{A}\right\} \tag{13}$$

By projecting Eq. (13) along the z-direction, we get:

$$J_z(z,t) = q\left\{\frac{\hbar}{2im}\left(\psi^*\frac{\partial\psi}{\partial z} - \psi\frac{\partial\psi^*}{\partial z}\right) - \frac{q}{m}|\psi|^2 A_z\right\} \tag{14}$$

The dimension of the current in (14) becomes [A] in the one-dimensional case. The above current, originated by quantum mechanical phenomena, can be modeled as a displacement current $\boldsymbol{J}_d$ along the nanotube, that pumps active energy into the electromagnetic field:

$$\mathbf{J}_d(\mathbf{r},t) = \frac{\partial\mathbf{D}(\mathbf{r},t)}{\partial t} \tag{15}$$

It is noted that the above Eq. (15) is just a working assumption (*Ansatz*) and other positions are possible. The $\boldsymbol{J}_d$ source current is sampled along the nanotube:

$$\mathbf{J}_d(\mathbf{r},t) = J_z(z,t)\,\mathbf{z} \quad J_z(z,t) = J^n(k) \tag{16}$$

Once the $J_d(k)$ distributed values of quantum mechanical current are known, the corresponding ${}_{TLM}E_z^{n+1}(k)$ equivalent distributed voltage density sources are calculated.

$$\begin{aligned} J_z(z,t) &= \varepsilon_0\varepsilon_r\frac{\partial E_z(z,t)}{\partial t} \quad \Rightarrow \\ E_z(z,t) &= \frac{1}{\varepsilon_0\varepsilon_r}\int_0^t J_z(z,t)\,dt \end{aligned} \tag{17}$$

These sources inject active energy into the EM field through the TLM chain of nodes z_j located on the nanotube discretized subregion $(z_1 < z_j < z_2)$.

2.5 Modeling of the Equivalent Sources

The modeling of the equivalent TLM active sources is depicted in Fig. 2. For each node belonging to the nanotube, the local $J_n(k)$ quantum-mechanical current is equivalent to injected pulses, impinging on to the nodes and scattered from them at the next time-step $t = (n+1)\Delta t$. In the model, we can refer either to equivalent current or voltage sources, depending on the chosen point of view. This conceptual equivalence is directly related to the Thévenin-Norton equivalent circuit transformation, as shown in Fig. 2 and reported below:

$$\begin{aligned} E_z^{n+1}(k) &= {}_{n+1}a^p_{\mu,x_0,y_0,k} = \frac{1}{\varepsilon_0\varepsilon_r(k)}\sum_{n_k=0}^{n} J_z^{n_k}(k)\Delta t_{TLM} \\ v(k) &= E_z^{n+1}(k)\,\Delta l \\ i(k) &= v(k)Y(k) \quad Y(k) = Z(k)^{-1} \end{aligned} \tag{18}$$

In Eq. (18), $Y = 1/Z = Y_{sz}$ is the Norton/Thévenin admittance/impedance, set equal to the source stub admittance/impedance, as explained in the following. The technique for introducing active source elements across a set of TLM nodes has been introduced in [24, 25]. If an active device is connected to the node, the node must be modified consequently. In general, assuming that the active sources can be oriented in any one of the three orthogonal directions, three equivalent generators and three stubs with appropriate admittances must be added to the node. In the present case, the active sources and the corresponding generators are z-oriented. The scattering matrix of the new node, taking now into account the interaction between 21 lines (12 link lines and 9 stubs), must be determined according to the laws of energy conservation, and assumes a form similar to that of the Symmetrical Condensed Node for the modeling of electrical losses. A full 21×21 matrix is therefore needed to model the material properties of the node in conjunction with the presence of the local lumped device. The coefficients of the new scattering matrix are reported in [25]. In the present case, the equivalent distributed sources are placed in the z-direction, the transverse (x_0, y_0) coordinate being fixed. According to the denomination of lines and ports of [18, 19, 20] Eq. (18) assumes the form:

$$\begin{aligned} J_z(z,t) &= \varepsilon_0\varepsilon_r\frac{\partial E_z(z,t)}{\partial t} \quad \Rightarrow \\ E_z(z,t) &= \frac{1}{\varepsilon_0\varepsilon_r}\int_0^t J_z(z,t)\,dt \end{aligned} \tag{19}$$

being ${}_na^p{}_{\mu,x_0,y_0}(k)$ the incident wave amplitude at the nodes (x_0, y_0, z), located at $(x_0\Delta x, y_0\Delta y, k\Delta z)$ in a Cartesian coordinate system; the p-index refers to port p in the Symmetrical Condensed Node scattering matrix. In the present case we have $p = 21$. The TLM grid is homogeneous with $\Delta x = \Delta y = \Delta z = \Delta x$. The μ index is set to z ($\mu = z$), being ($z_1 < z < z_2$) the polarization of the wave amplitude that corresponds to the local active source, as shown in Figs. 1 and 2. In Eq. (19), the value of the reflected wave amplitude ${}_{n+1}b_{\mu,x,y,k}$ is self-consistently set to zero, as we choose the Y_{sz} source stub admittance appropriately in order to decouple the incident and reflected wave amplitudes traveling in the device stub, as reported in [24, 25].

2.6 Boundary Conditions

In order to increase accuracy, the 3*D*-Maxwell domain, modeled by the Transmission Line Matrix method, is bounded by the exact boundary conditions provided by the Transmission Line Matrix- Integral Equation (TLM-IE) technique, [20].

For the 1*D*-Schrödinger domain, we first impose Dirichlet boundary conditions, meaning that the wave function $\psi(z,t)$, is set to zero at the boundaries: $\psi(z_1,t) = \psi(z_2,t) = 0$, for each t. Then, by following the guidelines of [26, 27, 28], we construct exact (matched) boundary conditions for the wave function $\psi(z,t)$ in $z = z_1$ and $z = z_2$. The former (Dirichlet) boundary conditions are suited for a nanotube that behaves as a resonator, the latter (matched) boundary conditions are properly related to absorbtion of charge by the source and/or drain electrodes in a transistor environment. In this case, the matched boundary conditions involve convolution integrals in the t time variable; these are:

$$\psi(z_1,t) = -\sqrt{\frac{\hbar}{2\pi m}}e^{i\frac{\pi}{4}} \int_0^t \frac{\partial \psi(z_1,t-\tau)}{\partial z} \frac{e^{-i\frac{V_p(z_1)}{\hbar}\tau}}{\sqrt{\tau}} d\tau \tag{20}$$

and

$$\psi(z_2,t) = -\sqrt{\frac{\hbar}{2\pi m}}e^{i\frac{\pi}{4}} \int_0^t \frac{\partial \psi(z_2,t-\tau)}{\partial z} \frac{e^{-i\frac{V_p(z_2)}{\hbar}\tau}}{\sqrt{\tau}} d\tau \tag{21}$$

Equations (20) and (21) are then discretized and inserted in the finite-difference algorithm (10) and (11).

2.7 Excitation

The structure can be excited: (i) by an external applied and/or impinging electromagnetic field, (ii) by an initial distribution $\psi(z,t=0)$ of the wave-function along the nanotube, (iii) by injection of charge from the electrodes, (iv) by a combination of the above conditions. In the present work, we deal with the case (ii). It is noted that the problem analyzed is 1*D* with respect to the Schrödinger equation, but full-3*D* with respect to Maxwell equations.

3 Results

We analyze the $3D$-structure of Fig. 1 representing a basic nanotube-transistor structure. An insulator, with $\varepsilon_r = 25$, completely fills the space between the nanotube, and the source-drain contacts. The electrodes are metallic plates with dimensions: $20 \times 20 \times 10\,\text{nm}$, being $10\,\text{nm}$ the width along the nanotube direction (z). A gate contact is placed below.

Let us consider an electron wavepacket, whose energy is centered near the bottom of the first band of a semiconducting nanotube (*16, 0*), defining the way the graphene sheet is rolled up (*chirality*). In particular, semi-conducting nanotubes are those whose chirality is given by index pairs of the (*n,0*) type. The first-band is assumed to be partly occupied by electrons diffused from the lateral metallic contacts. In fact, electrons are allowed to enter the nanotube due to the chosen profile of the potential $V_p(z)$ of Fig. 2, which can follow from a typical voltage configuration of the electrodes [10]. The TLM grid is homogeneous: $\Delta x = \Delta y = \Delta z = \Delta L \sim 1.5$ nm, with $\Delta L = 1/10\lambda_{elec}$; being λ_{elec} the electron-wavelength associated to its E_{elec} spectral width in energy, with $E_{elec} = 0.1\,\text{eV}$. The resulting grid consists of $(50 \times 50 \times 150)$ nodes. The time-step is $\Delta t = 2.5 * 10^{-3}\,fs$. We consider $n_{step} = 200000$ time-steps, corresponding to a time-slot of $T_{\max} \sim 0.5\,\text{ps}$.

3.1 Exciting Wavepacket

The exciting wavepacket $\psi(z,t=0)$ has a gaussian spatial shape centered at $z_k = z_{k_c}$. We may associate a spectral width (in energy) to the launched wavepacket by simply employing the relation between energy and wave-number for a steady state solution of the Schrödinger equation (at $t = 0$ the EM field is set to zero everywhere):

$$\psi_R\left(z_k = k\Delta l, t=0\right) = \psi_0 \exp\left(\frac{-\left(\left(z_k - z_{k_c}\right)\Delta l\right)^2}{2\sigma^2}\right)$$
$$k_{\max} = \frac{2\pi}{\lambda_{elec}} = \left(\frac{2mE_{elec}}{\hbar^2}\right)^{\frac{1}{2}} \qquad \sigma = \frac{\lambda_{elec}}{2\pi} = \frac{1}{k_{\max}} \tag{22}$$

In Eq. (22), $\Delta k = k_{\max}$ is the width of the wavepacket in the wave-number domain, corresponding to the electron E_{elec} energy and m the electron “effective mass”, set to $0.06 * m_e$, being $m_e = 9.109 * 10^{-31}$ [kg] the electron free mass, [22]. The use of the effective mass approximation is consistent with the choice of such a sharp pulse (in energy). Furthermore, this choice ensures that the contribution of sub-bands, seen as different channels for carrier transport having different effective masses, can be neglected. No external EM field is applied at $t = 0$.

3.2 Injected Charge

The total charge injected in the initial ($t = 0$) electron wavepacket is chosen according to typical values used for nanotubes embedded in a realistic device [12]. This charge is defined as the spatial integral of the linear density of charge diffused along the nanotube and depends on the operative conditions, such as (i) the applied external voltages, (ii) the geometrical parameters (electrode dimensions, nanotube size). The applied external voltages provide direct control on the electronic transmission properties of the nanotube channel. Moreover, the longer and larger is the nanotube, the higher is the total charge trapped between the electrodes. A typical value of the total charge for the 60 nm nanotube of Fig. 1 is estimated according to [10]. Therefore, we normalize the initial wavepacket in order to deal with a charge Q_0 that is about one order of magnitude greater then the electron charge: $Q_0 \sim 10_{q,\ q} = 1.602 * 10^{-19}$ [C].

$$q \int_{-\infty}^{+\infty} |\psi(\mathbf{r},0)|^2 d\mathbf{r} = q \int_{z_{t_1}}^{z_{t_2}} |\psi(z,0)|^2 dz = Q_0 \tag{23}$$

With respect to the case analyzed in [10], the Schrödinger equation (1) is solved for just one type of charge, the electron, because in the case of Fig. 1, the external bias is chosen in such a way as not to allow hole propagation.

The potential profile for electrons and the initial charge Gaussian wavepacket $|\psi(z,t=0)|^2$ are shown in Fig. 3. Ohmic contacts are assumed at the nanotube

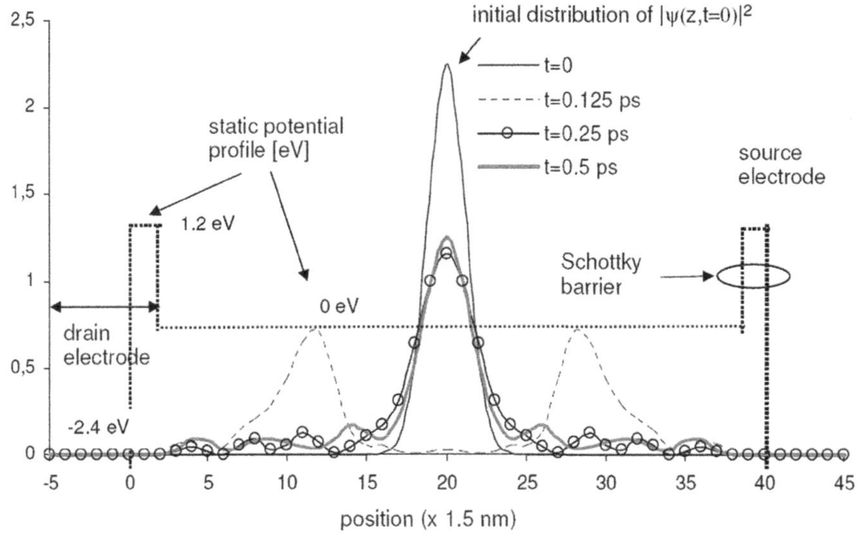

Fig. 3 Space evolution of the $|\psi(z,t)|^2$ charge wavepacket for $t = 0$ (initial distribution), $t = 0.125$, $t = 0.25$, $t = 0.5$ ps, respectively. Case of $A_z(z,t) = \phi(z,t) = 0$. The distribution of the static potential profile [eV] with the Schottky barriers is also reported

terminations. (the origin of the z-axis is the border of the left-side electrode): we observe the presence of Schottky barriers ($z = 0$, $z = 60\,\text{nm}$), detected by the electrons at the nanotube terminations. These barriers are modelled, as a first approximation, by rectangular functions. It has to be noted that they cause the electron wavepacket to be better confined within the nanotube cavity. The amplitude of the launched wavepacket has to be multiplied by the ψ_0 normalization constant, as derived from (23). We consider Dirichlet boundary conditions, by imposing $\psi(z_1,t) = \psi(z_2,t) = 0$, for each t. We first analyze the case of absence of interaction between the wavepacket propagation and the EM field, by setting $A_z(z,t) = \phi(z,t) = 0$ in Eqs. (8) and (9).

Figure 3 shows the spatial evolution of the Gaussian wavepacket $|\psi(z,t)|^2$ along the nanotube domain, for different time samples: for $t = 0$, $t = 0.125$, $t = 0.25$, $t = 0.5\,\text{ps}$. We then analyze the full wave coupling between the wavepacket propagation and the EM field, ($A_z(z,t) \neq 0$, $\phi(z,t) \neq 0$).

In Fig. 4, we report the comparison between the $|\psi(z,t)|^2$ wavepacket propagation in the case of full-wave coupling with the electromagnetic field and the previous case of Fig. 3 (no EM coupling).

In Fig. 5, we show the time-space evolution of the $|\psi(z,t)|^2$ $[1/m]$ charge density (normalized to the q electron charge). Figure 6 plots the time-space evolution of the corresponding $\boldsymbol{J}(z,t)$ [A] quantum-mechanical current. While approaching the end of the time interval, the charge begins to penetrate the potential barriers, located around $z = 0$ and $z = 60\,\text{nm}$.

In Figs. 7 and 8 we show the time-space evolution of the E_x, E_z field components, respectively, in the transversal plane containing the nanotube. It is noted that the E_z-field follows the behavior of the current (14), whereas the E_x-field is led by the charge (7), consistently with the Gauss law.

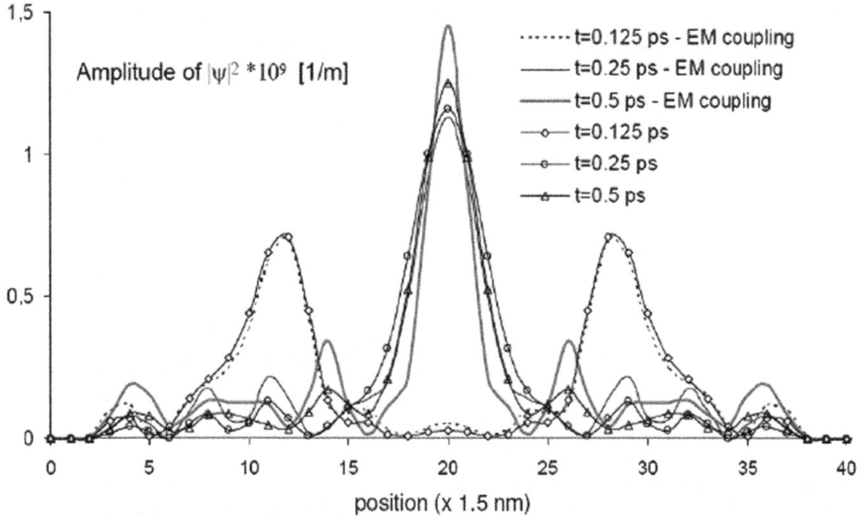

Fig. 4 Space evolution of $|\psi(z,t)|^2$ for $t = 0.125$, $t = 0.25$, $t = 0.5\,\text{ps}$. Comparison between the case of $A_z(z,t) = \phi(z,t) = 0$ and the case of $A_z(z,t) \neq 0$, $\phi(z,t) \neq 0$ (coupling with the EM field)

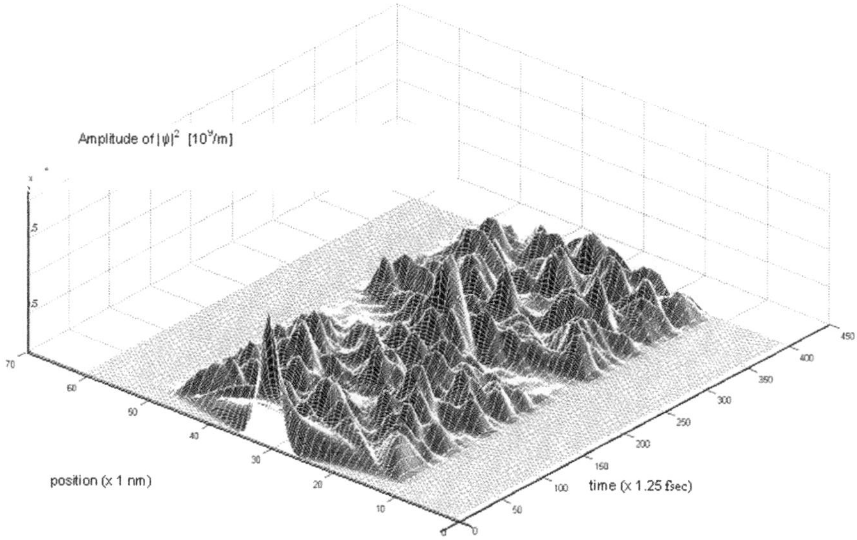

Fig. 5 Time-space evolution of the $|\psi(z,t)|^2$ charge density

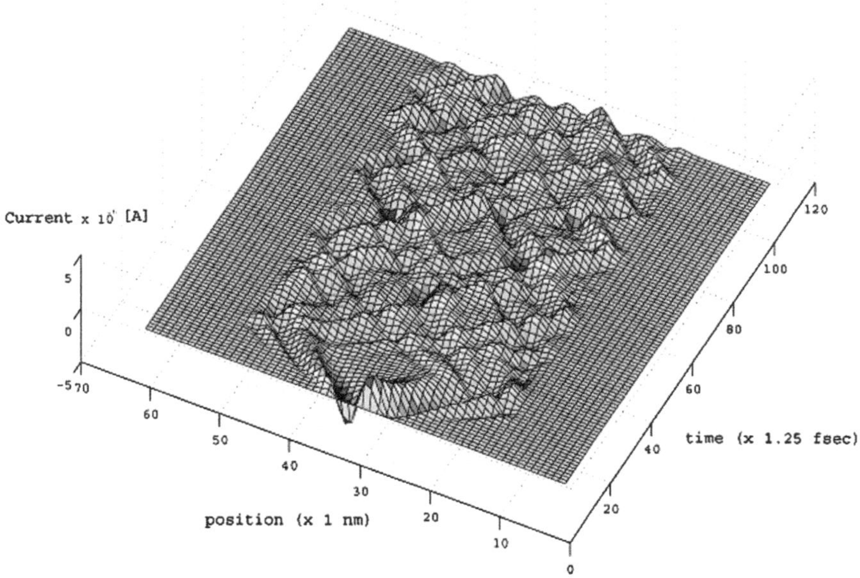

Fig. 6 Time-space evolution of the $\boldsymbol{J}(z,t)$ quantum mechanical current

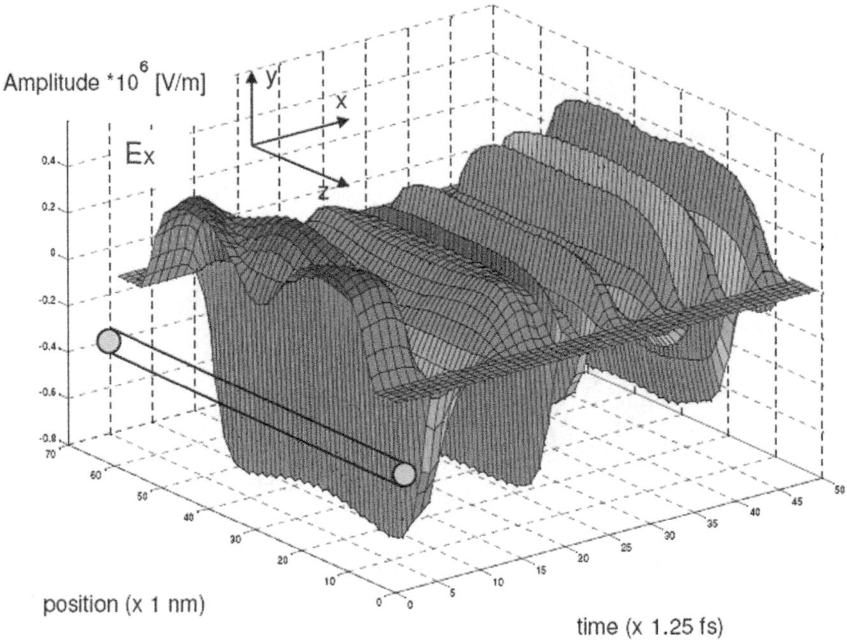

Fig. 7 Time-space evolution of the E_x-field

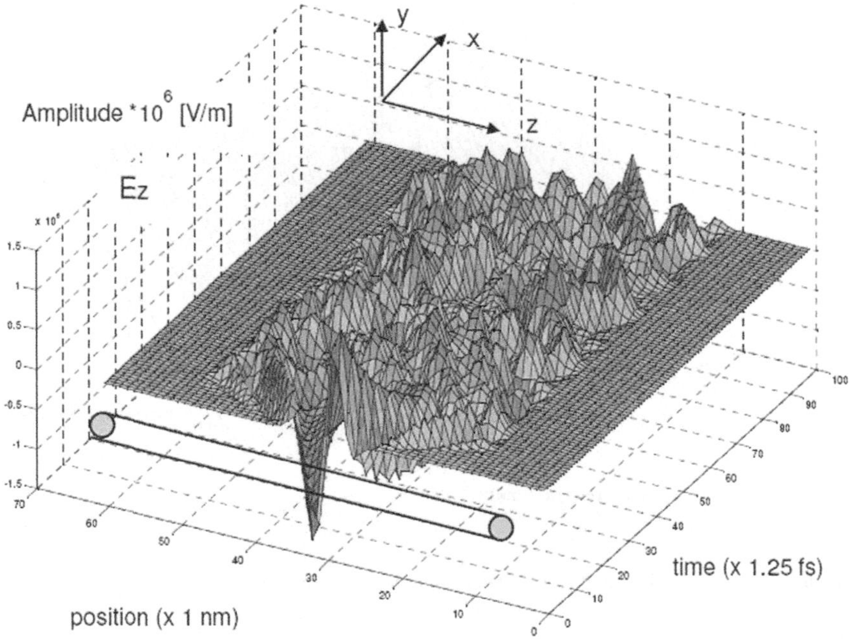

Fig. 8 Time-space evolution of the E_z-field

We now analyze a CNT with a V_{ds} bias: drain-voltage>source-voltage so that $V_{ds} = 0.3$ [V]. The dimensions of the CNT structure are the same as before, apart for the length, that is now of 30 nm. As before, $E_{elec} = 0.1$ eV; we now choose $\Delta L = 1/16\lambda_{elec}$, leading to $\Delta t = 1.65 * 10^{-3}\,fs$. We consider $n_{step} = 100000$ time-steps, corresponding to a time-slot of $T_{\max} \sim 0.16$ ps.

For the wave function $\psi(z,t)$, we apply the exact boundary conditions (20) and (21) in $z = z_1$ and $z = z_2$, respectively. These matched boundary conditions are properly related to absorbtion of charge by the source-drain electrodes in a transistor structure. In Fig. 9 we show the evolution of the $|\psi(z,t)|^2$ wavepacket and the bias potential profile. As expected, the positive charge tends to concentrate towards the source electrode and the current profile evolves consistently. The corresponding $\boldsymbol{J}(z,t)$ [A] current dynamic is reported in Fig. 10.

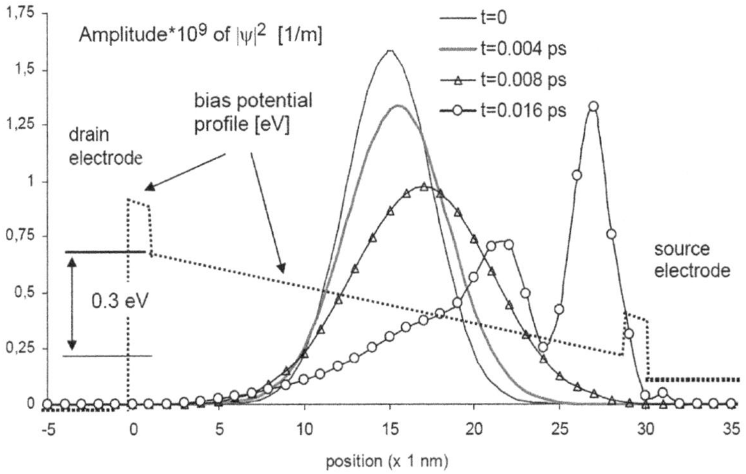

Fig. 9 Wavepacket propagation ($|\psi(z,t)|^2$) for $t = 0$, $t = 0.004$, $t = 0.008$, $t = 0.016$ps, in the case of $V_{ds} = 0.3$ [eV] bias potential

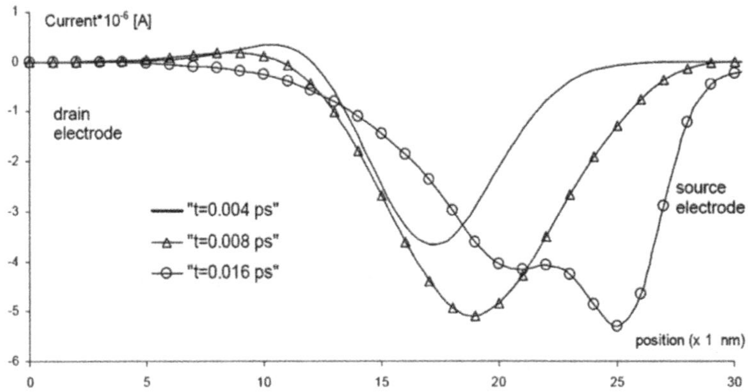

Fig. 10 Current density [A] corresponding to the wavepacket of Fig. 9

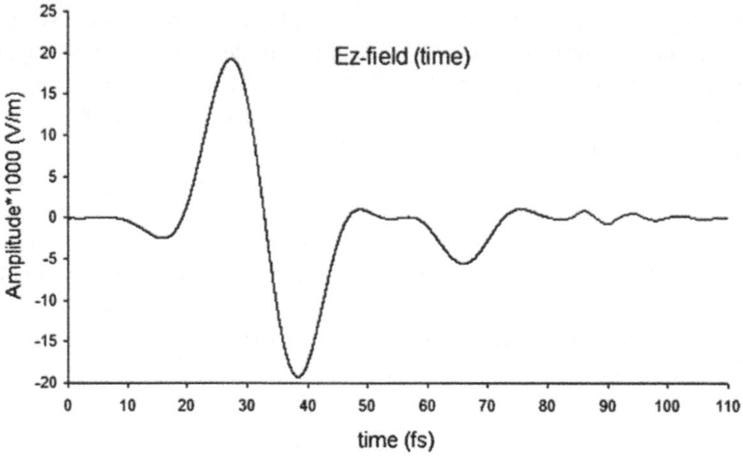

Fig. 11 Time evolution of the E_z-field [V/m] at a point close (2 nm) to the CNT

In Fig. 11 we show time evolution of the E_z-field component at a point placed $3\,nm$ close to the nanotube, in the dielectric region. Fig. 12 plots the correspondent frequency response (normalized amplitude vs. frequency in [THz]). We observe that the spectrum extends up to frequencies corresponding to $\lambda = c/f$ optical wavelengths, being c the light velocity and $\varepsilon_r = 25$.

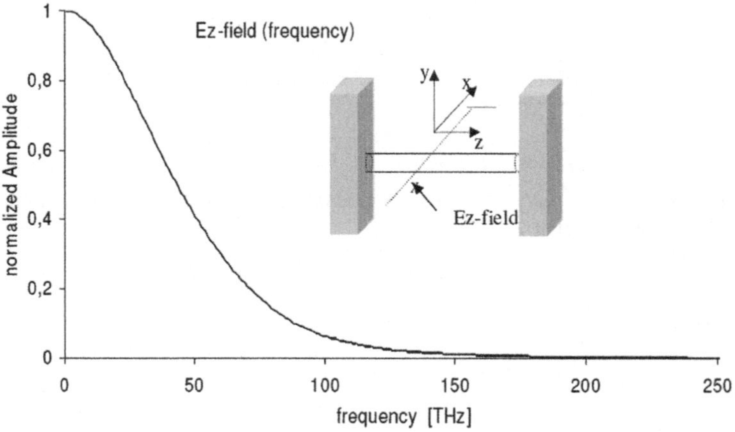

Fig. 12 Normalized frequency response [THz] for the field of Fig. 11

4 Semi-Analytical Approach in the Frequency Domain

4.1 Single-Channel Transport

A general model operating in frequency domain is currently proposed in the literature in order to analyze the properties of carrier transport [29] in carbon nanotube-field effect transistors [22, 12]. A typical device geometry, used to model the nano-FET behavior, is shown in Fig. 13. Source and drain contacts are placed at the nanotube terminations; typically, a cylindrical gate contact is assumed around the nanotube, so that cylindrical symmetry is preserved and the potential calculation can be performed analytically by means of a Green's function approach. An insulator, of relative permittivity $\varepsilon = 25$, completely fills the space between the nanotube and the gate contact; the nanotube is considered to be filled by air.

To the aim of obtaining the electrical properties of nano-FETs, we have to solve the system of coupled Poisson and Schrödinger equations, given below:

$$\begin{cases} \dfrac{\partial^2 V}{\partial \rho^2} + \dfrac{1}{\rho}\dfrac{\partial V}{\partial \rho} + \dfrac{\partial V}{\partial z^2} = \dfrac{-Q}{\varepsilon} \\ \dfrac{\partial^2 \Psi_{h,e}}{\partial z^2} = -\dfrac{2m}{\hbar^2}\left(E - U_{h,e}\right)\Psi_{h,e} \end{cases} \tag{24}$$

here, V is the electrostatic potential within the outer cylinder, $\Psi_h(\Psi_e)$ is the z-dependent wave-function of a hole (electron) of energy E, traveling under the effect of a local potential energy $U_h(U_e)$, Q is the nanotube charge linear-density corresponding to the carriers diffused from drain and source contacts. The boundary conditions for Eq. (24) are provided by V_g (gate voltage), V_s (source voltage) and V_d (drain voltage). Depending on the metal used and on the electrode geometry, the resulting metal-nanotube band offset may give rise either to a Schottky-type device, where both electrons and holes see an energy barrier of about $Eg/2$, or to Ohmic type nano-FETs, which generally exhibit better performances, since holes see no barrier [29]. Charge diffusion is governed by Schrödinger equation and by the local electrostatic potential, shifting the nanotube band-structure; the spatial charge distribution is recursively assumed as source term for the Poisson equation. In synthesis,

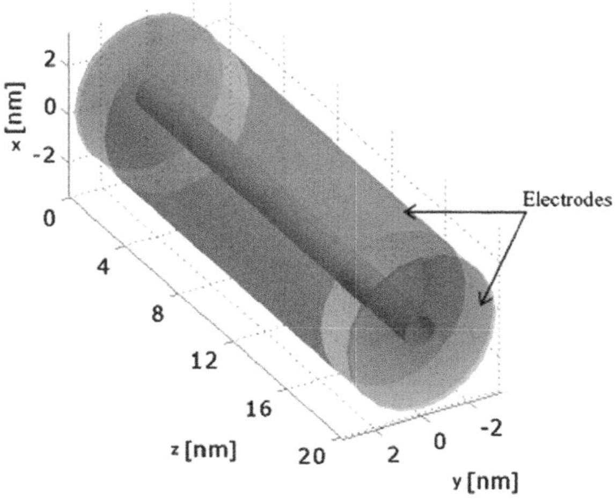

Fig. 13 Device Geometry of a CNT-FET

V_g modulates the total current by controlling the transmission probability for carriers through the channel (differently from MOSFETs, which are charge controlled devices).

4.2 Multi-Channel Transport

Multi-channel transport takes place either in sw-CNTs, where several sub-bands contribute to the charge diffusion, or in mw-CNTs, where not only different sub-bands but also different walls contribute to the current. This multichannel effect is expected to increase as the external voltage and/or the CNT diameter increases. Figure 14 shows qualitatively the band-structure of a mw-CNT-FET under external bias. We report the generic m-th wall of the CNT, with just the first sub-band, for the sake of clarity.

In the following, different bands will be denoted by the apex n, while the walls are indicated by m. Each sub-band and each wall of the CNT are formally considered as if they extended up to the metal regions. In other words, a complete correspondence is established between the channels of metal and CNT: these channels differ only in the potential energy seen by the transported carriers. In [29], metal carriers are assumed to have the same effective mass as CNT carriers. In this work, we use this assumption for all the channels: the effective masses of carriers are simply extended from the CNT to the metal. However, in the CNT the energy of a carrier of the channel (n,m) depends on the n-th branch of the dispersion curves of the m-th wall

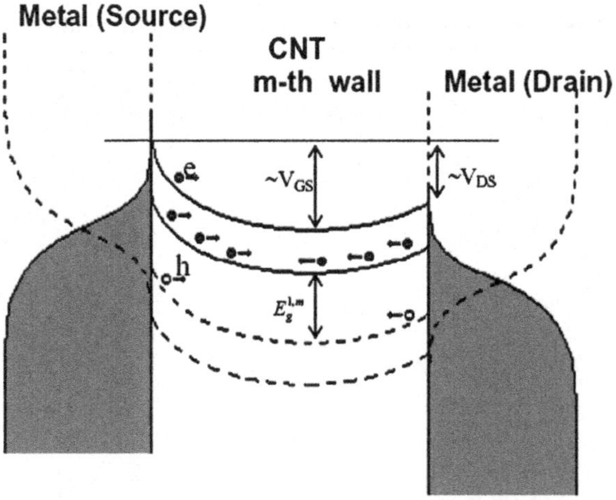

Fig. 14 Band diagram of the junction metal-CNT-metal: charges incident from metal electrodes are weighted by the Fermi function. The conduction and valence bands of the m-th CNT wall, together with the first sub-band for electrons (*solid lines*) and holes (*dashed lines*), are reported

while in the metal, on the other hand, carrier energies are all referred to the bottom of the conduction band.

As the main task of the present approach is extending the analysis to multi-channel transport, we maintain some approximations that are usually found in the literature, often without being properly highlighted: (i) each carrier wavefunction is characterized by a definite effective mass: although the potential energy varies along the CNT, being related to the solution of the Poisson equation, it does not produce a distortion, but only a local shift, of the CNT dispersion curves; (ii) consistently with the previous point, the polarization induced by the external voltage, i.e. the shielding effect due to the distortion of the electron and hole wavefunctions [11], is neglected as we consider small applied electric fields. For the same reasons, the effects of band gap reduction and splitting of degenerate states are neglected [11]; (iii) the wavefunctions of carriers transported by different channels do not couple at the metal-CNT interfaces: the only discontinuity being given by the CNT-metal band offset. In the present transmission-line approach, this kind of discontinuity is modeled by means of an ideal transformer, representing an abrupt change of waveguide potential energy. The Schrödinger equation becomes:

$$\frac{\partial^2 \Psi_{h,e}^{n,m}}{\partial z^2} = -\frac{2m_{h,e}^{n,m}}{\hbar^2}\left(E - U_{h,e}^{n,m}(z)\right)\Psi_{h,e}^{n,m} \tag{25}$$

We still keep distinct all quantities related to electrons and holes. The potential energy appearing in (25) is defined as:

$$\begin{cases} U_e^{n,m}(z) = \widetilde{U}_e^{n,m}(z) - eV(R+\delta_m, z) \\ U_h^{n,m}(z) = \widetilde{U}_h^{n,m}(z) + eV(R+\delta_m, z) \end{cases} \tag{26}$$

with

$$\begin{cases} \widetilde{U}_e^{n,m}(z) = E_{vac} - e\chi_T^{n,m} \\ \widetilde{U}_h^{n,m}(z) = \widetilde{U}_e^{n,m}(z) - E_g^{n,m} \end{cases} \tag{27}$$

where R is the radius of the smallest of the CNT walls, $Evac$ is the vacuum energy, $\chi_T^{n,m}$ is the electron affinity for the carriers of the (n,m)-channel, $E_g^{n,m}$ is the n-th energy gap of the m-th wall and V is the electrostatic potential satisfying the Poisson equation, evaluated in correspondence of the radius $(R+\delta_m)$ of the m-th wall. The boundary conditions for V are given by

$$\begin{cases} V(R_G, z) = V_g - \Phi_g \\ V(R+\delta_m, 0) = V_s - \Phi_s \\ V(R+\delta_m, L) = V_d - \Phi_d \end{cases} \tag{28}$$

where R_G is the radius of the gate electrode and Φ_g, Φ_d, Φ_s, are the work functions of gate, drain and source respectively. Each nanotube wall has been treated as if it were an independent nanotube. As a consequence, the dispersion curves of the m-th wall are computed by means of a TB (Tight Binding) approach, neglecting the interaction with the other walls. Since the dispersion curves are symmetric in energy with respect to the center of the fundamental band gap, the parameters of Eq. (27) can be defined as shown in Fig. 15. Figure 27 also shows the electron affinity, appearing in (28). Its value stems directly from the assumption that, in absence of external bias, the Fermi level, hence the work function, is the same for the metal and the CNT [29]: in this work the assumption is extended to all the CNT walls. A different choice may be made just by redefining Φ and χ in (27) and (28).

The effective mass $m_e^{n,m}$ ($m_h^{n,m}$) is derived from the parabolic approximation of the n-th branch above (below) the band gap of the dispersion curves of the m-th wall. The contributions of the (n,m)-channel to the linear charge and to the current are computed by solving separately the respective Schrödinger equations (25). However, even though the wavefunctions of different channels are not spatially coupled, they are coupled through the Poisson equation, since they all contribute to the source charge:

$$
\begin{aligned}
q_e^{n,m} &= \int \left|\Psi_e^{n,m,D}(z)\right|^2 + \left|\Psi_e^{n,m,S}(z)\right|^2 dE \\
q_h^{n,m} &= \int \left|\Psi_h^{n,m,D}(z)\right|^2 + \left|\Psi_h^{n,m,S}(z)\right|^2 dE
\end{aligned}
\qquad (29)
$$

$$
Q = -\frac{e}{2\pi} \frac{\delta(\rho - R - \delta_m)}{\rho} \sum_{n,m} \left(q_e^{n,m} - q_h^{n,m}\right) \qquad (30)
$$

As for the charge, the total current is provided by the sum of the contributions of each wall with its own sub-bands

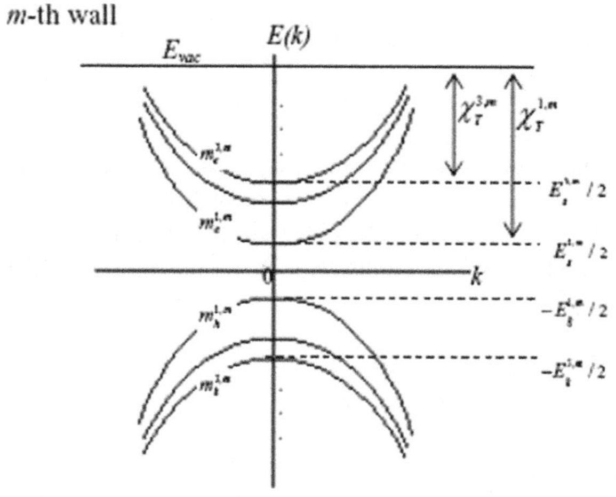

Fig. 15 Dispersion curves (energies not in scale) of the m-th wall of a mw-CNT

$$I = \sum_{n,m} \left(I_h^{n,m} + I_e^{n,m} \right) \tag{31}$$

In (31), $I_{h,e}^{n,m}$ is the current of the (n,m)-channel (Landauer-Büttiker formula):

$$I_{h,e}^{n,m} = \frac{4e}{h} \int \left(f_{h,e}^S - f_{h,e}^D \right) T_{h,e}^{n,m} \, dE \tag{32}$$

where $T_h^{n,m}$ $(T_e^{n,m})$ is the transmission probability through the (n,m)-channel for holes (electrons). and $f_{s,d}$ is the Fermi function at the source and drain.

4.3 Transmission Line Approach

With reference to (25), we start by making the positions:

$$\begin{cases} \Psi_{h,e}^{n,m} = \left(V_{h,e}^{n,m} \right)^+ e^{-j\beta_{h,e}^{n,m} z} + \left(V_{h,e}^{n,m} \right)^- e^{j\beta_{h,e}^{n,m} z} \\ \dfrac{\partial \Psi_{h,e}^{n,m}}{\partial z} = \left(I_{h,e}^{n,m} \right)^+ e^{-j\beta_{h,e}^{n,m} z} + \left(I_{h,e}^{n,m} \right)^- e^{-j\beta_{h,e}^{n,m} z} \\ \qquad = \dfrac{\left(V_{h,e}^{n,m} \right)^+}{jZ_{h,e}^{n,m}} e^{-j\beta_{h,e}^{n,m} z} - \dfrac{\left(V_{h,e}^{n,m} \right)^-}{jZ_{h,e}^{n,m}} e^{j\beta_{h,e}^{n,m} z} \end{cases} \tag{33}$$

From (33), it can be realized that the charge diffusion through the (n,m)-th channel is described by the telegrapher's equations, written for $V_{h,e}^{n,m}$ and $I_{h,e}^{n,m}$. The characteristic impedance and the propagation constant of the (n,m)-th transmission line are spatially variable, as they depend on the local potential energy U:

$$\beta_{h,e}^{n,m} = \sqrt{E - U_{h,e}^{n,m}} \qquad Z_{h,e}^{n,m} = 1/\beta_{h,e}^{n,m} \tag{34}$$

The potential energy U can be well approximated by a staircase function. Each constant step is modeled by means of a transmission line with fixed impedance and propagation constant. Then, the Schrödinger equation is solved as a cascade of transmission lines for each channel (n,m). The initial $(z < 0)$ and final regions $(z > L)$, with constant potential energy, correspond to the Source and Drain electrodes. By following the notation of [23] and by extending it to the case of n sub-bands and m walls, the wavefunction at source and drain becomes:

$$\Psi_{h,e}^{n,m} = \begin{cases} A_{h,e}^{n,m} \mathrm{e}^{-j\beta_{h,e}^{n,m,S} z} + B_{h,e}^{n,m} \mathrm{e}^{j\beta_{h,e}^{n,m,S} z} & \mathrm{z < 0 (source)} \\ C_{h,e}^{n,m} \mathrm{e}^{-j\beta_{h,e}^{n,m,D} z} + D_{h,e}^{n,m} \mathrm{e}^{j\beta_{h,e}^{n,m,D} z} & \mathrm{z > L (drain)} \end{cases} \tag{35}$$

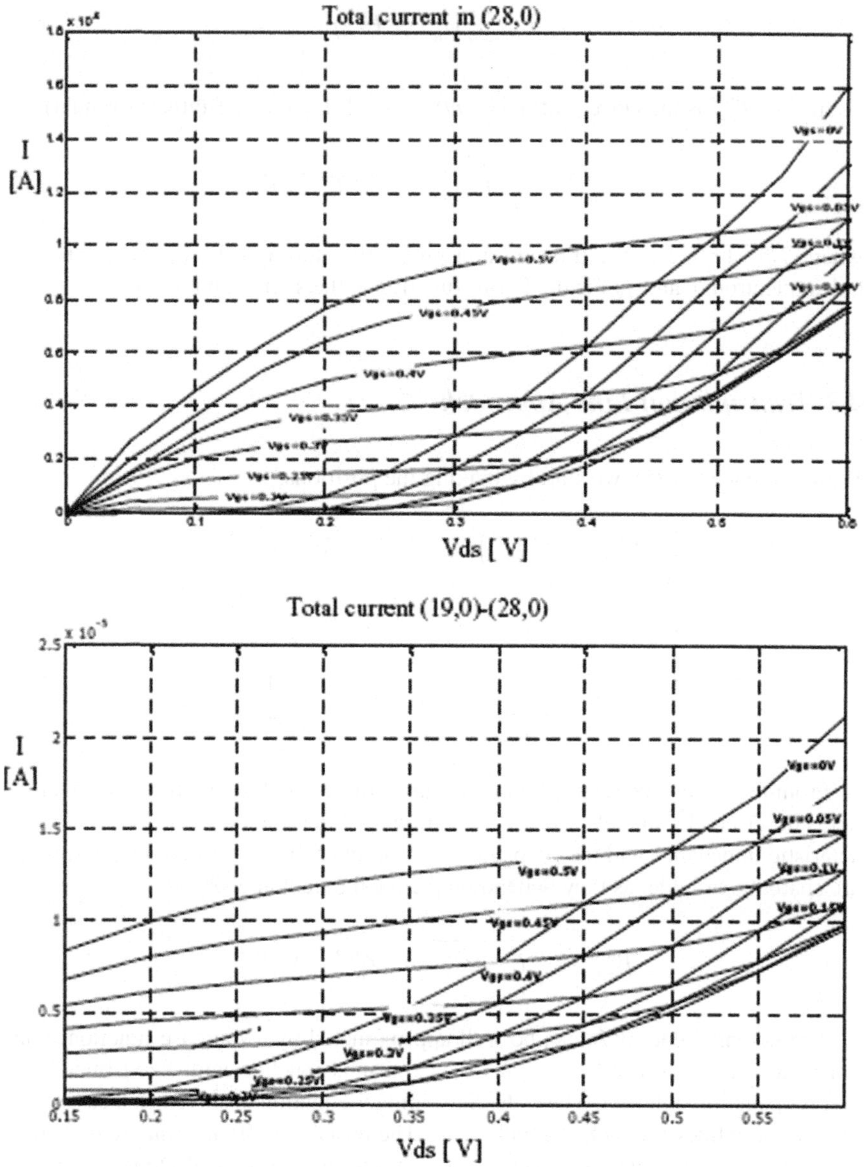

Fig. 16 Currents in a (28, 0) sw-CNT and in a (19, 0), (28, 0) dw-CNT. $V_{ds} = V_d - V_s$, $V_{gs} = V_g - V_s$

$A_{h,e}^{n,m}$ and $D_{h,e}^{n,m}$ are the normalization constants of the incident waves coming from Source and Drain respectively. In order to make a choice consistent with the Landauer equation [29], they are fixed as:

$$\left(A_{h,e}^{n,m}\right)^2 = \frac{2m_{h,e}^{n,m}}{\pi\hbar^2}\frac{f_{h,e}^{n,m,S}}{\beta_{h,e}^{n,m,S}} \qquad \left(D_{h,e}^{n,m}\right)^2 = \frac{2m_{h,e}^{n,m}}{\pi\hbar^2}\frac{f_{h,e}^{n,m,D}}{\beta_{h,e}^{n,m,D}} \tag{36}$$

Here is an example of the obtained results

In the Fig. 16 we plotted I vs V_{ds} for many fixed values of V_g.

5 Electromagnetic Solvers Applied to Carbon Nanotubes

The electronic properties of carbon nanotubes are strictly linked to the "dispersion curves", which describe the propagation of electronic waves within the material. The "Tight Binding" (TB) perturbation technique is widely used in the literature since it is the easiest way to address this problem: the sum of the atomic potentials of each carbon atom is just assumed as the Hamiltonian operator. In a recent work [11] we asked ourselves whether to define the potential V as a superposition of well-potentials instead of ion-potentials. By a proper choice of depth and width of these wells, we were able obtain the expected behavior of the dispersion curves, at least for a specified energy range (for example the energies around the band-gap of a semiconducting nanotube). In this way, we accept some kind of simplification for what concerns the potential profile, but we avoid all further approximation made by perturbation models.

A fundamental feature of the proposed model consists in the possibility to make directly use of electromagnetic solvers to the aim of solving the Schrödinger equation. In fact, the resulting nanotube lattice is very easy to design and to implement on the computer. In detail, the nanotube will be modelled by an ideal two-dimensional lattice, whose width is equal to the its circumference, by switching to a curvilinear coordinate-system on its surface. The formal coincidence between Maxwell and Schrödinger equations, in the particular case of a planar lattice, justifies theoretically the use of electromagnetic solvers:

$$\nabla^2 E_y = -\frac{(2\pi f)^2}{c^2}\varepsilon(x,z)E_y \tag{37}$$

where E_y is the electric field component orthogonal to the lattice plane and ε spatial distribution of the dielectric constant. The situation is depicted in Fig. 17:

The proposed model can be seen as a photonic crystal, where the dielectric distribution plays the role of potential energy (by comparison with the Schrödinger equation) and the electric field plays the role of wave-function. In this way, we provide a useful prototype for the solution of several problems of particular relevance.

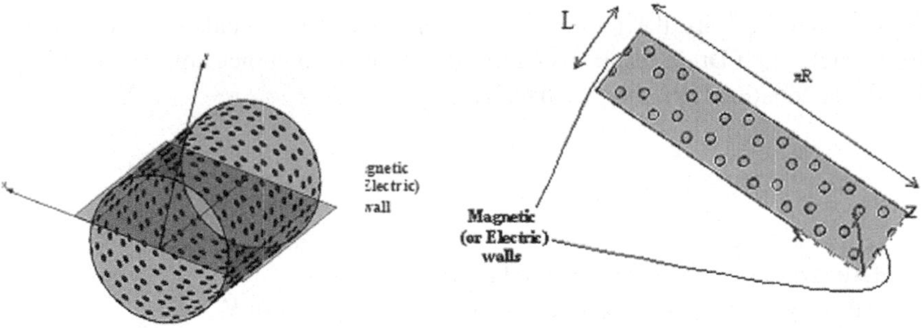

Fig. 17 (16, 0) CNT and half 2D unit cell

For instance, the contact between the nanotube and the metallic electrode may be modeled and simulated by means of a change of the dielectric profile of the prototype itself, in correspondence of the contact region: the resulting device will show hybrid metallic-semiconducting properties. Analogously, basing on the analogy between potential energy and dielectric constant, the effect of an external applied electric field to the nanotube electronic properties can be analyzed (the electric field simply modifies the local potential energy): here we mainly focus our attention on band gap reduction and distortion of the electronic density of states (DOS). It is also important to mention an important feature of the present approach: the possibility to avail the electronic wave functions in numerical format. In fact, it becomes easy to manipulate them, in order, for example, to calculate the matrix elements related to the optical absorption (or gain): thus, a direct way is provided to analyze the nanotube response at optical frequencies. The well known precision, flexibility, efficiency of electromagnetic simulators, even commercial ones, makes particularly attractive their application to the study of nano-devices, not only with reference to electromagnetic and optical problem, for which they were developed and optimized, but also to quantum problems.

The above mentioned example of band gap distortion and hole/electron wavefunctions polarization due to external voltage applied (electric field transverse with respect to the CNT axis) is shown in the following Fig. 18, which reports the dependence of the band gap of a (13, 0) CNT on the applied electric field.

The even solution (magnetic wall), whose wavefunction has relative maxima at $x = 0$ (see Fig. 19) in correspondence of the potential maximum, is more affected by the electric field than the odd solution: the even field drops faster than the odd one by 15–17%. Figure 19 shows the amplitude of the electron wave-function of a (13, 0) CNT at the bottom of the conduction band, for a local electric field of about 2.5 [V/nm].In the even case the band gap is equal to zero, whereas, in the odd case, the conduction and valence bands are still slightly separated. The even solution of the wave-function appears to be more polarized by the electric field with respect to the odd solution.

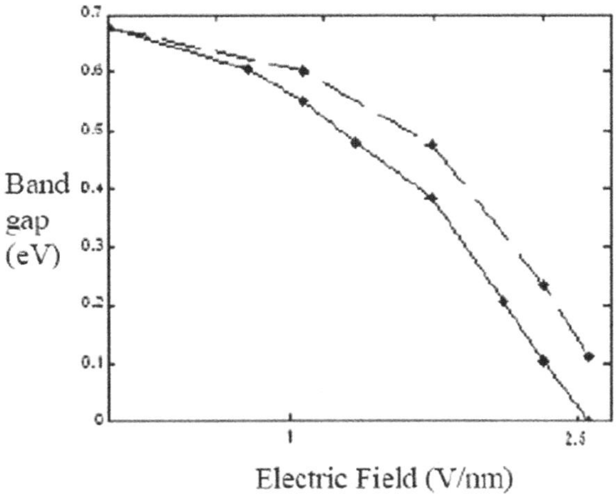

Fig. 18 Band gap distortion

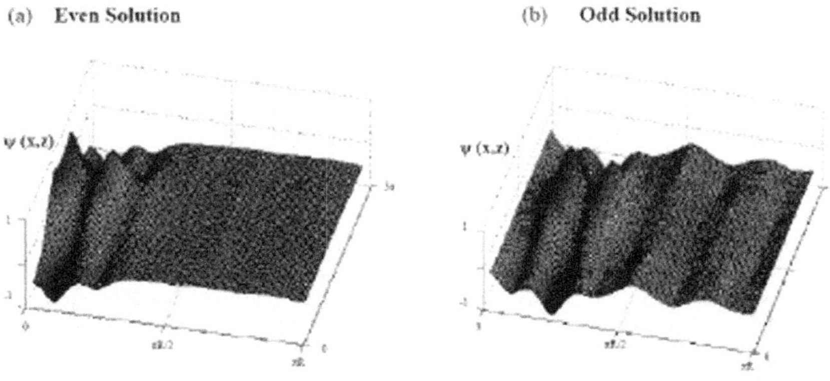

Fig. 19 Distortion of the even (**a**) and odd (**b**) wavefunctions

6 Conclusions

In this paper, we provide new simulation tools for emergent nanodevices. We introduce a novel numerical technique whereby Maxwell's equations are coupled to the Schrödinger's equation and solved. It consists of an iterative process in which the formers are solved by the Transmission Line Matrix method and, simultaneously, the latter is solved by a Finite Difference scheme. We analyze the space-time dynamics of an electron wavepacket and the corresponding self-generated EM-field. We then present a self-consistent analysis of multi-channel transport in multi-wall CNTs. The contribution of the sub-bands to charge diffusion is shown to be not negligible, especially for large diameters and/or high applied external voltages.

Finally, we show that quantum-mechanical properties of CNTs, such as band gap and effective mass, can be described by means of a "potential well" model.

Owing to the formal analogy between Maxwell and Schrödinger equations in low-dimensional systems, e.m. simulators can be directly used in order to analyze the carrier transport in CNTs.

References

1. S. Li, Z. Yu, S. F. Yen, W. C. Tang, and P. J. Burke, "Carbon nanotube transistor operation at 2.6 ghz," *Nano Lett.*, vol. 4, pp. 753–756, 2004.
2. J. P. Clifford, D. L. John, L. C. Castro, and D. L. Pulfrey, "Electrostatics of partially gated carbon nanotube fets," *IEEE Trans. Nanotechnol.*, vol. 3, pp. 281–286, Jun. 2004.
3. P. J. Burke, "An rf circuit model for cnts," *IEEE Trans. Nanotechnol.*, vol. 2, pp. 55–58, Mar. 2002.
4. P. Kim and C. M. Lieber, "Nanotube nanotweezers," *Science*, vol. 286, pp. 2148–2150, 1999.
5. G. Pirio, P. Legagneux, D. Pribat, M. C. K. B. K. Teo, G. A. J. Amaratunga, and W. I. Milne, "Fabrication and electrical characteristics of carbon nanotube field emission microcathodes with an integrated gate electrode," *Nanotechnology*, vol. 13, pp. 1–4, Feb. 2002.
6. G. W. Hanson, "Current on an infinitely-long carbon nanotube antenna excited by a gap generator," *IEEE Trans. Microw. Theory Tech.*, vol. 54, no. 1, pp. 76–81, Jan. 2006.
7. H. Sanada, M. Suzuki, and N. Nagai, "Analysis of resonant tunneling using the equivalent transmission-line model," *IEEE Trans. Quantum Electonics*, vol. 33, pp. 731–741, May 1997.
8. A. V. Maslov and C. Z. Ning, "Modal gain in a semiconductor nanowirelaser with anisotropic bandstructure," *IEEE J. Quantum Electron.*, vol. 40, no. 10, pp. 1389–1397, Oct. 2004.
9. T. Rozzi and D. Mencarelli, "Application of algebraic invariants to full-wave simulators - rigorous analysis of the optical properties of nanowires," *IEEE Trans. Microw. Theory Tech.*, vol. 54, no. 2, pp. 797–803, Feb. 2006.
10. T. Rozzi, D. Mencarelli, A. D. Donato, and M. Farina, "Self-consistent analysis of carbon nanotube (cnt) transistors: state-of-the-art and critical discussion," *In Proceedings of the 7th International Conference on RF MEMs and RF Microsystems*, Orvieto, Italy, June 2006, pp. 59–61.
11. D. Mencarelli, T. Rozzi, L. Maccari, A. D. Donato, and M. Farina, "Standard electromagnetic simulators for the combined electromagnetic/quantum-mechanical analysis of carbon nanotubes," *Phys. Rev. B*, vol. *75, 085402*, 2007.
12. M. Pourfath, H. Kosina, B. H. Cheong, W. J. Park, and S. Selberherr, "The effect of device geometry on the static and dynamic response of carbon nanotube field effect transistors," *In Proceedings of 5th Conference on Nanotechnology*, 2005.
13. P. L. M. Euen, M. S. Fuhrer, and H. Park, "Single-walled carbon nanotube electronics," *IEEE Trans on Nanotechnol.*, vol. 1, no. 1, pp. 78–85, Mar. 2002.
14. A. Taflove, *Computational Electrodynamics–the Finite-difference Time-domain Method.* Boston, MA: Artech House, 1995.
15. P. B. Johns, "A symmetrical condensed node for the tlm method," *IEEE Trans. Microw. Theory Tech.*, vol. 35, no. 4, pp. 370–377, Apr. 1987.
16. H. Jin and R. Vahldieck, "Direct derivation of tlm symmetrical condensednode and hybrid symmetrical condensed node from maxwell's equations using centered differencing and averaging," *IEEE Trans. Microw. Theory Tech.*, vol. 42, no. 12, pp. 2554–2562, Dec. 1994.
17. T. M. B. V. Trenkic and C. Christopoulus, "Optimization of tlm schemes based on the general symmetrical condensed node," *IEEE Trans. Microw. Theory Tech.*, vol. 45, no. 3, pp. 457–465, Mar. 1997.
18. M. Krumpholz and P. Russer, "A field theoretical derivation of tlm," *IEEE Trans. Microw. Theory Tech.*, vol. 12, no. 9, pp. 1660–1668, Sept. 1994.
19. W. Hoefer, "The transmission-line-matrix method-theory and applications," *IEEE Trans. Microw. Theory Tech.*, vol. 35, no. 4, pp. 882–893, Oct. 1985.

20. L. Pierantoni, A. Massaro, and T. Rozzi, "Accurate modeling of te/tm propagation and losses of integrated optical waveguide," *IEEE Trans. Microw. Theory Tech.*, vol. 53, no. 6, pp. 1856–1862, June 2005.
21. P. Enders and D. D. Cogan, "Tlm routines for the paraxial wave equation and the time-dependent schroedinger equation," *In Proceed. of Micr. Symp. Digest, 1998 MTT Intern. Symp.*, Aug. 1995, pp. 137–140.
22. A. Soriano, E. A. Navarro, J. A. Portì, and V. Such, "Analysis of the finite difference time domain technique to solve the schroedinger equation for quantum devices," *J. Appl. Phys.*, vol. 95, no. 12, pp. 811–829, June 2004.
23. J. H. Davis, *The Physics of Low-dimensional Semiconductor*. Cambridge Univercity Press, Cambridge, MA 1995.
24. P. Russer, W. J. R. Hoefer, and P. P. M. So, "Modeling of nonlinear active regions in tlm," *IEEE Microw. and Guided Lett.*, vol. 1, no. 1, pp. 10–13, June 1991.
25. L. Cascio and W. J. R. Hoefer, "Modification of the 3d-tlm scattering matrix to model nonlinear devices in graded and heterogeneous regions," *In Proceedings of the 41st IEEE Conference on Decision and Control*, Las Vegas, Nevada, USA, June 1998, pp. 897–900.
26. M. Mangin-Brinet, J. Carbonell, and C. Gignoux, "Exact boundary conditions at finite distance for the time-dependent schrödinger equation," *Phys. Rev. A*, vol. 57, no. 5, pp. 3245–3255, May 1998.
27. M. E. A. Arnoldy and I. Sofronov, "Discrete transparent boundary conditions for the schrödinger equation: fast calculation, approximation, and stability," *Math. Comput. Modell.*, vol. 1, no. 3, pp. 501–556, Dec 2003.
28. A. Zisowsky and M. Ehrhardt, "Discrete transparent boundary conditions for parabolic systems," *Comm. Math. Sci.*, vol. 1, no. 3–4, pp. 294–309, March 2005.
29. P. J. S. P. D. L. John, L. C. Castro, and D. L. Pulfrey, "A schrödinger-poisson solver for modelling carbon nanotube fets," *Nanotechnology*, vol. 3, pp. 65–68, 2004.

Recent Advances in the Combination of the Unscented Transform (UT) with the Transmission Line Modeling Method (TLM)

Leonardo R.A.X de Menezes, Ajibola Ajayi, Christos Christopoulos, Phillip Sewell and Geovany A. Borges

Abstract This work presents some of the latest improvements in the combination of the Unscented Transform (UT) with Transmission Line Modeling Method (TLM) for the description of statistical uncertainty in electromagnetic simulations. These advances include: the correct approximation of different distributions, an effective general approach to the UT-TLM combination and their use in time and frequency domains. The work concludes with examples which are compared to Monte Carlo simulations and closed form solutions. Our results are in very good agreement, and were obtained at a fraction of computational cost.

1 Introduction

Electromagnetic full-wave simulation software is usually based on a deterministic approach of the problems. This means that one has to know with absolute accuracy all geometric and electric parameters of the simulation domain. This is the case for modeling techniques such as Finite Difference Time Domain (FDTD) and Transmission Line Modeling (TLM) techniques [1].

Leonardo R.A.X de Menezes
Dep. de Eng. Elétrica - Universidade de Brasília, CEP 70910-919 - Brasília - DF, Brazil, e-mail: leonardo@ene.unb.br

Ajibola Ajayi
The George Green Institute for Electromagnetics Research, University Park, Nottingham, NG7 2RD, UK, e-mail: eezks@gwmail.nottingham.ac.uk

Christos Christopoulos
The George Green Institute for Electromagnetics Research, University Park, Nottingham, NG7 2RD, UK, e-mail: eezks@gwmail.nottingham.ac.uk

Phillip Sewell
The George Green Institute for Electromagnetics Research, University Park, Nottingham, NG7 2RD, UK, e-mail: eezks@gwmail.nottingham.ac.uk

Geovany A. Borges
Dep. de Eng. Elétrica - Universidade de Brasília, CEP 70910-919 - Brasília - DF, Brazil

P. Russer, U. Siart (eds.), *Time Domain Methods in Electrodynamics*,

However, this knowledge may not be always available [2]. In certain problems, some parameters may be known only up to certain accuracy due to manufacturing or measurement tolerances. In other problems, the model uses random variables to represent the result of several deterministic parameters assembled in a random way (caused by complex environments). All these effects introduce uncertainty in the original problem. The consequence of this uncertainty may or may not be negligible or not, depending on the complexity of the problem and the accuracy of the expected response.

In the case where the electrical parameters are a function of other variables, it may be advantageous to include uncertainty in the simulation domain. This is the case when parameters change as a function of humidity, pressure, heat or even time. Another set of problems where it is interesting to include uncertainty is in cases of position uncertainty. One example is the modeling of biological effects due to mobile phones: the distance and orientation of the antenna in respect to the head can be modeled as two or more random variables. Therefore, there are several cases in which the scope of the calculated solution could be improved if random variables are introduced in the computational domain.

This work uses a discrete specification of the probability distribution function to model uncertainty in TLM simulations. The main idea is described in [3, 4] and it is based on the Unscented Transform (UT) [5] and the Moment Design Technique (MDT) [6]. The main scope of the work is the accurate modeling of different probability density functions using the UT-TLM approach.

The combination of the UT and TLM techniques provides accurate and simple forms to represent the effect of normal zero mean random variables. In this work the applicability of UT-TLM is enhanced by the calculation of sigma points and weights for different probability density functions. The accuracy of the approach is demonstrated by a comparison of the results in frequency and time domains. Moreover, the theory of the UT is presented in the light of Gaussian quadrature integration techniques, and a close relationship between interpolating polynomial and the sigma points is presented.

2 Theory

The UT was developed by Julier and Uhlman in 1997 [5]. The idea is to approximate a nonlinear mapping by a set of selected points (sigma points). Both expected value and variance of the mapping are available through a weighted average of the mapped values at the sigma points. The method is quite similar to the MDT [6]. The technique uses the moments of the probability distribution function to develop a selected set of points. In MDT, these points are called design values. In the UT approach they are called sigma points. One possible interpretation of the UT is a discrete approximation of the continuous probability density function $w(\hat{u})$ by a discrete distribution w_i. Therefore, one has two distributions: a continuous and a discrete one, as shown in Fig. 1.

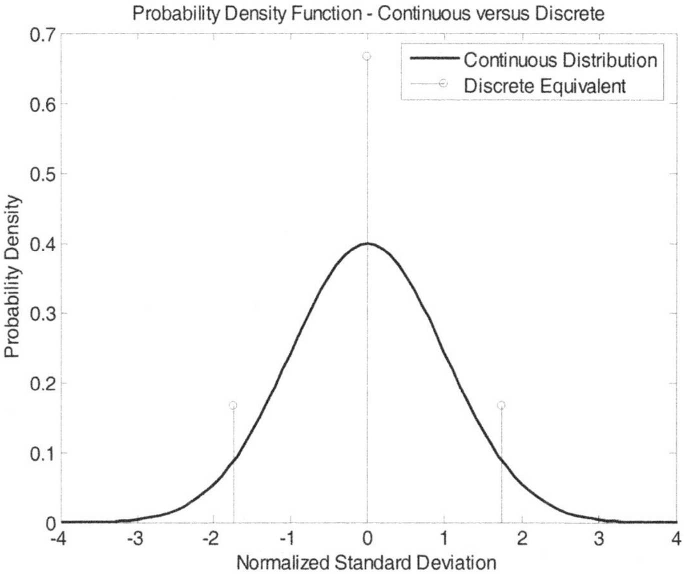

Fig. 1 Representation of the continuous normal distribution and the discrete approximation

The approximation is performed so that the mapping of the two distributions yields the same moments after the nonlinear mapping:

$$E\{\hat{u}^k\} = \int_{-\infty}^{\infty} \hat{u}^k w(\hat{u})d\hat{u} = \sum_i w_i S_i. \quad (1)$$

The weights w_i and sigma points S_i are fully available once the necessary moments are calculated. From (1) and considering a zero mean distribution, the second order sigma points are calculated using the following set of equations:

$$\begin{aligned} E\{\hat{u}\} &= \sum_{i=1}^{2} w_i S_i = 0, & E\{\hat{u}^2\} &= \sum_{i=1}^{2} w_i S_i^2 = \sigma^2, \\ E\{\hat{u}^3\} &= \sum_{i=1}^{2} w_i S_i^3 = \gamma_1 \sigma^3, & E\{\hat{u}^4\} &= \sum_{i=1}^{2} w_i S_i^4 = (\gamma_2 + 3)\sigma^4. \end{aligned} \quad (2)$$

In above, σ^2 is the variance, γ_1 is the skew of the probability distribution and γ_2 is its excess kurtosis. The solution of (2) is:

$$\begin{aligned} S_1 &= \tfrac{1}{2}\left[\gamma_1 - \sqrt{4(\gamma_2+3) - 3\gamma_1^2}\right]\sigma & S_2 &= \tfrac{1}{2}\left[\gamma_1 + \sqrt{4(\gamma_2+3) - 3\gamma_1^2}\right]\sigma \\ w_1 &= \frac{2}{\left[\gamma_1 - \sqrt{4(\gamma_2+3)-3\gamma_1^2}\right]\sqrt{4(\gamma_2+3)-3\gamma_1^2}} & w_2 &= \frac{2}{\left[\gamma_1 + \sqrt{4(\gamma_2+3)-3\gamma_1^2}\right]\sqrt{4(\gamma_2+3)-3\gamma_1^2}} \end{aligned} \quad (3)$$

These results are useful for any zero mean distribution. The calculated UT parameters (sigma points and weights) are used in the computation of the statistical parameters of the final mapped distribution. Therefore, in the case of TLM, the solution of uncertainty propagation is the equivalent to the nonlinear mapping $G(U + \hat{u})$ of a Random Variable (RV).

Once the sigma points are known, it is straightforward to apply them to the non-linear mapping. Using (2) with the results of (4) or (5), it is possible to calculate the expected value $\bar{G}$ and the variance σ_G^2 as follows:

$$\bar{G} = E\{G(\bar{U}+\hat{u})\} = \sum_i w_i G(\bar{U}+S_i), \tag{4}$$

$$\sigma_G^2 = E\left\{ \left[G(\bar{U}+\hat{u}) - \bar{G}\right]^2 \right\} = \sum_i w_i \left[G(\bar{U}+S_i) - \bar{G}\right]^2. \tag{5}$$

The combination of (2) with the results from (4) or (8) also allows the calculation of skew γ_{G1} and kurtosis γ_{G2}:

$$\gamma_{G1}\sigma_G^3 = \sum_i w_i \left[G(\bar{U}+S_i) - \bar{G}\right]^3, \tag{6}$$

$$(\gamma_{G2}+3)\sigma_G^4 = \sum_i w_i \left[G(\bar{U}+S_i) - \bar{G}\right]^4. \tag{7}$$

The moments are correctly calculated after the nonlinear mapping using a small set of sigma points. The interpretation that UT is an approximation for the continuous distribution also means that (4) is the discrete approximation to

$$\sum_i w_i G(\bar{U}+S_i) = \int_{-\infty}^{\infty} G(\hat{u}) w(\hat{u}) d\hat{u}. \tag{8}$$

Equation (8) shows that the UT expression for the mean is an approximation to the integral of the function $G(\hat{u})$ weighted by a window function $w(\hat{u})$. This is actually the formulation for the Gaussian quadrature integration scheme [7]. The advantage looking the UT as an integration scheme is that the weights and sigma points are now easily calculated from the roots of the interpolation polynomial. Naturally, the polynomial is dependent on the window function $w(\hat{u})$. Table 1 summarizes the polynomials and their corresponding probability functions.

Naturally, the expression of UT as an integration scheme suggests that other schemes could be used instead of Gaussian quadrature. However, the optimal

Table 1 Probability density functions and corresponding polynomials

	Probability density function	Polynomial
A:	$w(\hat{u}) = \begin{cases} 0.5 & \lvert\hat{u}\rvert < 1 \\ 0 & \text{otherwise} \end{cases}$ (Uniform)	Legendre
B:	$w(\hat{u}) = \begin{cases} \frac{1}{\pi\sqrt{1-\hat{u}^2}} & \lvert\hat{u}\rvert < 1 \\ 0 & \text{otherwise} \end{cases}$	Chebyshev (1st kind)
C:	$w(\hat{u}) = \begin{cases} e^{-\hat{u}} & \hat{u} > 0 \\ 0 & \text{otherwise} \end{cases}$ (Exponential)	Laguerre
D:	$w(\hat{u}) = \frac{1}{\sqrt{2}} e^{-\frac{\hat{u}^2}{2}}$ (Gaussian)	Hermite

placing of abscissas given by the quadrature provides the best numerical estimate for the integral [7].

The order N of the polynomial has a direct effect on the calculation of higher order moments (6) and (7). The higher moment approximation improves as the order of the polynomial increases. As a rule of thumb, the order of the polynomial results in good estimates of the moments up to that order. A 5th order polynomial will give good estimates of the four main statistical moments (mean, variance, skew and kurtosis [2]). Table 2 shows the weights and sigma points for the different probability density functions. As expected, the weights and sigma points calculated with (3) are exactly the same provided in Table 2 with $N = 3$. The relationships described in Tables 1 and 2 can be summarized with the use of Rodrigues Formula [8] and the weight calculation [7]:

$$w_i = \left(\left. \frac{dp_n}{dx} \right|_{x=x_i} \right)^{-1} \int_{-\infty}^{\infty} w(x) \frac{p_n(x)}{x - x_i} dx, \tag{9}$$

with

$$p_n(x) = \frac{1}{w(x)} \frac{d^n(w(x)Q^n(x))}{dx^n}.$$

In this case x_i will be the zeros of the polynomial. The combination of UT and TLM is straightforward. The moments are calculated considering TLM as a nonlinear mapping function. This is shown considering the basic formulation of TLM with global matrices. This approach provides a more clear representation of

Table 2 Probability density functions (PDF) and corresponding weights and sigma points

PDF	Unscented transform parameters										
	N	Weights					Sigma points				
A	2	0.500	0.500				−0.577	0.577			
	3	0.278	0.444	0.278			−0.775	0	0.775		
	4	0.174	0.326	0.326	0.174		−0.861	−0.340	0.340	0.861	
	5	0.118	0.239	0.284	0.239	0.118	−0.906	−0.538	0	0.538	0.906
B	2	0.500	0.500				−0.707	0.707			
	3	0.333	0.333	0.333			−0.866	0	0.866		
	4	0.250	0.250	0.250	0.250		−0.9238	−0.3827	0.3827	0.9238	
	5	0.200	0.200	0.200	0.200	0.200	−0.951	−0.5878	0	0.5878	0.951
C	2	0.854	0.146				0.586	3.414			
	3	0.711	0.279	0.010			0.416	2.294	6.290		
	4	0.6031	0.3574	0.0389	0.0005		0.323	1.746	4.537	9.395	
	5	0.5218	0.3987	0.0759	0.0036	0.00002	0.264	1.413	3.596	7.086	12.641
D	2	0.5	0.5				−1	1			
	3	0.167	0.666	0.167			−1.732	0	1.732		
	4	0.107	0.337	0.337	0.107		−2.334	−0.742	0.742	2.334	
	5	0.011	0.222	0.5333	0.222	0.011	−2.857	−1.356	0	1.356	2.857

random variables in TLM. The time-domain global formulation represents TLM as the iterated matrix equation:

$$[V^r]_k = [S][V^i]_k, \tag{10}$$
$$[V^i]_{k+1} = [C][V^r]_k. \tag{11}$$

The combination of the equations yields the global state equation

$$[V^i]_{k+1} = [C][S][V^i]_k. \tag{12}$$

The set of random variables $[\hat{u}]$ are part of the description of the problem (boundaries or media). Therefore, they are either represented in the global scattering and/or connection matrices. A compact representation combines all the set $[\hat{u}]$ in a single matrix $[A(\hat{u})]$. Therefore (12) becomes the matrix state equation:

$$[V^i]_{k+1} = [A(\hat{u})][V^i]_k. \tag{13}$$

Equation (13) shows that the state in time step $k+1$ is dependent on the state in time step k. However, it is possible to further simplify this equation. The simplification involves the elimination of the dependency of the previous time step. This is possible only for impulsive excitation, which results in

$$[V^i]_{k+1} = [A(\hat{u})]^{k+1}[V^i]_0. \tag{14}$$

The vector of initial voltages $[V^i]_0$ is not a random variable. Therefore, the mapping $[A(\hat{u})]^{k+1}$ represents the effect of the complete set of random variables. The representation of TLM by (13) and (14) is equivalent, although there are some differences regarding how to represent random variables in each one (especially regarding covariance between variables). The same formulation is possible in frequency domain TLM or eigenvalue problems. This description shows that for all random variables in the problem, the TLM procedure acts as a nonlinear mapping:

$$[V^i]_{k+1} = G(\hat{u}). \tag{15}$$

Equation (15) shows that TLM acts as a nonlinear mapping of a set of random variables. Therefore, TLM results can be used with Eqs. (4), (5), (6) and (7) to obtain the desired statistical parameters.

3 Results

The accuracy of the technique can be measured by comparing with the exact results in a closed form problem. A simple test is the computation of the resonant frequencies of a 2D cavity of 1 by 0.5 m. The cavity is filled with a dielectric of relative permittivity of 2.01. The TLM simulation was performed in the frequency domain. In this experiment the spatial discretization is 1/24 m and the permittivity standard deviation was 0.15. This problem was calculated with different probability density functions. In the setup a TLM simulation was performed for each sigma point and a new calculated frequency was stored. After the simulation of all sigma points, the moments were calculated using (4), (5), (6) and (7) and the frequency values. The results calculated with 5 sigma points are summarized in Table 3.

Table 3 Comparison between exact results and UT-TLM (5 sigma points)

PDF	Mean (MHz)		σ (MHz)		Skew		Kurtosis	
	exact	TLM	exact	TLM	*exact*	*TLM*	*exact*	*TLM*
Gaussian (D)	237.0804	236.7167	8.9509	8.9535	0.3463	0.3470	0.2526	0.2530
Uniform (A)	237.0778	236.7136	8.8947	8.8904	0.1351	0.1354	−1.1764	−1.1761
Exponential (C)	228.5928	228.2341	7.2938	7.2835	−1.537	−1.531	2.9391	2.8601

Naturally, the results with 3 sigma points are less accurate regarding the kurtosis and the skew. This effect can be seen in Table 4. Therefore, it is clear that in this case three sigma points is enough for estimates of the expected value and the variance, but for higher moments it is best to use five sigma points. This observation remains valid for most problems concerning the calculation of resonant frequencies.

Table 4 Comparison between exact results and UT-TLM (3 sigma points)

PDF	Mean (MHz)		σ (MHz)		Skew		Kurtosis	
	exact	TLM	exact	TLM	*exact*	*TLM*	*exact*	*TLM*
Gaussian (D)	237.0804	236.7167	8.9509	8.9515	0.3463	0.3342	0.2526	0.0373
Uniform (A)	237.0778	236.7151	8.8947	8.9051	0.1351	0.1341	−1.1764	−1.1984
Exponential (C)	228.5928	228.2276	7.2938	7.2906	−1.537	−1.560	2.9391	2.6604

The second example is also the calculation of resonant frequencies of a cavity with a floating dielectric. The dielectric was placed in the center of the cavity, which had the same dimensions as the previous case. The width of the dielectric was 0.333 m. and the height was 0.167 m. In this case, there is no closed form solution. Therefore it is necessary to use Monte Carlo to obtain the estimates for the moments. Since Monte Carlo took nearly a hundred thousand repetitions, results were only obtained for the normal distribution. Both skew and kurtosis still had some convergence problems even with such a number of Monte Carlo repetitions. The results are shown in Table 5.

The third example is a set of time domain simulations for different distributions. The time-domain simulation is more challenging than the frequency domain case.

Table 5 Comparison between Monte Carlo and UT-TLM computations (5 sigma points) in the floating dielectric case for different permittivity standard deviations

Permittivity standard deviation	σ		Skew	Kurtosis
	UT-TLM	Monte Carlo	UT-TLM	UT-TLM
0.05	2.4205	2.4160	0.07445	0.011022
0.1	4.8498	4.8404	0.14932	0.044353
0.15	7.2968	7.2822	0.22501	0.10077
0.2	9.7707	9.7501	0.30192	0.18151
0.25	12.2805	12.2534	0.3804	0.28813

The reason is the multiple interactions of the random variable in time. Even for one random variable, this interaction does exist (it couples to itself after a certain amount of time steps). The computational domain was a TEM guide with 1 by 30 cells. The spatial discretization Δl was 1 mm. The guide was filled with air and had two 5 mm dielectrics of average relative permittivity 8 separated by a gap of 10 mm. Both dielectrics had a variance of 0.4. The number of time steps was 500. The input point was at 1 Δl, and the output was at 2 Δl. The sampled field quantity was the electric field, which is proportional to the voltage. The results were compared to the result of 40000 Monte Carlo expected value and standard deviation. Both uniform and normal distributions were simulated. In the first set of simulations the problem was closed with electric boundaries. The result is the increased coupling between electric fields of earlier and later times (due to the random variable). The effect of this is a gradual increasing discrepancy as time marches (both in the expected value and the variance). Figure 2 shows the comparison of the expected value calculated using Monte Carlo and the combination of the UT-TLM. Figure 3 shows the comparison of the standard deviation calculated with Monte Carlo and UT-TLM.

The effect of the multiple interactions is quite visible in Fig. 3. After 100 time steps, the standard deviation tends to deviate from the Monte Carlo calculations. This effect is caused by the interaction of the random variable with itself after some time steps have been calculated. One way to minimize this effect is to increase the order of the UT approximation. This will demand more simulations and the necessary time will increase. This effect is clearer in Fig. 4. In this case, the random variable has a normal zero mean distribution.

The same problem was simulated considering an exponential distribution with standard deviation of 0.4. The results for the calculated standard deviation are shown in Fig. 5.

The comparison between Figs. 3, 4 and 5 shows a similar behavior for the initial time steps of the standard deviation (although with different magnitudes). The curves start to differ only after a few time steps.

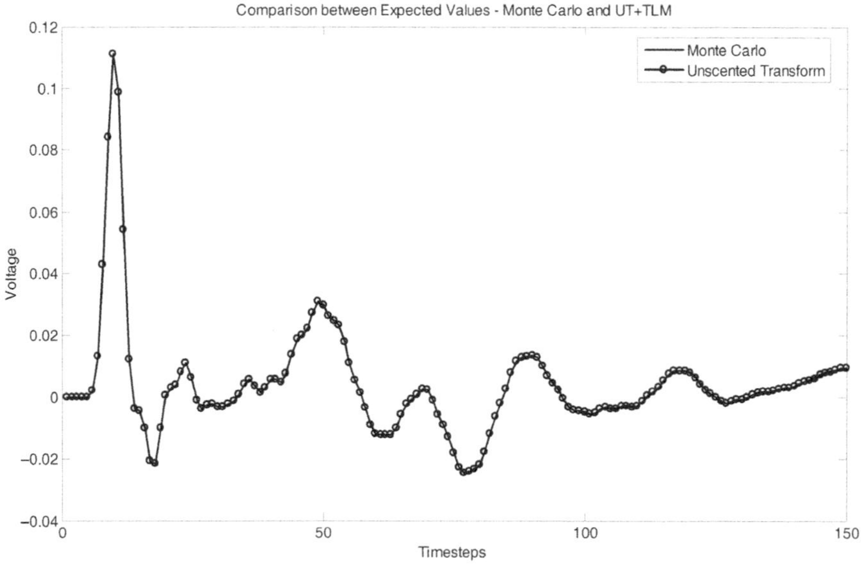

Fig. 2 Expected value of voltage at the output node calculated with Monte Carlo and the Unscented Transform (3 sigma points). The random variable has a standard deviation of 0.4 and a zero mean uniform distribution

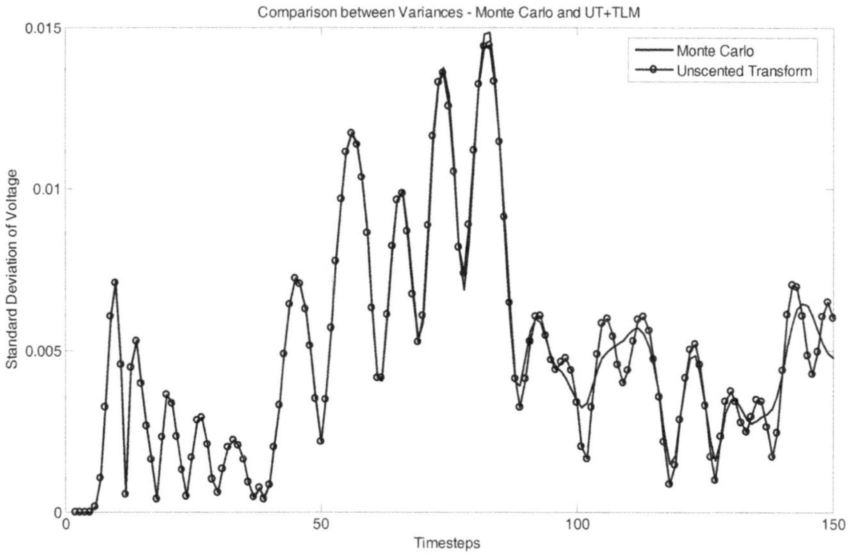

Fig. 3 Standard deviation of voltage at the output calculated with Monte Carlo and the Unscented Transform (3 sigma points). The random variable has a standard deviation of 0.4 and a zero mean uniform distribution

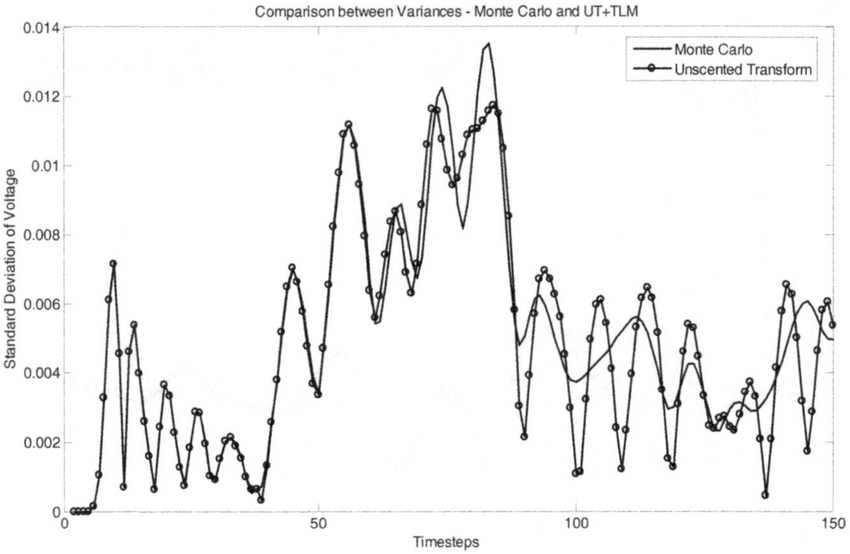

Fig. 4 Standard deviation of voltage at the output calculated with Monte Carlo and the Unscented Transform (3 sigma points). The random variable has a standard deviation of 0.4 and a zero mean normal distribution

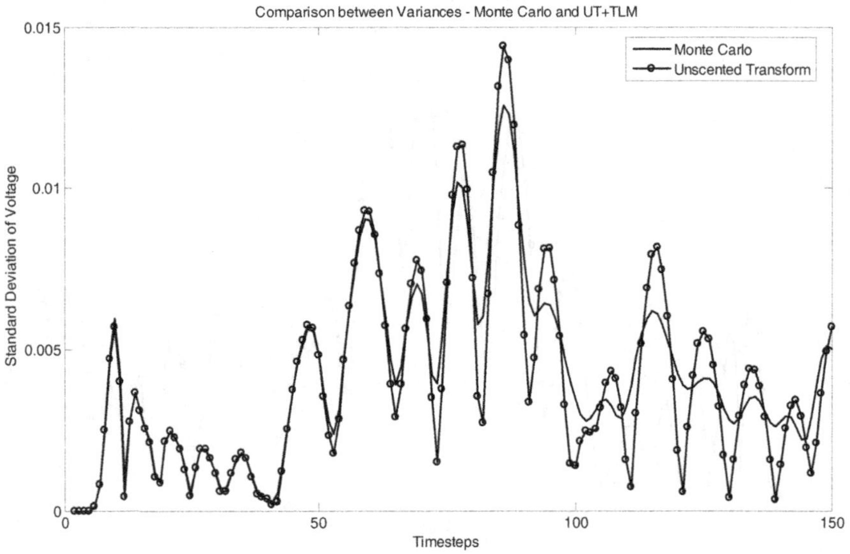

Fig. 5 Standard deviation of voltage at the output calculated with Monte Carlo and the Unscented Transform (3 sigma points). The random variable has a standard deviation of 0.4 and an exponential distribution

4 Conclusions

This paper presented the combination of the UT and TLM to model uncertainty of different probability density functions with electromagnetic solvers. Although the results were calculated using a TLM simulator, the approach is readily applicable to FDTD, FEM and MoM. The application of the procedure does not require any changes to the solver. It is only used as a pre-processing and post-processing stage. In TLM, the combination with the UT shows accurate results both in time and frequency domains. The simulation of the full sigma point set is not only faster than the Monte Carlo approach, but also it uses much less computational requirements. The calculation of sigma points and weights is similar to a Gaussian quadrature procedure, and so this work presented new forms to calculate the necessary parameters to the desired accuracy. The UT-TLM was used with different distributions and showed good agreement with the analytical and Monte Carlo-based results.

References

1. Itoh T. *Numerical Techniques for Microwave and Millimeter-Wave Passive Structures.* John Wiley, New York, 1989.
2. Papoulis A. *Probability, Random Variables, and Stochastic Processes.* McGraw-Hill, New York, 1991.
3. de Menezes L.R.A.X., Ajayi A., Christopoulos C., Sewell P., and Borges G.A. Efficient computation of stochastic electromagnetic problems using unscented transforms. Submitted to IET Science, Measurement & Technology on May 23th, 2007.
4. de Menezes L.R.A.X. and Borges G.A. Uncertainty propagation in transmission line modeling (tlm) method. In *Frontiers on Applied Computational Electromagnetics. Face 2006 Digest*, volume 1, pages 13–16, 2006.
5. Julier S.J. and Uhlmann J.K. Unscented filtering and nonlinear estimation. *Proceedings of the IEEE*, 92(3):401–422, March 2004.
6. Zhang J. The calculating formulae, and experimental methods in error propagation analysis. *IEEE Trans. om Reliability*, 55(2):169–181, June 2006.
7. Eric W.W. Gaussian quadrature. From MathWorld–A Wolfram Web Resource: `http://mathworld.wolfram.com/GaussianQuadrature.html`, last checked July 24, 2007.
8. Eric W.W. Rodrigues representation. From MathWorld–A Wolfram Web Resource: `http://mathworld.wolfram.com/RodriguesRepresentation.html`, last checked July 24, 2007.

Bandwidth Optimization using Transmission Line Matrix Modeling and System Identification

Nikolaus Fichtner, Uwe Siart, Yury Kuznetsov, Andrey Baev and Peter Russer

The optimization of electromagnetic structures demands for a fast and efficient field solver in combination with a flexible and stable optimization method. In this work we present the time domain based transmission line matrix (TLM) method in conjunction with the genetic algorithm (GA) for the design optimization. The proposed technique is advantageous compared to conventional frequency domain methods when several frequency points are considered. The efficiency and flexibility of the combined TLM–GA optimization method is shown on the example of a low profile planar patch antenna. A further reduction of the total simulation and optimization time is achieved by using system identification (SI) procedures. Employing a SI the total TLM simulation time can be reduced. A new criterion based on the stability of the physical poles is introduced which allows to truncate the time response without loss of information. The SI procedure determines the pole position in the complex plane and antenna structures can be represented both by means of lumped element networks and distributed parameter models.

Nikolaus Fichtner
Institute for High-Frequency Engineering, Technische Universität München, Arcisstr. 21, 80333 München, Germany, e-mail: fichtner@tum.de

Uwe Siart
Institute for High-Frequency Engineering, Technische Universität München, Arcisstr. 21, 80333 München, Germany, e-mail: uwe.siart@tum.de

Yury Kuznetsov
Moscow Aviation Institute (State Technical University) Volokolamskoe sh. 4, A-80, GSP-3, Moscow, 125993, Russia, e-mail: kuznetsov@mai-trt.ru

Andrey Baev
Moscow Aviation Institute (State Technical University) Volokolamskoe sh. 4, A-80, GSP-3, Moscow, 125993, Russia, e-mail: baev@mai-trt.ru

Peter Russer
Institute for High-Frequency Engineering, Technische Universität München, Arcisstr. 21, 80333 München, Germany, e-mail: russer@tum.de

P. Russer, U. Siart (eds.), *Time Domain Methods in Electrodynamics*,

1 Introduction

In this article we present a optimization technique based on the time domain transmission line matrix (TLM) method in conjunction with the genetic algorithm (GA) and a system identification (SI) procedure. Design optimization is an important issue in the development of new systems and services and demand for a flexible and efficient optimization routine. Especially the need for faster data transmission and the combination of several services requires increased operation bandwidths of already established structures. Time domain (TD) based methods are advantageous in this case as they can provide a broad frequency information in a single simulation run.

The TLM method introduced by P.B. Johns [1] in the early 1970s is based on the analogy between the electromagnetic field and a mesh of transmission lines and is for that reason a descendant of the seminal contribution of G. Kron [2]. TLM can be considered as a space and time discrete analog of the Huygens' principle of wave propagation. A mesh of transmission lines which are connected at scattering junctions is used to model the dynamics of the electromagnetic field propagation. In the 2D case modes of TE and TM type can be modeled by series and parallel connection of the transmission lines. A extension is this principle for the 3D case is known as expanded node [3]. However, the expanded mode is difficult to implement and the electric and magnetic field components are separated by half the unit cell size which complicates the application of boundary conditions. The symmetrical condensed node (SCN) again introduced by P.B. Johns [4] in 1987 resolved the aforementioned problems. The SCN is easy to implement and exhibits excellent stability properties. Also of numerous offspring's, such as the hybrid symmetrical condensed node and the symmetrical supercondensed node [5] the SCN is probably the most widely used TLM unit cell. In the present work the open source TLM simulation package YATPAC will be used [6] which uses the SCN node as well.

The genetic algorithms (GA) emerged in the early 1990s as a new global stochastic optimization method applicable in many electromagnetic design problems [7, 8]. As the optimization of antenna structures includes a large number of degrees of freedom accompanied by severe design constraints the GA procedure turns out to be suitable. The GA routine can be paraphrased by a natural selection and evolution process of the problem domain (electromagnetic structure plus initial and boundary conditions), which is mapped on a string of bits, the so–called genes. Several genes constitute a chromosome, representing the particular designs. For every chromosome a fitness value (also known as cost function) is computed which is a measure for the performance and capability of the structure under investigation. Based on the respective fitness values the chromosomes are modified which results in new designs. Iterative repetition of the outlined steps leads to a stochastic search in a definite function domain that can be described as survival of the fittest. The GA optimization procedure itself is decoupled from the considered electromagnetic structure or the optimization goal only acting on the genes and chromosomes of the coded problem domain. Thus the GA turns out to be extremely flexible and is used in a large area of electromagnetic design optimizations [7].

GA's have been used in the past in conjunction with the method of moments (MoM) for the optimization of antenna structures with high success [9, 10]. In this article the impedance bandwidth of planar antenna structures are considered. For the analysis of the matching in the desired frequency range time domain (TD) methods are preferable because the frequency response and thus the matching behavior of the antenna structure is obtained in a single simulation run.

The application of CAD tools for the numerical treatment of electromagnetic problems demands (usually) for large computer resources as well as for a correct interpretation of the simulation results. A meaningful interpretation must therefore be independent of the used input signals or the used numerical method. One of the universal methods for describing general multiport structures is by means of the scattering matrix representation. The coefficients of the scattering matrix describe the reflection respectively transmission of incident and scattered waves at all ports of the microwave circuit. By defining a characteristic impedance the scattering matrix representation can be transformed into impedance, admittance or ABCD representation [11].

A analytic expression for the impulse response or its Fourier transform the transfer function at the ports of a microwave circuit can be found from numerical simulations. System identification (SI) in that sense describes the estimation of the model parameters through observation of the system response exited by some known input signal. The SI procedure may be used in parallel to the numerical simulation of the microwave structure. By using a time domain based CAD tool a simultaneous estimation of the model parameters may lead to a considerable reduction of the total simulation time. If the recorded time response at any port of the circuit contains sufficient information in a sense that the estimated model is complete, the further time response is given by the model parameters and the simulation can be stopped.

Passive and linear distributed circuits can be described by a equivalent circuit consisting of lumped elements and transformers. Such circuits are fully described by a system of linear ordinary differential equations (ODE) with constant coefficients. The ODE system is adequate if the behavior of the electromagnetic field inside the structure can be characterized by quasi electrostatic or quasi magnetostatic fields similar to the case of lumped elements of the low frequency equivalent circuits. The visualization of the electromagnetic fields obtained from a full wave time domain simulation of a microwave structure highlights the presence of positions with stable electric and magnetic fields concentrations.

A different behavior is expected for the early time part of the electromagnetic field inside the microwave circuit. This part will be dominated by multiple reflections and interferences as the signals propagates through the structure. Contrary to the quasi static parts of the electromagnetic fields the reflections are mainly due to a direct interaction of the electric and magnetic fields with the structure, where the early time part consists of much higher frequencies.

In practice two different processes during the interaction of a incident electromagnetic signal with the microwave circuit are observed. One process is characterized by a long signal response composed of several damped oscillations with stable frequencies while the other process has relatively short duration with nearly random and unstable fluctuations. This behavior of the output signals corresponds to a model first presented by C.E. Baum in 1976 and is known as the singularity expansion method (SEM) [12]. Later on L.B. Felsen explained the nature of the early time part of the scattered ultra wide band (UWB) electromagnetic signal by "the hierarchy of reflected and diffracted fields of the geometrical theory of diffraction (GTD)" [13].

The interaction of the incident and scattered waves at different ports respectively at the same port of a multiport microwave circuit may be characterized by a model consisting of two parts: The first part consists of a *pole model*. The Laplace transform of a linear ODE with zero initial conditions (no previously impressed energy inside the microwave circuit) can be written as a rational function with poles and zeros. The second part consist of the *entire* or *non–pole model*. The time border between this two parts of the impulse response can be found empirically or directly from the numerical simulation. In practice the a priori prediction of the duration of the early time part is challenging and only possible for a small number of simple structures. At this point we want to emphasize the importance of a accurate late time border estimation because of the possible decaying of some exponential terms in the vicinity of the late time border.

Another important issue is the correct choice of the pole model order P which gives the number of physical poles. The conventional choice of the pole model order is based on the relative mean square error (RMSE) criterion. The RMSE is a measure of the modeled impulse response approximating with respect to the simulated impulse response. The application of the RMSE criterion may lead to excessive number of poles whereas only a fraction of them represents the physical nature of the structure and the remaining poles (unphysical poles) only improve the approximation and are numerical artifacts.

In this work we will present a new *stability criterion* for the late time border computation and the choice of the pole model order. The criterion is based on the fact that the pole positions can be found from only a impulse response segment with finite duration. A variation of the segment position inside the late time part of the impulse response results in stable positions of the estimated poles if the order of the model corresponds to the number of the physical poles [14]. The realization of the stability criterion is based on the comparison of the different sets of poles with some predefined reference set of poles. The difference between the estimated pole sets is obtained from a signature comparison method [15] which maps each pole set into a single point in a multidimensional vector space and determines the distances between different pole sets.

2 A combined TLM-GA Optimization Technique

The performance of a combined TLM and GA optimization routine is demonstrated on a planar patch antenna. The patch antenna is a low profile antenna with a rather narrow bandwidth. The relative bandwidth $\Delta f/f_0$ of patch antennas lies typically

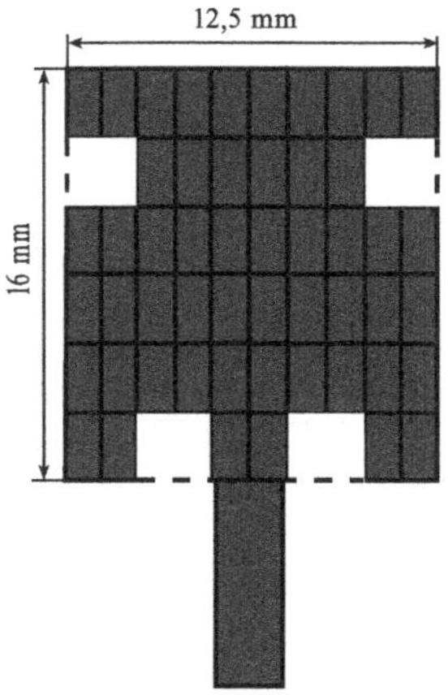

(a)

$$\begin{bmatrix} 1&1&1&1&1&1&1&1&1&1 \\ 0&0&1&1&1&1&1&1&0&0 \\ 1&1&1&1&1&1&1&1&1&1 \\ 1&1&1&1&1&1&1&1&1&1 \\ 1&1&1&1&1&1&1&1&1&1 \\ 1&1&0&0&1&1&0&0&1&1 \end{bmatrix}$$

(b)

Fig. 1 (**a**) Top metallization of the initial patch antenna structure (*dashed line*) with discretized patch elements. Removed patch elements are indicated by white areas. (**b**) Corresponding chromosome set

in the range of a few percent [16]. The initial patch antenna, shown in Fig. 1a, has a width of 12.5 mm and a length of 16 mm. The permittivity of the antenna substrate is $\varepsilon_r = 2.2$ and the substrate height is 1 mm. The antenna is feed by a microstrip line with 50 Ω impedance. The antenna patch dimensions where chosen such, that the first resonance for the TM_{010} mode is at 6 GHz and the second resonance for the TM_{001} mode is at 12 GHz (for the patch to be in the xy-plane). The low substrate height suppresses the excitation of undesired surface waves and the low substrate permittivity provides improved efficiency and larger bandwidth [16].

In several previously published approaches the enhancement of the matching bandwidth by removing appropriate elements of the antenna patch is demonstrated [9, 10]. This procedure is advantageous compared to a modification of the substrate height and/or permittivity, because the 50 Ω feed impedance is maintained and the modified structure can easily be manufactured. The first step in any GA optimization routine is the coding (or mapping) of the given antenna design into a string of genes. For this purpose the antenna patch is discretized into small rectangular elements, as shown in Fig. 1a. A straight forward mapping scheme is introduced. To every patch element a single bit is assigned: "1" if the element is present and "0" if the element is absent, i.e., the metallization is removed. This procedure is illustrated in Fig. 1. Several arbitrary chosen patch elements are removed from the initial patch in Fig. 1a, indicated by the dashed line, and the corresponding chromosome is shown in Fig. 1b.

Thus the complete design of the top patch needed for the TLM simulation is coded into a single chromosome. Parameters such as the substrate height and ε_r remain constant and need not be coded in the chromosome because these parameters aren't affected by the GA and are predefined in the TLM input file.

The next step in the GA optimization run is the generation of several different chromosomes. This can be achieved through a random bit generator or manually. Randomly generated chromosomes however can lead to immoderate long optimization times with low convergence. Hence the manual creation of several

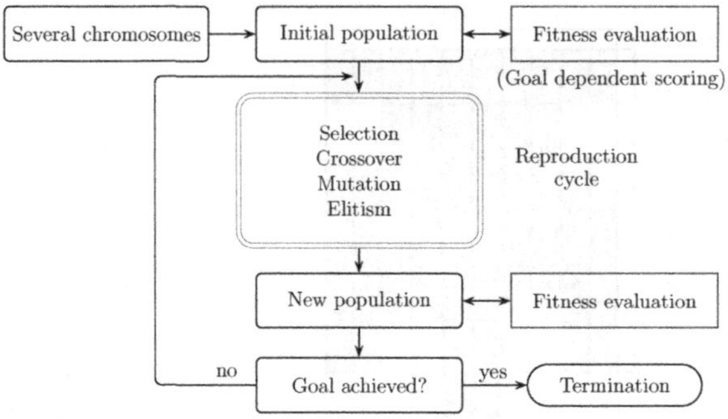

Fig. 2 Block diagram of the implemented genetic algorithm iteration scheme [7]

chromosomes – respectively antenna designs – to represent a reasonable initial guess is preferable. The generated chromosomes constitute the initial population, cp. Fig. 2. The number of members in the population is a crucial parameter in any GA optimization. The more chromosomes are used the longer a single iteration takes but the better is the convergence to the specified goal. A direct correlation of the population number to the convergence rate hasn't been presented up to now [7]. With a number of 10 members in a single population an acceptable convergence rate was achieved. As can be seen from the GA iteration scheme in Fig. 2 the next task is the fitness evaluation. A fitness value is assigned to every member of the initial population. The return loss is a suitable quantity for the fitness evaluation. Frequency domain based methods as e. g. MoM become impractical for problems where the return loss at several frequencies is needed. To give a quantitative example, we setup a TLM simulation in YATSIM of the initial patch antenna structure with a total of 240.000 TLM unit cells and run the simulation for 10.000 time steps. The simulation was performed on a P4 (Dual core), 2.8 GHz CPU with 2 GB RAM. The total TLM simulation took 250 s. For comparison a MoM simulation with the Matlab antenna toolbox [17] performed on the same machine and the same space discretization took approximately 60 s. However, the TLM simulation with a Fast Fourier Transformation (FFT) of the incoming and reflected time signals at the antenna port provides full information of the return loss over all frequencies of interest (here 2–15 GHz). Compared to the respective MoM simulation the TLM simulation enables a faster GA optimization as soon as more than 5 distinct frequencies are considered.

In the fitness evaluation the return loss $a_\mathrm{L} = 20\log|S_{11}|$ is sampled at several frequencies f_m and then to each return loss value a penalty value j is assigned. The penalty values j and the corresponding return loss ranges are given in Table 1. The total fitness is then obtained from the total penalty value

$$J = \sum_{m=1}^{M} j\{a_L\}. \tag{1}$$

Table 1 Penalty lookup table for fitness evaluation

Return loss					Penalty value j
$-4\,\text{dB}$	$<$	a_L	$\leq$	$0\,\text{dB}$	2000
$-8\,\text{dB}$	$<$	a_L	$\leq$	$-4\,\text{dB}$	400
$-10\,\text{dB}$	$<$	a_L	$\leq$	$-8\,\text{dB}$	100
$-12\,\text{dB}$	$<$	a_L	$\leq$	$-10\,\text{dB}$	10
$-15\,\text{dB}$	$<$	a_L	$\leq$	$-12\,\text{dB}$	5
$-20\,\text{dB}$	$<$	a_L	$\leq$	$-15\,\text{dB}$	2
		a_L	$\leq$	$-20\,\text{dB}$	0

The values of j were chosen in such a manner that several sampled return loss values close to the $-10\,\text{dB}$ line yield a better fitness (a lower total penalty value J) than one deep but narrow notch. Hence the optimization acts on the whole specified frequency range and so the penalty value J provides a measure for the antenna performance in the considered frequency band represented by a single number.

With the information about the performance of every chromosome the reproduction cycle is started. First suitable parent chromosomes are selected. Different strategies exist for the selection. We used the proportionate selection (also known as roulette wheel selection), which is the most commonly used in GA optimizations [7]. There the individuals are selected with a probability given by

$$p_l = \frac{\sum_i J_i - J_l}{\sum_i J_i}, \tag{2}$$

where J_l is the total penalty value of the considered individual. In the crossover operation the selected parents are mate together with a probability p_{cross}. At a randomly chosen position in the chromosome the chromosome is split into 2 parts. Now a new child chromosome is build from the first part of parent 1 and the second part of parent 2. A second child chromosome is obtained if the order of parts is exchanged. The objective of the crossover operation is to combine the best attributes of the parents to the new generated child. The mutation operation considers every gene of a child chromosome and changes its value for a given probability p_{mut}. Typical values for p_{mut} are between 0.01 and 0.1 [9]. This step is necessary to introduce random mutation of the structure and to avoid the algorithm to get stuck in local minima of the total penalty value. The elitism operation simply keeps the best individual of a population unchanged and transfers it to the next generation. Consequently the GA guarantees convergence because the best individual of a new generated population is at least equal to the best individual of its preceding population or even better. The described steps of the reproduction cycle are repeated until a complete new population is generated. Again the fitness value for every member of the new population is computed from a TLM simulation with FFT and equation (1). Now a goal for the optimization must be defined. For the presented bandwidth optimization the goal respectively termination criteria is set to a minimum total penalty value J in the considered frequency spectrum. However to avoid unnecessary long optimization times a maximum number of iterations should be defined.

3 The Description of a Linear Microwave Circuit

Any computer–aided design (CAD) tool for modeling and optimization of electromagnetic (EM) structures is based on one of the numerous numerical method for the computation of the approximate field solution (in some cases more then one methods are used in parallel which are then called hybrid methods). If the transmission–line matrix (TLM) method is used for the modeling of general three–dimensional structures such as antennas or microwave filters, the field solutions are obtained in time domain (TD). Because any linear EM structure can be represented as a linear multiport, every port can be described by a set of incident waves $a_{ij}(t)$ and scattered waves $b_{ij}(t)$ in the TD.

Without loss of generality we will from now on only consider one–port devices, so $a_{ij}(t)$ and $b_{ij}(t)$ reduce to $a(t)$ and $b(t)$. An example of a microwave one–port is shown in Fig. 3. In this case the voltage and current relations are fully described by the microwave circuit. In the numerical treatment of EM structures the field or circuit quantities are only given at a finite set of samples. In the case of TLM we obtain for a constant sampling time T a ordered sequence, i.e. $a[n] = a(nT)$, of wave pulses. By defining a characteristic impedance Z_0 we can express the incident and scattered waves in terms of the generalized port voltage and current and vice versa. Using the convention for the current and voltage orientation as indicated in Fig. 3 we get [11]

$$a[n] = \frac{1}{2\sqrt{Z_0}}\left(v[n] + Z_0 i[n]\right), \quad b[n] = \frac{1}{2\sqrt{Z_0}}\left(v[n] - Z_0 i[n]\right). \tag{3}$$

Each arbitrary time discrete sequence of samples $x[n]$ has a time discrete Fourier transform (DTFT) defined by

$$X(f) = \sum_{n=0}^{\infty} x[n] \cdot e^{-j2\pi f n T}, \tag{4}$$

where the frequency f is continuous and the sequence $x[n]$ is assumed to be causal. In the following text $x[n]$ and its transforms are arbitrary functions and are replaced by $a[n]$, $b[n]$ and $h[n]$ if required. In every numerical analysis only a finite number of samples is considered. A truncated sequence of samples in TD is given by $x_N[n]$ with N elements. A discrete and truncated spectrum of a time discrete sequence $x_N[n]$ is obtained from the discrete Fourier transform (DFT)

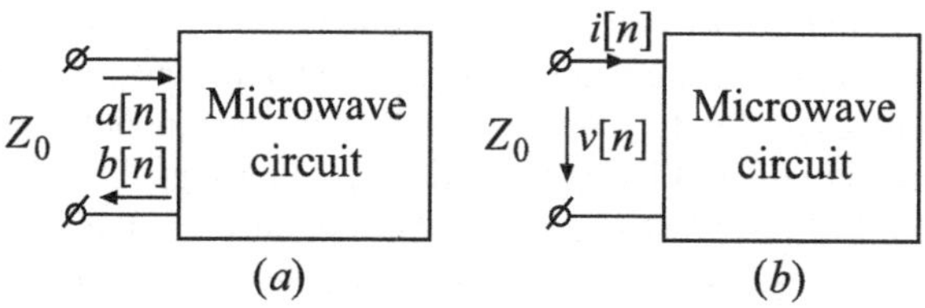

Fig. 3 Microwave one–port in (**a**) scattering and (**b**) impedance representation

$$X_N[m] = \sum_{n=0}^{N-1} x_N[n] \cdot W_N^{nm}, \tag{5}$$

with $W_N^{nm} = \exp(-j2\pi nm/N)$. The relation between the incident and a scattered waves in TD is characterized by the impulse response $h[n]$ of the microwave circuit corresponding to the chosen port. In general the scattered wave is given by

$$b[n] = a[n] * h[n] = \sum_{k=0}^{\infty} h[k] \cdot a[n-k], \tag{6}$$

where "$*$" denotes the discrete convolution for causal and arbitrary long sequences. If the duration of the output sequence is limited by N elements than a periodic discrete convolution can be used

$$b_N[n] = a_N[n] \otimes h_N[n] = \sum_{k=0}^{N-1} h_N[k] \cdot a_N[n-k]. \tag{7}$$

Here all three quantities are assumed to be periodically continued outside the observation interval $n \in [0, N-1]$. In this case the truncated quantities are related to the infinite sequences as follows:

$$x[n] = \sum_{l=-\infty}^{\infty} x_N[n + l\,N] \tag{8}$$

where l is an integer number. The relation between the incident and scattered waves in frequency domain is given by the transfer function $H(f)$ which relates the spectra of the scattered waves $B(f)$ and the incident waves $A(f)$ obtained from the DTFT (4)

$$H(f) = \frac{B(f)}{A(f)}. \tag{9}$$

In case of truncated sequences we have the corresponding transfer function $H_N[m]$

$$H_N[m] = \frac{B_N[m]}{A_N[m]}. \tag{10}$$

Another useful representation of the transfer function is given by the z-transform of the time domain sequences $a[n]$ and $b[n]$

$$\tilde{A}(z) = \sum_{n=0}^{\infty} a[n] \cdot z^{-n}, \qquad \tilde{B}(z) = \sum_{n=0}^{\infty} b[n] \cdot z^{-n}, \tag{11}$$

and hence

$$\tilde{H}(z) = \frac{\tilde{B}(z)}{\tilde{A}(z)}. \tag{12}$$

with $z = e^{sT}$. An alternative description of the microwave circuit with respect to the chosen port is the port impedance which can be defined in the frequency domain by any of the above mentioned transforms. In the continuous frequency domain we thus have

$$Z(f) = \frac{V(f)}{I(f)}, \tag{13}$$

where $V(f)$ and $I(f)$ are the DTFT's of the corresponding generalized voltage and current sequences $v[n]$ and $i[n]$. The transfer function $H(f)$ and the port impedance $Z(f)$ are related to each other by

$$Z(f) = Z_0 \frac{1 + H(f)}{1 - H(f)}. \tag{14}$$

and

$$H(f) = \frac{Z(f) - Z_0}{Z(f) + Z_0}, \tag{15}$$

The transfer function $H(f)$ and its time domain analogue the impulse response $h[n]$ can be used to describe the scattering properties of the electromagnetic structure at any port of a multiport circuit. For that reason the first step of any system identification procedure is the extraction of the impulse response $h[n]$ or the transfer function $H(f)$ of the microwave circuit from the observed quantities.

4 Transfer Function and Impulse Response of the Microwave Circuit

The estimation of the transfer function or the impulse response can be implemented either in frequency or in time domain. In general the transfer function can be computed from (9). In case of time and frequency discrete quantities of finite length we need to apply Eq. (10). The number N of samples in the time domain is equal to the number of samples in the frequency domain with a frequency spacing given by $f_s = 1/T$ where T is the TLM time step. This time step is given from the TLM discretization length as $T = \Delta l / 2c_0$ where the length Δl is chosen such, that the condition $\Delta l < \lambda_{\text{min}}/10$ is fulfilled. With λ_{min} the smallest wavelength of interest is denoted in the TLM simulation domain. All wavelengths with $\lambda < \lambda_{\text{min}}$ have a considerable dispersion error and should not be considered in any analysis. Furthermore the TLM discretization length must be small enough to model all details of the electromagnetic structure accurately. In all practical cases considered so far the TLM cutoff frequency $f_s/2$ is much larger than the Nyquist frequency given by the bandwidth of the used input signal $A(f)$. Therefore the TLM signals $a_N[n]$ and $b_N[n]$ are highly oversampled. The oversampling in frequency domain results in a large number of samples with zero contribution in the spectrum $A_N[m]$. From Eq. (10) we see that the transfer function $H_N[m]$ exhibits singularities in that case.

To overcome the singularities we can exclude the samples of $A_N[m]$ which are close enough to zero from the division in (10). This can be done in several ways. The most simple and straight forward is a decimation procedure. The implementation of this technique needs the choice of a decimation number L. This number denotes the cutoff frequency of the incident wave spectrum. The frequency bandwidth of the incident wave is defined by the shape of the input excitation. The typical choice for the excitation signal is a Gaussian pulse $a_N[n]$, shown in Fig. 4. The Gaussian pulse in the continuous time domain has the form

$$a(t) = A_0 \cdot \exp\left\{-\pi\left(\frac{t-T_G}{\tau}\right)^2\right\}. \tag{16}$$

The spectrum of the Gaussian pulse in the continuous frequency domain has again a Gaussian form

$$A(f) = A_0\tau \cdot \exp\left(-\pi f^2\tau^2\right) \cdot \exp(-j2\pi f T_G). \tag{17}$$

The discretization of the input wave results in a truncated pulse in the discrete time domain

$$a_N[n] = A_0 \cdot \exp\left[-\pi\left(\frac{n-N_G}{G}\right)^2\right] \tag{18}$$

with $N_G = T_G/T$, $G = \tau/T$ and $N \gg N_G$, G. The spectrum of the input wave in discrete frequency domain is given by

$$A_N[m] = \begin{cases} \frac{A_0}{G} \cdot e^{-\pi\left(\frac{mG}{N}\right)^2} \cdot e^{-j2\pi\frac{mN_G}{N}} & \text{for} \quad m = 0, 1, \ldots, \frac{N}{2}-1 \\ \frac{A_0}{G} \cdot e^{-\pi\left(\frac{N-m}{N}\cdot G\right)^2} \cdot e^{-j2\pi\frac{N-m}{N}N_G} & \text{for} \quad m = \frac{N}{2}, \ldots, N-1 \end{cases} \tag{19}$$

An example of the excitation pulse and its spectrum is shown in Fig. 4. The choice of the decimation number L can be defined by the ratio of the sampling rate f_s and the cutoff frequency of the excitation wave f_c

$$L = \left\lceil \frac{f_s}{f_c} \right\rceil = \frac{N}{M}, \tag{20}$$

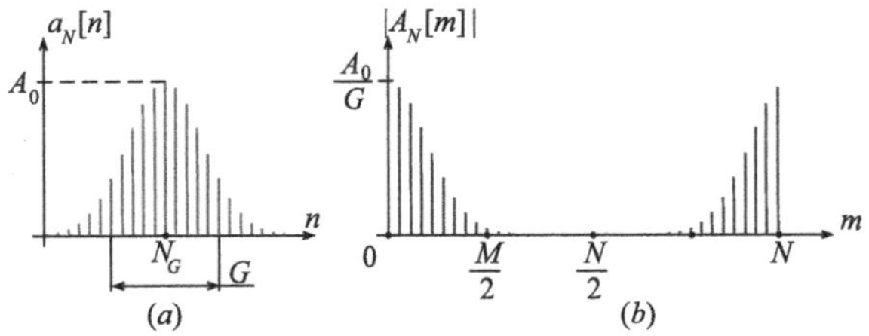

Fig. 4 Gaussian signal in (**a**) in discrete time domain and (**b**) in discrete frequency domain

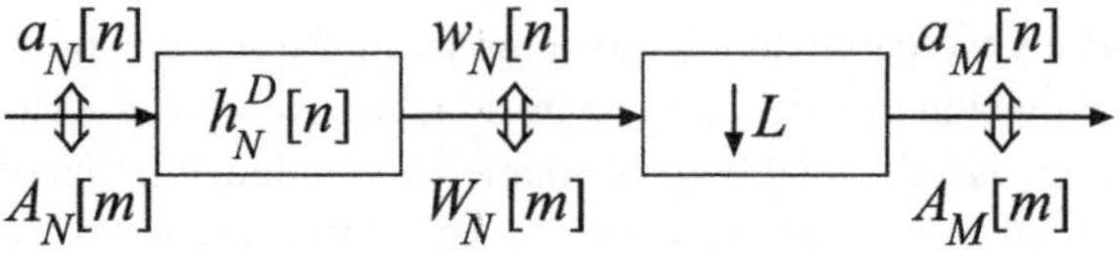

Fig. 5 Block diagram for the decimation procedure

where $\lceil u \rceil$ denotes the smallest integer greater than or equal to u. The decimated excitation wave $a_M[n]$ can be obtained from the initial input signal by two the succeeding steps as outlined in Fig. 5.

The first step is a digital lowpass filtering which prevents aliasing effects throughout the decimation process. The frequency characteristic of the ideal lowpass filter has the form

$$H_N^D[m] = \begin{cases} 1 & \text{for} \quad m = 0,\, 1,\, \ldots,\, \frac{M}{2} \quad \text{and} \quad m = N - \frac{M}{2},\, \ldots, N-1 \\ 0 & \text{for} \quad m = \frac{M}{2}, \frac{M}{2}+1,\, \ldots,\, N - \frac{M}{2} - 1 \end{cases} \tag{21}$$

with M is assumed to be an even integer. The corresponding spectrum is obtained from the inverse discrete Fourier transform (IDFT) of $H_N^D[m]$ by

$$h_N^D[n] = \frac{1}{N} \sum_{m=0}^{N-1} H_N^D[m] \cdot W_N^{-nm} \tag{22}$$

The spectrum $H_N^D[m]$ and impulse response $h_N^D[n]$ of the ideal lowpass filter are shown in Fig. 6a.

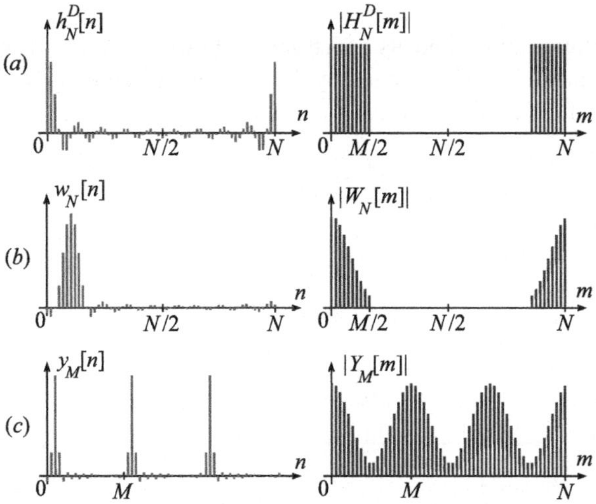

Fig. 6 Decimation procedure for Gaussian excitation pulse. Characteristics of ideal lowpass filter (**a**), filtered signal (**b**) and decimated signal (**c**)

The filtered output signal $w_N[n]$ is obtained from a periodic convolution by

$$w_N[n] = a_N[n] \otimes h_N^D[n] = \sum_{k=0}^{N-1} h_N^D[k] \cdot a_N[n-k], \tag{23}$$

where the spectrum $W_N[m]$ is given by the DFT of $w_N[n]$ or by the product

$$W_N[m] = A_N[m] \cdot H_N^D[m]. \tag{24}$$

The filtered output signal $w_N[n]$ and its spectrum $W_N[m]$ are shown in Fig. 6b. Finally the decimated signal $y_M[n]$ is

$$y_M[n] = w_{N/L}\left[\frac{N}{M} \cdot n\right] = w_M[L \cdot n] \tag{25}$$

as denoted by the second step in the decimation scheme in Fig. 5. The decimated signal $y_M[n]$ and its spectrum $Y_M[m]$ are shown in Fig. 6c. The same decimation procedure (23) and (25) is also applied to the output waveform $b_N[n]$ of the multiport by

$$v_N[n] = b_N[n] \otimes h_N^D[n] = \sum_{k=0}^{N-1} h_N^D[k] \cdot b_N[n-k], \tag{26}$$

$$x_M[n] = v_{N/L}\left[\frac{M}{N} \cdot n\right] = v_M[L \cdot n] \tag{27}$$

and finally Eq. (10) can be written in the decimated form

$$H_M[m] = \frac{X_M[m]}{Y_M[m]}. \tag{28}$$

The inverse discrete Fourier transform (IDFT) of Eq. (28) gives the decimated impulse response of the microwave circuit

$$h_M[n] = \frac{1}{M} \sum_{m=0}^{M-1} H_M[m] \cdot W_M^{mn}. \tag{29}$$

Besides the Gaussian pulse (16) other excitation signals are possible. In many problems only a specific frequency range with minimal frequency larger than zero is desired. The typical excitation form for this problem is a modulated Gaussian pulse with center frequency f_0 as given by

$$a(t) = A_0 \cdot \exp\left\{-\pi\left(\frac{t-T_G}{\tau}\right)^2\right\} \cdot \cos\left[2\pi f_0(t-T_G)\right]. \tag{30}$$

Another example for a excitation signal is a superposition of the decaying exponentials with complex frequencies $s_i = \sigma_i + j2\pi f_i$ placed inside the desired frequency band of the microwave circuit

$$a(t) = \Re\left\{\sum_i A_i \cdot e^{-s_i t} \cdot u(t)\right\}, \tag{31}$$

where $u(t)$ is a Heaviside step function. Such a excitation can easily be transformed into the Laplace domain which is helpful for the extraction of the transfer function [18].

All this additional types of excitation functions have in common the existence of some frequencies with a zero spectral components inside the decimated region. It is obvious that the existence of the nearly zero quantities of the spectrum will result in singularities of the transfer function and will cause numerical instabilities in (28).

To overcome the numerical instabilities a singular value decomposition (SVD) to the matrix representing the excitation signal is applied [19]. The scattered signal can be represented by a periodic convolution of the decimated excitation signal and the impulse response of the microwave circuit

$$b_M[n] = a_M[n] \otimes h_M[n] = \sum_{k=0}^{M-1} h_M[k] \cdot a_M[n-k]. \tag{32}$$

We can express (32) in a matrix equation

$$\mathbf{b} = \mathbf{a} \cdot \mathbf{h}, \tag{33}$$

which yields

$$\begin{pmatrix} b_m[0] \\ b_m[1] \\ \vdots \\ b_M[M-1] \end{pmatrix} = \begin{pmatrix} a_m[0] & a_m[M-1] & \cdots & a_m[1] \\ a_m[1] & a_m[0] & \cdots & a_m[2] \\ \vdots & \vdots & \ddots & \vdots \\ a_m[M-1] & a_M[M-2] & \cdots & a_M[0] \end{pmatrix} \cdot \begin{pmatrix} h_m[0] \\ h_m[1] \\ \vdots \\ h_M[M-1] \end{pmatrix} \tag{34}$$

Equation (33) can be solved for $\mathbf{h}$ by inversion of the excitation matrix $\mathbf{a}$. The main problem is that this matrix is ill–conditioned. The condition number Q of the excitation matrix $\mathbf{a}$ gives a measure of the accuracy and the sensitivity to coefficient variations. The condition number Q of the excitation matrix $\mathbf{a}$ can be defined by the singular values

$$Q = \frac{(\sigma_i)_{\max}}{(\sigma_i)_{\min}}, \tag{35}$$

where the σ_i's are the singular values of the matrix $\mathbf{a}$.

The condition number of the excitation matrix estimates the coefficient variations of $\mathbf{h}$ as a result of the coefficient variations of $\mathbf{b}$. Therefore we have

$$\frac{||\Delta\mathbf{h}||}{||\mathbf{h}||} \leq Q\frac{||\Delta\mathbf{b}||}{||\mathbf{b}||}, \tag{36}$$

where $\Delta\mathbf{h}$ and $\Delta\mathbf{b}$ are the absolute deviations of the impulse response $\mathbf{h}$ and the response signal $\mathbf{b}$ respectively. The symbol $||\bullet||$ denotes the root–mean–square value which is the norm of the corresponding vectors. For large Q the matrix $\mathbf{a}$ is ill–conditioned and even small variations of the response vector can result in very large deviations in the estimated impulse response sequence $\mathbf{h}$.

There are many different reasons for a matrix to be ill–conditioned. In the present case our investigated electromagnetic structure constitutes a linear system. So in the spectrum of the output signal $B_M[m]$ only spectral components will be present that were excited by the input signal $A_M[m]$. Other frequencies would only appear in the case of non–linear structures. However, the spectrum of the excitation signal must have some decay towards $f = f_{min}$ and $f = f_{max}$ (see e.g. the Gaussian pulse in Fig. 7). Therefore some spectral components of $A_M[m]$ tend also to zero and the matrix in (34) becomes ill–conditioned.

The solution of Eq. (33) can be obtained by a SVD of the excitation matrix $\mathbf{a}$ [20]

$$\mathbf{a} = \mathbf{U}\boldsymbol{\Sigma}\mathbf{V}^T, \tag{37}$$

where $\mathbf{U}$ and $\mathbf{V}$ are orthogonal $M \times M$ matrices, $\boldsymbol{\Sigma}$ is a diagonal matrix with elements which are in decreasing order the singular values σ_i, $i = 1,\ldots,M$ and $\mathbf{V}^T$ denotes the transpose of matrix $\mathbf{V}$.

In accordance with the SVD of the excitation matrix $\mathbf{a}$ the solution to Eq. (33) can be written as

$$\mathbf{h} = \mathbf{a}^{-1}\mathbf{b} = \mathbf{V}\boldsymbol{\Sigma}^{-1}\mathbf{U}^T\mathbf{b}, \tag{38}$$

where $\boldsymbol{\Sigma}^{-1}$ is the inverse of the diagonal matrix with the elements $1/\sigma_i$. Furthermore we rewrite Eq. (38) as

$$\mathbf{h} = \mathbf{V}\boldsymbol{\eta}, \tag{39}$$

where $\boldsymbol{\eta}$ is a vector with the elements

$$\eta_i = \mathbf{a}^{-1}\mathbf{b} = \frac{1}{\sigma_i}(\mathbf{U}^T\mathbf{b})_i, \tag{40}$$

$(\mathbf{U}^T\mathbf{b})_i$ is i-th element of the vector $\mathbf{U}^T\mathbf{b}$. If the matrix $\mathbf{a}$ is ill–conditioned, then small singular values σ_i will lead to a increase of the corresponding elements of the vector $\boldsymbol{\eta}$ and hence in $\mathbf{h}$ too. Physically the columns of the matrix $\mathbf{V}$ represent the discrete time signals with negligible spectral components in the excitation signal. Therefore the matrix $\mathbf{V}$ cannot be used for the estimation of the impulse response $\mathbf{h}$. On the other side from matrix $\mathbf{V}$ we can determine easily the singular values

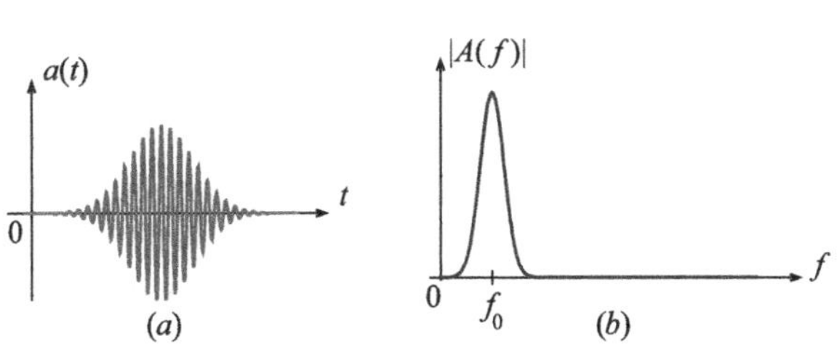

Fig. 7 The decimated band-selective signal (**a**) and its spectrum (**b**)

σ_i with $P \leq i \leq M$ that can be set to zero. P is the number of physical poles for the characterization of the late time part which will be described later in the text. Finally the approximation of the impulse response **h** will give a solution to Eq. (33) in the sense of a total least square approach [21].

Some different type for the excitation in the TLM modeling of microwave structures is presented in [22], where a time discrete Dirac impulse was used for the excitation $a_N[n] = \delta_N[n - n_0]$ with a constant spectrum for all m. But the problem of the transfer function extraction still exists in this case. The reason is the oversampling of the time domain signals by the full wave TLM simulation of the EM structure. The oversampling in frequency domain gives a huge number of unphysical samples in the response spectrum $B_N[m]$. So the estimated transfer function $H_N[m]$ which is equal to $B_N[m]$ is unphysical in the high frequency region starting from some cutoff frequency. In TLM (and also FDTD) method the highest frequency up to which the model is valid is related to the spatial discretization in the simulation domain. We assume that all signal parts with wavelengths shorter then $\Delta l/10$ are not reasonable and do not contribute to a correct physical solution. An effective way to solve the problem of the transfer function extraction in this case is a windowing $B_N[m]$ in the frequency domain with subsequent decimation of the impulse response in time domain. An example of the trapezoidal window $H_N^W[m]$ in frequency domain and its time domain representation is shown in Fig. 8a. The product of $B_N[m]$ with $H_N^W[m]$ according to (24) gives a physical representation of the transfer function $W_N[m]$ in Fig. 8c. It is important to remind that the decimation number L defined by (20) need to be an integer number where M is an even number. The resultant impulse response $x_M[n]$ defined by (26) can be used as a representation of the modeled microwave circuit in TD.

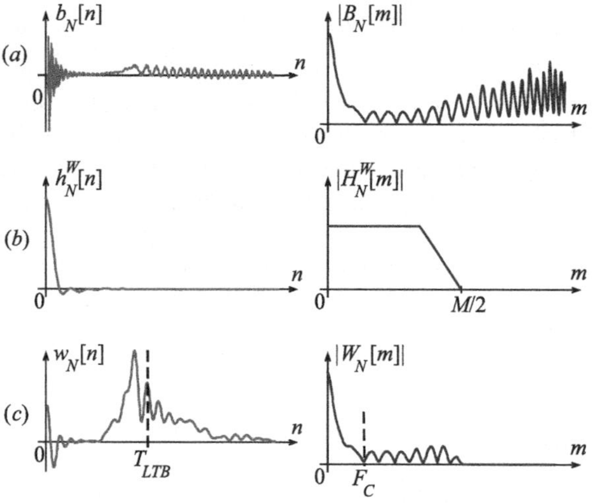

Fig. 8 Scattered waveform in time and frequency domain (**a**). Trapezoidal window (**b**) and the windowed transfer function (**c**) of the microwave circuit [22]

5 The Structure of the Microwave Circuit Model

The system identification procedure can be defined as an algorithm for the parameter estimation of the model corresponding to the microwave circuit. The model is a physical model that describes the physical process of the EM field interaction inside the circuit. The singularity expansion method as proposed by C.E. Baum [12] assumes a subdivision of the scattering transfer function into two parts as follows

$$H(f) = H_{ENT}(f) + H_{POLE}(f), \tag{41}$$

where $H_{ENT}(f)$ is an entire part of the transfer function and $H_{POLE}(f)$ is the pole part. The entire part is responsible for the high frequency process inside the EM structure concerning the propagation of the EM waves resulting in reflections, delays and other transformations of the initial waves. This processes according to L.B. Felsen [13] can be described by the geometrical theory of diffraction.

The pole part of the transfer function is responsible for the low frequency processes inside the EM structure of the quasi–static exchange of the electric and magnetic energy stored at different positions in the microwave circuit. This processes can be described by linear ODEs with constant coefficients. Linear dynamic systems describe the relation between input $a(t)$ and output $b'(t)$ signals by a differential equation of the form

$$\begin{aligned} &\frac{d^P b'(t)}{dt^P} + \alpha_{P-1}\frac{d^{P-1}b'(t)}{dt^{P-1}} + \cdots + \alpha_0 b'(t) \\ &\quad = \beta_M \frac{d^Q a(t)}{dt^Q} + \beta_{Q-1}\frac{d^{Q-1}a(t)}{dt^{Q-1}} + \cdots + \beta_0 a(t), \end{aligned} \tag{42}$$

where P is the order of the dynamic system and α_i and β_i are the constant coefficients. Such differential equations are fully describing circuits with lumped elements and representing linear electric circuits within network theory concepts.

The Laplace domain representation of the dynamic system is a rational function in the complex s–plane

$$\begin{aligned} B'(s) &= \frac{\beta_Q s^Q + \beta_{Q-1}s^{Q-1} + \cdots + \beta_0}{s^P + \alpha_{P-1}s^{P-1} + \cdots + \alpha_0} \cdot A(s) \\ &= \frac{\beta_Q (s - s_{z1}) \cdots (s - s_{zQ})}{(s - s_{p1}) \cdots (s - s_p P)} \cdot A(s) \\ &= H_{POLE}(s) \cdot A(s). \end{aligned} \tag{43}$$

$B'(s)$ and $A(s)$ are the Laplace transforms of the corresponding TD signals, s_{pi} are the poles or natural frequencies and s_{zi} are the zeros of the system. P is the order of the dynamic system.

Another type of a linear system represents a superposition of time delayed and weighted copies of the input signal

$$b''(t) = \sum_{k=1}^{K} D_k \cdot a(t - T_k) = h_{ENT}(t) * a(t) \quad (44)$$

with the Laplace transform

$$B''(s) = A(s) \sum_{k=1}^{K} D_k \cdot e^{-sT_k} = H_{ENT}(s) \cdot A(s), \quad (45)$$

where T_k are the time delays and D_k are the corresponding weighting coefficients. This equation can be applied for the multi–path propagation of the electromagnetic signals inside some channel with reflections and fading. The combination of these two linear system models can be used as model a of the passive electromagnetic structure or the microwave circuit. Of course, this assumption needs some verification, but first we can explain our approach as follows: The transfer function $H_{ENT}(s)$ does not contain any poles. System identification for this part includes the estimation of the time delays T_k. The resultant scattered signal $B(s)$ is a superposition of the two parts represented by Eqs. (43) and (45)

$$B(s) = B'(s) + B''(s) = A(s) \cdot [H_{POLE}(s) + H_{ENT}(s)] = A(s) \cdot H(s). \quad (46)$$

The TD representation of the scattered wave is given by the inverse Laplace transform of (46)

$$b(t) = b'(t) + b''(t) = a(t) * [h_{POLE}(t) + h_{ENT}(t)], \quad (47)$$

with

$$h_{POLE}(t) = \sum_{i=1}^{P} C_i \cdot e^{-s_{pi}t}, \quad (48)$$

$$h_{ENT}(t) = \sum_{k=1}^{K} D_k \cdot a(t - T_k), \quad (49)$$

C_i are the residues of the pole transfer function at the corresponding natural frequencies s_{pi}. We can make the conclusion that for the correct implementation of the SI procedure it is necessary to separate this two parts of the transfer function.

The main feature of the frequency domain characteristic is the presence of a cutoff frequency f_{LTB} which determines the border of the pole part $H_{POLE}(f)$ of the entire transfer function. For frequencies greater than f_{LTB} only the entire part of the transfer function is present. Contrary to the frequency domain the pole part of the impulse response in time domain has infinite duration. The entire part of the impulse response $h_{ENT}(t)$ has finite duration and the time border t_{LTB} divides the whole impulse response into two parts. The time where both parts of the impulse response are present is identified by the *early–time* part contrary to the remaining *late–time* part of the impulse response.

It is obvious that the calculation of the cutoff frequency f_{LTB} and the late time border T_{LTB} is a important issue in the system identification process. After the calculation of the late–time border the parameter estimation process for the pole and the

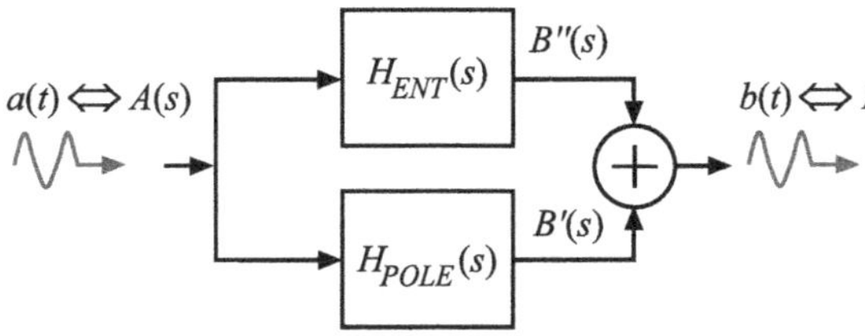

Fig. 9 Structure of the circuit SI model

entire part can be implemented separately. Figure 9 shows the schematic drawing of the model which links a pair of the incident and scattered electromagnetic signals at different ports or at the same port of the multiport microwave circuit.

6 Stability Criterion

In conventional methods the order of the pole model is chosen such that a appropriate approximation quality is achieved. The relative mean square error (RMSE) between the estimated pole model and the simulation data provides a measure for the quality of the approximation. However, the poles extracted by the Matrix Pencil Method [23] are subdivided in physical and numerical or unphysical poles. The physical poles describe the linear system whereas the numerical poles only minimize the RMSE without any physical significance. The new stability criterion is based on the stability of the physical poles when the location of the observation window within the late–time part is changed. It turns out that at different observation frames in the late–time part only the positions of the physical poles are stable whereas the numerical poles may swirl around in the complex plane. For the realization of the physical pole stability criterion different pole sets are compared by means of the signature comparison technique. Herein a set of P poles in the complex plane, given by the point $\mathbf{Z}_i = \{z_1, \ldots, z_P\}$ in the P-dimensional space is used. The distance between two points $\mathbf{Z}_1 = z_p^{(1)}$ and $\mathbf{Z}_2 = z_p^{(2)}$ is defined as

$$\Delta Z = \left|\left|\mathbf{Z}_1 - \mathbf{Z}_2\right|\right| = \left[\sum_{p=1}^{P} \left|z_p^{(1)} - z_p^{(2)}\right|^2\right]^{1/2} . \tag{50}$$

The correct dimension range P is important for every pole of one set must have a corresponding pole in the other set with low deviation. This condition per se leads to the continuity of the location of the physical poles and the pole order. The minimum value for all possible combinations is given by

$$\Delta Z_{\min} = \min_{d} \{\Delta Z_d\} , \tag{51}$$

where d is a combination index of the two dimensions.

7 Results and Verification

7.1 Patch Antenna Optimization

The combined TLM–GA optimization method presented in section 2 is tested on the patch antenna structure shown in Fig. 1a. The initial patch antenna exhibits its first two resonances at 6 and 12 GHz with a relative bandwidth of $f_{010} = 7.8\%$ and $f_{001} = 12\%$ respectively. For the optimization the frequency range was specified from 5 to 10 GHz. In this range $M = 10$ samples for the fitness evaluation (1) where taken. The settings for the GA are $p_{\text{cross}} = 0.75$ and $p_{\text{mut}} = 0.08$. The goal is set to minimize the fitness value J in the specified frequency range. Therefore the maximum number of iterations for the GA is limited to 50 runs.

In order to obtain a symmetrical radiation characteristic the patch should be symmetric with respect to its center axis which is centered along the microstrip feed line. Therefore patch elements being mutual images with respect to the symmetry axis are treated in the same manner. Removing one patch element from the right half immediately removes its counterpart from the left half and vice versa. In that way the symmetry of the optimized patch is ensured. Beyond this no further constraints are imposed in the optimization run and all patch elements can be affected by the GA.

In Fig. 10 the return loss of the optimized and the initial patch antenna are shown. The chart on the right side of Fig. 10 shows the patch of the antenna after the GA optimization. The relative bandwidth for the optimized patch antenna is now increased up to 61% for a center frequency of 7.5 GHz if the limit for the return loss is set to −10 dB. Hence a large increase of the operating bandwidth is achieved. The removed patch elements close to the microstrip feed improve the matching over the desired frequency range and the holes in the patch give rise to several additional resonance frequencies as can be seen in Fig. 10. In [24] the same patch antenna was optimized in the range from 10 to 15 GHz.

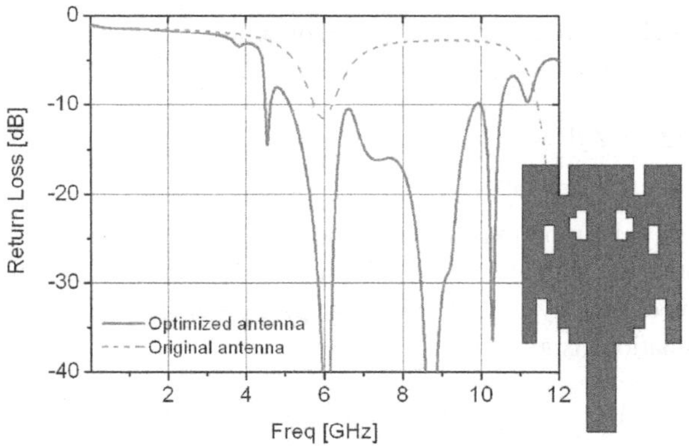

Fig. 10 Structure and return loss of optimized (*straight red line*) and initial (*dashed green line*) patch antenna. The optimization bandwidth is 5–10 GHz

The presented antenna is not yet optimized for UWB pulse transmission because of its strongly resonant behavior of the return loss in the operating frequency range. Therefore the group delay properties may cause unacceptable dispersion during transmission of extremely narrow time bursts [15]. Nevertheless the optimized patch antenna can be used as a common antenna for several narrow frequency bands and services located in the optimized frequency range.

The implementation of the TLM–GA optimization requires for a fast manipulation of the electromagnetic structure. The used open source TLM simulator YATPAC [6] provides a plain text based input file (TLM structure file) for the simulation kernel, containing all information about geometry, material, excitation and boundaries of the electromagnetic structure. Because the TLM simulation domain is subdivided into cubic unit cells each cube in the TLM domain can be addressed separately in the TLM structure file. A direct relation in form of a connection matrix is established, referring the single patch elements, respectively the genes, to the corresponding lines in the input file. So an efficient linkage of the electromagnetic structure to the rather abstract coding of a chromosome is established. This procedure is similar to the well known direct impedance matrix manipulation (DMM) used in MoM based GA optimizers [9].

7.2 System Identification

The impulse response of the non–optimized (initial) and the bandwidth optimized patch antenna has been estimated by the proposed SI method presented in Sects. 3, 4, 5 and 6. The resulting pole constellation in both cases is shown in Fig. 11.

Figure 12 shows the corresponding amplitude characteristics. It can be seen that the optimized antenna has additional resonant frequencies. These provide a larger impedance bandwidth compared to the initial antenna with two resonances as can be

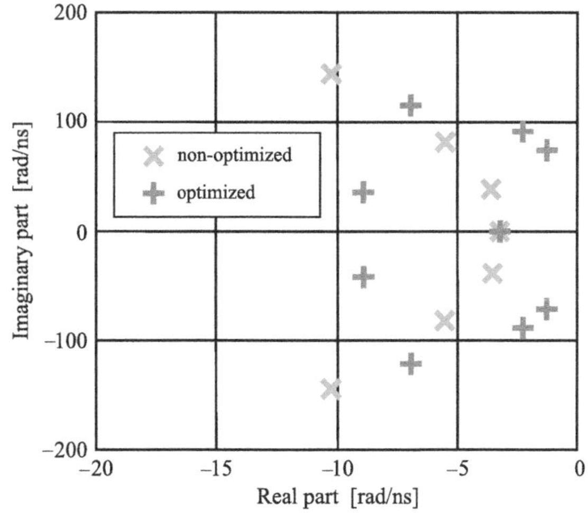

Fig. 11 Pole diagram estimated from the patch antenna impulse response

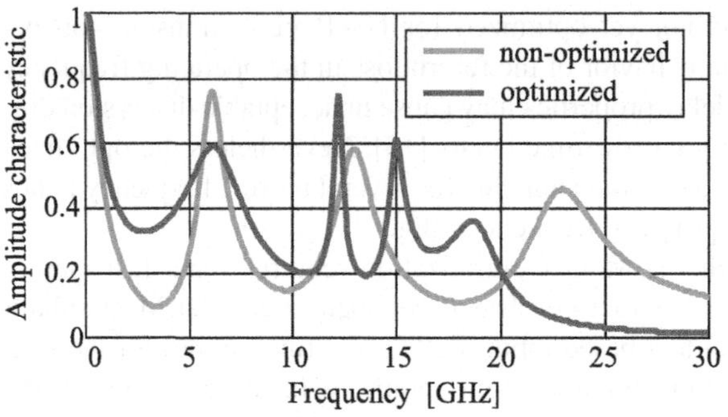

Fig. 12 Modeled patch antenna amplitude spectral response

seen from Fig. 10. It has to be mentioned that additional resonances may destroy the group delay properties. Therefore the impedance bandwidth criterion used here applies well to wideband antennas for several narrowband services. However, antennas optimized in that way may not be suitable for ultrawide band pulse transmission.

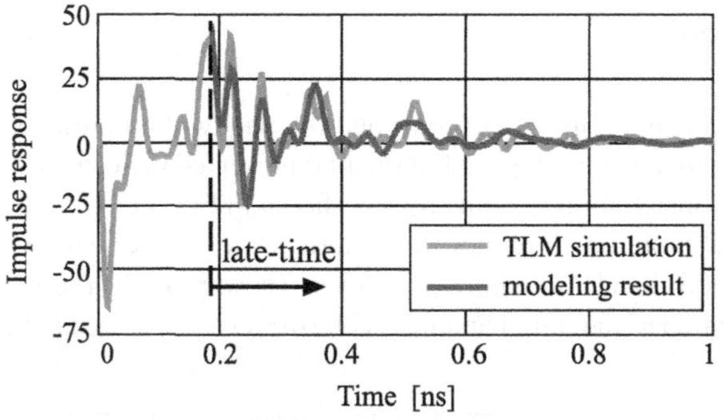

(a) Non-optimized antenna

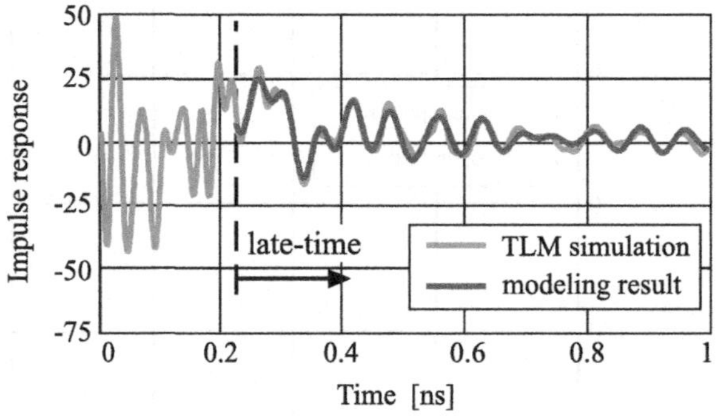

(b) Optimized antenna

Fig. 13 Patch antenna impulse response: TLM simulated and reconstructed by system identification

The initial TLM impulse response and the modeled late–time part are shown together in Fig. 13. Figure 13a shows the result for the non–optimized antenna, Fig. 13b shows the same for the optimized antenna.

The estimated model parameters can be used to establish network oriented equivalent models of the microwave structures by means of network synthesis. As an example the antenna input impedance derived from the all–pole model is shown in Fig. 14.

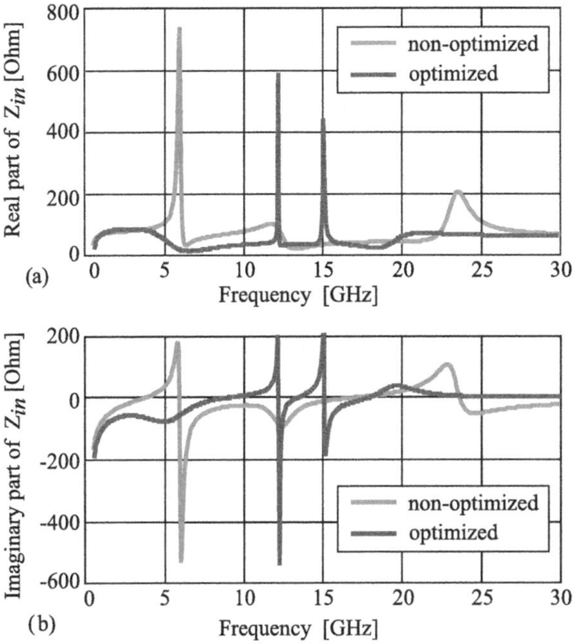

Fig. 14 Modeled patch antenna input impedance

8 Conclusions

Broadband electromagnetic systems are efficiently modeled by means of time–domain simulations. The coupling between a GA optimization and a time–domain EM simulator requires effective coding of the electromagnetic structure. We introduced a fast coding scheme for patch antenna structures in the open–source TLM simulator YATSIM [6]. The potential of a combined TLM–GA optimization method is demonstrated considering the example of narrow band patch antennas. Furthermore a system identification procedure is used to establish network oriented models of the antennas from parts of the impulse response thus avoiding the need to simulate the total duration of the scattered signal. A combination of time–domain electromagnetic modeling and system identification has therefore the potential to reduce computation time considerably.

References

1. P.B. Johns and R.L. Beurle (1971) Numerical Solution of 2-Dimensional Scattering Problems using a Transmission-Line Matrix, Proc. IEE, vol. 118, pp. 1203–1208, Sept. 1971.
2. G. Kron (1944) Equivalent Circuit of the Field Equations of Maxwell. Proceedings of the IRE, vol. 32, no. 5, pp. 294–299, May 1944.
3. W.J.R. Hoefer (1985) The Transmission Line Matrix Method – Theory and Applications, IEEE Trans. Microw. Theory Techn., vol. 33, no. 10, pp. 882–893, Oct. 1985.
4. P.B. Johns (1987) A Symmetrical Condensed Node for the TLM Method, IEEE Trans. Microw. Theory Techn., vol. 35, no. 4, pp. 370–377, Apr. 1987.
5. C. Christopoulos (1995) The Transmission-Line Modeling Method TLM, IEEE Press, New York.
6. Yatpac Homepage, May 2007, http://www.yatpac.org.
7. Y. Rahmat-Samii and E. Michielssen (1999) Electromagnetic Optimization by Genetic Algorithms. Wiley Series in Microwave and Optical Engineering, Wiley & Sons New York.
8. J.M. Johnson and Y. Rahmii-Samii (1997) Genetic Algorithms in Engineering Electromagnetics, IEEE Antenn. Propag. Mag., vol. 39, pp. 7–25, Aug. 1997.
9. J.M. Johnson and Y. Rahmii-Samii (1999) Genetic Algorithms and Method of Moments (GA/MOM) for the Design of Integrated Antennas, IEEE Trans. Antenn. Propag., vol. 47, No. 10, pp. 1606–1614, Oct. 1999.
10. B. Biscontini (2006) A Novel Antenna for Ultra Wide Band (UWB) Intelligent Antenna Systems, Int. Microwave Symposium Digest, San Francisco, pp. 2023–2026, June 2006.
11. P. Russer (2006) Electromagnetics, Microwave Circuit and Antenna Design for Communication Engineering, 2nd Ed., Boston, Artech House.
12. C.E. Baum (1976) The Singularity Expansion Method, In: Transient Electromagnetic Fields, Edt.: L.B. Felsen, Springer, New York.
13. L.B. Felsen (1985) Comments on Early Time SEM, IEEE Trans. Antenn. Prop., vol. 33, pp. 118–119, Jan. 1985.
14. Y. Kuznetsov, A. Baev, T. Shevgunov, M. Zedler and P. Russer (2005) Transfer Function Representation of Passive Electromagnetic Structures, Int. Microwave Symposium Digest, Los Angeles, pp. 183–186, June 2004.
15. A. Baev, Y. Kuznetsov and A. Aleksandrov (2003) Ultra Wideband Radar Target Discrimination Using the Signatures Algorithm. In: 33rd European Microwave Conference, Munich, Oct. 2003.
16. C.A. Balanis (1997) Antenna Theory – Analysis and Design, 2nd Ed. John Wiley & Sons New York.
17. S.N. Makarov (2002) Antenna and EM Modeling with Matlab, 1st Ed. John Wiley & Sons New York.
18. Y. Kuznetsov, A. Baev, T. Shevgunov, P. Lorenz and P. Russer (2006), Application of the Stability Criterion to the Passive Electromagnetic Structures Modeling, Proc. of the 36th Euro. Microwave Conf., pp. 13–16, Sept. 2006.
19. E.J. Rothwell and S. Weimin (1990) Time Domain Deconvolution of Transient Radar Data, IEEE Trans. Antennas Prop., vol. 38, pp. 470–475, Apr. 1990.
20. A. Baev, Y. Kuznetsov, F. Coccetti and P. Russer (2004) The Ultra Wideband Transfer Function Representation of Complex Three-Dimensional Electromagnetic Structures. In: 34-th European Microwave Conference, Amsterdam, Oct. 2004.

21. T.K. Sarkar and J. Rahman (1995) Deconvolution and Total Least Squares in Finding the Impulse Response of an Electromagnetic System from Measured Data, IEEE Trans. on Antenn. Prop., vol. 43, pp. 416–421, Apr. 1995.
22. A. Baev, Y. Kuznetsov, P. Lorenz and P. Russer (2007) Network Oriented Modeling of Passive Microwave Structures. In: The IEEE Region 8 EUROCON 2007. Computer as a Tool, Warsaw, Sept. 2007.
23. Y. Hua and T. Sarkar (1990) Matrix pencil method for estimating parameters of exponentially damped/undamped sinusoids in noise, IEEE Trans. Antenn. Prop., vol. 38, pp. 814–824, May 1990.
24. N. Fichtner, U. Siart and P. Russer (2007) Antenna Bandwidth Optimization using Transmission Line Matrix Modeling and Genetic Algorithms. In: Int. Symp. on Signals, Systems and Electronics (ISSSE), Montréal, Québec, July 30–August 2 2007.

Study of Single and Dual Band Wearable Metallic Button Antennas for Personal Area Networks (PANs)

Benito Sanz-Izquierdo, Fengxi Huang, John C. Batchelor and Mohammed I. Sobhy

Abstract A small WLAN/Bluetooth antenna with the appearance of a metallic button is reported for use with wearable computer systems and covert communications. A single band version is discussed in addition to a dual band design that also covers the HiperLAN/2 band. No external matching circuits are required when the antenna is fed with a 50 Ω coaxial line. A lumped element equivalent circuit is presented for the dual band design together with parametric data curves to inform the design process. A 3 layer human body is included in the simulation and measurements are given to demonstrate the on-body performance. The antenna is disguised as a small metallic button, having the appearance of the waist button used in denim jeans.

Keywords Wearable antennas · dual band PAN · WLAN and HiperLAN antennas

1 Introduction

This paper describes a small antenna suitable for worn WLAN/Bluetooth and HiperLAN/2 Personal Area Network (PAN) and mobile computing applications.

Benito Sanz-Izquierdo
Department of Electronics, The University of Kent, Canterbury, Kent, UK, CT2 7NT, e-mail: b.sanz@kent.ac.uk

Fengxi Huang
Formerly with the Department of Electronics, The University of Kent, Canterbury, Kent, UK, CT2 7NT

John C. Batchelor
Department of Electronics, The University of Kent, Canterbury, Kent, UK, CT2 7NT, e-mail: j.c.batchelor@kent.ac.uk

Mohammed I. Sobhy
Department of Electronics, The University of Kent, Canterbury, Kent, UK, CT2 7NT, e-mail: m.i.sobhy@kent.ac.uk

Numerous wearable devices and systems have been developed over the last decade with the aim of simplifying the human-computer interface, reducing individual device sizes and removing the need for physical connections [1]. For instance; eye glasses with integrated cameras, MP3 players sewn directly into clothing and wearable sensor systems [2]. There is an ongoing trend towards personal communications and mobile and pervasive computing. As a result, the wireless community has shown increasing interest in recent years to create antennas suitable for mounting on the person. Research published on the subject includes [3, 4, 5] which present linearly polarized fabric based patch antennas and [6] which describes a circularly polarized design. The subject of human body-antenna interaction has also been studied and the performance of a PIFA placed close to a human body is investigated in [7] while [8] presents a study to identify suitable antenna sites on a wearer using a simple monopole and coil antenna. Finally, on-body propagation channels are characterized in [9] and [10] for narrowband and Ultra Wide Band (UWB) channels respectively.

Although a variety of different antenna designs are described above, the unifying factor is a desire to make the antenna unobtrusive to the point of invisibility. However, an alternative design philosophy to the covert fabric based antennas proposed in [3, 4, 5, 6] is to disguise the antenna rather than hide it.

The antenna described in this paper is a compact rigid metallic structure over a small ground plane. The antenna has the appearance and size of a conventional waist button used in denim jeans. The rationale for designing an antenna of this structure is as follows: (a) clothing fabric is not used as an integral substrate which fills a resonant cavity as could be the case in a microstrip patch. This alleviates the performance consistency issues associated with substrate compression, poor electrical characterization of textile material and variable losses caused by moisture uptake which will be significant in a worn textile. (b) The small surface area of the antenna means that it is not required to be flexible so that surface distortion will not alter the input impedance and radiation patterns. (c) The antenna does not require a large flat ground plane and (d) the structure is relatively simple to incorporate into clothing by conventional construction methods, and it is even possible to fit retrospectively. A microstrip feed line on a flexible substrate is proposed as a method of excitation.

The paper is organized as follows. Section II describes the structure and performance of the Single Band Metallic Button Antenna (SBMBA) giving an S11 curve. Section 3 presents the characteristics of the SBMBA and establishes how the topology affects the operation. Section 4 contains the current distribution over the structure and Sect. 5 presents the measured radiation patterns. Section 6 introduces a dual band version of the SBMBA giving structure and performance simulated and measured on and off the human body, while Sect. 7 describes the DBMBA design and operation. Section 8 contains a parametric study of the DBMBA offering design insight and Sect. 9 gives a lumped element equivalent circuit of the DBMBA for use in system modelling and offering the future possibility of closed form design equations. Section 10 presents the radiation patterns and on-body performance of the DBMBA and Sect. 11 concludes the paper.

2 Single Band Metallic Button Antenna Structure and Performance

The structure of the Single Band Metallic Button Antenna (SBMBA) is shown in Fig. 1 and the dimensions are tabulated in Table 1. The structure consists of discs 1 and 3 of diameters d_1 and d_3 respectively which are separated by cylinder 2 of diameter d_2 and height h_2. The antenna has similar dimensions to a standard jeans button and is mounted on a flexible Velcro® substrate of height h_1 which is backed by a metallic ground plane of 5.0×5.0 cm. Two 50 Ω feed types were used in this study: (i) a coaxial cable connecting to disc 1 through the ground plane and (ii) a side feeding microstrip line (width 7 mm) attached to the edge of disc 1. The Velcro® substrate offers good flexibility but strong resistance to compression and has a relative permittivity, ε_r, measured to be approximately 1.3. Measured and simulated S11 curves for the SBMBA are given in Fig. 2 where it can be seen that the antenna is resonant at 5 GHz. This matched band is primarily due to the capacitance between discs 1 and 2 and the inductance associated with current flow on cylinder 2. All simulations were carried out by CST Microwave Studio and the measurement in Fig. 2 confirms the SBMBA is matched at the HiperLAN/2 band.

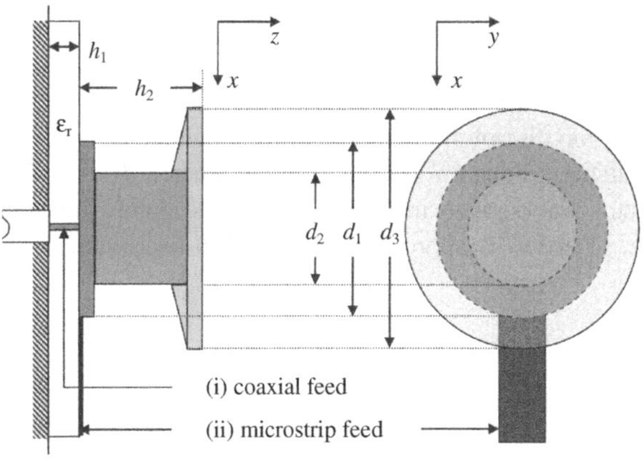

Fig. 1 Elevation and plan of Single Band Metallic Button Antenna

Table 1 Principle dimensions of the SBMBA (Fig. 1)

Dimension	Value
d_1	9.2 mm
d_2	6.8 mm
d_3	16 mm
h_1	1.8 mm
h_2	7.9 mm
ε_r	1.4

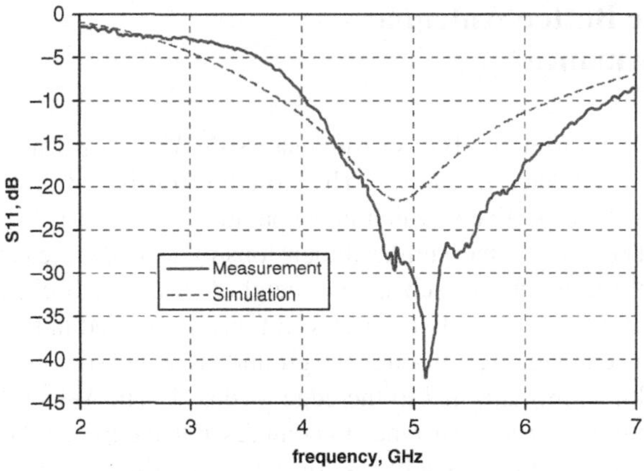

Fig. 2 Measured and simulated S11 of SBMBA

3 SBMBA Characteristics

Altering the principal dimensions of the SBMBA gives insight into the operation and design process. The antenna comprises a microstrip patch as the base disc with a diameter d_1 of 11 mm. In isolation from the rest of the structure and on a substrate of 1.8 mm height and relative permittivity 1.3, the disc is naturally resonant at 11 GHz. The resonant frequency of the structure falls to 5.5 GHz with the addition of the cylinder of diameter d_2 and the top disc of diameter d_3. For the complete structure, Fig. 3 shows the change in matched frequency as the top disc diameter, d_3, is varied from 7 to 16 mm and the data is also given in Table 2. The best bandwidth and match are obtained at 5 GHz, where d_3 is approximately equal to one quarter of a wavelength.

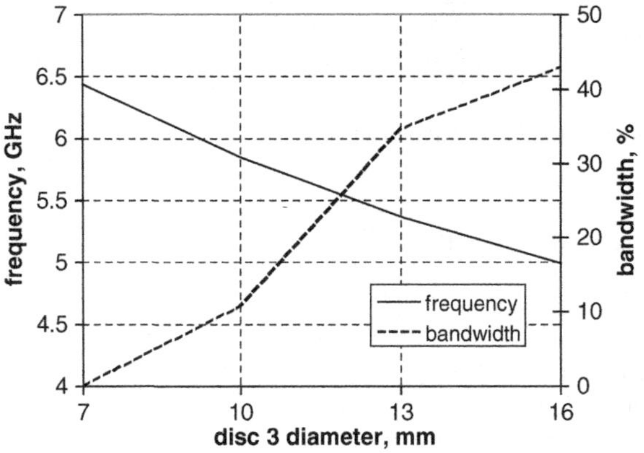

Fig. 3 Simulated SBMBA resonances as a function of upper disc diameter, d_3

Table 2 Resonant frequencies and electrical dimensions (D_1 and D_3 varied independently)

Diameter, mm	Disc 1 ($d_3 = 16$ mm)		Disc 3 ($d_1 = 9.2$ mm)	
	Frequency, GHz	Circumference/ lambda	Frequency, GHz	Diameter/ lambda
7	6.2	0.45	6.4	0.15
10	5.0	0.52	5.9	0.20
13	4.5	0.62	5.4	0.23
16	4.3	0.72	5.0	0.26

The base disc diameter d_1 also influences the resonant frequency and Fig. 4 illustrates that the mode changes from 6.3 to 4.3 GHz as d_1 increases. The frequency is tabulated in Table 2 together with disc circumference referenced to wavelength. The mode is best matched at 5.0 GHz where the circumference of the base disc is approximately one half wavelength. The height of the central cylinder, h_2, also has a significant effect on the resonance of the SBMBA, Fig. 5. The frequency decreases as h_2 increases which lengthens the current path along the structure and reduces the capacitance between discs 1 and 3. The cylinder height is small compared to wavelength, being approximately 0.1 λ at the normal button height h_2 of 7.9 mm.

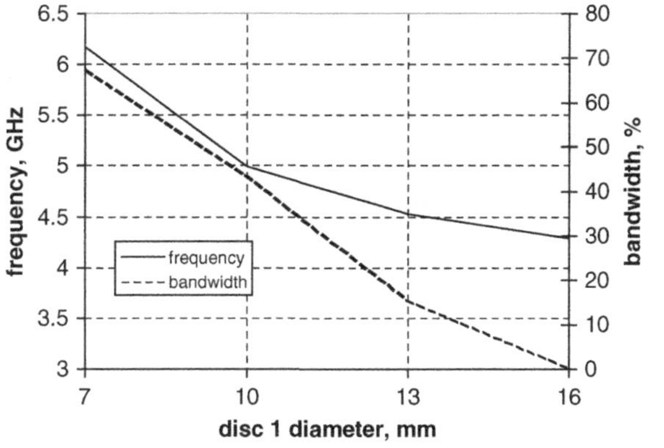

Fig. 4 Simulated SBMBA resonance as a function of base disc diameter, d_1

The diameter of the central cylinder, d_2, affects the resonance frequency of the SBMBA as shown in Fig. 6 where the frequency is plotted against the cylinder diameter. The antenna is loaded by the capacitance between the top and base plates which is reduced as the cylinder diameter increases. The maximum matched bandwidth of 43% is offered at a cylinder diameter of 7 mm where the circumference is 0.36 λ. Near the operating frequency, the mode varies with cylinder diameter at a rate of 0.22 GHz/mm.

The principal dimensions of the SBMBA are tabulated in Table 3 to give a design basis for the antenna.

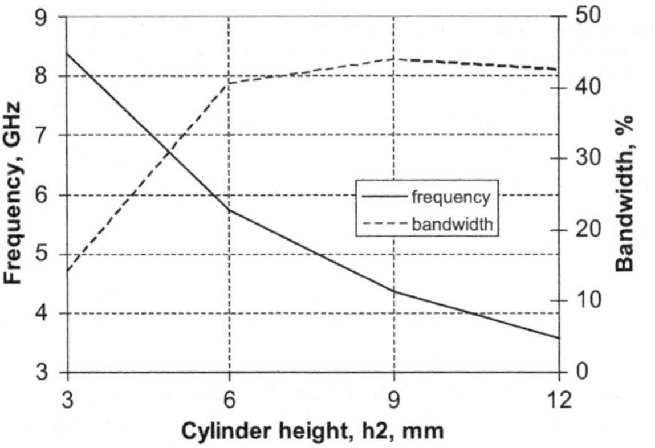

Fig. 5 SBMBA cylinder height vs resonant frequency

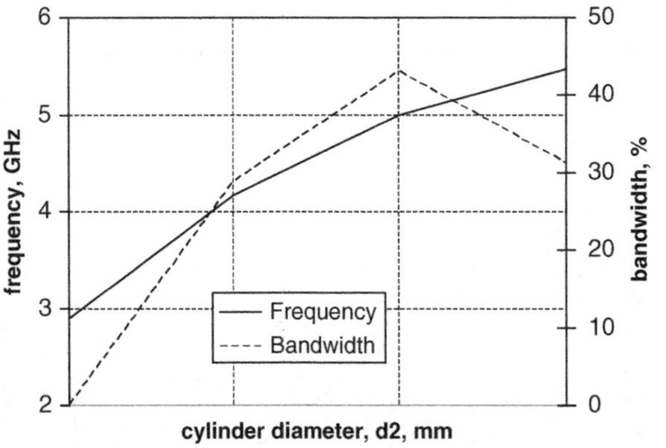

Fig. 6 Resonant frequency as a function of cylinder diameter, d_2

Table 3 Principle electrical dimensions of the SBMBA

Dimension	Electrical length
Disc 1 circumference, πd_1	0.52
Cylinder diameter, d_2	0.12
Disc 3 diameter, d_3	0.26
Substrate height, h_1	0.03
Cylinder height, h_2	0.12

4 Current Distribution of the SBMBA

The current distribution at 5 GHz on the SBMBA with optimal dimensions, given in Table 1 is shown in Fig. 7 where the substrate and ground plane have been omitted for clarity. The current variation observed around the circumference of the cylinder is caused by a small ground plane which is not shown here. The cylinder diameter is sufficiently large for there to be some radial variation in current around its circumference. The primary radiating element is the centre cylinder and the current is polarized along the z-axis.

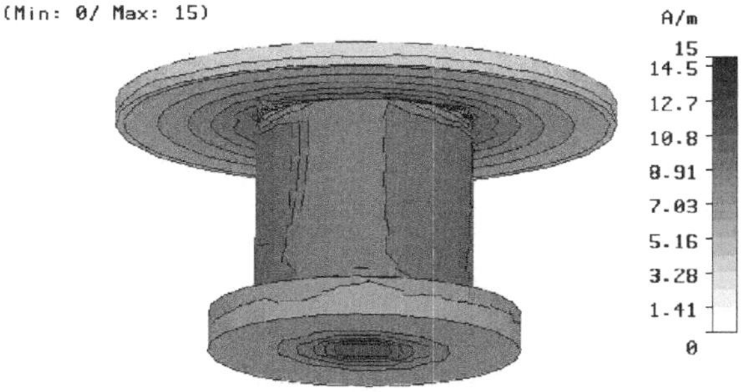

Fig. 7 SBMBA current distribution at 5.5 GHz resonance

5 Radiation Patterns and Gain of the SBMBA

The radiation patterns were measured in an anechoic chamber, with the SBMBA on a ground plane of 5 cm by 5 cm ($0.92 \times 0.92\ \lambda$). Feeding was via a coaxial feed through the ground plane. The radiation patterns are shown in Fig. 8 at 5.5 GHz and are as for a monopole on a finite ground plane.

No variation in pattern shape was discernable across the HiperLAN/2 band of 5.15–5.725 GHz. The gain was measured to be 3.8 dBi.

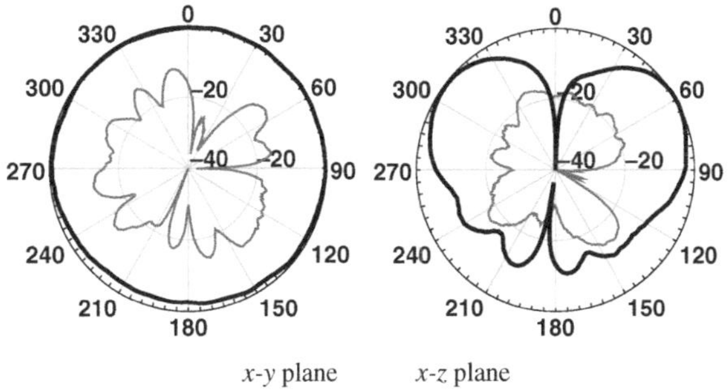

x-y plane *x-z* plane

Fig. 8 Measured radiation patterns of SBMBA. *Heavy line*: Co-polarization, *thin line*: Cross Polarization

6 Dual Band Metallic Button Antenna Structure and Performance

It was established in Sect. 2 that the SBMBA supports a mode with a frequency determined by the capacitance between the upper and lower discs, 3 and 1 respectively. The frequency is controlled by altering the ratio of the diameter between discs 1 and 3 and the central cylinder. Unfortunately, changing the top disc diameter alters the appearance of the button and may not be desirable. It is possible to create a dual band version of the antenna while maintaining the appearance and size of a standard button and a proposed dual Band Metallic Button Antenna (DBMBA) design is shown in Fig. 9. The top of the DBMBA consists of 2 discs of diameters d_3 and d_4. Disc 4 is shorted to disc 1 by a coaxially placed via pin in the cylinder. Honeycomb spacer is used between discs 3 and 4 to give mechanical rigidity with a relative permittivity close to 1. The lower resonance frequency is determined by the capacitance between the two top discs 3 and 4 which are loaded by a short circuited coaxial line of length, h_2. The impedance of the short circuited coaxial line is given by:

$$Z = jZ_0 \tan \frac{2\pi}{\lambda} h_2 \tag{1}$$

where Z_0 is the characteristic impedance of the coaxial line and is equal to:

$$Z_0 = \frac{60}{\sqrt{\varepsilon_r}} \ln \left(\frac{d_{2\,inner}}{d_5} \right) \tag{2}$$

where ε_r is the relative permittivity within the coaxial waveguide and is equal to 1, and d_{2inner} refers to the inner radius of cylinder 2 and is 3 mm. At 2.4 GHz the inductance of the shorted coaxial line is 1.9 nH. There is also a series inductance due to the inner via that extends through disc 3 and connects with disc 4. If the

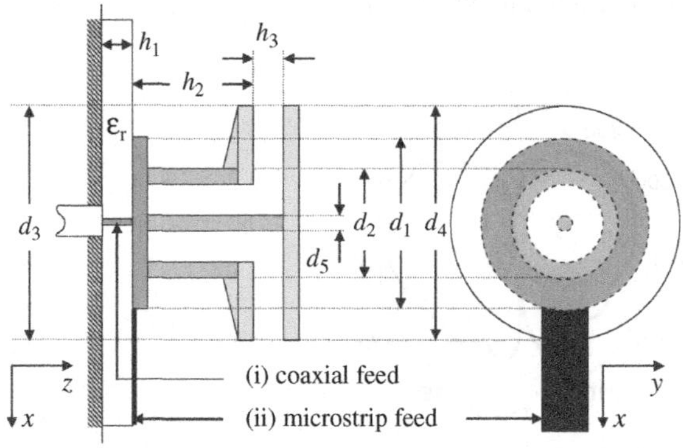

Fig. 9 Cross section and plan of Dual Band Metallic Button Antenna, DBMBA

field distribution across the coaxial aperture is neglected the probe inductance can be calculated according to the method reported in [11]:

$$L_{probe} = -\frac{\mu_0 h_3}{2\pi}\left[\ln\left(\frac{\pi d_5}{\lambda}\right) + 0.5772\right] \quad (3)$$

where μ_0 is the permeability of free space.

The original mode at 5 GHz due to the capacitance between discs 1 and 3 is largely unaffected by the addition of disc 4 and the inner via. Fig. 10 shows the measured and simulated S11 curves for the DBMBA of Fig. 9 with the dimensions given in Table 4. S11curves are given in Fig. 10 for both on and off the human body where the body is simulated as a 3 layer tissue model [12] placed immediately below to the antenna ground plane. The three body layers were: skin of 1 mm thickness; fat of 3 mm thickness; and muscle of 44 mm thickness. The dimension of the simulated body was $120 \times 120 \times 40\,\text{mm}^3$. The electrical values of the tissues were assumed to be non-dispersive [12] and were taken at 4 GHz which was the centre point between the bands. The relative permittivity, ε_r, and conductivity, σ, of the tissues were obtained from [13] and were as follows: skin: $\varepsilon_r = 38$, $\sigma = 2.34$ S/m; fat : $\varepsilon_r = 5.1$, $\sigma = 0.18$ S/m; muscle : $\varepsilon_r = 50.8$, $\sigma = 3$ S/m. It can be seen that attaching the antenna to a human body has not significantly altered its response at the lower band. The upper bandwidth falls from 27 to 20% which is still sufficient to cover the HiperLAN2 band.

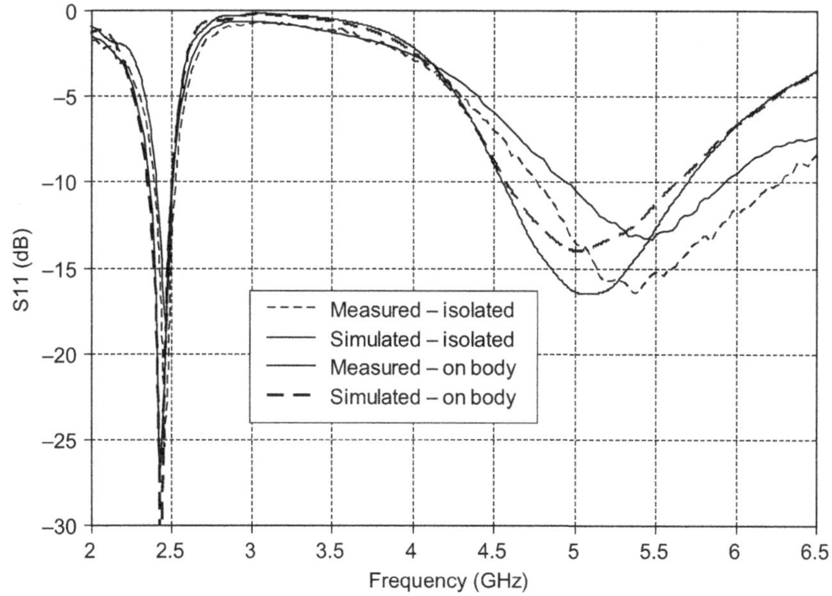

Fig. 10 Measured and simulated S11 for DBMBA on and off-body

Table 4 Dimensions of DBMBA (Fig. 9)

	Dimension	Electrical length, lower band	Electrical length, upper band
Disc 1 diameter, d_1	10.0 mm	0.081 λ	0.198 λ
Cylinder 2 diameter, d_2	7.0 mm	0.056 λ	0.137 λ
Disc 3 diameter, d_3	16 mm	0.129 λ	0.315 λ
Disc 4 diameter, d_4	16 mm	0.129 λ	0.315 λ
Inner via diameter, d_5	2 mm		
Substrate height, h_1	1.8 mm		
Cylinder height, h_2	8.3 mm	0.064 λ	0.154 λ
Gap height, h_3	3.0 mm		
Substrate relative permittivity	1.4	–	–
Ground plane size	50 × 50 mm	$(0.40\ \lambda)^2$	$(0.98\ \lambda)^2$

7 DBMBA Design and Operation

The surface currents of the DBMBA with the dimensions of Table 4 are shown in Figs. 11a and b at 2.5 and 5 GHz respectively. Fig. 11a shows that at the lower band the top plates 3 and 4 are strongly capacitive and current flows down the central via to the base disc 1. The inductance of the inner via and the coaxial cylinders is highly significant at the lower band. At the upper band the most significant fields are between disc 3 and disc 1 (Fig. 11b) as described in Sect. 3. Simulation demonstrated the presence of the ground did not significantly affect the resonance frequency of the lower mode as the capacitance between discs 3 and 1 dominates the capacitance to ground. The cylinder height, h_2, tunes both the upper and lower bands. Further detail will be given in the next section.

8 DBMBA Parametric Study

A parametric study was conducted to give design insight for the DBMBA. In all cases, the structure considered is based on the dimensions of Table 4. Table 5 tabulates the maximum percentage frequency variation and difference in fractional bandwidth caused by parametric sweeps of the DBMBA dimensions. Note, maximum changes are given and these do not necessarily occur at the extremes of the parametric ranges.

These results are summarized in Table 6 to give an oversight of the DBMBA tuning process. The table uses the following ranges of change: $5\% \leq$ not significant ('–'); $5\% <$ moderate $\leq 25\%$; $25\% <$ significant. Examination of the table demonstrates that the DBMBA antenna can be tuned in the upper band by d_3 and in the lower band by d_4. These dimensions are relatively uncoupled in their effect which makes the tuning process straightforward. For second order tuning, base disc diameter d_1 and cylinder diameter d_2 have a moderate tuning effect on the upper frequency and a significant effect on the bandwidth. The diameter of the inner via d_5

(a)

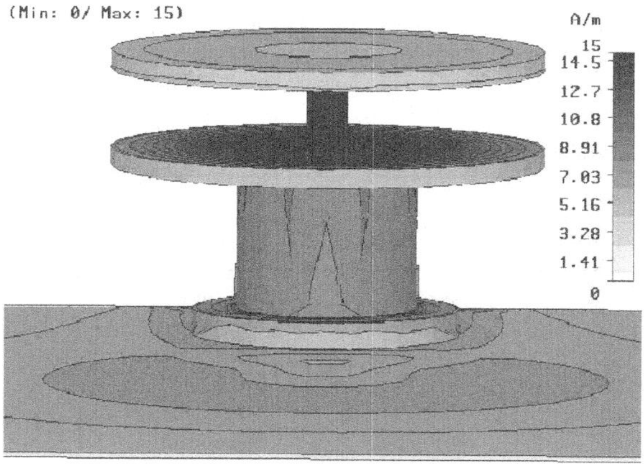

(b)

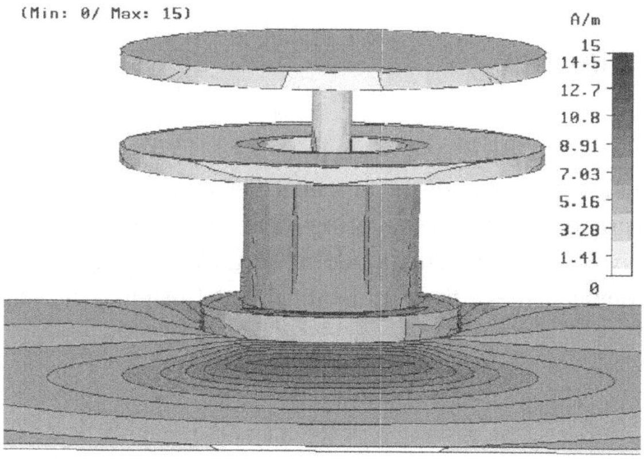

Fig. 11 (**a**) Surface currents of DBMBA at 2.5 GHz. (**b**) DBMBA surface currents at 5 GHz

Table 5 DBMBA parametric variation

	Parameter range, mm	Lower Frequency change (%)	Lower Bandwidth Range (%)	Upper Frequency Change (%)	Upper Bandwidth Range (%)
d_1	10.0 mm	+2	0–8	−25	57–0
d_2	7.0 mm	−12	0–3	+15	22–34
d_3	16 mm	−3	0–5	−35	0–38
d_4	16 mm	−49	18–0	−4	27–33
d_5	2 mm	+37	0–11	+5	38–29
h_2	1.8 mm	−49	9–0	−64	0–34
h_3	8.3 mm	−4	0–10	+14	45–28

has a significant effect on the lower band frequency and moderately alters the bandwidth. The cylinder height, h_2 significantly affects both bands. Table 6 indicates with * where a band has become matched as a result of the parametric variation. Note the lower band match is sensitive to change in all dimensions.

Table 6 DBMBA parametric study summary

	Effect on			
	Lower frequency	Lower bandwidth	Upper frequency	Upper bandwidth
d_1	–	Moderate*	Moderate	Significant*
d_2	Moderate	–*	Moderate	Moderate
d_3	–	–*	Significant	Significant*
d_4	Significant	Moderate*	–	Moderate
d_5	Significant	Moderate*	–	Moderate
h_2	Significant	Moderate*	Significant	Significant*
h_3	–	Moderate*	Moderate	Moderate

* Indicates that band becomes matched ($S11 < -10\,dB$) over the parametric variation.

9 Equivalent Circuit of DBMBA

An equivalent lumped circuit of the DBMBA is shown in Fig. 12. The circuit is of 4th order and reflects the dual band nature of the antenna. The component values have been generated by capturing the measured scattering parameters of the antenna and using them to generate a system model. This model is then transformed to the topology indicated and component values are optimized to fit the original curves.

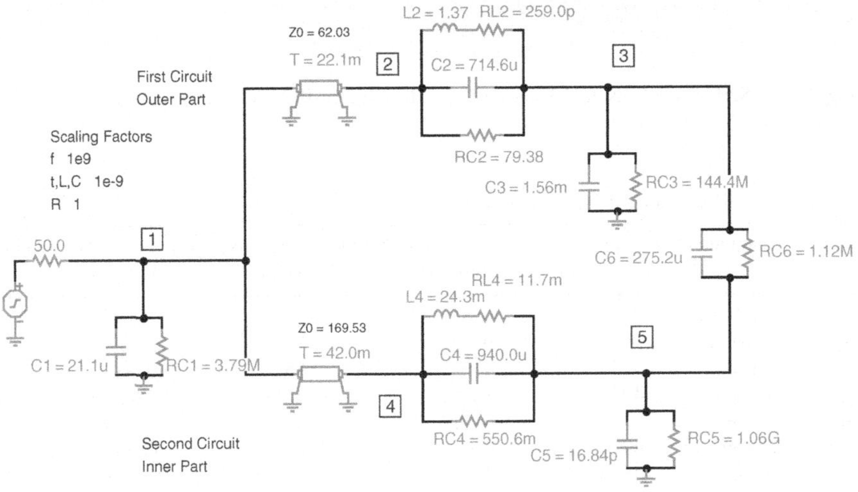

Fig. 12 Lumped element 4th order equivalent circuit of the DBMBA

The upper and lower sections of the circuit provide the upper and lower resonances respectively. A very close fit between the original measurement and the frequency response of the equivalent circuit can be seen in Fig. 13. The short length of transmission line in the equivalent circuit of Fig. 12 is to remove a phase offset present in the measurement.

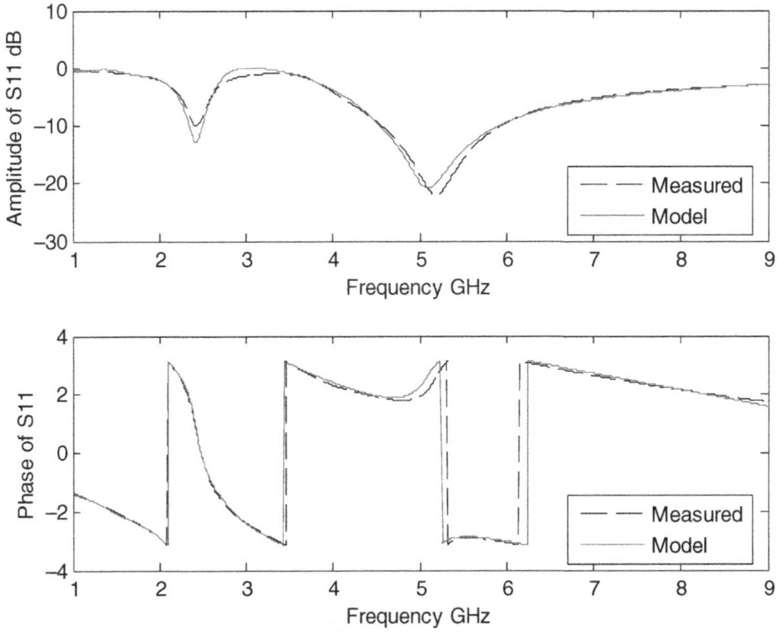

Fig. 13 Measured and circuit modelled S11 of the DBMBA

10 DBMBA Radiation Patterns and Gain

The radiation patterns were measured on a live adult male and are shown in Fig. 14 with a comparison to off-body measurement. The antenna is proposed for use in Personal Area Networks (PANs) where communication is necessary both to devices on-body and to external, off-body, network access nodes. With the DBMBA mounted as a waist height front trouser button there is good coverage at the front of the user's body providing connection to other worn devices. The plane cutting the user's waist shows there to be reasonable forwards coverage with a centre null of around 20 dB and there is a deep null in the rear direction. The measured gains for the antenna with a coaxial feed and not mounted on a body were: 2.2 dB at 2.45 GHz; 3.6 dB at 5.2 GHz and 3.8 dB at 5.5 GHz. Placing the DBMBA at waist height on a human caused the gains to become: 0.5 dB at 2.45 GHz; 2.7 dB at 5.2 GHz; and 1.8 dB at 5.5 GHz. The side feed caused the gains to be higher than simulated due to asymmetry in the patterns.

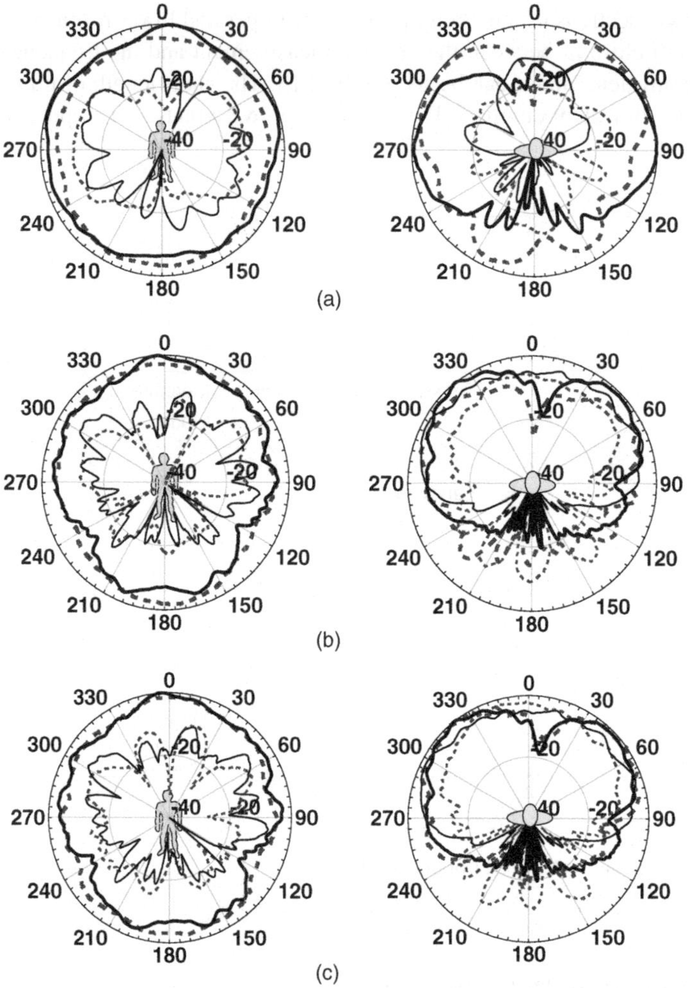

Fig. 14 DBMBA Radiation patterns. (**a**) 2.45 GHz, (**b**) 5.2 GHz, (**c**) 5.5 GHz. *LHS*: *x-y* plane, *RHS*: *x-z* plane. In all cases DBMBA is mounted as a front trouser button at waist height. *Solid line* shows on-body measurement and *broken line* represents isolated antenna. *Heavy line*: co-polarization, *thin line*: cross polarization

11 Conclusions

A new antenna proposed for use in Personal Area Networks (PANs) has been introduced with the appearance and dimensions of a standard jeans button. The antenna has been realized in both single and double band versions covering the WLAN and HiperLAN/2 bands. The antenna tuning is strongly determined by the capacitance between the various discs that make up the structure and frequencies can be controlled by varying the disc diameters or separations according to the parametric study presented. The radiation currents mainly flow axially along the central post of the structure giving it the radiation characteristics of a monopole on a small ground plane. The measured scattering parameters of the antenna are captured and used to create an equivalent circuit of lumped elements which accurately describes the frequency response. This model will be developed to allow closed form design of the DBMBA in the future.

References

1. Barfield, W. and Thomas, C. eds., "Fundamentals of Wearable Computers and Augmented Reality", Lawrence Erlbaun, 2000. ISBN 0-8058-2902-4.
2. Lukowicz, P., Buren, T.V., Junker, H., Stager, M., "WearNET: A Distributed Multi-Sensor System for Context Aware Wearables", In Proc. 4th Int. Conf. On Ubiquitous Computing, Goteborg, Sweden, 2002.
3. Salonen, P., Kim, J., and Rahmat-Samii, Y., "Dual-Band E-Shaped Patch Wearable Textile Antenna", IEEE Proc. AP-S, Washington DC, 2005.
4. Ouyang, Y., Karayianni, E., and Chappell, W.J., "Effect of Fabric Patterns on Electrotextile Patch Antennas", IEEE Proc. AP-S, Washington DC, 2005.
5. Tanaka, M. and Jang, J.-H., "Wearable Microstrip Antenna", IEEE Antenn. Propag. Int. Symposium, Vol. 2, June 2003, pp. 704–707.
6. Klemm, M., Locher, I., and Troster, G., "A Novel Circularly Polarized Textile Antenna for Wearable Applications", 34th European Microwave Conference (EuMC), Amsterdam, Netherlands, October 2004.
7. See, T.S.P. and Chen, Z.N., "Effects of Human Body on Performance of Wearable PIFAs and RF Transmission", IEEE Proc. AP-S, Washington DC, 2005.
8. Kamarudin, M.R., Nechayev, Y.I., and Hall, P.S., "Performance of Antennas in the On-Body Environment", IEEE Int. Workshop on Antenna Technology, IWAT, pp.17–20, Singapore, March 2005.
9. Nachayev, Y., Hall, P., Constantinou, C.C., Hao, Y., Owadally, A., and Parini, C.G., "Path loss Measurements of On-Body Propagation Channels", in Proc. 2004, Int. Symp. Antenn. Propag., Sendai, Japan, pp. 745–748.
10. Alomainy, A., Hao, Y., Parini, C.G., and Hall, P.S., "Comparison Between Two Different Antennas for UWB On-Body Propagation Measurements", IEEE Antenn. Wirel. Propag. Lett., Vol. 4, 2005, pp. 31–34.
11. Zheng, J.-X. and Chang, D., 'End-Correction Network of a Coaxial Probe for Microstrip Patch Antennas', IEEE Trans. Antennas Propagat., Vol. 39, No. 1, pp. 115–118, 1991.
12. Klemm, M., Kovcs, I.Z., Pedersen, G.F., and Troster, G., "Novel Small-Size Directional Antenna for UWB WBAN/WPAN Applications", IEEE Trans. Antenn. Progag., Vol. 53, No. 12, December 2005, pp. 3884–3896.
13. 'Calculation of the Dielectric Properties of Body Tissues in the Frequency Range 10Hz–100GHz', Nello Carrara, Italian National Research Council, Institute for Applied Physics, Florence, http://niremf.ifac.cnr.it/tissprop/htmlclie/htmlclie.htm.

Fast and Efficient Methods for Circuit-based Automotive EMC Simulation

Martin L. Zitzmann and Robert Weigel

Abstract This paper presents a novel approach to eliminate the most serious bottlenecks in BEM-based EMC simulation by introducing fast and efficient modeling and solving techniques. A suitable modeling of 3D electromagnetic effects by an equivalent circuit representation can be accomplished by the well-established *PEEC* method. Typically, the resulting system matrices can be very large, dense and ill-conditioned. To enable the analysis of real-life problems, an acceleration has to be applied to the modeling as well as the solving process. Within the scope of this work, a methodology was developed to realize an overall EMC simulation process with significantly reduced complexity in terms of CPU time and storage demands for the underlying equation system. Various methods for model extraction and solving were investigated, including *Hierarchical matrices* ($\mathcal{H}$-matrices) and the well-known *Algebraic multigrid* (AMG) approach, allowing simulations with almost optimal complexity, in principle. These methods fulfill the accuracy demands and are suitable for parallel implementation. The generality, robustness, flexibility and efficiency of the proposed methods are shown by means of numerical results from a typical automotive application.

1 Introduction

Electromagnetic compatibility (EMC) deals with the interaction between electronic systems and their electromagnetic (EM) environment. Developing electronic components, some EMC aspects have to be taken into consideration to guarantee overall functionality in a complex electromagnetic environment. The complexity of problems arising in automotive EMC analysis increases with the number of electronic

Martin L. Zitzmann
BMW Group, Development Ignition Systems, Hufelandstrasse 4, 80788 Munich, Germany, e-mail: Martin-Ludwig.Zitzmann@bmw.de

Robert Weigel
Institute for Electronics Engineering, University of Erlangen-Nuremberg, Cauerstrasse 9, 91058 Erlangen Germany, e-mail: weigel@lfte.de

P. Russer, U. Siart (eds.), *Time Domain Methods in Electrodynamics*,

systems assembled in modern cars. Components containing integrated circuits and controlled sensors and actuators are influencing factors for the EMC quality. Disturbances like electronic noise on power and ground planes which is known as *simultaneous switching noise* (SSN) are transfered via the cable harnesses to the whole system. To avoid cost intensive re-design cycles in the development process, numerical simulation techniques help to identify EMC relevant problems in the very early design stage. Whereas cable harness simulations base on a 2D TL (*transmission line*) apporach, for PCB simulations the 3D PEEC (*partial element equivalent circuit*) method [1] is applied. The PEEC method enables a circuit representation of the electric field integral equation (EFIE) and therefore a connection between the field domain and the circuit domain. Circuit simulators compute voltages and currents in *lumped* circuit elements as resistors, capacitors and inductors. A quasi-stationary consideration can often be sufficient performing signal integrity (SI) analysis. Therefore the interconnects can be modeled using lumped circuit elements. If the signal rise time is comparable to the round-trip propagation time, however, a lumped model is not sufficient and retardation effects have to be included [2]. To enable an efficient and fast system simulation containing complex models of various types, a flexible and robust solver strategy is necessary. This paper describes a fast and efficient modeling and solver strategy for circuit-based EMC simulation which takes influences on the matrix properties into account and uses high-sophisticated mathematical methods to enhance the overall simulation process. Hierarchical methods allow to accelerate the modeling as well as the simulation process tremendously. Hierarchical matrices provide a possibility to extract parasitic effects in an efficient way. The application of a specific $\mathcal{H}$-matrix arithmetic enables a very efficient computation of preconditioners for iterative solving techniques. Combined with the well-known algebraic multigrid approach, this methodology can help to decrease the effort in today's circuit-based EMC simulation drastically. The presented methods are also suitable to realize a hybrid simulation strategy for an automotive EMC simulation on system level.

This paper is organized as follows. In Sect. 2 a short introduction to the theoretical aspects of PEEC modeling and the MNA formulation is given, providing the basis for circuit-level EM simulation. Section 3 presents a new extraction and compression method for parasitics from a BEM approach which can be combined together with iterative solvers in Sect. 4. Section 5 describes the application of multigrid methods to PEEC system matrices and Sect. 6 some parallelization aspects for hierarchical methods. Numerical results are given in Sect. 7. In the following small bold letters represent vectors, capital bold letter denote matrices and normal letters represent scalar variables.

2 PEEC-based Circuit Simulation

The PEEC method enables a circuit interpretation of the EFIE full-wave formulation in 3D for a combined analysis of electromagnetic and circuit problems. Power systems, VLSI systems, EMI and ESD are typical problems, which are studied by means of circuit simulators based on the solution of ordinary differential equations.

2.1 PEEC Model Derivation

A complete theoretical derivation of the PEEC method is given in [1]. This integral equation based *boundary element method* (BEM) approach is used for 3D modeling of complex conducting structures as PCB traces and leads to dense matrices, in general. It explains how a conducting object can be transformed into a linear circuit. A discretization of the conducting medium into volume and surface cells is necessary to approximate current and charge densities, respectively. The *method of moments* (MoM) [3] is applied togehter with GALERKIN's method [4] to the EFIE. Constant pulse functions are used as fundamental basis functions. For an analysis how such effects influence simulation results using PEEC models, the reader is referred to [2]. The PEEC formulation proved to be equivalent to a MoM formulation but it exhibits some important advantages. The resulting circuit representation can be analyzed in the time domain (TD) as well as in the frequency domain (FD) using conventional SPICE simulators [5]. It allows to include dielectrics and retardation effects, it is compatible with non-orthogonal discretizations [6, 4], it enables the application of various methods of model order reduction and it is very flexible for extensions by additional circuit elements. In contrast with MoM, PEEC is a *full-spectrum* method, valid from DC to the maximum admissible frequency defined by the discretization. For accuracy and validity aspects of PEEC models in the TD and the FD refer to [7].

2.2 MNA Formulation of Linear Networks

A given linear circuit like a PEEC model with n_n nodes and n_b branches can be described mathematically by an *adjacency matrix* $\boldsymbol{A}_a \in \mathbb{Z}^{n_b \times n_n}$ [8], representing the network's topology. Each row only contains two nonzero entries $\{-1,+1\}$ and the row-sums are zero. With this adjacency matrix, KIRCHHOFF's current and voltage laws and the branch constitutive relations (BCR) characterize the electrical behavior of a network completely. The BCR contain the mathematical models of the electrical behavior of circuit primitives (e.g. $u = Ri$) [9].

Combining KIRCHHOFF's laws with the BCR results in the *modified nodal analysis* (MNA) formulation [9, 8]

$$\boldsymbol{H}\boldsymbol{x} + \boldsymbol{N}\frac{\partial \boldsymbol{x}}{\partial t} = \boldsymbol{b} \tag{1}$$

with

$$\boldsymbol{H} = \begin{bmatrix} \boldsymbol{G} & \boldsymbol{A}_l^T \\ -\boldsymbol{A}_l & \boldsymbol{0} \end{bmatrix}, \quad \boldsymbol{N} = \begin{bmatrix} \boldsymbol{C} & \boldsymbol{0} \\ \boldsymbol{0} & \boldsymbol{L} \end{bmatrix}. \tag{2}$$

In (2) $\boldsymbol{H}, \boldsymbol{N} \in \mathbb{R}^{n \times n}$, $\boldsymbol{x} = [\boldsymbol{u}_n^T \ \boldsymbol{i}_b^T]^T \in \mathbb{R}^n$ represents the vector of node voltages and branch currents, $\boldsymbol{b} = [\boldsymbol{b}_1^T \ \boldsymbol{b}_2^T]^T$ represents the vector of current sources and $\boldsymbol{G}$, $\boldsymbol{C}$, $\boldsymbol{L}$ the symmetric positive definite conductance, capacitance and inductance matrices. The topology of the inductive elements is represented by the inductive incidence matrix $\boldsymbol{A}_l$. The output equation is given by $\boldsymbol{u}_p = \boldsymbol{B}^T \boldsymbol{x}$ with the port voltages $\boldsymbol{u}_p$ and the number of ports n_T. Each column of the port connectivity matrix $\boldsymbol{B} \in \mathbb{N}^{n \times n_T}$ contains zeros and one single $+1$.

The applied solver strategy for circuit simulation is based on the solution of large systems of linear equations with n degrees of freedom (DOF) of the form

$$\boldsymbol{A}\boldsymbol{x} = \boldsymbol{b} \tag{3}$$

with $\boldsymbol{A} = \boldsymbol{H} + s\boldsymbol{N}$ in FD (s represents the complex frequency variable) to reach a given accuracy of the residuals $\|\boldsymbol{A}\tilde{\boldsymbol{x}} - \boldsymbol{b}\|_2 \leq \varepsilon$ for an approximation $\tilde{\boldsymbol{x}}$. In SPICE the build-up process of the system matrix $\boldsymbol{A}$ can be carried out very fast by using matrix stamping [9] for circuit primitives. If a system to be simulated consists of models of different modeling strategies, the individual model properties have to be taken into account. They have a direct impact to the systems condition and thus on time and memory requirements of the solution process.

2.3 Circuit-based EMC Simulation

For the analysis of PEEC models, a conventional SPICE-like circuit simulator can be used. The fundamental EMC simulation workflow is visualized in Fig. 1. Based on a geometrical design, a discretization is necessary, building the basis for

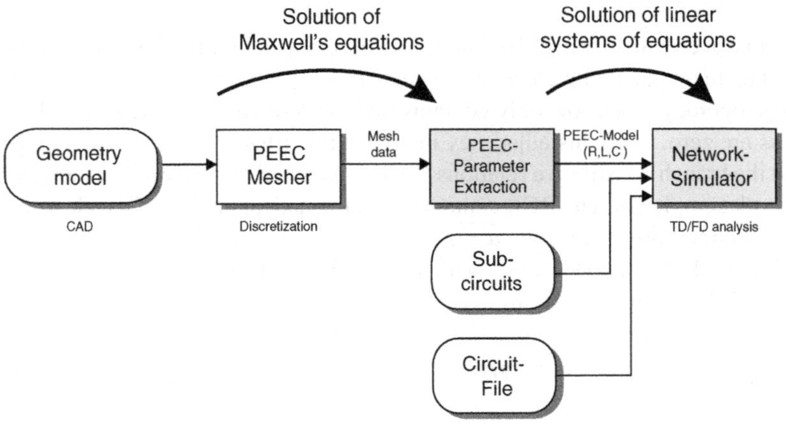

Fig. 1 Circuit-based EMC simulation workflow

a numerical computation. By solving Maxwell's equations (refer to Sect. 2.1), the PEEC parameter extraction (modeling) can be accomplished, resulting in a so-called (R,L,C)PEEC model. Together with optional subcircuits, the network simulator solves the network equations in the TD or the FD, corresponding to the simulation options predefined in the circuit file. The modeling as well as the solving process represent the major two bottlenecks of the computational performance of the total EMC simulation workflow. As described in the following sections, hierarchical techniques can be used to handle BEM-based element matrices with high accuracy and efficiency and therefore to avoid these shortcomings.

3 Fast and Efficient Extraction of Parasitic Effects

A fast and accurate computation of matrix elements represents one major issue to realize an efficient simulation process. This is especially important for parasitic matrices originating from electromagnetic boundary integral equation (BIE) approaches as the PEEC method. BEM-based matrices as inductance matrices and potential coefficient matrices, whose elements resulting from the application of an integral operator of the form

$$\mathscr{L}u(\boldsymbol{r}) = \int_{\Omega} g(\boldsymbol{r},\boldsymbol{r}')u(\boldsymbol{r}')\,\mathrm{d}\boldsymbol{r}', \tag{4}$$

defined on domain $\Omega \in \mathbb{R}^3$ with kernel function $g : \mathbb{R}^3 \times \mathbb{R}^3 \rightarrow \mathbb{R}$, are dense, in general. These matrices comprise $n = \mathscr{O}(h^{-2})$ degrees of freedom for geometric meshing with meshsize h thus exhibiting $\mathscr{O}(h^{-4})$ matrix entries [10] in 2D. Both the TD and the FD analysis need to compute large numbers of interactions between basis functions for each and every simulation step, leading to enormous CPU time and memory requirements. This can form a severe limitation to the number of DOF in practical relevant problems. Fast methods are necessary to reduce the complexity of basic arithmetic operations. The basic idea of fast computation methods for BEM-based matrices is, to take advantage of the smoothness of the integral operator's kernel function far away from the singularity. Smooth functions can be described by series expansions, in general. A good approximation can be obtained by applying a truncation to the series expansion. The *fast multipole method* (FMM) enables fast model computations with a complexity of $\mathscr{O}(n^{1.5})$ for matrix dimension n, in principle. For a detailed view to the FMM for PEEC-based circuit simulation, the reader is referred to [11, 12]. In a multilevel context, the FMM allows to reduce the storage as well as the computational complexity to $\mathscr{O}(n \log^{\kappa} n)$, where $\kappa = 4$ holds for some special kernel functions and $\kappa \sim 6,7$ for general kernels [13]. The disadvantage of the FMM is, that the expansions can not be applied globally to the whole matrix but only to matrix subblocks belonging to smooth areas of the kernel function. A more advantageous method for circuit-based EM modeling is described in the following subsection.

3.1 Hierarchical Matrices

Hierarchical matrices ("$\mathcal{H}$-matrices") help to represent the matrix as a decomposition of matrix subblocks. They provide a purely mathematical approach to describe dense matrices by so-called "*data-sparse*" matrices. Data-sparse matrices are sparse in a sense, that only few data are necessary for representation [14]. Based on this kind of description, it is possible to apply approximative algorithms for matrix arithmetic with almost optimal complexity. Together with PEEC matrices one exploits the smoothness of GREEN's function

$$G(\boldsymbol{r},\boldsymbol{r}') = \frac{1}{4\pi\|\boldsymbol{r}-\boldsymbol{r}'\|_2} \tag{5}$$

far from the origin. Green's function represents the most common kernel function in computational EM analysis. This matter of fact makes it possible, to approximate the integral operator of the occuring terms up to a tolerance of $\mathcal{O}(n^{-\kappa}),\ \kappa > 0$ by a data-sparse $\mathcal{H}$-matrix of dimension $n \times n$ with almost linear complexity. Compared to the FMM, $\mathcal{H}$-matrices offer the advantage that the kernel function has not to be known explicitely. It is sufficient that the kernel proceeds smooth far away from the singularity. Consequently, the approximation has to be applied locally in each subblock. A problem occurs with blocks containing a singularity. These so-called *nearfield* blocks can effectively be treated by quadrature rules. In multipole methods the so-called *farfield* blocks are computed "on-the-fly", which makes it necessary only to store the nearfield. Using $\mathcal{H}$-matrices, the kernel function expansions enable a description of farfield blocks by matrices with low rank. It can be shown, that the rank for these low-rank blocks is bounded by the number of terms in the expansion. By storing the farfield subblocks as low-rank matrices, one can take advantage of the possibility to apply special optimization algorithms combined with minimal storage requirements [15]. *Adaptive recompression* techniques allow an approximation with "optimal" rank, comprising an optimization with regard to minimal storage requirements. The FMM shows the disadvantage that the implementation needs a complete reconding of the matrix-vector multiplication [16]. Moreover, $\mathcal{H}$-matrix techniques allow to apply a special *$\mathcal{H}$-matrix arithmetic* which makes it possible to realize efficient direct as well as iterative methods [17, 18]. Together with $\mathcal{H}$-matrices, it is possible to store nearfield parts as well as farfield parts of the matrix in an efficient way with only low storage requirements and to apply modern recompression algorithms for a further optimization. $\mathcal{H}$-matrix techniques require slightly more memory compared to FMM-based methods, in general, because the farfield part has to be stored. In contrast, $\mathcal{H}$-matrices are more efficient in the computation of matrix-vector multiplications since only elementary operations as multiplications and additions of floating point values have to be executed. The purely algebraic approach of $\mathcal{H}$-matrices, together with a suitably defined $\mathcal{H}$-matrix arithmetic, exhibits the possibility to construct very efficient preconditioners for iterative methods. This represents a bonus, leading to a maximum of flexibility. To enable an optimal asymptotic efficiency for very large systems, so-called *$\mathcal{H}^2$-matrices* [19] can be useful. $\mathscr{H}^2$-matrices are a variant of $\mathscr{H}$-matrices and apply techniques from the field of multigrid methods (refer to Sect. 5). $\mathscr{H}^2$-matrices enable optimizations similar to $\mathscr{H}$-matrices, however, the securing of special structural properties leads to a higher algorithmic complexity. Basic algorithms and numerical results for verification are presented in [19].

3.2 $\mathscr{H}$-Matrix-based fast Model Extraction

In $\mathscr{H}$-matrix-based model computation, a fast and accurate approximation of a parasitics matrix is aspired, comparable to the FMM. Controlled by a suitable "*heuristics*" only required matrix entries have to be computed, leading to a complexity reduction from $\mathscr{O}(n^2)$ to $\mathscr{O}(n \log n)$ [20]. Furthermore, $\mathscr{H}$-matrices provide additional advantageous features to enhance efficiency and flexibility, compared to the FMM. First of all, a block clustering structure (empty $\mathscr{H}$-matrix) has to be arranged as described in [18], solely based on the geometric information concerning the grid nodes. A graphical illustration for an influence extraction is presented in Fig. 2. The interactions between the clusters can be classified into nearfield and farfield, implying a suitably chosen "admissibility condition". Such a separation leads to the fact, that for cluster pairs offering a certain distance to each other, the singularity in the boundary integral kernels not need to be considered [21]. In a second step, the farfield low-rank blocks, representing the interactions, have to be filled data-sparse. This can be realized intelligently using the *ACA algorithm (adaptive cross approximation)*[1] [22] which needs no explicit knowledge about the kernel function, unlike together with the FMM. The approximation purely bases on algebraic transformations where the heuristics only requests such entries for computation, which are necessary for filling the blocks. The so-called "η-admissibility" [17] has

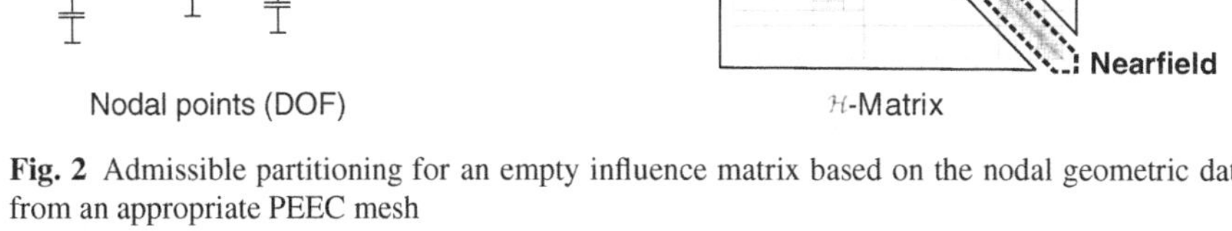

Fig. 2 Admissible partitioning for an empty influence matrix based on the nodal geometric data from an appropriate PEEC mesh

[1] From literature [22] it is well known, that under suitable conditions for BEM matrices and applying the single layer potential, for asymptotically smooth kernels an approximation of the truncated SVD can be obtained by ACA with time complexity $\mathscr{O}(nk)$, where k denotes the reduced rank [23].

to be checked for each matrix-block in the empty $\mathscr{H}$-matrix. If the admissibility condition is fulfilled, only few entries have to be computed. In contrast, inadmissible blocks are computed as full matrices with complexity $\mathscr{O}(n^2)$. In principle, the ACA algorithm can be understood as a function that returns two rank-k matrices $\boldsymbol{A}_k,\, \boldsymbol{B}_k \in \mathbb{C}^{n\times k}$ such that

$$\|\boldsymbol{A} - \boldsymbol{A}_k \boldsymbol{B}_k^T\|_2 < \varepsilon \tag{6}$$

holds for a given accuracy $\varepsilon > 0$ and the matrix $\boldsymbol{A}$ corresponding to an admissible block. Only $\mathscr{O}(n\log n) \ll \mathscr{O}(n^2)$ matrix entries A_{ij} for an index-pair (i,j) have to be computed, with indices i,j originating from the index set $\mathscr{I}$. According to [24], a relative approximation error of $\mathscr{O}(\varepsilon)$ can be achieved using $k = \log(\varepsilon)^2$. Although the ACA algorithm needs no expansion of the kernel function, it may malfunction together with not asymptotic smooth kernel functions also in the modified version[2]. While the ACA is absolutely adequate for the quasi-stationary PEEC case, problems can occur in the retarded case. A remedy can be achieved by applying an enhanced version of the ACA, the so-called *HCA algorithm (hybrid cross approximation)* [24].

4 Hierarchical Matrices Based Solvers for PEEC

Utilizing iterative methods to obtain an accurate solution of $\boldsymbol{Ax} = \boldsymbol{b}$ inevitably leads to square complexity w.r.t. CPU time requirements even for sparse matrices. In the context of BEM-based PEEC matrices the intention of this section is, to realize iterative solutions with almost linear complexity w.r.t. CPU time and memory requirements for dense matrices. This can be accomplished by approximating the PEEC-based system matrix $\boldsymbol{A}$ by a *data-sparse* $\mathscr{H}$-matrix $\boldsymbol{A}_{\mathscr{H}}$, as described in the previous section. The approximative solution $\tilde{\boldsymbol{x}}$ of the equation system $\boldsymbol{A}_{\mathscr{H}}\boldsymbol{x} = \boldsymbol{b}$ must fulfill the accuracy condition $\|\boldsymbol{x} - \tilde{\boldsymbol{x}}\|_2 \leq \varepsilon$ for a given $\varepsilon > 0$. Based on the *formatted $\mathscr{H}$-matrix arithmetic* [17], efficient preconditioning techniques can be realized. Implementation issues for an appropriate matrix setup and suitable preconditioning for an efficient PEEC-based circuit simulation have to be taken into account.

4.1 Iterative Solution Strategy

Using conventional methods, a complexity of $\mathscr{O}(n_s n^2)$ for a number of n_s simulation steps and dense matrix dimension n [11] can be obtained. The complexity of the matrix-vector multiplication can be reduced to $\mathscr{O}(n_s n \log^2(n))$ by applying a multilevel FMM approach. Where the nearfield interactions have to be computed with

[2] *Remark*: The ACA is not an exact method, since it does not produce the optimal rank k. Compared to the truncated SVD which results in the optimal approximation, however, the ACA produces almost optimal results in practice with only a fraction of the cost.

complexity $\mathcal{O}(n^2)$, the farfield can effectively be treated using multipole approaches or hierarchical matrix techniques. Hierarchical methods can be well suited also for original sparse matrices since the inverse of a sparse matrix is dense, in general. If the approximate inverse can be calculated with a prescribed accuracy, a bounded number of iterations of Krylov-subspace methods can be obtained [25]. For the following theoretical investigations a circuit simulation in the TD or the FD can be understood as the task to solve a sequence of systems of the form [25]

$$\boldsymbol{A}_N \boldsymbol{x}_N = \boldsymbol{b}_N, \quad N = 1, \ldots, n_s \tag{7}$$

with an invertible system matrix $\boldsymbol{A}_N \in \mathbb{R}^{n \times n}$. Applying a left preconditioning procedure with a preconditioner $\boldsymbol{A}_{\mathcal{H},\text{pre}}$, the resulting linear equation system in $\mathcal{H}$-notation can be given in the form

$$\left(\boldsymbol{A}_{\mathcal{H},\text{pre}}\right)^{-1} \boldsymbol{A}_{\mathcal{H}} \boldsymbol{x} = \left(\boldsymbol{A}_{\mathcal{H},\text{pre}}\right)^{-1} \boldsymbol{b}. \tag{8}$$

The convergence rate is defined by the spectral radius of $\mathbf{1} - \left(\boldsymbol{A}_{\mathcal{H},\text{pre}}\right)^{-1} \boldsymbol{A}_{\mathcal{H}}$. An upper bound for the spectral radius is given by the norm $\|\mathbf{1} - \left(\boldsymbol{A}_{\mathcal{H},\text{pre}}\right)^{-1} \boldsymbol{A}_{\mathcal{H}}\|_2$ which should be < 1.

4.2 $\mathcal{H}$-Matrix-based Adaptive Preconditioning

A suitable preconditioner for data-sparse network matrices can effectively be computed in $\mathcal{H}$-matrix arithmetic. Within the context of $\mathcal{H}$-matrices, the conventional matrix operations as matrix additions or matrix multiplications have to be replaced by the *formatted* operations [25]. Compared to the $\mathcal{H}$-Inverse-preconditioning, the $\mathcal{H}$-*LU*-preconditioning comprises the same complexity and usually produces less fill-ins. Both approaches, however, are used for practical relevant circuit simulations. For investigations and analysis in the scope of this work, the $\mathcal{H}$-*LU*-decomposition is used without sacrifying general validity. The hierarchical *LU*-decomposition is based on a block-decomposition of the matrix and a gradual forward and backward substitution. Summarizing, this approach is more efficient than computing the inverse of the matrix. To realize an efficient $\mathcal{H}$-preconditioner $\boldsymbol{M}$, the approximate decomposition

$$\boldsymbol{M}^{-1} \simeq (\boldsymbol{L}_{\mathcal{H}} \boldsymbol{U}_{\mathcal{H}})^{-1} = \boldsymbol{U}_{\mathcal{H}}^{-1} \boldsymbol{L}_{\mathcal{H}}^{-1} \tag{9}$$

with the lower and upper triangular $\mathcal{H}$-matrices $\boldsymbol{L}_{\mathcal{H}}$ and $\boldsymbol{U}_{\mathcal{H}}$ can be applied. The $\mathcal{H}$-matrices have to be constructed with $\varepsilon_{\mathcal{H},\text{pre}} \gg \varepsilon_{\mathcal{H}}$, such that (9) represents a decomposition, accurate enough for preconditioning. For circuit simulation purposes w.r.t. efficiency, the iterative algorithms GMRes and BiCGstab are recommended, requiring specific forward/backward substitutions and matrix-matrix multiplications to compute the approximate inverse of $\boldsymbol{M}$. The preconditioner computation usually represents the dominating effort in an iterative solving step. Moreover, together with circuit simulations in the TD as well as the FD, the spectral properties of the system matrix only slightly vary during simulation progress. Consequently, monitoring the convergence behavior of the iterative solution process from one step to another can be advantageous to exactly determine the right time for a recomputation or update of the preconditioner instead of recomputing in each step. So, a very efficient $\mathscr{H}$-matrix-based "adaptive" preconditioning can be realized.

4.3 $\mathscr{H}$-Matrices for the Retarded Case

To utilize the benefits from the application of $\mathscr{H}$-matrices also together with retarded PEEC models [2], it is necessary to map the proposed premises and methods, described in the previous sections, to the retarded case. The "*retarded*" Green's function in 3D

$$G(\boldsymbol{r},\boldsymbol{r}') = \frac{e^{j\beta\|\boldsymbol{r}-\boldsymbol{r}'\|_2}}{4\pi\|\boldsymbol{r}-\boldsymbol{r}'\|_2} \tag{10}$$

represents the kernel function originating from the HELMHOLTZ equation. While a separable expansion of the kernel function for the FMM is possible, the aim is to find a method which is numerically stable and moreover exhibits almost optimal complexity w.r.t. time and memory requirements. From the literature [23], it is well known that for the high-frequency regime in 2D the number of the boundary elements n is proportional to the wave number β and in 3D n is proportional to β^2. With the meshsize h, $\beta\ h \overset{!}{=} \text{const}$ must hold. If a required accuracy is fixed, length k of the kernel functions separable expansion can be expected to be proportional to β and hence strongly dependent on n. This length of expansion is constant for a fixed accuracy since for the low-frequency domain β is a constant w.r.t. n. Most of the FMM techniques, comprising a complexity of $\mathscr{O}(n \log^{\kappa} n)$ for some small constant κ, suffer from numerical instabilities at low frequencies in treating the retarded Green's functions [23, 26]. The approach proposed by BANJAI and HACKBUSCH [23] approximates the matrices resulting from the low-frequency domain by $\mathscr{H}$-matrices, while for matrices from the high-frequency regime a combination of a $\mathscr{H}$-matrix and a $\mathscr{H}^2$-matrix is used. However, while recompression techniques are applicable in a straightforward manner for the low-frequency regime, additional research is necessary for the high-frequency range. In 3D, the fundamental solution of the integral operator respresents the zero-order spherical Hankel function of the first kind. From basic $\mathscr{H}$-matrix theory it can be observed, that in the high-frequency regime, where βh is a constant, $\mathscr{H}$-matrix techniques can not be applied. As a remedy, $\mathscr{H}^2$-matrices can be used. Starting from a separable expansion of the corresponding retarded kernel function and a truncation, the $\mathscr{H}^2$-matrix construction by construction of the cluster basis, the transfer operator and the corresponding coefficient matrix can be applied in a straightforward manner.

5 Multigrid Solvers for PEEC

In the literature it is well known that the number of iteration steps, required to achieve convergence, increases with the problem size [27] using conventional preconditioned iterative solvers. *Multilevel approaches*, if suitably applied, are "*scalable*", i.e. the convergence rate is independent of the problem size and the number of iteration steps stays almost constant[3]. Very suited problem classes for multilevel methods are, e.g. discrete elliptic matrices or mixed elliptic-hyperbolic matrices. In general, the maximum efficiency of multilevel methods unfolds with very large problem dimensions. Multilevel methods are often called "*multigrid*" (MG) methods, which can achieve linear complexity, in principle. The two fundamental principles of multigrid are *smoothing* and *coarse-level correction*. To realize a very robust and efficient solution process, the advantages of MG and Krylov one-level methods can be combined in a *multigrid preconditioned Krylov method*. PEEC matrices from a BEM approach, however, need special treatment in an AMG context [28]. The superiority of AMG particularly results from the robustness and applicability for *unstructured* grids in 2D and 3D as well and problems with anisotropic or even discontinuous coefficients. Furthermore, AMG allows for an efficient parallel implementation.

5.1 Motivation

Although one-level iterative solvers exhibit a complexity of $\mathcal{O}(n^\kappa)$, $(\kappa > 1)$ w.r.t. computational work [29] for matrix dimension n, they can show a bad convergence behavior for large and ill-conditioned systems of equations. The reason for this lies in the fact that high-frequency error components are much better reduced than low-frequency error components[4]. Furthermore, it is obvious that for the treatment of lower-frequency contributions of the iteration error a coarse grid is sufficient, while highly-oscillating contributions are treated well on a fine grid [28]. It can be observed that the error components are damped by the iteration matrix. The spectral radius of the iteration matrix is uniformly bounded by a constant smaller than one, leading to the efficiency of multigrid methods. With decreasing meshsize, more and more high frequencies in the error occur, resulting in a mesh-dependence of the overall efficiency of the iterative method [30]. Due to the capability of better reducing high-frequency error components than low-frequency error components, simple iteration schemes as *Jacobi* or *Gauß-Seidel* are also called *"smoothers"*. The basic intention of multilevel approaches is, to restrict the remaining relatively smooth error to a coarser grid, where it can be effectively reduced (so-called *"coarse-grid*

[3] An algorithm is numerically scalable if its complexity in terms of computational time and memory requirements proceeds linearly with the problem size [31].

[4] While small eigenvalues exhibit low-frequency eigenfunctions, large eigenvalues correspond to high-frequency eigenfunctions.

correction"). Consequently, the mesh-dependency of the convergence behavior can be removed and optimal complexity of $\mathcal{O}(n)$, w.r.t both computational work and memory requirements can be achieved [29], in principle.

5.2 AMG Preconditioners for PEEC Matrices

Although, by applying multigrid methods one can achieve high efficiency for various problems, slowly converging eigenvectors of the iteration matrix can lead to bad convergence results. Multigrid algorithms tend to be problem specific and less robust than preconditioned Krylov iterative methods. In order to enhance the robustness and efficiency of multigrid methods, Krylov methods can be included. In this context, multigrid is used as a preconditioner in a Krylov-subspace method, often leading to a more robust, efficient and scalable algorithm. As mentioned in the introduction of this section, multigrid solvers are scalable if applied suitably, in contrast to conventional preconditioned iterative solvers. Multigrid preconditioning means that the iteration matrix of the multigrid method, which is not explicitely needed, plays the role of the preconditioner [10]. The preconditioning steps are substituted by "solving" approximately appropriate systems by the aforementioned multigrid method. Since a multigrid cycle is designed to reduce the error, it can also be used as an approximate inverse of the system matrix, and therefore as a preconditioner for an iterative solver. If the multigrid cycle is very effective, classical multigrid already converges very quickly.

AMG strategies are applicable for various circuit simulation purposes as shown in [32]. In the field of effective power grid reduction [33] sparse system matrices occur, suitable for AMG-based circuit simulation for large-scale power grid transient and FD analysis. The AMG components are constructed directly from the circuit MNA matrices as described in detail in [32]. A significant memory reduction can be obtained, since no geometric information is needed. Assuming smooth voltage variations, experimental results show a significant reduction in problem size without sacrifying the accuracy. Electromagnetic effects (e.g. cross-talking) prevent the resulting system matrices from being M-matrices[5]. Such circuit simulation matrices often exhibit very large condition numbers. To achieve high efficiency of AMG also together with BEM-based PEEC matrices, a special treatment of the fully-populated matrices has to be accomplished [7]. The character of the underlying boundary integral operator has to be taken into account for an efficient component computation. For the single layer potential operator with Green's function as kernel, one has to consider that high-frequency eigenfunctions correspond to small eigenvalues [28]. AMG was originally designed for differential-based approaches. Together with PEEC matrices, therefore, one has to construct a smoother for *pseudo-differential* operators of order minus one, as mentioned in the context with the SCHWARTZ

[5] M-matrices often occur from discrete elliptic problems.

kernel theorem [34][6]. To avoid the problem resulting from the conversely behavior of pseudo-differential operators of negative order, a certain inner product can be applied. In fact, the MG reduction process is equivalent to the linear preconditioned iterative scheme. To reduce the algorithmic complexity of a matrix-vector multiplication for BEM-based circuit simulation matrices, two different approaches can be applied – *truncated reluctance matrix* based PEEC and PEEC based on the *adaptive cross approximation* [7].

6 Parallelization Aspects for Hierarchical Methods

As described in the previous sections, multigrid-based methods as well as $\mathscr{H}$-matrix techniques are able to offer high efficiency together with iterative solution methods for circuit simulation purposes implying almost optimal complexity w.r.t. CPU-time and memory requirements. An additional speed-up, however, often can only be achieved by applying *parallelization* techniques. Only the most powerful and sequentially fastest algorithms can be considered to realize optimal parallel solvers for complex industrial applications. A decomposition of the domain of interest into subdomains respresents a common parallelization strategy in numerical field computation. Orthogonal (rectangular) meshes easily can be partitioned into boxes or cubes, enabling a balanced workload. For unstructured meshes, however, an equal amount of work has to be assigned to each processor. Furthermore, a minimization of the communication overhead between the processors has to be aspired by essentially minimizing the surface area of the subdomain. The adoption of parallel computing is only efficient if the benefit, resulting from the speed-up of the computations, outperforms the communication overhead between the processors and the work is distributed equally to the processors [35]. As a precondition for an adequate achievable speed-up, the operations to be executed must allow a simultaneous processing in addition to an equally distributed workload. The subproblems assigned to each processor are still coupled. This makes a data-transfer between the processors necessary. These data-transfer to be minimized by ensuring uniform distribution of the load, is a high-dimensional discrete optimization problem, a so-called *load-balancing problem*. Iterative methods are easier to implement on parallel computers than sparse direct methods [36]. In iterative algorithms usually matrix-vector multiplication operations are parallelized since they can be implemented on high-performance computers in an easy manner. Especially the application of the CSR/CSC matrix storage format [36, 37, 38] turns out to be advantageous for parallel computations. To implement the computation of the matrix-vector product $\boldsymbol{y} = \boldsymbol{A}\boldsymbol{x}$ in parallel using the CSR/CSC-format, it has to be taken into account that each component of the resulting vector $\boldsymbol{y}$ can be computed independently from the remaining as the dot-product of the ith row of the matrix with vector $\boldsymbol{x}$.

[6] One reason for the ill-conditioning of BEM matrices can also be found in the character of the single layer potential operator, to be represented as a differential operator of order minus one.

For a parallelization of algebraic multigrid algorithms the application of a domain decomposition approach proved to be useful. In contrast to geometric multgrid methods the coarsening hierarchy and the corresponding coarse-grid operators in AMG develop dynamically from one level to another why a parallelization is not trivial, in general [39]. As a main part, the original sequential coarsening strategy has to be replaced by its parallel equivalent. For the parallelization of $\mathscr{H}$-matrices, shared as well as distributed memory systems are suitable. For a parallel $\mathscr{H}$-matrix approximation, *online scheduling* techniques are preferable, since the cost for computing can not be determined in advance. For a parallel $\mathscr{H}$-matrix-vector multiplication, it can be advantageous to split the multiplication into two phases. So, access conflicts can be avoided. A parallel $\mathscr{H}$-matrix inversion can be accomplished, based on the recursive application of the SCHUR complement. For a more detailed view refer to [36].

7 Numerical Results

For simulation as well as measurement purposes an automotive specific demonstrator board with dimensions $160 \times 150\,\text{mm}^2$ is considered whose layout is given in Fig. 3. The board consists of four layers with dielectric medium FR4 ($\varepsilon_r = 4.2$) and stems from a typical application in automotive industry. The fully assembled board contains a microcontroller TC1796 and several power drivers which are driven application-specifically. Diverse operation modes are possible. Measurements are possible via provided connectors. An interface for uC-programming is available.

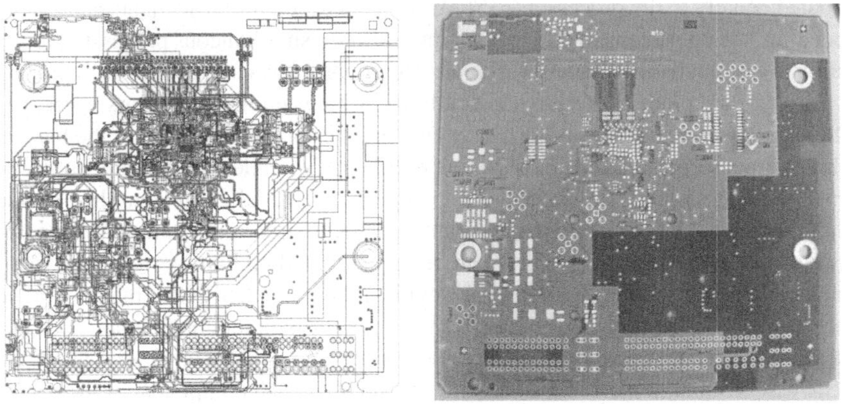

(a) (b)

Fig. 3 MISEA demonstrator board [40]. (**a**) CAD-layout for simulation purposes; (**b**) PCB for measurement purposes (without components)

The investigated design is part of a research project named “MISEA” [40]. A SPICE compatible simulation model of the board is provided by conventional model extraction. It reflects the respective transmission behavior of the power supply traces of the microcontroller as well as for the power drivers and the transmission behavior of the signal traces of the power drivers, respectively.

To show the correspondence of measurement and simulation[7] results a reluctance-based model of a relevant signal trace is considered. A suitably chosen meshsize guarantees the validity of the model in the TD as well as in the FD. Simulations based on a suitably chosen parameter setting agree well with physical measurements. Both model extraction and simulation in the conventional case require a large amount of time. Therefore the total analysis process cannot be carried out in less than one or two days. A first approach for simulation acceleration can be enabled by applying simple matrix sparsification by *truncation* with varying sparsification factor δ_C, δ_K for capacitances and reluctances. For a detailed definition of the truncation procedure, refer to [41, 7]. The partial reluctance K represents the inverse of the partial inductance and enables a sparse matrix formulation while guaranteeing stable circuit models in the contrast to a truncation of partial inductance matrices [42, 43, 7]. The shielding effect of partial reluctances leads to the fact, that mutual reluctance terms drop off much faster than the mutual inductance terms. By applying the truncation with a tolerance limit δ_K to small coupling values in the partial reluctance matrix, the inductive effects can be captured more effectively compared to partial inductances. Although the complexity of the modeling process remains almost constant or slightly increases, respectively, the simulation can be accelerated by a speed-up factor of ≈ 5 compared to the full inductance-based approach. The simulation results for $\delta_C = \delta_K = 0.05\%,\ 0.08\%,\ 0.15\%,\ 0.2\%$ for the S-parameters S_{11} and S_{12} are given in Fig. 4, applying a standard direct solver (GMATRIX). The comparison to measurement results shows good accordance for $\delta_C = \delta_K = 0.05\%,\ 0.08\%,\ 0.15\%$, while a higher tolerance limit directly results in an unaccaptable loss in accuracy. Although a truncation enables a simple sparsification method, it is not efficient enough to realize accurate simulations for large industrially relevant problems. A further enhancement can be accomplished by accelerating model extraction as well as simulation using hierarchical techniques. Applying $\mathcal{H}$-matrix-based computation of parasitics, a modeling speed-up by an additional factor of ≈ 7 can be achieved. Compared to conventional techniques an increased modeling accuracy can be obtained. Therefore, the obtained curves agree much better with the measurement results than in the conventional case with a maximal deviation between measurement and simulation results of 1 dB. An implementation of the proposed algorithms was applied. A gain with a factor of ≈ 20 can be achieved concerning storage requirements by using the $\mathcal{H}$-matrix storage format instead of

[7] *Remark*: Good accordance of experimental and simulation results can only be achieved by preserving strictly controlled conditions. Only under these circumstances simulation and measurement results coincide. Already marginal deviations concerning the arrangement, the wiring or the placement of control modules can cause a significant deviation in the numerical results [44].

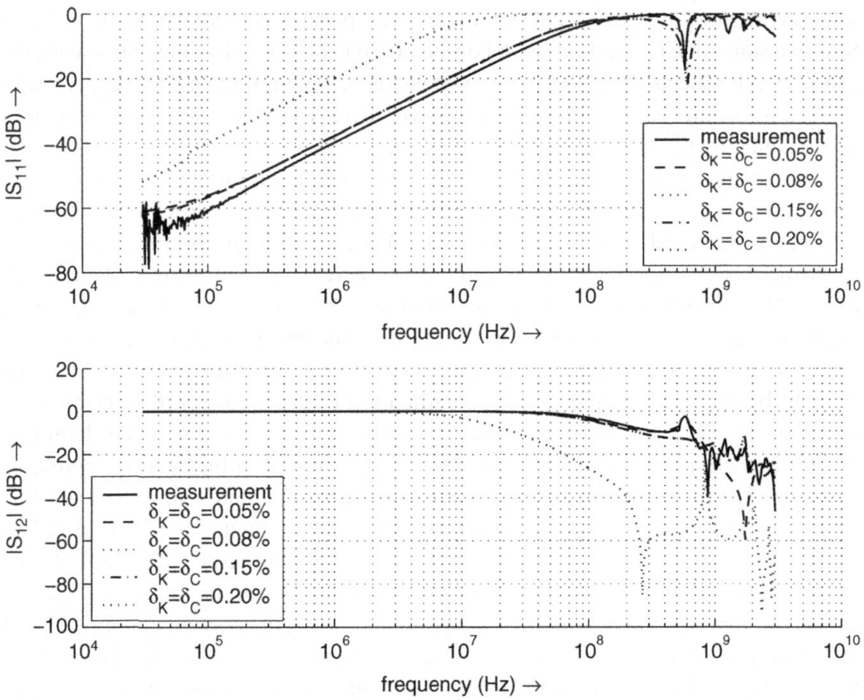

Fig. 4 Comparsion between scattering parameter simulation and measurement results. Due to symmetry reasons only S_{11} and S_{12} are given for varying tolerance limits δ_C, δ_K

the conventional coordinate sparse matrix storage format. The simulation process can be accelerated up to a factor of ≈ 10 by applying an AMG solver, in accordance with the numerical results. While dense PEEC models represent the most critical case for multigrid solvers, an even higher gain can be achieved for TL models. The implementation of AMG in existing codes is a crucial task, due to the fact that multigrid-based solvers usually have to be adapted problem specifically. Using $\mathcal{H}$-matrix techniques, the proposed simulation can be carried out with a speed-up factor of ≈ 5 by applying a one-level iterative solver with $\mathcal{H}$-preconditioning, based on a transformation into an equivalent real-valued system[8] [36]. This also explains increased storage requirements compared to the direct solver case. Adaptive preconditioning, however, enables a speed-up by a factor of ≈ 10. A further gain can be predicted using a combination of hierarchical matrix techniques and algebraic multigrid methods. This is the subject of a current research project in joint work together with the Fraunhofer Institute for Algorithms and Scientific Computing (SCAI) and the Max Planck Institute for Mathematics in the Sciences (MPI) Leipzig, Germany.

[8] An extension of the used $\mathcal{H}$-matrix library to be able to treat complex values is still under work.

8 Conclusion

In the course of this paper, a fast and accurate modeling and simulation methodology has been introduced, suitable for electromagnetic field problems as they are encountered in automotive applications. The BEM-based PEEC modeling approach proved to be useful for the analysis of the electromagnetic behavior of 3D interconnects and traces on PCBs. As the main shortcomings in EMC simulation in the context of a BEM discretization approach, the memory requirements as well as the total simulation times severely limit the performance and therefore the managable problem sizes. The application of direct methods together with dense matrices leads to a time complexity of $\mathcal{O}(n^3)$ and a storage complexity of $\mathcal{O}(n^2)$ with n DOF. Consequently, only small problems could be solved in the past. Even simple iterative solution algorithms usually can only reduce the complexity to $\mathcal{O}(n^{1.5})$ by ensuring convergency. Model order reduction (MOR) techniques often proved to be not efficient enough to enable simulations of real-life problems on system level. The two existing bottlenecks are the modeling as well as the solving processes. Both parts have to be optimized to realize a simulation of industrially relevant problem sizes. This work introduces some promising modern hierarchical techniques for an accelerated model extraction and circuit-level simulation. Hierarchical approaches enable a complexity reduction to $\mathcal{O}(n\log n)$ w.r.t time as well as storage requirements, in principle. To check this ability also in the automotive environment, the presented methods and algorithms are tested by means of real-life problems.

An efficient model extraction approach based on hierarchical matrices is introduced in Sect. 3. This technique overcomes some serious limitations of the FMM and enables a purely mathematical approach to transform dense BEM-based matrices to a so-called *data-sparse* format. The introduced data-sparse representation of element matrices as well as complete network matrices enables a very memory-economical description. Consequently, the memory bottleneck can be eliminated. Numerical results, performed by means of relevant examples in the context of a PEEC-based model extraction, show a significant acceleration of the element evaluation process by a factor of > 100. In particular, computation times for matrix inversions in the hierarchical format can be drastically reduced compared to a full matrix inversion. The storage requirements can be reduced by a factor of $8\ldots50$. Losses in accuracy are vanishing, compared to the accuracy of the conventional PEEC approach. An additional performance improvement can be realized by novel methods as adaptive refinement techniques, adaptive recompression and so-called $\mathcal{H}^2$-matrices. Apart from being dense, BEM-based circuit simulation network matrices usually exhibit large condition numbers, resulting from the different types of element characteristics involved. Fundamental details are presented to realize efficient preconditioned Krylov-subspace iterative methods w.r.t. the application at hand.

Another special focus of this paper has been on the application of $\mathcal{H}$-matrices in a PEEC context in combination with iterative solvers as described in Sect. 4. We introduced *$\mathcal{H}$-matrix arithmetics* to realize an efficient preconditioning for PEEC-based network matrices for simulations in the TD and the FD. An implementation of iterative one-level solvers such as GMRes and BiCGstab in a $\mathcal{H}$-matrix-based framework pointed out to be straightforward using the HLIB [45] and lead to significant gains w.r.t. time and memory requirements. Numerical results show very robust, fast and accurate simulation behavior for industrially relevant PEEC examples. The proposed algorithms work well together with very large, dense and ill-conditioned PEEC-based network matrices. Although this approach unfolds its highest efficiency together with dense matrices, a significant improvement can also be observed for sparse matrices. In combination with different preconditioners, a speed-up by a factor of ≈ 32 in the TD was achieved. The residuals are sufficiently small ($\approx 10^{-10}$) to ensure accurate system simulation. The storage requirements can be reduced up to a factor of ≈ 4 compared to the conventionally used standard direct solver. Especially the proposed adaptive preconditioning [7], enables an additional speed-up by a factor of ≈ 4. In general, efficiency, robustness and flexibility are drastically increased and the limits for managable problem sizes are extended.

To enable fast and accurate circuit simulations especially together with sparse system matrices, the functionality of an AMG approach was introduced in Sect. 5. This very powerful hierarchical approach enables significant speed-up by a factor of ≈ 10 in the TD as well as the FD by using the standard AMG library SAMG [37] together with reluctance-based sparsified PEEC models as well as together with TL models. The achievable residuals applying GMRes or BiCGstab are very small, usually leading to very accurate results. Although AMG is well-established for various kinds of numerical problems (e.g. computational fluid dynamics (CFD) or semiconductor simulation) together with circuit simulation approaches it requires some algorithmic modifications to enable accurate and robust system simulations. As a fallback solution, it is meaningful to provide a direct solver to guarantee a stable and accurate solution if the iterative solvers fail. A combination of multigrid solvers and data-sparse PEEC matrices seems to be most promising to realize highly-efficient EM analysis techniques. This new and novel approach was first introduced in [28] and further extended to PEEC matrices in [7]. Applying the $\mathcal{H}$-matrix stamping procedure, the PEEC model directly can be provided in the $\mathcal{H}$-matrix format, eliminating the memory bottleneck. The computational complexity with $\mathcal{O}(n \log n)$ moreover leads to a fast solution process. Due to the almost optimal complexity of hierarchical techniques presented so far, a further increase in performance often can only be obtained by applying *parallelization techniques* (refer to Sect. 6).

To demonstrate the applicability and efficiency of the presented methodology, a comprehensive analysis of an automotive specific subject was accomplished in Sect. 7. With respect to the requirements for an automotive EMC simulation on system level, challenges and remedy methods are presented. The applied hierarchical techniques facilitate the inclusion of models of different types in the context of a *"hybrid"* solver approach [46]. As a basic requirement for real-life simulations, the bottlenecks concerning time and memory consumptions can be avoided.

Possible Extensions and Future Work Although the presented modeling and simulation methodology already allows EMC analysis with high efficiency and accuracy, it comprises some still outstanding scopes for further investigations and improvements. To exploit the total computational power of the proposed hierarchical approaches, a more suitable $\mathscr{H}$-network-matrix construction containing filling and post-processing has to be developed. The inclusion of PEEC models as well as TL models must be possible and the interfaces have to be constructed in a way to enable a continuous data exchange between modeling and simulation. As a first step, a simple $\mathscr{H}$-stamping for the elementwise and blockwise addition of matrix entries was presented in [7]. A more powerful approach, however, is subject of a current research project together with the Max-Planck-Institute Leipzig and the Fraunhofer Institute SCAI. The first step to enable a suitable automatic multi-level solver switching strategy will be introduced in [47].

The implementation of $\mathscr{H}^2$-matrices for an iterative solution is still under work. Due to the fact, that the preconditioner computation represents the most complex part in the iterative solving process, the application of $\mathscr{H}^2$-matrices can help to significantly reduce the complexity w.r.t. time as well as storage requirements. Although modern quadrature rules exhibit powerful techniques to treat complex and singular numerical integrals, problems are inevitable in plate arrangements with very small distances between the plates, as it occurs frequently in PCB analysis. *Local mesh refinement* techniques can help to overcome these shortcomings. Accuracy problems as well as occuring simulation artifacts can be avoided or at least moderated (refer to [48, 7]). The proposed hierarchical techniques proved to be well suited for a parallel implementation. While an "outer" parallelization of the FD circuit analysis is still under work, the "inner" parallelization for the simulator and inner as well as outer parallelizations for the modeling are still open problems. Last but not least additional research work concerning a combination of hierarchical techniques and MOR is supposed to be promising.

Acknowledgements Special thanks go to Dr. Tanja Clees from the Fraunhofer Institute for Algorithms and Scientific Computing (SCAI) for her support and for many fruitful discussions in the field of electrical engineering and mathematics.

References

1. A. E. Ruehli, "Equivalent Circuit Models for Three-Dimensional Multiconductor Systems," *IEEE Trans. Microw. Theory Tech.*, vol. MTT-22, no. 3, pp. 216–221, March 1974.
2. H. Heeb and A. E. Ruehli, "Retarded Models for PC Board Interconnects - or How the Speed of Light Affects Your SPICE Circuit Simulation," in *Proc. of the IEEE International Conference on Computer-Aided Design*, November 1991, pp. 70–73.
3. R. F. Harrington, *Field Computation by Moment Methods*. Macmillan Publishing Company, 1968, reprint: IEEE Press, New York, 1993.
4. J. Ekman, "Electromagnetic Modeling Using the Partial Element Equivalent Circuit Method," Ph.D. dissertation, Luleå University of Technology, Luleå, Sweden, 2003.
5. A. Görisch and G. Wollenberg, "Simulation von PEEC Modellen mit SPICE," in *Proc. of the EMV, Düsseldorf, Germany*, 2000, pp. 181–188, (In German).

6. A. E. Ruehli, G. Antonini, J. Esch, J. Ekman, A. Orlandi, and A. Mayo, "Nonorthogonal PEEC Formulation for Time- and Frequency Domain EM and Circuit Modeling," *IEEE Trans. Electromagn. Compat.*, vol. 45, no. 2, pp. 167–176, 2003.
7. M. L. Zitzmann, "Fast and Efficient Methods for Circuit-based Automotive EMC Simulation," Ph.D. dissertation, University of Erlangen-Nuremberg, Germany, 2007.
8. T.-H. Chen, C. Luk, and C. C.-P. Chen, "INDUCTWISE: Inductance-Wise Interconnect Simulator and Extractor," *IEEE Trans. Computer-Aided Design*, vol. 22, no. 7, pp. 884–894, July 2003.
9. J. Vlach and K. Singhal, *Computer-Methods for Circuit Analysis and Design*, 2nd ed. Van Nostrand Reinhold, New York, 1994.
10. M. E. Verbeek, "Iterative Solvers and Preconditioning for Electromagnetic Boundary Integral Equations," Ph.D. dissertation, University of Utrecht, Netherlands, 2001.
11. G. Antonini, "Fast Multipole Method for Time Domain PEEC Analysis," *IEEE Trans. Mobile Comput.*, vol. 2, no. 4, pp. 275–287, October-December 2003.
12. G. Antonini and A. E. Ruehli, "Fast Multiple and Multifunction PEEC Methods," *IEEE Trans. Mobile Comput.*, vol. 2, no. 4, pp. 288–298, October-December 2003.
13. S. Börm and S. A. Sauter, "Classical BEM with Linear Complexity," *MathComp.*, vol. 74, pp. 1139–1177, 2005.
14. M. Lintner, "Lösung der 2D Wellengleichung mittels hierarchischer Matrizen," Ph.D. dissertation, Technical University of Munich, Germany, 2002, (In German).
15. L. Grasedyk, "Adaptive Recompression of $\mathscr{H}$-Matrices for BEM," *Computing*, vol. 74, pp. 205–223, 2004.
16. M. Bebendorf and R. Kriemann, "Fast Parallel Solution of Boundary Integral Equations and Related Problems," Max Planck Institute for Mathematics in the Sciences, preprint 10/2004. [Online]. Available: url: http://www.mis.mpg.de/preprints/
17. W. Hackbusch, "A Sparse Matrix arithmetic Based on $\mathscr{H}$-Matrices. Part I: Introduction to $\mathscr{H}$-Matrices," *Computing*, vol. 62, pp. 89–108, 1999. [Online]. Available: url: http://www.mis.mpg.de/preprints/
18. W. Hackbusch, "A Sparse $\mathscr{H}$-Matrix Arithmetic, Part II: Application to Multi-Dimensional Problems," *Computing*, vol. 64, pp. 21–47, 2000. [Online]. Available: url: http://www.mis.mpg.de/preprints/
19. S. Börm, "Approximation of Integral Operators by $\mathscr{H}^2$-Matrices with Adaptive Bases," *Computing*, vol. 74, pp. 249–271, 2005.
20. M. Bebendorf, "Project Homepage." [Online]. Available: url: http://www.math.uni-leipzig.de/bebendorf/projekte.html
21. U. Langer and D. Pusch, "Sparse Algebraic Multigrid Methods for Large Scale Boundary Element Equations," Johannes Kepler University of Linz, Austria, 2003, Special Research Program SFB F013.
22. M. Bebendorf, "Approximation of Boundary Element Matrices," *Numer. Math.*, vol. 86, no. 4, pp. 565–589, 2000.
23. L. Banjai and W. Hackbusch, "$\mathscr{H}$- and $\mathscr{H}^2$-Matrices for Low and High Frequency Helmholtz Equation," Max Planck Institute for Mathematics in the Sciences, preprint 17/2005. [Online]. Available: url: http://www.mis.mpg.de/preprints/
24. S. Börm and L. Grasedyk, "Hybrid Cross Approximation of Integral Operators," Max Planck Institute for Mathematics in the Sciences, preprint 68/2004. [On-line]. Available: url: http://www.mis.mpg.de/preprints/
25. M. Bebendorf, "Approximate Inverse Preconditioning of FE Systems for Elliptic Operators with non-smooth Coefficients," Max Planck Institute for Mathematics in the Sciences, preprint 7/2004. [Online]. Available: url: http://www.mis.mpg.de/preprints/
26. W. C. Chew, J. M. Jin, E. Michielssen, and J. Song, *Fast and Efficient Algorithms in Computational Electromagnetics*. ARTECH House Norwood, MA 2001.
27. W. Hackbusch, "Multi-Grid Methods for FEM and BEM Applications," Max Planck Institute for Mathematics in the Sciences, preprint 72, 2003. [Online]. Available: url: http://www.mis.mpg.de/preprints/

28. D. Pusch, "Efficient Algebraic Multigrid Preconditioners for Boundary Element Methods," Master's thesis, Johannes Kepler University of Linz, Austria, 2003.
29. T. Clees, "AMG Strategies for PDE Systems with Applications in Industrial Semiconductor Simulation," Ph.D. dissertation, University of Köln, Germany, 2005.
30. M. Bollhöfer, "Geometrische Mehrgitterverfahren und Algebraische Mehrgitterverfahren - Eine Einführung," Lecture notes, Technical University of Berlin, Germany, 2001, (In German).
31. V. E. Henson, "An Algebraic Multigrid Tutorial," Center of Applied Scientific Computing, Lawrence Livermore National Laboratory Livermore, CA, 1999.
32. H. Su, E. Acar, and S. R. Nassif, "Power Grid Reduction Based on Algebraic Multigrid Principles," in *Proc. of the International Design Automation Conference, Anaheim, CA, USA*, June 2003.
33. S. R. Nassif and J. N. Kozhaya, "Fast Power Grid Simulation," in *Proc. of the International Design Automation Conference, Los Angeles, CA, USA*, June 2000.
34. R. Beals, "Exact Fundamental Solutions," *Journées Équations aux dérivées partielles*, June 1998.
35. A. Schüller, K. Stüben, C.-A. Thole, and U. Trottenberg, "Paralleles Rechnen und schnelle Simulation," *Der GMD-Spiegel*, no. 1/2, pp. 9–11, 2000, (In German).
36. Y. Saad, *Iterative Methods for Sparse Linear Systems*. PWS Publishing Company Boston, MA, 1996.
37. K. Stüben and T. Clees, *SAMG User's Manual v. 22c*, Fraunhofer Institute SCAI, Schloss Birlinghoven, Sankt Augustin, Germany, May 2005. [Online]. Available: url: http://www.scai.fhg.de/303.0.html
38. K. Stüben, *SAMG - Data Structure and File Format Specification*, Fraunhofer Institute SCAI, Sankt Augustin, Germany, February 2005.
39. A. Krechel and K. Stüben, "GMD Report 71: Parallel Algebraic Multigrid Based on Subdomain Blocking," GMD - Forschungszentrum Informationstechnik GmbH, Germany, Tech. Rep., 1999.
40. The MISEA Project. [Online]. Available: url: *http://misea.de*
41. M. L. Zitzmann, T. Clees, and R. Weigel, "Iterative Methods for Reluctance based PEEC Models," in *Proc. of the 17th International Zurich Symposium on Electromagnetic Compatibility, Singapore 2006, February 27th - March 3rd*, 2006, pp. 81–84.
42. A. Devgan, H. Ji, and W. Dai, "How to Efficiently Capture On-Chip Inductance Effects: Introducing a New Circuit Element K," in *Proc. of the International Conference on Computer-Aided Design*, November 2000, pp. 150–155.
43. H. Zheng, B. Krauter, M. Beattie, and L. Pileggi, "Window-Based Susceptance Models for Large-Scale RLC Circuit Analyses," in *Proc. of the International Conference on Design, Automation and Test in Europe*, 2002, pp. 628–633.
44. F. Canavero, J.-C. Kedzia, P. Ravier, and B. Scholl, "Numerical Simulation for Early EMC Design of Cars," in *Proc. of the CEM 2000 Symposium, Brugge, Belgium*, 2000.
45. S. Börm and L. Grasedyk, "HLIB – A Library for $\mathcal{H}$- and $\mathcal{H}^2$-matrices," Max Planck Institute for Mathematics in the Sciences. [Online]. Available: url: http://www.hlib.org/
46. M. L. Zitzmann, R. Grillmair, T. Clees, and R. Weigel, "Hybrid Solver Strategies in Automotive EMC Simulation," in *Proc. of the 17th International Zurich Symposium on Electromagnetic Compatibility, Singapore 2006, February 27th - March 3rd*, 2006, pp. 340–343.
47. T. Clees, T. Samrowski, M. L. Zitzmann, and R. Weigel, "An Automatic Multi-Level Solver Switching Strategy for PEEC-Based EMC Simulation," in *Proc. of the 18th International Zurich Symposium on Electromagnetic Compatibility, Munich 2007*, (accepted).
48. J. Ekman and G. Antonini, "On Characterizing Artifacts Observed in PEEC Based Modeling," in *Proc. of the IEEE International Symposium on Electromagnetic Compatibility, Santa Clara, CA, USA*, August 2004.
49. R. Achar and M. S. Nakhla, "Simulation of High-Speed Interconnects," *Proc. IEEE*, vol. 89, no. 5, pp. 693–728, May 2001.

50. J. H. Bramble, Z. Leyk, and J. E. Pasciak, "The Analysis of Multigrid Algorithms for Pseudo-Differential Operators of Order Minus One," *Math. Comp.*, vol. 63, no. 208, pp. 461–478, 1994.
51. K. Stüben and T. Clees, *SAMG User's Manual v. 22c*, Fraunhofer Institute SCAI, Schloss Birlinghoven, Sankt Augustin, Germany, May 2005. [Online]. Available: url: http://www.scai.fhg.de/303.0.html
52. R. Kriemann, "Parallel $\mathscr{H}$-Matrix Arithmetics on Shared Memory Systems," Max Planck Institute for Mathematics in the Sciences, preprint 29/2004. [Online]. Available: url: http://www.mis.mpg.de/preprints/

Equivalent Circuit (EC) FDTD Method for Dispersive Materials: Derivation, Stability Criteria and Application Examples

A. Rennings, A. Lauer, C. Caloz and I. Wolff

1 Introduction

Although many computational methods are currently available to analyze electromagnetic problems of high complexity, further progress is necessary to unify the different approaches and to improve the robustness of some algorithms. This chapter aims at reaching these two goals. It presents an equivalent circuit (EC) finite difference time domain (FDTD) method which both represents a complete circuital formulation of conventional FDTD [1] based on the electric and magnetic fields Yee scheme and provides guaranteed stability criteria for dispersive media.

A circuital formulation of FDTD offers several benefits. Since its update values are voltages and currents (as opposed to electric and magnetic fields), it facilitates the integration of electronic components, such as for instance diodes and transistors, into the simulation. In addition, it provides increased computational speed thanks to a reduced number of multiplications in the update equations [2, 3]. Finally, it may be easily generalized to arbitrary dispersive media via simple manipulations of the mesh immittances.

While no convenient explicit guaranteed stability criterion for dispersive media is available in the framework of the conventional FDTD technique, the EC FDTD provides such a criterion. This is a distinct advantage for both emerging metamaterials [4], which are inherently dispersive, and for various natural materials [5], such as metals at the optical frequencies, ferrimagnetic and ferromagnetic materials and biological tissues, which exhibit various types of more or less complex dispersion responses.

A. Rennings
IMST GmbH, D-47475 Kamp-Lintfort, Germany, e-mail: A.Rennings@ieee.org

A. Lauer
IMST GmbH, D-47475 Kamp-Lintfort, Germany

C. Caloz
École Polytechnique, Montréal, H3T 1J4, Québec, Canada, e-mail: christophe.caloz@polymtl.ca

I. Wolff
IMST GmbH, D-47475 Kamp-Lintfort, Germany

The chapter is organized as follows. Section 2 introduces a compact and convenient notation for the method, derives the EC FDTD method first for the simplest case of non-dispersive materials and then extends it for the case of dispersive materials, including the cases of Drude, Lorentz and Debye media. Section 3 develops within the framework of the EC FDTD a discretized and circuital version of the Poynting theorem both for the time-continuous and time-discrete cases. Upon this foundation and following a Lyapunov energy approach, Sect. 4 establishes a guaranteed stability criterium, which may be particularized to arbitrary types of dispersion responses. Application examples of metamaterial and natural material dispersive structures and devices are next presented in Sect. 5. Finally, Sect. 6 outlines the main results.

2 Derivation of the Method

2.1 Notation of the Discrete Calculus

Here a compact and convenient notation for the EC FDTD method is introduced. Additionally to conventional mathematical notations here all operators including matrices are denoted by upright characters (like $\mathrm{S}[\cdot]$), and vectorial/tensorial objects are denoted by capital bold characters (vector field $\boldsymbol{E}$ or matrix $\mathbf{A}$).

2.1.1 Discrete Fields

Each of the six continuous scalar fields of the form $F_{\text{conti}}(\boldsymbol{x},t)$, $\boldsymbol{x} \in \mathbb{R}^3$, $t \in \mathbb{R}$ of the electromagnetic field are represented by six *discrete fields* of the form $F^n_{\text{disc}}(\boldsymbol{X})$, $\boldsymbol{X} \in \mathbb{Z}^3$, $n \in \mathbb{Z}$, where $\boldsymbol{X}$ is the three-dimensional spatial index vector and n is the time index. The index vector $\boldsymbol{X}$ is usually not given [F^n_{disc} and not $F^n_{\text{disc}}(\boldsymbol{X})$], since it is assumed that all discrete fields depend on $\boldsymbol{X}$. Whereas the time index n is always given for the time discrete case (conventional technique with upper index).

2.1.2 Modulo-Three Notation and Variable Indices

For the sake of simpler and compacter notation, component and spatial direction indices are numbered in a modulo-three sense, e.g. $F_0 \equiv F_x$, $F_1 \equiv F_y$, $F_2 \equiv F_z$ and $F_3 \equiv F_0$, $F_4 \equiv F_2$, i.e., the effective index is always element of the set $\{0, 1, 2\}$.

In 3D figures the direction indices $d \in \{0,\ 1,\ 2\}$ are explicitly given (as absolute numbers), whereas in 2D figures (projections of a 3D arrangement along the Cartesian axes) variable indices $(d,\ d+1,\ d+2)$ are used. By inserting $\{0,\ 1,\ 2\}$ for d one gets the projection onto a desired plane. The variable indices are also utilized to specify vectorial equations in a scalar manner. Again, by inserting $\{0, 1, 2\}$ for d the specific component is obtained. The fact that d is element of the set $\{0, 1, 2\}$ is not repeated in each equation with variable indices, it is part of the general assumptions outlined in this section.

2.1.3 Discrete Spatial-Shift and Index-Shift Operators

In finite difference schemes the application of spatial operators onto discrete fields is usually specified by using index changes together with appropriate weighting factors. Quite often the resulting equations are cumbersome. In order to circumvent this, we introduce shift operators S_d and its inverse S_d^{-1}. Together with the null $0[\cdot]$ and the identity element $\mathrm{I}[\cdot]$ the family of shift operators generate an operator space $\mathbb{O}$, which in algebraic terms is called a commutative ring. The shift operators are defined via their action onto the discrete fields. It is

$$\mathrm{S}_d F(\boldsymbol{X}) = F(\boldsymbol{X} + \boldsymbol{E}_d), \; \{\mathrm{S}_d^{-1} + \mathrm{S}_{d+1}\} F(\boldsymbol{X}) = F(\boldsymbol{X} - \boldsymbol{E}_d) + F(\boldsymbol{X} + \boldsymbol{E}_{d+1}), \quad (1)$$

with the index unit vectors $\boldsymbol{E}_0 = (1, 0, 0)^{\mathrm{T}}$, $\boldsymbol{E}_1 = (0, 1, 0)^{\mathrm{T}}$ and $\boldsymbol{E}_2 = (0, 0, 1)^{\mathrm{T}}$. In Eq. (1) the summation in $\mathbb{O}$ is also shown. If a summation of several operators is applied to a field, braces $\{\ldots\}$ are used around the operator sum as in Eq. (1). The application of S_d to a field that does not depend on X_d yields the unchanged field (identity operation). This rule is important for the application of shift-operators to fields that depend only on one component of the index vector, like the dimensions of the mesh cells. For instance Δ_d is only a function of X_d. Thus expressions like $\mathrm{S}_d \Delta_d$ can be simplified to $\mathrm{S}\Delta_d$.

The following example should clarify the multiplication in $\mathbb{O}$:

$$\alpha\, \mathrm{S}_d\, \mathrm{S}_{d+2}^{-1} F(\boldsymbol{X}) = \mathrm{S}_{d+2}^{-1}\, \mathrm{S}_d\, \alpha F(\boldsymbol{X}) = \alpha\, F(\boldsymbol{X} + \boldsymbol{E}_d - \boldsymbol{E}_{d+2}) \quad (2)$$

Due to inhomogeneous material distributions the simple multiplication with a constant factor in $\mathbb{O}$ needs to be extended to a multiplicative operator $\mathrm{M}(\boldsymbol{X})$ that depends on $\boldsymbol{X}$. In this case the commutativity of the multiplication in $\mathbb{O}$ is lost, since e.g.

$$\mathrm{S}_d\,[\mathrm{M}(\boldsymbol{X}) F(\boldsymbol{X})] = \mathrm{M}(\boldsymbol{X} + \boldsymbol{E}_d) F(\boldsymbol{X} + \boldsymbol{E}_d) \neq \mathrm{M}(\boldsymbol{X}) \mathrm{S}_d F(\boldsymbol{X}). \quad (3)$$

If an operator is applied to more than the field directly behind it, brackets $[\ldots]$ are used, for instance $\mathrm{S}_d\,[\mathrm{M}(\boldsymbol{X}) F_d(\boldsymbol{X}) + F_{d+2}(\boldsymbol{X})]$.

The index-shift operator R and its inverse R^{-1} are defined by

$$\mathrm{R} F_d = F_{d+1}, \quad \mathrm{R}^{-1} F_d = F_{d-1}. \quad (4)$$

This operator is used in Sect. 3 where a discrete Poynting theorem will be derived. R is used for re-formulations and simplification of terms that are summed up over all three directions $d \in \{0, 1, 2\}$. The overall sum does not change if these operators are applied due to the modulo-3 sense of the indices.

2.2 Basic EC FDTD for Non-Dispersive Materials

The basic EC FDTD scheme is derived from an approximation of Maxwell's equations in integral form, including electric loss σ_e, magnetic loss σ_m, and a current density source $\boldsymbol{J}_\mathrm{src}$ (soft excitation),

$$\oint_{\partial A} \boldsymbol{H} \cdot \mathrm{d}\boldsymbol{\ell} = +\frac{\mathrm{d}}{\mathrm{d}t} \iint_A \varepsilon \boldsymbol{E} \cdot \mathrm{d}\boldsymbol{A} + \iint_A \sigma_\mathrm{e}\, \boldsymbol{E} \cdot \mathrm{d}\boldsymbol{A} - \iint_A \boldsymbol{J}_\mathrm{src} \cdot \mathrm{d}\boldsymbol{A}, \tag{5}$$

$$\oint_{\partial A} \boldsymbol{E} \cdot \mathrm{d}\boldsymbol{\ell} = -\frac{\mathrm{d}}{\mathrm{d}t} \iint_A \mu \boldsymbol{H} \cdot \mathrm{d}\boldsymbol{A} - \iint_A \sigma_\mathrm{m}\, \boldsymbol{H} \cdot \mathrm{d}\boldsymbol{A}, \tag{6}$$

where all material constitutive parameters $(\varepsilon, \mu, \sigma_\mathrm{e}, \sigma_\mathrm{m})$ are considered *non-dispersive*, i.e. independent of frequency (but possibly inhomogeneous).

These space-time continuous equations are *spatially discretized* using the well-known *Yee cell* [1], which is depicted in Fig. 1. Within the Yee cell, the electric field is sampled at the edges of the main mesh subcell while the magnetic field is sampled at the edges of the dual mesh subcell. The field samples within the Yee cell are located by the index vector $\boldsymbol{X} = (X_0, X_1, X_2)^\mathrm{T}$, according to the notations of Sect. 2.1.

The Yee discretization details of the proposed EC FDTD formulation are shown in Fig. 2. The magnetic field samples are located normally at the *center* of the faces of the main mesh subcells and are *symmetrically surrounded* by four adjacent tangential electric field samples. In contrast, in the general case of a non-uniform mesh, the electric field samples are *not* located symmetrically with respect to the surrounding magnetic field samples.

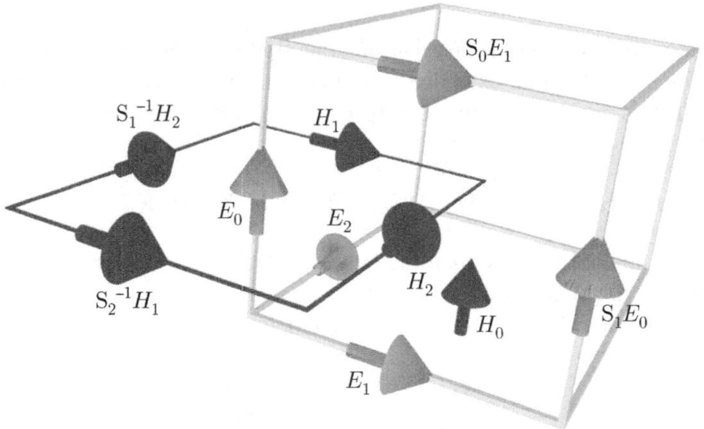

Fig. 1 Yee cell, which consists of a *main mesh* subcell (*parallelepipedic structure* in the figure) and a *dual mesh* subcell (*staggered parallelepipedic structure*, of which only one face is shown in the figure). The Yee cell includes six unshifted field samples. Additionally four positively or negatively shifted field samples are shown, which are required for the approximation of Eqs. (5) and (6)

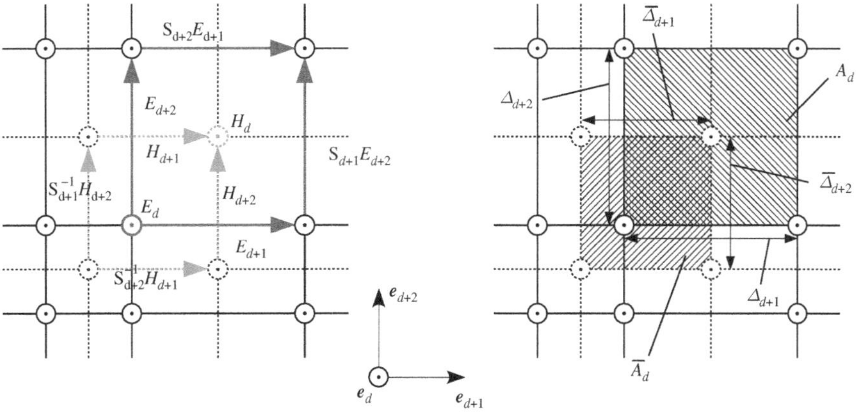

Fig. 2 Details of the Yee cell (Fig. 1) discretization. Used components and corresponding subcell dimensions for the spatial discretization of Eqs. (5) and (6) leading to Eqs. (8) and (9)

Here in this scheme the electric *and also* the magnetic material parameters are approximated by *main mesh "blocks" with piecewise constant constitutive parameters*[1], which allows a natural meshing of inhomogeneous materials, since a change in the material parameters is mostly due to an object boundary. In the case of a non-uniform mesh, the dimensions of the mesh cells vary across space, and the edge length Δ_d is only a function of the component X_d of the index vector $\boldsymbol{X}$. According to the notations of Sect. 2.1, the dimensions of the dual mesh subcell, $\overline{\Delta}_d$, are related to the dimensions of the main mesh subcell, Δ_d, by

$$\overline{\Delta}_d = \frac{1}{2}\left(\mathrm{S}^{-1}\Delta_d + \Delta_d\right) = \frac{1}{2}\left\{\mathrm{S}^{-1}+1\right\}\Delta_d, \quad \forall \quad d \in \{0,1,2\}. \tag{7}$$

Evaluating Maxwell's Eqs. (5) and (6) on all faces of the main and dual mesh subcells, as represented in Fig. 2, yields the (still time-continuous) spatially discretized generic equations

$$\oint_{\partial\overline{A}_d} \boldsymbol{H}\cdot \mathrm{d}\boldsymbol{\ell} \approx \left\{1-\mathrm{S}^{-1}_{d+1}\right\} H_{d+2}\,\overline{\Delta}_{d+2} - \left\{1-\mathrm{S}^{-1}_{d+2}\right\} H_{d+1}\,\overline{\Delta}_{d+1} = \tag{8}$$

$$+\left(\varepsilon_{d,\mathrm{eff}}\,\frac{\mathrm{d}}{\mathrm{d}t}\,E_d + \sigma^{\mathrm{e}}_{d,\mathrm{eff}}\,E_d - J^{\mathrm{src}}_d\right)\overline{\Delta}_{d+1}\overline{\Delta}_{d+2} \approx \iint_{\overline{A}_d} \ldots \cdot \mathrm{d}\boldsymbol{A},$$

$$\oint_{\partial A_d} \boldsymbol{E}\cdot \mathrm{d}\boldsymbol{\ell} \approx \left\{\mathrm{S}_{d+1}-1\right\} E_{d+2}\,\Delta_{d+2} - \left\{\mathrm{S}_{d+2}-1\right\} E_{d+1}\,\Delta_{d+1} = \tag{9}$$

$$-\left(\mu_{d,\mathrm{eff}}\,\frac{\mathrm{d}}{\mathrm{d}t}\,H_d + \sigma^{\mathrm{m}}_{d,\mathrm{eff}}\,H_d\right)\Delta_{d+1}\Delta_{d+2} \approx \iint_{A_d} \ldots \cdot \mathrm{d}\boldsymbol{A},$$

[1] Usually the magnetic material parameters are approximated by *dual* mesh blocks with piecewise constant constitutive parameters.

where the four effective constitutive parameters $\varepsilon_{d,\mathrm{eff}}$, $\mu_{d,\mathrm{eff}}$, $\sigma^{\mathrm{e}}_{d,\mathrm{eff}}$, $\sigma^{\mathrm{m}}_{d,\mathrm{eff}}$ will be determined now.

The effective *electric* parameters are approximated by averaging the fluxes $\iint_{\overline{A}_d}\{\varepsilon,\sigma_{\mathrm{e}}\}\boldsymbol{E}\cdot\mathrm{d}\boldsymbol{A}$ on each face $\overline{A}_d$ (Fig. 2) of the *dual* mesh subcells, leading to

$$\varepsilon_{d,\mathrm{eff}} \doteq \frac{\frac{1}{4}\left\{1+\mathrm{S}_{d+1}^{-1}+\mathrm{S}_{d+2}^{-1}+\mathrm{S}_{d+1}^{-1}\mathrm{S}_{d+2}^{-1}\right\}\left[\varepsilon\;\Delta_{d+1}\;\Delta_{d+2}\right]}{\frac{1}{4}\left\{1+\mathrm{S}_{d+1}^{-1}+\mathrm{S}_{d+2}^{-1}+\mathrm{S}_{d+1}^{-1}\mathrm{S}_{d+2}^{-1}\right\}\left[\Delta_{d+1}\;\Delta_{d+2}\right]}, \tag{10}$$

for $\varepsilon_{d,\mathrm{eff}}$, as illustrated in Fig. 3(a), and $\sigma^{\mathrm{e}}_{d,\mathrm{eff}}$ is evaluated in the same manner.

The magnetic parameters are constant in each A_d–transversal direction (where A_d is defined by the directions $d+1$ and $d+2$, Fig. 2, right) of the *main* mesh subcells, and therefore no useful averaging quantity may defined from the fluxes $\iint_{A_d}\{\mu,\sigma_{\mathrm{m}}\}\boldsymbol{H}\cdot\mathrm{d}\boldsymbol{A}$. In contrast, two different magnetic materials may exist in the longitudinal direction (d), and therefore the effective *magnetic* parameters may be approximated by averaging the magneto-motive force[2] $H_d\overline{\Delta}_d$ as illustrated in Fig. 3(b)

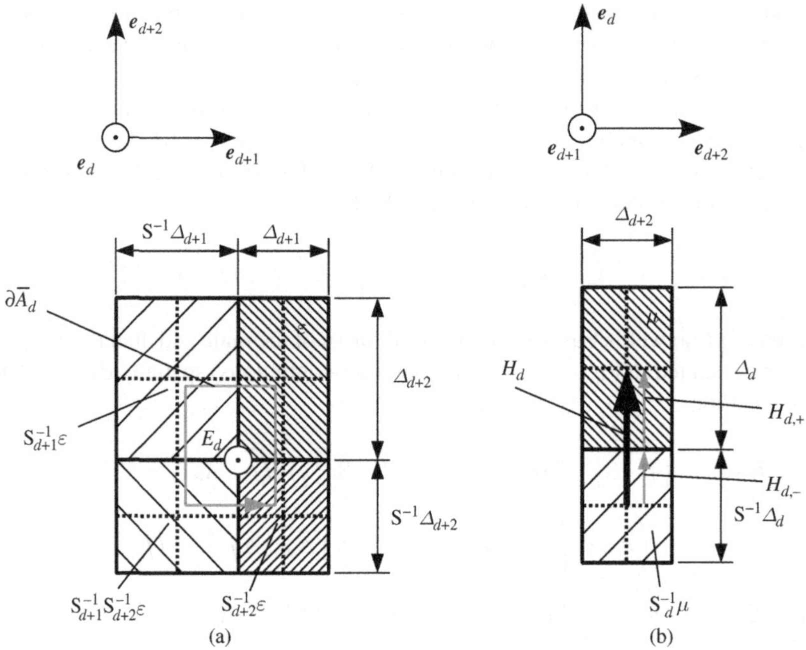

Fig. 3 Averaging process for the approximation of the effective constitutive parameters. (**a**) Local virtual capacitance filled with 4 different materials to define $\varepsilon_{d,\mathrm{eff}}$. (**b**) Local virtual inductance filled with 2 different materials to define $\mu_{d,\mathrm{eff}}$

[2] The magnetic flux density B_d is continuous here.

$$H_d\overline{\Delta}_d = H_{d,+}\frac{1}{2}\Delta_d + H_{d,-}\frac{1}{2}\mathrm{S}_d^{-1}\Delta_d = B_d\frac{1}{2}\left\{1+\mathrm{S}_d^{-1}\right\}\left[\frac{\Delta_d}{\mu}\right] \doteq B_d\frac{\overline{\Delta}_d}{\mu_{d,\mathrm{eff}}}, \tag{11}$$

from which

$$\mu_{d,\mathrm{eff}} \doteq \frac{\overline{\Delta}_d}{\frac{1}{2}\left\{1+\mathrm{S}_d^{-1}\right\}\left[\frac{\Delta_d}{\mu}\right]} = \frac{\frac{1}{2}\left\{1+\mathrm{S}_d^{-1}\right\}\Delta_d}{\frac{1}{2}\left\{1+\mathrm{S}_d^{-1}\right\}\left[\frac{\Delta_d}{\mu}\right]}, \tag{12}$$

for $\mu_{d,\mathrm{eff}}$, and $\sigma^{\mathrm{m}}_{d,\mathrm{eff}}$ is evaluated in the same manner.

Inserting the so-defined effective constitutive parameters into Eqs. (8) and (9) yields the complete space-discretized Maxwell's equations

$$\begin{aligned}\left\{1-\mathrm{S}_{d+1}^{-1}\right\}&H_{d+2}\,\overline{\Delta}_{d+2} - \left\{1-\mathrm{S}_{d+2}^{-1}\right\}H_{d+1}\,\overline{\Delta}_{d+1} = -\overline{\Delta}_{d+1}\overline{\Delta}_{d+2}\,J_d^{\mathrm{src}}\\ &+\frac{1}{4}\left\{1+\mathrm{S}_{d+1}^{-1}+\mathrm{S}_{d+2}^{-1}+\mathrm{S}_{d+1}^{-1}\mathrm{S}_{d+2}^{-1}\right\}\left[\varepsilon\,\Delta_{d+1}\,\Delta_{d+2}\right]\frac{1}{\Delta_d}\frac{\mathrm{d}}{\mathrm{d}t}E_d\Delta_d\\ &+\frac{1}{4}\left\{1+\mathrm{S}_{d+1}^{-1}+\mathrm{S}_{d+2}^{-1}+\mathrm{S}_{d+1}^{-1}\mathrm{S}_{d+2}^{-1}\right\}\left[\sigma_{\mathrm{e}}\,\Delta_{d+1}\,\Delta_{d+2}\right]\frac{1}{\Delta_d}E_d\Delta_d,\end{aligned} \tag{13}$$

$$\begin{aligned}\left\{\mathrm{S}_{d+1}-1\right\}&E_{d+2}\,\Delta_{d+2} - \left\{\mathrm{S}_{d+2}-1\right\}E_{d+1}\,\Delta_{d+1} =\\ &-\frac{\Delta_{d+1}\Delta_{d+2}}{\frac{1}{2}\left\{1+\mathrm{S}_d^{-1}\right\}\left[\frac{\Delta_d}{\mu}\right]}\frac{\mathrm{d}}{\mathrm{d}t}H_d\overline{\Delta}_d + \frac{\Delta_{d+1}\Delta_{d+2}}{\frac{1}{2}\left\{1+\mathrm{S}_d^{-1}\right\}\left[\frac{\Delta_d}{\sigma_{\mathrm{m}}}\right]}H_d\overline{\Delta}_d,\end{aligned} \tag{14}$$

where the terms $E_d\Delta_d \doteq v_d$ may be interpreted as the electric voltages along the edges of the main mesh subcells and the terms $H_d\overline{\Delta}_d \doteq i_d$ may be interpreted as the "magnetic voltages" along the edges of the dual mesh subcells.[3]

This circuit analogy and inspection of Eqs. (13) and (14) naturally lead to the definition of the equivalent *lumped elements*

$$C_d = \frac{1}{4}\left\{1+\mathrm{S}_{d+1}^{-1}+\mathrm{S}_{d+2}^{-1}+\mathrm{S}_{d+1}^{-1}\mathrm{S}_{d+2}^{-1}\right\}\left[\varepsilon\,\Delta_{d+1}\,\Delta_{d+2}\right]\frac{1}{\Delta_d}, \tag{15}$$

$$G_d = \frac{1}{4}\left\{1+\mathrm{S}_{d+1}^{-1}+\mathrm{S}_{d+2}^{-1}+\mathrm{S}_{d+1}^{-1}\mathrm{S}_{d+2}^{-1}\right\}\left[\sigma_{\mathrm{e}}\,\Delta_{d+1}\,\Delta_{d+2}\right]\frac{1}{\Delta_d}, \tag{16}$$

which are interpreted as the local capacitance and conductance, and

$$L_d = \frac{\Delta_{d+1}\Delta_{d+2}}{\frac{1}{2}\left\{1+\mathrm{S}_d^{-1}\right\}\left[\frac{\Delta_d}{\mu}\right]}, \quad R_d = \frac{\Delta_{d+1}\Delta_{d+2}}{\frac{1}{2}\left\{1+\mathrm{S}_d^{-1}\right\}\left[\frac{\Delta_d}{\sigma_{\mathrm{m}}}\right]}, \tag{17}$$

which are interpreted as the local inductance and resistance. To complete the circuit analogy, a lumped current source i_d^{src} is defined

---

[3] Alternatively the term $H_d\overline{\Delta}_d \doteq i_d$ can be interpreted as corresponding loop current in the face A_d of the main mesh cell (Fig. 4), which is related to the orthogonal magnetic field H_d component by the right-hand rule.

$$i_d^{\mathrm{src}} = \frac{1}{4}\{1+\mathrm{S}_{d+1}^{-1}+\mathrm{S}_{d+2}^{-1}+\mathrm{S}_{d+1}^{-1}\mathrm{S}_{d+2}^{-1}\}\,[\Delta_{d+1}\,\Delta_{d+2}]\,J_d^{\mathrm{src}}. \tag{18}$$

With these local lumped elements Ampere's (5) and Faraday's (6) laws take the very compact form

$$\{1-\mathrm{S}_{d+1}^{-1}\}\,i_{d+2}-\{1-\mathrm{S}_{d+2}^{-1}\}\,i_{d+1}+i_d^{\mathrm{src}} = +C_d\,\frac{\mathrm{d}}{\mathrm{d}t}v_d+G_d\,v_d, \tag{19}$$

$$\{\mathrm{S}_{d+1}-1\}\,v_{d+2}-\{\mathrm{S}_{d+2}-1\}\,v_{d+1} = -L_d\,\frac{\mathrm{d}}{\mathrm{d}t}i_d-R_d\,i_d. \tag{20}$$

For each $d \in \{0,\ 1,\ 2\}$ Eq. (19) represents the Kirchhoff's current law for a current node and Eq. (20) represents the Kirchhoff's voltage law for a voltage loop. This "circuitization" of the Yee FDTD discretization of space may be represented by the equivalent circuit cell shown in Fig. 4.

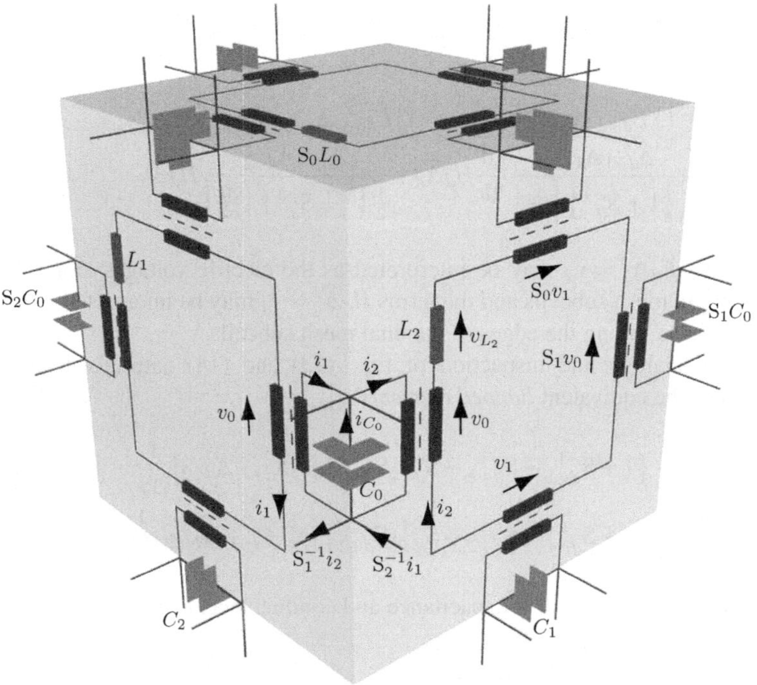

Fig. 4 3D equivalent circuit cell for a non-dispersive material with (*shunt*) capacitances located at the edges of the main mesh cells (*e-nodes*) and (*series*) inductances (represented by the *rectangle in the loop*) located at the faces of the main mesh cells. The 1:1-transformers are neccessary to ensure a well defined loop current with a *one-to-one* correspondance to the orthogonal magnetic field component. The direction of the voltages v_d is positive in the parallel Cartesian direction $\boldsymbol{e}_d$ and the one for the loop currents i_d is positive with respect to the corresponding magnetic field component (*right-hand rule*). The values $i_{C_0} = C_0\,\frac{\mathrm{d}}{\mathrm{d}t}v_0$ and $v_{L_2} = L_2\,\frac{\mathrm{d}}{\mathrm{d}t}i_2$ are used as an abbreviation within the figure

The final step in the derivation of the EC FDTD scheme consists in *time-discretize* by finite difference the previous equations in order to set up the update equations. Defining a time step Δt and using the traditional leap-frog scheme, the voltage values are sampled at time multiples of Δt while the current values are sampled between these instances, i.e. at odd multiples of $\Delta t/2$. This leads to

$$\{1-\mathrm{S}_{d+1}^{-1}\}\, i_{d+2}^{n-\frac{1}{2}} - \{1-\mathrm{S}_{d+2}^{-1}\}\, i_{d+1}^{n-\frac{1}{2}} = +C_d\,\frac{v_d^n - v_d^{n-1}}{\Delta t} + G_d\,\frac{v_d^n + v_d^{n-1}}{2} - i_{\mathrm{src},d}^{n-\frac{1}{2}}, \tag{21}$$

$$\{\mathrm{S}_{d+1}-1\}\, v_{d+2}^n - \{\mathrm{S}_{d+2}-1\}\, v_{d+1}^n = -L_d\,\frac{i_d^{n+\frac{1}{2}} - i_d^{n-\frac{1}{2}}}{\Delta t} - R_d\,\frac{i_d^{n+\frac{1}{2}} + i_d^{n-\frac{1}{2}}}{2}, \tag{22}$$

and solving these equations for the latest voltages v_d^n and currents $i_d^{n+\frac{1}{2}}$ yield the final EC FDTD update equations for a non-dispersive material region of the computational domain

$$v_d^n = \frac{1-\frac{\Delta t G_d}{2C_d}}{1+\frac{\Delta t G_d}{2C_d}}\, v_d^{n-1} + \frac{\Delta t}{C_d}\,\frac{\{1-\mathrm{S}_{d+1}^{-1}\}\, i_{d+2}^{n-\frac{1}{2}} - \{1-\mathrm{S}_{d+2}^{-1}\}\, i_{d+1}^{n-\frac{1}{2}} + i_{\mathrm{src},d}^{n-\frac{1}{2}}}{1+\frac{\Delta t G_d}{2C_d}}, \tag{23}$$

$$i_d^{n+\frac{1}{2}} = \frac{1-\frac{\Delta t R_d}{2L_d}}{1+\frac{\Delta t R_d}{2L_d}}\, i_d^{n-\frac{1}{2}} - \frac{\Delta t}{L_d}\,\frac{\{\mathrm{S}_{d+1}-1\}\, v_{d+2}^n - \{\mathrm{S}_{d+2}-1\}\, v_{d+1}^n}{1+\frac{\Delta t R_d}{2L_d}}. \tag{24}$$

2.3 *Extended Method for General Dispersive Materials*

In Sect. 2.2 the basic EC FDTD scheme for *non-dispersive* materials is derived. This has led to the emergence of the local impedance $Z(\omega) = \mathrm{j}\omega L \propto \mathrm{j}\omega\mu$ constituted of the inductance L (with resistance R for loss) and local admittance $Y(\omega) = \mathrm{j}\omega C \propto \mathrm{j}\omega\varepsilon$ constituted of a capacitance C (with conductance G for loss), which are characteristic of non-dispersive media in the transmission line framework and correspond to the material non-dispersive constitutive parameters.

However, many problems in electrodynamics involve dispersive media, including *metamaterials* and *various natural materials*, for which application examples are presented in Sects. 5.1 and 5.2, respectively. In order to handle such problems, we generalize now the EC FDTD scheme to an *extended* EC FDTD scheme for dispersive materials. This extended scheme fundamentally consists in adding to the equivalent impedance $Z(\omega)$ and admittance $Y(\omega)$ the appropriate lumped elements to account for the specific dispersions considered, $\tilde{\mu}(\omega) = Z(\omega)/(\mathrm{j}\omega\Delta_{\mathrm{m}})$ and $\tilde{\varepsilon}(\omega) = Y(\omega)/(\mathrm{j}\omega\Delta_{\mathrm{e}})$ [the lengths Δ_{m} and Δ_{e} are specified later]. This transformation will require additional update equations. The present section specifically focuses on *double-Drude* dispersion, which is the dispersion encountered in the so-called composite right/left-handed (CRLH) metamaterials [4]. Dispersions of the Lorentz and Debye types are also discussed.

The relation between the generalized-dispersive admittance and the corresponding dispersive function $\tilde{\varepsilon}(\omega) = \varepsilon_\infty + \Delta\varepsilon'(\omega) - \mathrm{j}\sigma_e/\omega - \mathrm{j}\varepsilon''(\omega)$ [where $\varepsilon_\infty \doteq \tilde{\varepsilon}(\omega \to \infty)$] reads

$$\tilde{\varepsilon}_{d,\mathrm{eff}}(\omega) = \frac{Y_d(\omega)}{\mathrm{j}\omega} \frac{\Delta_d}{\frac{1}{4}\{1+\mathrm{S}_{d+1}^{-1}+\mathrm{S}_{d+2}^{-1}+\mathrm{S}_{d+1}^{-1}\mathrm{S}_{d+2}^{-1}\}\,[\Delta_{d+1}\,\Delta_{d+2}]} = \frac{Y_d(\omega)}{\mathrm{j}\omega\Delta_{\mathrm{e},d}}, \quad (25)$$

where $\Delta_{\mathrm{e},d}$ represents the formfactor of an artificial subcell with cross section $\overline{A}_d$ of the dual mesh and the height Δ_d of the main mesh and has the dimension of a length. Similarly, the dispersive function $\tilde{\mu}(\omega) = \mu_\infty + \Delta\mu'(\omega) - \mathrm{j}\sigma_\mathrm{m}/\omega - \mathrm{j}\mu''(\omega)$ [where $\mu_\infty \doteq \tilde{\mu}(\omega \to \infty)$] can be determined from its corresponding generalized-dispersive impedance via

$$\tilde{\mu}_{d,\mathrm{eff}}(\omega) = \frac{Z_d(\omega)}{\mathrm{j}\omega} \frac{\frac{1}{2}\{1+\mathrm{S}_d^{-1}\}\,\Delta_d}{\Delta_{d+1}\Delta_{d+2}} = \frac{Z_d(\omega)}{\mathrm{j}\omega\Delta_{\mathrm{m},d}}, \quad (26)$$

where $\Delta_{\mathrm{m},d}$ represents the formfactor of the artificial cell with cross section A_d of the main mesh subcell and height $\overline{\Delta}_d$ of the dual mesh subcell.

2.3.1 Drude Dispersion

Let us now particularize these generalized dispersive relations to the important case of *Drude* materials, which include for instance noble metals at optical frequencies (ε-Drude) or CRLH metamaterials (both ε and μ-Drude). In this case, the relevant admittance Y_d includes an inductance $L_{\mathrm{L},d}$ in parallel with the capacitance C_d as represented in Fig. 5(a) and the relevant impedance Z_d includes a capacitance $C_{\mathrm{L},d}$ in series with the inductance L_d, [4], as represented in Fig. 5(b). In terms of losses, we consider the most general situation, i.e., in series to each inductance there is a resistance, and parallel each capacitor a conductance consideres the losses. The general dispersive parameters in Eqs. (25) and (26) become then

$$\tilde{\varepsilon}_{\text{ext-Drude}}(\omega) = \frac{Y_{\text{ext-Drude}}(\omega)}{\mathrm{j}\omega\Delta_\mathrm{e}} = \frac{C}{\Delta_\mathrm{e}}\left\{1 - \frac{\frac{1}{L_\mathrm{L}C}}{\omega\left(\omega - \mathrm{j}\frac{R_\mathrm{L}}{L_\mathrm{L}}\right)}\right\} - \mathrm{j}\frac{G}{\omega\Delta_\mathrm{e}}$$

$$= \varepsilon_0\varepsilon_{\mathrm{r},\infty}\left\{1 - \frac{\omega_\mathrm{pe}^2}{\omega(\omega - \mathrm{j}\Omega)}\right\} - \mathrm{j}\frac{\sigma_\mathrm{e}}{\omega}, \quad (27)$$

$$\tilde{\mu}_{\text{ext-Drude}}(\omega) = \frac{Z_{\text{ext-Drude}}(\omega)}{\mathrm{j}\omega\Delta_\mathrm{m}} = \frac{L}{\Delta_\mathrm{m}}\left\{1 - \frac{\frac{1}{LC_\mathrm{L}}}{\omega\left(\omega - \mathrm{j}\frac{G_\mathrm{L}}{C_\mathrm{L}}\right)}\right\} - \mathrm{j}\frac{R}{\omega\Delta_\mathrm{m}}$$

$$= \mu_0\mu_{\mathrm{r},\infty}\left\{1 - \frac{\omega_\mathrm{pm}^2}{\omega(\omega - \mathrm{j}\Omega)}\right\} - \mathrm{j}\frac{\sigma_\mathrm{m}}{\omega}. \quad (28)$$

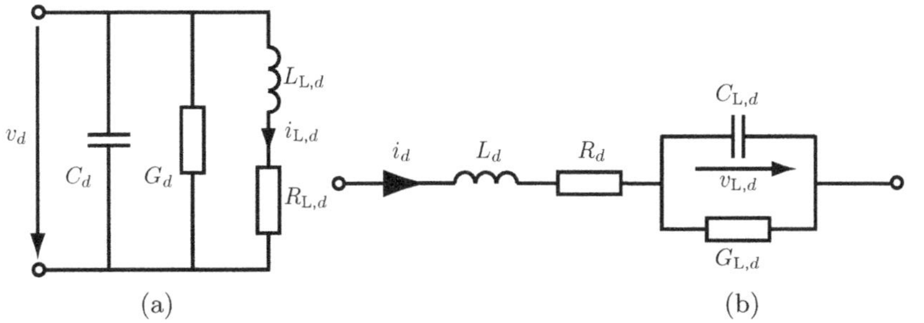

Fig. 5 Extended equivalent circuit models for Drude dispersive and lossy media. (**a**) Admittance $Y_{\text{ext-Drude}}(\omega)$ located at E-field nodes and related to the dispersive $\tilde{\varepsilon}_{\text{ext-Drude}}(\omega)$. (**b**) Impedance $Z_{\text{ext-Drude}}(\omega)$ located at H-field nodes and related to the dispersive $\tilde{\mu}_{\text{ext-Drude}}(\omega)$

Due to the presence of the additional Drude lumped elements, the time-continuous Ampere's and Faraday's circuit law system of Eqs. (19) and (20) must be extended to the system

$$\{1-\mathrm{S}_{d+1}^{-1}\}\,i_{d+2}-\{1-\mathrm{S}_{d+2}^{-1}\}\,i_{d+1}+i_{\mathrm{src},d}=+C_d\,\frac{\mathrm{d}}{\mathrm{d}t}v_d+G_d\,v_d+i_{\mathrm{L},d}, \tag{29}$$

$$v_d=+L_{\mathrm{L},d}\,\frac{\mathrm{d}}{\mathrm{d}t}i_{\mathrm{L},d}+R_{\mathrm{L},d}\,i_{\mathrm{L},d}, \tag{30}$$

$$\{\mathrm{S}_{d+1}-1\}\,v_{d+2}-\{\mathrm{S}_{d+2}-1\}\,v_{d+1}=-L_d\,\frac{\mathrm{d}}{\mathrm{d}t}i_d-R_d\,i_d-v_{\mathrm{L},d}, \tag{31}$$

$$i_d=+C_{\mathrm{L},d}\,\frac{\mathrm{d}}{\mathrm{d}t}v_{\mathrm{L},d}+G_{\mathrm{L},d}\,v_{\mathrm{L},d}, \tag{32}$$

where the two original equations include one more term and which has been increased by additional equations, as seen from the models of Fig. 5.

The final update equations representing the Drude counterparts of Eqs. (23) and (24) are finally obtained as

$$v_d^n=\frac{1-\dfrac{\Delta t G_d}{2C_d}}{1+\dfrac{\Delta t G_d}{2C_d}}\,v_d^{n-1}+\frac{\Delta t}{C_d}\,\frac{\{1-\mathrm{S}_{d+1}^{-1}\}\,i_{d+2}^{n-\frac{1}{2}}-\{1-\mathrm{S}_{d+2}^{-1}\}\,i_{d+1}^{n-\frac{1}{2}}-i_{\mathrm{L},d}^{n-\frac{1}{2}}+i_{\mathrm{src},d}^{n-\frac{1}{2}}}{1+\dfrac{\Delta t G_d}{2C_d}}, \tag{33}$$

$$i_{\mathrm{L},d}^{n+\frac{1}{2}}=\frac{1-\dfrac{\Delta t R_{\mathrm{L},d}}{2L_{\mathrm{L},d}}}{1+\dfrac{\Delta t R_{\mathrm{L},d}}{2L_{\mathrm{L},d}}}\,i_{\mathrm{L},d}^{n-\frac{1}{2}}+\frac{\Delta t}{L_{\mathrm{L},d}}\,\frac{v_d^n}{1+\dfrac{\Delta t R_{\mathrm{L},d}}{2L_{\mathrm{L},d}}}, \tag{34}$$

$$i_d^{n+\frac{1}{2}}=\frac{1-\dfrac{\Delta t R_d}{2L_d}}{1+\dfrac{\Delta t R_d}{2L_d}}\,i_d^{n-\frac{1}{2}}-\frac{\Delta t}{L_d}\,\frac{\{\mathrm{S}_{d+1}-1\}\,v_{d+2}^n-\{\mathrm{S}_{d+2}-1\}\,v_{d+1}^n+v_{\mathrm{L},d}^n}{1+\dfrac{\Delta t R_d}{2L_d}}, \tag{35}$$

$$v_{\mathrm{L},d}^{n} = \frac{1 - \dfrac{\Delta t G_{\mathrm{L},d}}{2C_{\mathrm{L},d}}}{1 + \dfrac{\Delta t G_{\mathrm{L},d}}{2C_{\mathrm{L},d}}} v_{\mathrm{L},d}^{n-1} + \frac{\Delta t}{C_{\mathrm{L},d}} \frac{i_d^{n-\frac{1}{2}}}{1 + \dfrac{\Delta t G_{\mathrm{L},d}}{2C_{\mathrm{L},d}}}. \tag{36}$$

2.3.2 Other Dispersive Media: Lorentz and Debye

Lorentz-type (one pole and one zero in ε or μ) dispersive materials may be analyzed by similar extended EC FDTD models [6]. Such Lorentz media are often encountered in natural materials (e.g. ferrimagnetic and ferromagnetic materials) and in metamaterials (e.g. split ring resonator structures). The circuit model for this type of material is shown in Fig. 6(a), and the extended EC FDTD scheme equations may be obtained following a procedure formally analogous to that for the Drude case.

As a final important example of dispersive media which can be analyzed by the extended EC FDTD scheme, let us consider the *Debye*-type dispersive case [typically characterized by the parameters $(\varepsilon_\infty, \sigma_\mathrm{e}, \Delta\varepsilon_n, \omega_n = 1/\tau_n)$], which are very common in biological systems. A Debye material typically exhibits a permittivity function of the form

$$\begin{aligned}\tilde{\varepsilon}_{\text{N-Debye}}(\omega) = \frac{Y_{\text{N-Debye}}(\omega)}{\mathrm{j}\omega\Delta_\mathrm{e}} &= \frac{1}{\Delta_\mathrm{e}}\left\{C_\infty - \mathrm{j}\frac{G_0}{\omega} + \sum_{n=1}^{N} \frac{\Delta C_n}{1 + \mathrm{j}\omega\Delta C_n R_n}\right\} \\ &= \varepsilon_\infty - \mathrm{j}\frac{\sigma_\mathrm{e}}{\omega} + \sum_{n=1}^{N} \frac{\Delta\varepsilon_n}{1 + \mathrm{j}\omega/\omega_n},\end{aligned} \tag{37}$$

which may include N poles and whose circuit model is shown in Fig. 6(b), while its permeability is usually constant and equal to μ_0 (non-magnetic medium). Again, the extended EC FDTD scheme equations may be obtained following a procedure formally analogous to that for the Drude case.

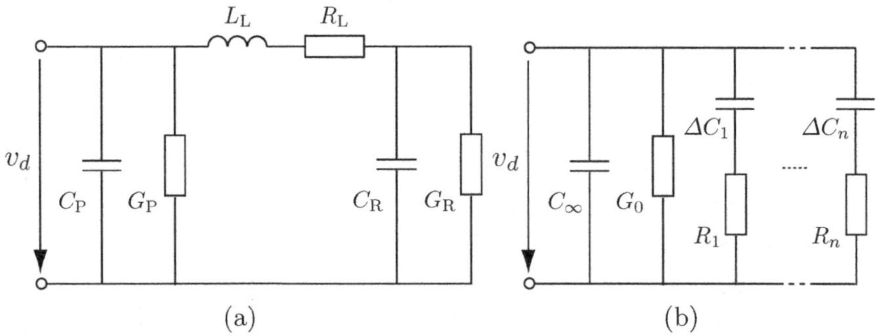

Fig. 6 Extended equivalent circuit admittance models for a Lorentz and a Debye dispersive and lossy media. (**a**) Extended lossy Lorentz model. (**b**) N-pole Debye circuit

3 Poynting Theorem for EC FDTD

This section presents the Poynting theorem for the EC FDTD scheme. It will show that the space-discrete Poynting theorem has exactly the same form as the space-continuous Poynting theorem in the time-continuous case but a slightly different form (with additional terms) in the time-discrete case. These results will be utilized in Sect. 4 to derive stability criteria. The procedure for the derivation of the space-discrete Poynting theorem is somewhat analogous to that for the space-continuous case. For generality, the derivations will be made for the double-Drude case. They can be easily particularized to the case of non-dispersive media or applied to other types of dispersive media.

3.1 Time-Continuous Case

In order to evaluate a Poynting quantity ($\boldsymbol{E} \times \boldsymbol{H}$, or scalar $E_d H_{d+1}$), one needs to consider perpendicular components of the E and H fields. Since the (Drude) current Eqs. (31) and (32) correspond to a H-field sample which is *parallel* to the E-field sample corresponding to the associated voltage Eqs. (29) and (30) (see Fig. 1, with $H_d \parallel E_d$), it is then necessary to consider, for one of the two fields, a different and perpendicular (i.e. *rotated*) component. Let us for instance keep the E_d sample unchanged and right-handedly *rotate* the H_d sample to select the appropriate perpendicular component, where the current and voltages considered are those orthogonal to each other; this is achieved by rotating or index-shifting ($d \rightarrow d+1$) Eqs. (31) and (32) by the operator R (Sect. 2.1).

The first step of the procedure consists then in multiplying Eqs. (29), (30), index-shifted (31) and index-shifted (32) with $v_d, i_{\mathrm{L},d}, (-i_{d+1})$ and $v_{\mathrm{L},d+1}$, respectively, and summing for each index-direction $d \in \{0,\ 1,\ 2\}$ over all the cells ($\boldsymbol{X} \in \Omega$) of the computational domain to determine its total power. This yields

$$\begin{aligned}
&\sum_{\boldsymbol{X}\in\Omega}\sum_{d=0}^{2}\Big(v_d\left\{1-\mathrm{S}_{d+1}^{-1}\right\}i_{d+2}-v_d\left\{1-\mathrm{S}_{d+2}^{-1}\right\}i_{d+1}+v_d\, i_{\mathrm{L},d}\\
&\qquad -i_{d+1}\left\{\mathrm{S}_{d+2}-1\right\}v_d+i_{d+1}\left\{\mathrm{S}_d-1\right\}v_{d+2}+v_{\mathrm{L},d+1}i_{d+1}\Big) \qquad (38)\\
&=\sum_{\boldsymbol{X}\in\Omega}\sum_{d=0}^{2}\Big(C_d\, v_d\frac{\mathrm{d}}{\mathrm{d}t}v_d+G_d\, v_d^2+v_d\, i_{\mathrm{L},d}-v_d\, i_{\mathrm{src},d}+L_{\mathrm{L},d}\, i_{\mathrm{L},d}\frac{\mathrm{d}}{\mathrm{d}t}i_{\mathrm{L},d}+R_{\mathrm{L},d}\, i_{\mathrm{L},d}^2+\\
&L_{d+1}i_{d+1}\frac{\mathrm{d}}{\mathrm{d}t}i_{d+1}+R_{d+1}i_{d+1}^2+v_{\mathrm{L},d+1}i_{d+1}+C_{\mathrm{L},d+1}v_{\mathrm{L},d+1}\frac{\mathrm{d}}{\mathrm{d}t}v_{\mathrm{L},d+1}+G_{\mathrm{L},d+1}v_{\mathrm{L},d+1}^2\Big).
\end{aligned}$$

The different terms of this relation may be rearranged as

$$\sum_{X\in\Omega}\sum_{d=0}^{2} P_{\mathrm{src},d} = \sum_{X\in\Omega}\sum_{d=0}^{2} v_d i_{\mathrm{src},d}$$
$$= \sum_{X\in\Omega}\sum_{d=0}^{2}\Bigg(G_d\, v_d^2 + R_{\mathrm{L},d}\, i_{\mathrm{L},d}^2 + R_{d+1}\, i_{d+1}^2 + G_{\mathrm{L},d+1}\, v_{\mathrm{L},d+1}^2$$
$$+ \frac{\mathrm{d}}{\mathrm{d}t}\left[\frac{1}{2}C_d v_d^2 + \frac{1}{2}L_{\mathrm{L},d} i_{\mathrm{L},d}^2 + \frac{1}{2}L_{d+1} i_{d+1}^2 + \frac{1}{2}C_{\mathrm{L},d+1} v_{\mathrm{L},d+1}^2\right]$$
$$- v_d i_{d+2} + v_d \mathrm{S}_{d+1}^{-1} i_{d+2} + v_d i_{d+1} - v_d \mathrm{S}_{d+2}^{-1} i_{d+1}$$
$$+ \mathrm{S}_{d+2} v_d i_{d+1} - v_d i_{d+1} - \mathrm{S}_d v_{d+2} i_{d+1} + \mathrm{R}^{-1}\left[v_d i_{d+2}\right]\Bigg) \tag{39}$$

where the following different *total powers of the 3D EC mesh within the computational domain* are identified: (i) left-hand side: *power delivered by the sources* ($\int_\Omega \boldsymbol{E}\cdot\boldsymbol{J}_{\mathrm{src}}\,\mathrm{d}v$ in the field-continuous case); (ii) right-hand side – first row: *power dissipated*; (iii) right-hand side – second row: *power deriving from the net reactive energy stored* in the inductances and capacitances ($\frac{\mathrm{d}}{\mathrm{d}t}[W_\mathrm{m} + W_\mathrm{e}]$); right-hand side - third and fourth row: *power radiated within the volume* ($\int_\Omega \mathrm{div}(\boldsymbol{E}\times\boldsymbol{H})\,\mathrm{d}v$ in the field-continuous case), where 4 out of the 8 products involving the terms $v_d i_{d+1}$ and $v_d i_{d+2}$ cancel out[4].

Furthermore, applying spatial shifting (operator S and its inverse S^{-1}), it is found that the radiated power terms *inside the volume* Ω cancel out so that only the terms *at the surface* $\partial\Omega$ of this volume contribute to the radiated power, leading to the substitution: $\sum_{X\in\Omega} \to \sum_{X\in\partial\Omega}$. This fact constitutes the *discretized version of Gauss theorem* ($\int_\Omega \mathrm{div}(\boldsymbol{E}\times\boldsymbol{H})\,\mathrm{d}v = \int_{\partial\Omega}(\boldsymbol{E}\times\boldsymbol{H})\cdot\mathrm{d}\boldsymbol{A}$). The previous relation may then be written in the form

$$\sum_{X\in\Omega}\sum_{d=0}^{2} P_{\mathrm{src},d} = \sum_{X\in\Omega}\sum_{d=0}^{2}\Bigg(P_{G_d} + P_{R_{\mathrm{L},d}} + P_{R_{d+1}} + P_{G_{\mathrm{L},d+1}}$$
$$+ \frac{\mathrm{d}}{\mathrm{d}t}\left[W_{C_d} + W_{L_{\mathrm{L},d}} + W_{L_{d+1}} + W_{C_{\mathrm{L},d+1}}\right]\Bigg)$$
$$+ \sum_{d=0}^{2}\Bigg\{\sum_{X_{d+1}}\sum_{X_{d+2}}\Big(\mathrm{S}_d v_{d+1} i_{d+2}\big|_{X_d=N_{+,d}} - \mathrm{S}_d v_{d+2} i_{d+1}\big|_{X_d=N_{+,d}}$$
$$+ v_{d+2}\mathrm{S}_d^{-1} i_{d+1}\big|_{X_d=N_{-,d}} - v_{d+1}\mathrm{S}_d^{-1} i_{d+2}\big|_{X_d=N_{-,d}}\Big)\Bigg\}, \tag{40}$$

which is the *EC FDTD discretized time-continuous version of the Poynting theorem*, stating that the power delivered by the sources is equal to the sum of the power

---

[4] To indicate the cross-cancellation of the 1st and 8th term, the 8th has been index-shifted by the operator R^{-1}.

transmitted through the surface[5] $\partial\Omega$, the power lost to heat in the volume Ω and the net reactive power stored in the volume Ω.

3.2 Time Discrete Case

Let us now derive the *time-discrete* counterpart of EC FDTD discretized Poynting theorem. This will lead to considerations of stability in the next section.

Multiplying the time-discretized versions of Eqs. (29), (30), (31) and (32) with $v_d^{n-\frac{1}{2}}$, $i_{\mathrm{L},d}^n$, $\left(-i_{d+1}^n\right)$ and $v_{\mathrm{L},d+1}^{n-\frac{1}{2}}$, respectively, yields

$$\begin{aligned}
&\sum_{X\in\Omega}\sum_{d=0}^{2}\Big(v_d^{n-\frac{1}{2}}\left\{1-\mathrm{S}_{d+1}^{-1}\right\}i_{d+2}^{n-\frac{1}{2}}-v_d^{n-\frac{1}{2}}\left\{1-\mathrm{S}_{d+2}^{-1}\right\}i_{d+1}^{n-\frac{1}{2}}+v_d^n i_{\mathrm{L},d}^n \\
&\qquad -i_{d+1}^n\left\{\mathrm{S}_{d+2}-1\right\}v_d^n+i_{d+1}^n\left\{\mathrm{S}_d-1\right\}v_{d+2}^n+v_{\mathrm{L},d+1}^{n-\frac{1}{2}}i_{d+1}^{n-\frac{1}{2}}\Big) \\
&=\sum_{X\in\Omega}\sum_{d=0}^{2}\left(C_d\,v_d^{n-\frac{1}{2}}\frac{v_d^n-v_d^{n-1}}{\Delta t}+G_d\,v_d^{n-\frac{1}{2}}\frac{v_d^n+v_d^{n-1}}{2}+v_d^{n-\frac{1}{2}}i_{\mathrm{L},d}^{n-\frac{1}{2}}-v_d^{n-\frac{1}{2}}i_{\mathrm{src},d}^{n-\frac{1}{2}}\right. \\
&\qquad +L_{\mathrm{L},d}\,i_{\mathrm{L},d}^n\frac{i_{\mathrm{L},d}^{n+\frac{1}{2}}-i_{\mathrm{L},d}^{n-\frac{1}{2}}}{\Delta t}+R_{\mathrm{L},d}\,i_{\mathrm{L},d}^n\frac{i_{\mathrm{L},d}^{n+\frac{1}{2}}+i_{\mathrm{L},d}^{n-\frac{1}{2}}}{2} \\
&\qquad +L_{d+1}\,i_{d+1}^n\frac{i_{d+1}^{n+\frac{1}{2}}-i_{d+1}^{n-\frac{1}{2}}}{\Delta t}+R_{d+1}\,i_{d+1}^n\frac{i_{d+1}^{n+\frac{1}{2}}+i_{d+1}^{n-\frac{1}{2}}}{2}+v_{\mathrm{L},d+1}^n i_{d+1}^n \\
&\qquad \left.+C_{\mathrm{L},d+1}\,v_{\mathrm{L},d+1}^{n-\frac{1}{2}}\frac{v_{\mathrm{L},d+1}^n-v_{\mathrm{L},d+1}^{n-1}}{\Delta t}+G_{\mathrm{L},d+1}\,v_{\mathrm{L},d+1}^{n-\frac{1}{2}}\frac{v_{\mathrm{L},d+1}^n+v_{\mathrm{L},d+1}^{n-1}}{2}\right).
\end{aligned}\tag{41}$$

Next partly replacing the time samples $v_d^{n-\frac{1}{2}}$, $i_{\mathrm{L},d}^n$, i_{d+1}^n and $v_{\mathrm{L},d+1}^{n-\frac{1}{2}}$ by their averages $\frac{1}{2}(v_d^n+v_d^{n-1})$, $\frac{1}{2}\left(i_{\mathrm{L},d}^{n+\frac{1}{2}}+i_{\mathrm{L},d}^{n-\frac{1}{2}}\right)$, $\frac{1}{2}\left(i_{d+1}^{n+\frac{1}{2}}+i_{d+1}^{n-\frac{1}{2}}\right)$ and $\frac{1}{2}\left(v_{\mathrm{L},d+1}^n+v_{\mathrm{L},d+1}^{n-1}\right)$, respectively, transforms Eq. (41) into the re-arranged relation

$$\begin{aligned}
\sum_{X\in\Omega}\sum_{d=0}^{2}P_{\mathrm{src},d}^{n-\frac{1}{2}}&=\sum_{X\in\Omega}\sum_{d=0}^{2}v_d^{n-\frac{1}{2}}i_{\mathrm{src},d}^{n-\frac{1}{2}}=\sum_{X\in\Omega}\sum_{d=0}^{2}\frac{1}{2}\left(v_d^n+v_d^{n-1}\right)i_{\mathrm{src},d}^{n-\frac{1}{2}} \\
&=\sum_{X\in\Omega}\sum_{d=0}^{2}\left(G_d\left(v_d^{n-\frac{1}{2}}\right)^2+R_{\mathrm{L},d}\left(i_{\mathrm{L},d}^n\right)^2+R_{d+1}\left(i_{d+1}^n\right)^2+G_{\mathrm{L},d+1}\left(v_{\mathrm{L},d+1}^{n-\frac{1}{2}}\right)^2\right. \\
&\qquad +\frac{1}{2}C_d\frac{\left(v_d^n\right)^2-\left(v_d^{n-1}\right)^2}{\Delta t}+\frac{1}{2}L_{\mathrm{L},d}\frac{\left(i_{\mathrm{L},d}^{n+\frac{1}{2}}\right)^2-\left(i_{\mathrm{L},d}^{n-\frac{1}{2}}\right)^2}{\Delta t}
\end{aligned}$$

[5] The indices values $N_{-,d}$ and $N_{+,d}$ in the last summation of Eq. (40) are the lower and upper bounds of the cuboidal domain Ω in the direction d. The inner two sums over X_{d+1} and X_{d+2} are due to the summation over the plane boundaries of the volume Ω.

$$
\begin{aligned}
&+\frac{1}{2}L_{d+1}\frac{\left(i_{d+1}^{n+\frac{1}{2}}\right)^2-\left(i_{d+1}^{n-\frac{1}{2}}\right)^2}{\Delta t}+\frac{1}{2}C_{\mathrm{L},d+1}\frac{\left(v_{\mathrm{L},d+1}^{n}\right)^2-\left(v_{\mathrm{L},d+1}^{n-1}\right)^2}{\Delta t}\\
&+\frac{1}{2}\Big[-v_d^n i_{d+2}^{n-\frac{1}{2}}-v_d^{n-1}i_{d+2}^{n-\frac{1}{2}}+v_d^n\mathrm{S}_{d+1}^{-1}i_{d+2}^{n-\frac{1}{2}}+v_d^{n-1}\mathrm{S}_{d+1}^{-1}i_{d+2}^{n-\frac{1}{2}}\\
&+v_d^n i_{d+1}^{n-\frac{1}{2}}+v_d^{n-1}i_{d+1}^{n-\frac{1}{2}}-v_d^n\mathrm{S}_{d+2}^{-1}i_{d+1}^{n-\frac{1}{2}}-v_d^{n-1}\mathrm{S}_{d+2}^{-1}i_{d+1}^{n-\frac{1}{2}}\\
&+\mathrm{S}_{d+2}v_d^n i_{d+1}^{n+\frac{1}{2}}+\mathrm{S}_{d+2}v_d^n i_{d+1}^{n-\frac{1}{2}}-v_d^n i_{d+1}^{n+\frac{1}{2}}-v_d^n i_{d+1}^{n-\frac{1}{2}}\\
&-\mathrm{S}_d v_{d+2}^n i_{d+1}^{n+\frac{1}{2}}-\mathrm{S}_d v_{d+2}^n i_{d+1}^{n-\frac{1}{2}}+v_{d+2}^n i_{d+1}^{n+\frac{1}{2}}+\mathrm{R}^{-1}\left[v_d^n i_{d+2}^{n-\frac{1}{2}}\right]\\
&-v_d^n i_{\mathrm{L},d}^{n+\frac{1}{2}}-v_d^n i_{\mathrm{L},d}^{n-\frac{1}{2}}-v_{\mathrm{L},d+1}^n i_{d+1}^{n-\frac{1}{2}}-v_{\mathrm{L},d+1}^{n-1}i_{d+1}^{n-\frac{1}{2}}\\
&+v_d^n i_{\mathrm{L},d}^{n-\frac{1}{2}}+v_d^{n-1}i_{\mathrm{L},d}^{n-\frac{1}{2}}+v_{\mathrm{L},d+1}^n i_{d+1}^{n+\frac{1}{2}}+v_{\mathrm{L},d+1}^n i_{d+1}^{n-\frac{1}{2}}\Big]\Bigg),
\end{aligned}
\tag{42}
$$

which includes 16 non-dispersive and 8 dispersive vi products, over which 4 non-dispersive and 4 dispersive cancel out[6]. As for the time-continuous case, applying spatial shifting (operator S and its inverse S^{-1}), it is found that some of the radiated power terms *inside the volume* Ω cancel out and only 4 surface terms remain. The previous relation may then be written in the form

$$
\begin{aligned}
\sum_{X\in\Omega}\sum_{d=0}^{2}P_{\mathrm{src},d}^{n-\frac{1}{2}}=\sum_{X\in\Omega}\sum_{d=0}^{2}\Bigg(&P_{G_d}^{n-\frac{1}{2}}+P_{R_{\mathrm{L},d}}^{n}+P_{R_{d+1}}^{n}+P_{G_{\mathrm{L},d+1}}^{n-\frac{1}{2}}+\frac{W_{C_d}^{n}-W_{C_d}^{n-1}}{\Delta t}\\
&+\frac{W_{L_{\mathrm{L},d}}^{n+\frac{1}{2}}-W_{L_{\mathrm{L},d}}^{n-\frac{1}{2}}}{\Delta t}+\frac{W_{L_{d+1}}^{n+\frac{1}{2}}-W_{L_{d+1}}^{n-\frac{1}{2}}}{\Delta t}+\frac{W_{C_{\mathrm{L},d+1}}^{n}-W_{C_{\mathrm{L},d+1}}^{n-1}}{\Delta t}\\
&+\frac{1}{2}\Big[-v_d^n\left\{1-\mathrm{S}_{d+2}^{-1}\right\}i_{d+1}^{n+\frac{1}{2}}+v_d^{n-1}\left\{1-\mathrm{S}_{d+2}^{-1}\right\}i_{d+1}^{n-\frac{1}{2}}\\
&+v_d^n\left\{1-\mathrm{S}_{d+1}^{-1}\right\}i_{d+2}^{n+\frac{1}{2}}-v_d^{n-1}\left\{1-\mathrm{S}_{d+1}^{-1}\right\}i_{d+2}^{n-\frac{1}{2}}\\
&-v_d^n i_{\mathrm{L},d}^{n+\frac{1}{2}}+v_d^{n-1}i_{\mathrm{L},d}^{n-\frac{1}{2}}+v_{\mathrm{L},d+1}^n i_{d+1}^{n+\frac{1}{2}}-v_{\mathrm{L},d+1}^{n-1}i_{d+1}^{n-\frac{1}{2}}\Big]\Bigg)\\
&+\sum_{d=0}^{2}\Bigg\{\sum_{X_{d+1}}\sum_{X_{d+2}}\Big(\mathrm{S}_d v_{d+1}^n i_{d+2}^n\big|_{X_d=N_{+,d}}-\mathrm{S}_d v_{d+2}^n i_{d+1}^n\big|_{X_d=N_{+,d}}\\
&+v_{d+2}^{n-\frac{1}{2}}\mathrm{S}_d^{-1}i_{d+1}^{n-\frac{1}{2}}\big|_{X_d=N_{-,d}}-v_{d+1}^{n-\frac{1}{2}}\mathrm{S}_d^{-1}i_{d+2}^{n-\frac{1}{2}}\big|_{X_d=N_{-,d}}\Big)\Bigg\},
\end{aligned}
\tag{43}
$$

---

[6] To indicate the cross-cancellation of the 1st and 16th term, the 16th has been index-shifted by the operator R^{-1}.

which is the time-discrete EC FDTD version of the Poynting theorem[7]. In contrast to its time-continuous counterpart [Eq. (40)], this relation does not simplify to a form exactly identical to the space-continuous Poynting theorem: 12 *vi* "time-discretization-compensatory" product terms (8 for the non-dispersive case and 4 additional ones for the double Drude material), due to the two different time steps $n-\frac{1}{2}$ and n involved, remain present in the power expressions.

4 Stability Criteria Based on a Lyapunov Energy

Time-discretized Maxwell's equations are subject to instability when the time step Δt is larger then the computational cell size Δ divided by the speed of light c, and multiplied by a factor depending on the dimension of the computational domain. Whereas in the case of *non-dispersive* media stability is ensured by following the simple Courant–Friedrichs–Lewy criterion $[\Delta t < \Delta/(\sqrt{3}c)]$ [1], no explicit guaranteed-stability criterion is available for *dispersive* media in FDTD schemes. The EC FDTD scheme provides such a criterium, which will be derived now applying a Lyapunov energy approach to the time-discrete Poynting theorem [Eq. (43)]. Here the volume Ω in the discrete Poynting theorem must be the whole computational domain in order to include the overall energy in the 3D EC network.

In order to simplify the derivation, we assume that the computational domain is truncated either by PECs, by PMCs or by lossy material for absorption (with PEC or PMC walls behind)[8], so that the last sum in Eq. (43) vanishes.

The *Lyapunov energy stability criterium* states that a system is stable if its energy: (1) is a positive definite scalar function, and (2) is constant or monotonically decreasing after all sources have been switched off [7].

Defining as a generalized energy expression the Lyapunov function

$$\begin{aligned} L^n_{\text{global}} = \sum_{X\in\Omega}\sum_{d=0}^{2}\Big(& W^n_{C_d} + W^{n+\frac{1}{2}}_{L_{\mathrm{L},d}} + W^{n+\frac{1}{2}}_{L_{d+1}} + W^n_{C_{\mathrm{L},d+1}} \qquad (44)\\ & + \frac{\Delta t}{2}\Big[-v^n_d\left\{1-\mathrm{S}^{-1}_{d+2}\right\} i^{n+\frac{1}{2}}_{d+1} + v^n_d\left\{1-\mathrm{S}^{-1}_{d+1}\right\} i^{n+\frac{1}{2}}_{d+2} - v^n_d i^{n+\frac{1}{2}}_{\mathrm{L},d} + v^n_{\mathrm{L},d+1} i^{n+\frac{1}{2}}_{d+1}\Big]\Big), \end{aligned}$$

the time-discrete Poynting theorem [Eq. (43)] may be reformulated as

$$\begin{aligned} \frac{L^n_{\text{global}} - L^{n-1}_{\text{global}}}{\Delta t} &= \sum_{X\in\Omega}\sum_{d=0}^{2}\left(P^{n-\frac{1}{2}}_{\mathrm{src},d} - P^{n-\frac{1}{2}}_{G_d} - P^n_{R_{\mathrm{L},d}} - P^n_{R_{d+1}} - P^{n-\frac{1}{2}}_{G_{\mathrm{L},d+1}} \right) \le 0 \\ & \text{for} \quad P^{n-\frac{1}{2}}_{\mathrm{src},d} = 0, \qquad (45) \end{aligned}$$

[7] The indices values $N_{-,d}$ and $N_{+,d}$ in the last summation of Eq. (43) are the lower and upper bounds of the cuboidal domain Ω in the direction d. The inner two sums over X_{d+1} and X_{d+2} are due to the summation over the plane boundaries of the volume Ω.

[8] Special absorbing boundary conditions of traveling-wave or perfectly matched layer type are not considered.

which means that the global energy of the system remains constant or decreases in time after the source has been switched off, and the second Lyapunov condition is thus fulfilled.

The demonstration of the first Lyapunov condition, namely the positive definiteness of the Lyapunov function, or equivalently the properness of L^n_{global} in Eq. (44) [7], is more intricate. The sought stability criterion will be obtained in terms of the maximum Δt delimiting the range where L^n_{global} is positive-definite.

Since the positive-definiteness of the global energy function L^n_{global} is difficult to establish, we decompose it in terms of *the local Lyapunov function* $L^n_{\text{local},d}(\boldsymbol{X})$

$$L^n_{\text{global}} = \sum_{\boldsymbol{X}\in\Omega} \sum_{d=0}^{2} L^n_{\text{local},d}, \tag{46}$$

which involves only local voltages and currents. Once the positive definiteness of this local function will have been shown, the positive definiteness of the global one will follow by superposition.

For the non-dispersive case each summand of the global Lyapunov energy function [Eq. (44)] includes the two orthogonal values v_d and i_{d+1} (via the energies W_{C_d} and $W_{L_{d+1}}$) *and* additionally the adjacent currents[9] i_{d+2}, and (the negatively shifted) $\mathrm{S}^{-1}_{d+2} i_{d+1}$ and $\mathrm{S}^{-1}_{d+1} i_{d+2}$. This local configuration is also used for the $L^n_{\text{local},d}$. But instead of just considering the energy of the the current i_{d+1} as in Eq. (44), the energies of all four surrounding currents are included. If this arrangement is summed up over all directions d and all $\boldsymbol{X}$ as defined in Eq. (46) the energies of the four surrounding currents would have been counted four times. Thus a weighting factor of 1/4 must be used for proper energy summation. The advantage of the mentioned re-arrangement of the energy terms is the fact that such a local Lyapunov function can be evaluated by a quadratic form. The positive definiteness of $L^n_{\text{local},d}$ is equivalent to the positive definiteness of the associated quadratic form matrix $\mathbf{A}$. This can be mostly done by applying Sylvester's criterion, which states that all of the leading principal minors of the square matrix must have a positive determinant for the matrix to be positive definite.

Within the following subsections three important practical cases are considered for stability analysis: the non-dispersive material, the plasmonic ε-Drude case and finally the double Drude metamaterial.

4.1 Stability Criterion for Non-Dispersive Materials

By using the advantageous computational configuration for $L^n_{\text{local},d}(\boldsymbol{X})$ where the admittance Y_d is at its center together with only one quarter (above-mentioned weighting factors) of the four adjacent and orthogonal impedances, the quadratic form

[9] These terms are due to the remaining 4 vi products.

including only the non-dispersive reactive EC elements[10] and the voltages/current vector $\boldsymbol{\Theta}_5$ reads[11]

$$
\begin{aligned}
2\,L^n_{\mathrm{local},d} &= \boldsymbol{\Theta}_5^{\mathrm{T}}\cdot\mathbf{A}_5\cdot\boldsymbol{\Theta}_5\\
&= \boldsymbol{\Theta}_5^{\mathrm{T}}\cdot
\begin{pmatrix}
C_d & -\frac{\Delta t}{2} & +\frac{\Delta t}{2} & +\frac{\Delta t}{2} & -\frac{\Delta t}{2}\\
-\frac{\Delta t}{2} & \frac{L_{d+1}}{4} & 0 & 0 & 0\\
+\frac{\Delta t}{2} & 0 & \frac{\mathrm{S}^{-1}_{d+2}L_{d+1}}{4} & 0 & 0\\
+\frac{\Delta t}{2} & 0 & 0 & \frac{L_{d+2}}{4} & 0\\
-\frac{\Delta t}{2} & 0 & 0 & 0 & \frac{\mathrm{S}^{-1}_{d+1}L_{d+2}}{4}
\end{pmatrix}
\cdot
\begin{pmatrix}
v^n_d\\
i^{n+\frac{1}{2}}_{d+1}\\
\mathrm{S}^{-1}_{d+2}i^{n+\frac{1}{2}}_{d+1}\\
i^{n+\frac{1}{2}}_{d+2}\\
\mathrm{S}^{-1}_{d+1}i^{n+\frac{1}{2}}_{d+2}
\end{pmatrix}
\end{aligned}
\tag{47}
$$

Sylvester's criterion demands that all of the leading principal minors of the square matrix must have a positive determinant in order to have a positive overall matrix, which leads to several stability criteria. Since the determinant of the overall matrix $\mathbf{A}_5$ is the most limiting one for the time step, we have for the maximal $\Delta t_{\text{non-disp}}$

$$
\Delta t_{\text{non-disp}} < \frac{2}{\max_{\mathbf{x}\in\Omega,\, d\in\{0,1,2\}}\sqrt{\frac{1}{C_d}\left(\frac{4}{L_{d+1}}+\frac{4}{\mathrm{S}^{-1}_{d+2}L_{d+1}}+\frac{4}{L_{d+2}}+\frac{4}{\mathrm{S}^{-1}_{d+1}L_{d+2}}\right)}}. \tag{48}
$$

For an uniform mesh the relation $\Delta t_{\text{non-disp}} = \sqrt{3/4}\,\Delta t_{\text{CFL}} \approx 87\%\Delta t_{\text{CFL}}$ holds, where Δt_{CFL} is the conventional time step based on the Courant-Friedrich-Levy (CFL) condition [1]. However, in non-uniform (graded) meshes, where the smallest cell is adjacent to a larger one (with at least 33% longer edges), the EC-based criterion yields a larger time step. This improvement depends on the size-ratio of directly adjacent cells. For extreme meshing situations an improvement factor $\Delta t_{\text{non-disp}}/\Delta t_{\text{CFL}}$ of 10 can easily be achieved. The ordinary CFL criteria has been derived for an uniform mesh and a homogeneous material distribution, therefore it seems to be too conservative in highly graded meshes, since the smallest cell size has to be used in the formula [1].

4.2 Stability Criterion for Plasmonic (ε-Drude) Materials

For a plasmonic (ε-Drude) medium, which has been deployed in the optical coupler example to be presented in Sect. 5.2, the admittances at the edges of the EC unit cell [Fig. 4] are extended by a LH inductance L_{L} in parallel, and the quadratic form reads

[10] All elements with an index L are set to zero.

[11] The index 5 should indicate the number of components of this vector.

$$\boldsymbol{\Theta}_6^{\mathrm{T}} \cdot \mathbf{A}_6 \cdot \boldsymbol{\Theta}_6 =$$

$$\boldsymbol{\Theta}_6^{\mathrm{T}} \cdot \begin{pmatrix} C_d & -\frac{\Delta t}{2} & -\frac{\Delta t}{2} & +\frac{\Delta t}{2} & +\frac{\Delta t}{2} & -\frac{\Delta t}{2} \\ -\frac{\Delta t}{2} & L_{\mathrm{L},d} & 0 & 0 & 0 & 0 \\ -\frac{\Delta t}{2} & 0 & \frac{L_{d+1}}{4} & 0 & 0 & 0 \\ +\frac{\Delta t}{2} & 0 & 0 & \frac{\mathrm{S}_{d+2}^{-1} L_{d+1}}{4} & 0 & 0 \\ +\frac{\Delta t}{2} & 0 & 0 & 0 & \frac{L_{d+2}}{4} & 0 \\ -\frac{\Delta t}{2} & 0 & 0 & 0 & 0 & \frac{\mathrm{S}_{d+1}^{-1} L_{d+2}}{4} \end{pmatrix} \cdot \begin{pmatrix} v_d^n \\ i_{\mathrm{L},d}^{n+\frac{1}{2}} \\ i_{d+1}^{n+\frac{1}{2}} \\ \mathrm{S}_{d+2}^{-1} i_{d+1}^{n+\frac{1}{2}} \\ i_{d+2}^{n+\frac{1}{2}} \\ \mathrm{S}_{d+1}^{-1} i_{d+2}^{n+\frac{1}{2}} \end{pmatrix}, \tag{49}$$

which yields by Sylvester's criterion (6×6-matrix determinant) the maximal possible Δt for an ε-Drude material

$$\Delta t_{\varepsilon\text{-Drude}} < \frac{2}{\max_{\mathbf{X} \in \Omega,\, d \in \{0,1,2\}} \sqrt{\frac{1}{C_d} \left(\frac{1}{L_{\mathrm{L},d}} + \frac{4}{L_{d+1}} + \frac{4}{\mathrm{S}_{d+2}^{-1} L_{d+1}} + \frac{4}{L_{d+2}} + \frac{4}{\mathrm{S}_{d+1}^{-1} L_{d+2}} \right)}} \tag{50}$$

Inspection of Eq. (50) confirms that a smaller time step limit must be used compared to the non-dispersive case $[\Delta t_{\varepsilon\text{-Drude}} < \Delta t_{\text{non-disp}}]$ due to the additional term $1/L_{\mathrm{L},d}$ in the denominator. This fact makes the novel criterion even more important for stable time domain simulation of plasmonic structures.

4.3 Stability Criterion for Double Drude Materials

For a double Drude (ε-Drude and μ-Drude) medium, the admittances at the edges of the EC unit cell are extended by a LH inductance L_{L} in parallel and the impedances in the faces of the EC unit cell [Fig. 4] are extended by a LH capacitance C_{L} in series. In this case the voltage/current vector reads

$$\begin{aligned} \boldsymbol{\Theta}_{10}^{\mathrm{T}} \doteq \Big(& v_d^n, i_{\mathrm{L},d}^{n+\frac{1}{2}}, i_{d+1}^{n+\frac{1}{2}}, \mathrm{S}_{d+2}^{-1} i_{d+1}^{n+\frac{1}{2}}, i_{d+2}^{n+\frac{1}{2}}, \mathrm{S}_{d+1}^{-1} i_{d+2}^{n+\frac{1}{2}}, \\ & v_{\mathrm{L},d+1}^n, \mathrm{S}_{d+2}^{-1} v_{\mathrm{L},d+1}^n, v_{\mathrm{L},d+2}^n, \mathrm{S}_{d+1}^{-1} v_{\mathrm{L},d+2}^n \Big), \end{aligned} \tag{51}$$

i.e. the equation $|\mathbf{A}_{10}|(\Delta t) > 0$ must be solved for Δt. In principle this might be possible, but the result will be very cumbersome. Therefore the special case

$$\begin{aligned} \frac{1}{\omega_{\mathrm{se},d}^2} & \doteq L_{d+1}\, C_{\mathrm{L},d+1} \qquad (52) \\ & = \mathrm{S}_{d+2}^{-1} L_{d+1}\, \mathrm{S}_{d+2}^{-1} C_{\mathrm{L},d+1} = L_{d+2}\, C_{\mathrm{L},d+2} = \mathrm{S}_{d+1}^{-1} L_{d+2}\, \mathrm{S}_{d+1}^{-1} C_{\mathrm{L},d+2} \end{aligned}$$

is considered here instead, due to a simplified determination of the stability criterion. Solving the simplified determinant equation yields the maximal possible Δt for a double-Drude material

$$\Delta t_{\text{double-Drude}} < \frac{2}{\max_{\boldsymbol{x}\in\Omega,\ d\in\{0,1,2\}} \sqrt{\omega_{\text{sh},d}^2 + \omega_{\text{se},d}^2 + \omega_{\text{mix},d}^2}} < \frac{2}{\max_{\boldsymbol{x}\in\Omega,\ d\in\{0,1,2\}} \omega_{\text{double-Drude},d}} \tag{53}$$

with

$$\omega_{\text{double-Drude},d}^2 = \frac{1}{2}\Big(\omega_{\text{sh},d}^2 + \omega_{\text{se},d}^2 + \omega_{\text{mix},d}^2 + \sqrt{\omega_{\text{sh},d}^4 + \omega_{\text{se},d}^4 + \omega_{\text{mix},d}^4 + 2\left(\omega_{\text{se},d}^2\,\omega_{\text{mix},d}^2 + \omega_{\text{sh},d}^2\,\omega_{\text{mix},d}^2 - \omega_{\text{sh},d}^2\,\omega_{\text{se},d}^2\right)}\Big) < \omega_{\text{sh},d}^2 + \omega_{\text{se},d}^2 + \omega_{\text{mix},d}^2 \tag{54}$$

where

$$\omega_{\text{sh},d}^2 = \frac{1}{L_{\text{L},d}\,C_d}, \tag{55}$$

$$\omega_{\text{mix},d}^2 = 16\,\omega_{\text{sh},d}^2\,\omega_{\text{se},d}^2\,L_d\,\overline{C}_{\text{L,se},d} \quad \text{with} \tag{56}$$

$$\overline{C}_{\text{L,se},d} = \frac{1}{4}\left(C_{\text{L},d+1} + \text{S}_{d+2}^{-1}C_{\text{L},d+1} + C_{\text{L},d+2} + \text{S}_{d+1}^{-1}C_{\text{L},d+2}\right). \tag{57}$$

The simplified denominator $[\omega_{\text{sh},d}^2 + \omega_{\text{se},d}^2 + \omega_{\text{mix},d}^2]$ in equation (53) yields a slightly lower limit for the time step, but is easier to compute. The term is an upper estimate for the complicated $\omega_{\text{double-Drude},d}$. Inspection of Eq. (53) reveals that a smaller time step limit must be used compared to the non-dispersive and also the ε-Drude case $[\Delta t_{\text{double-Drude}} < \Delta t_{\varepsilon\text{-Drude}} < \Delta t_{\text{non-disp}}]$, due to the additional term $\omega_{\text{se},d}^2$ in the denominator of Eq. (53).

5 Simulation Examples

This section presents examples of dispersive material structures and devices computed by the EC FDTD method. The dispersive materials used are *metamaterials* in Sect. 5.1 and *natural materials* in Sect. 5.2.

5.1 Applications based on Dispersive Metamaterials

A *metamaterial* is an artificial structure which is electromagnetically homogeneous (i.e. possesses a structural unit cell much smaller than the guided wavelength,

$p \ll \lambda_g$) and which exhibits unusual properties not readily found in nature, such as for instance a negative refractive index (NRI) [4]. Metamaterials are generally dispersive media. This section presents a spatial-frequencial refractive filter and an open surface-plasmonic resonator. In addition, it discusses a few other possible metamaterial applications.

5.1.1 Spatial-Frequencial Refractive Filter

The spatial-frequencial refractive filter is shown in Fig. 7. It consists of a composite right/left-handed (CRLH) [4] wedge placed in vacuum. This wedge is excited by a two-tone Gaussian (plane) wave with the lower frequency (f_l) in its left-handed range and the higher frequency (f_h) in its right-handed range. As a consequence, the lower frequency part of the signal experiences negative refraction and is therefore refracted toward the left, while the higher frequency part of the signal experiences positive refraction and is therefore refracted toward the right. The structure presented is specifically designed such that, with a 11.3° inclined-plane wedge, the output beams refract under the angles ±30° with respect to the direction of the incident wave. The structure operates thus as a spatial ($\pm\theta$) – frequencial (f_l/f_h) filter. Such a device may be used as quasi-optical diplexer.

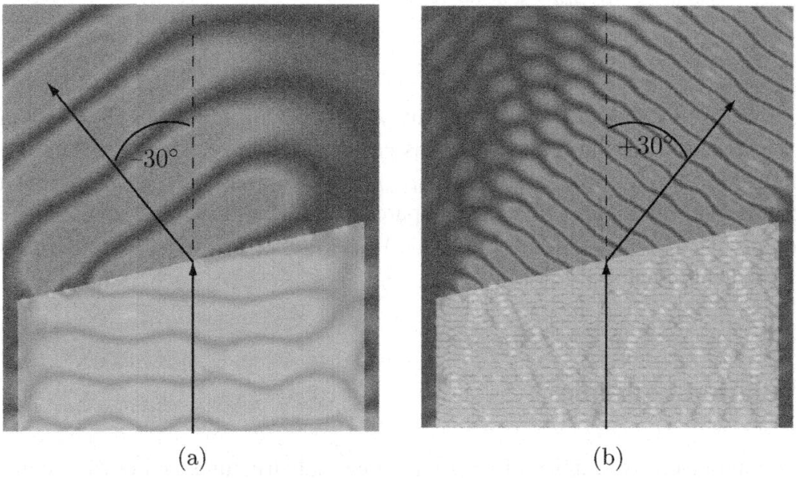

Fig. 7 Spatial-frequencial refractive filter, consisting of a CRLH wedge excited by a two-tone signal. (**a**) Negatively refracted lower-frequency beam, for which $n(4.21\,\text{GHz}) = -1.63$. (**b**) Positively refracted higher frequency beam, for which $n(12.56\,\text{GHz}) = +3.37$

5.1.2 Open Surface-Plasmonic Resonator

The open surface-plasmonic resonator is shown in Fig. 8(a). It consists of four NRI metamaterial (MTM) triangular regions with $n = -1$ interleaved with comple-

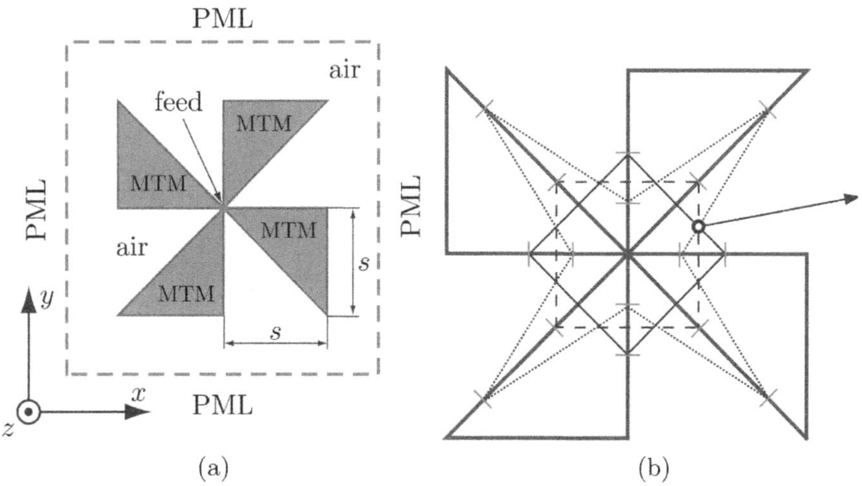

Fig. 8 Open surface plasmonic resonator. The metamaterial (MTM) regions are made of negative refractive index (NRI) medium with $n = -1$ while the complementary regions are vacuum ($n = +1$). The structure is bounded in the xy-plane by 4 perfect matched layers (PMLs) for absorption and in the z-direction by two perfect electric conductor (PEC) walls (bottom and top) for TM_z operation. (**a**) Top view of the structure. (**b**) Ray-tracing description, for a source located at the position of the black circle (3 rays at different angles are shown).

mentary vacuum ($n = +1$) regions, and this arrangement is bounded by perfectly matched layer (PML) absorbing boundaries on the sides and by perfect electric (PEC) walls at the bottom and at the top to ensure TM_z operation.

Figure 8(b) shows a naive ray-tracing description of the expected behavior of the resontator: due to the negative and iso-density refraction occurring at each interface, the electromagnetic wave radiated by a source within the structure should be "trapped", as indicated, despite the fact that no walls surround the structure to confined the energy as in a conventional cavity resonator (hence the term "open" resonator). In addition, it is known that the evanescent components (corresponding to the near-field radiation contributions) of a wave incident at the interface between a right-handed and a NRI medium excite *surface plasmons* along the interface [8], as suggested in Fig. 9(a). The time snapshots of Figs. 9(b), (c), (d), (e) and (f) illustrate the refractive traping and surface plasmonic behavior of the resonator. Note that only a small amount of energy is radiated through the open boundaries. This type of "energy traps" may lead to high-quality factor resonators, due to the absence of lossy metal walls, if purely dielectric NRI media are used.

5.1.3 Other Metamaterial Structures

Many other types of metamaterials structures may be easily modeled by the proposed EC FDTD method. CRLH metamaterials are double Drude (only one zero

Fig. 9 Open surface-plasmonic resonator. (**a**) Surface plasmon propagation, and distribution of the vertical component of the electric field $[E_z(x,y)]$ at the resonance frequency f_{res} at the phase instants (**b**) 0°, (**c**) 45°, (**d**) 90°, (**e**) 135°, (**f**) 180°. Here, the excitation is located at the center of the structure

in both ε and μ) media [$\varepsilon(\omega)p = C_{\text{R}} - 1/(\omega^2 L_{\text{L}})$ and $\mu(\omega)p = L_{\text{R}} - 1/(\omega^2 C_{\text{L}})$]. Other metamaterials are double Lorentz (one pole and one zero in both ε and μ) [6] or mixed Drude-Lorentz as the thin wire and split ring resonator structure which was the first experimentally verified NRI metamaterial [9] (Fig. 10), and more com-

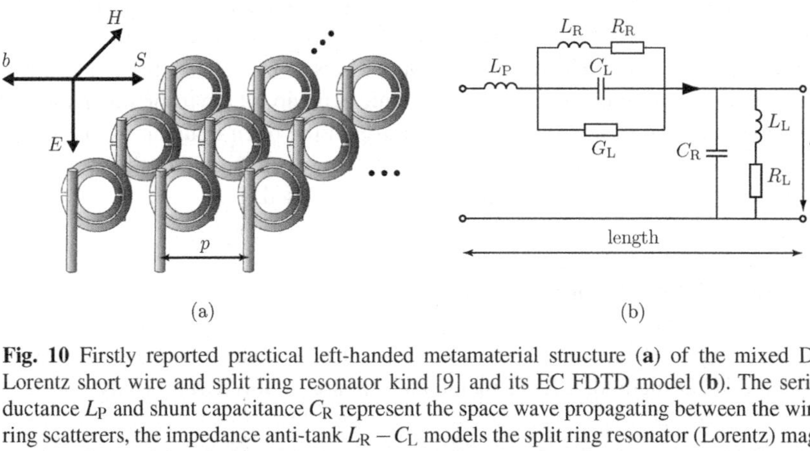

Fig. 10 Firstly reported practical left-handed metamaterial structure (**a**) of the mixed Drude-Lorentz short wire and split ring resonator kind [9] and its EC FDTD model (**b**). The series inductance L_{P} and shunt capacitance C_{R} represent the space wave propagating between the wire and ring scatterers, the impedance anti-tank $L_{\text{R}} - C_{\text{L}}$ models the split ring resonator (Lorentz) magnetic dipolar effect and the admittance inductance L_{L} models the wire (Drude) electric dipolar effect

plex possibilities exist. Corresponding dispersive EC FDTD models are straightforwardly available in the proposed EC FDTD scheme.

5.2 Applications Based on Natural Dispersive Materials

Many natural materials are dispersive. This includes for instance plasmas, ferromagnetic and ferroelectric materials, noble metals at optical frequencies and biological tissues. The EC FDTD method, by including various circuit-based dispersive (and also lossy) models both for ε and for μ (Drude, Lorentz, Debye, and variations thereof) can ideally analyze structures and devices made of such materials. This section presents the example of an optical coupled surface plasmon subwavelength power divider and discusses the applicability of the method to biological tissues.

5.2.1 Optical Coupled Surface Plasmon Subwavelength Power Divider

The optical power divider to be discussed here [10] is essentially constituted of a thin metallic Ag film within SiO_2 ($\varepsilon_{SiO_2} = 2.09$) and supporting two surface (in fact interface) surface plasmons coupling to each other. It is appropriate to first provide an accurate description of the plasmonic response of Ag at optical frequencies.

It is well known that noble metals, such as silver (Ag), exhibit at optical frequencies a Drude-type permittivity response

$$\tilde{\varepsilon}_{\text{Drude}}(\omega) = \varepsilon_0 \left\{ 1 - \frac{\omega_{\text{p}}^2}{\omega(\omega - \text{j}\Omega)} \right\}. \tag{58}$$

This expression with the parameters $\omega_{\text{p}} = 12.9 \times 10^{15}\text{rad/s}$, $\Omega = 8.0 \times 10^{13}$ rad/s was used for instance in [11] for the design of a plasmonic contra-directional coupler. Since this simple model fits experimental data [12] only in a very narrowband region around 500 THz, we propose the following extended model

$$\tilde{\varepsilon}_{\text{ext-Drude}}(\omega) = \varepsilon_0 \varepsilon_{\text{r},\infty} \left\{ 1 - \frac{\omega_{\text{p}}^2}{\omega(\omega - \text{j}\Omega)} \right\} - \text{j}\frac{\sigma_0}{\omega}, \tag{59}$$

which includes the two additional parameters $\varepsilon_{\text{r},\infty}$ (asymptotic term) and σ_0 (conductive term) to better fit experimental data in the frequency range of interest. The complex permittivity $\varepsilon_{\text{r}} = \varepsilon_{\text{r}}' - \text{j}\varepsilon_{\text{r}}''$ obtained from this model for Ag with $\varepsilon_{\text{r},\infty} = 3.9$, $\omega_{\text{p}} = 7.0 \times 10^{15}\text{rad/s}$, $\Omega = 2.3 \times 10^{13}\text{rad/s}$ and $\sigma_0 = 8.0 \times 10^3\text{S/m}$, yielding the best fit to the measured values of [12] in the frequency range of 450 THz–850 THz, is shown in Fig. 11.

The optical coupled surface plasmon subwavelength power divider is presented in Fig. 12 along with its field distribution. It consists of a tapered metal (Ag) film embedded in a dielectric (SiO_2) material, with the coupling region in the center

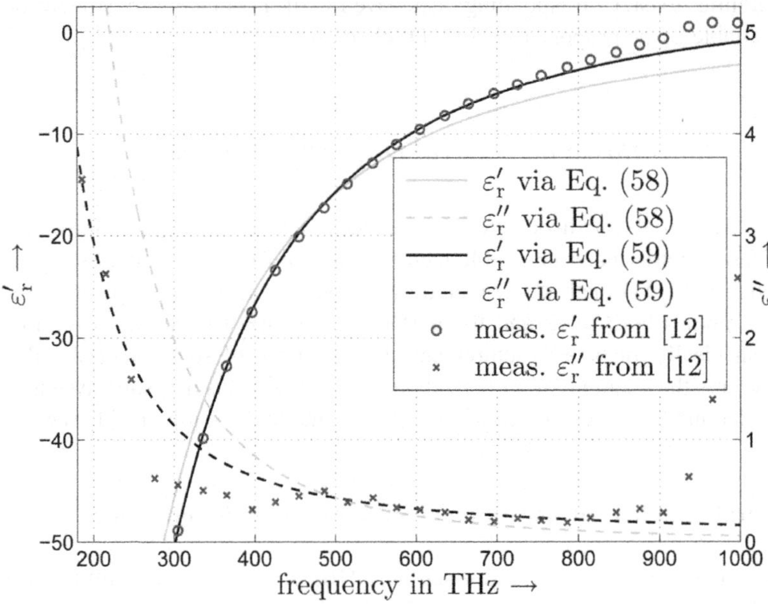

Fig. 11 Measured complex permittivity $\varepsilon_r = \varepsilon_r' - j\varepsilon_r''$ from [12] compared to the data obtained by the simple Drude model given in Eq. (58) and an extended Drude function given in Eq. (59)

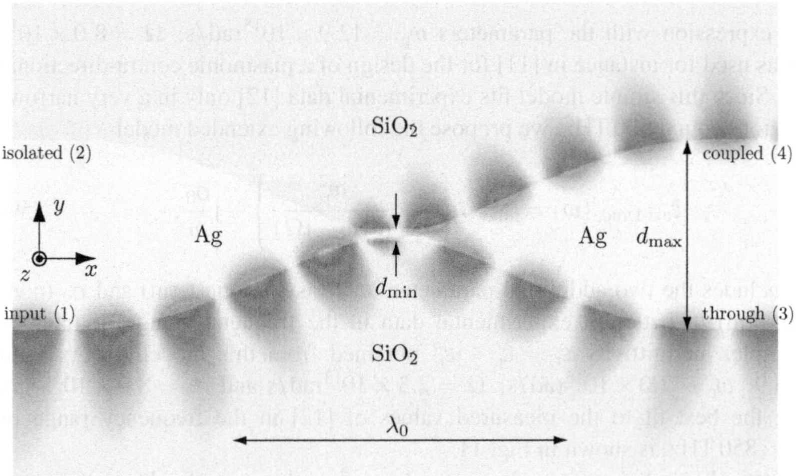

Fig. 12 Optical coupled surface plasmon subwavelength 3-dB/quadrature power divider (*longitudinal section view*) with intensity plot of the driven-mode TE_z magnetic field $H_z(x, y;\, f = 770\,\text{THz})$

and the input/output regions on either sides. The field distribution shown indicates 3-dB/quadrature splitting of the incoming guided surface plasmon. The dimensions are $d_{\min} = 15\,\text{nm}$, $d_{\max} = 240\,\text{nm}$, $\ell_{\text{narrow}} = 50\,\text{nm}$ and $\ell_{\text{taper}} = 375\,\text{nm}$ yielding an overall length of the device, including the tapering sections, of 800 nm. The coupling region of the device is therefore of ultra-compact subwavelength size, since the free-space wavelength λ_0 at the operation frequency of 770 THz is 390 nm.

5.2.2 Ultra Wide-Band Exposure of Human Tissues

Human tissues illuminated by ultra wide-band electromagnetic fields generally exhibit a Debye-type response of the form given by Eq. (37), where the frequencies ω_n are typically determined by experimental and empirical techniques. Table 1 lists typical 2-pole corresponding parameters for the brain, for muscles and for bones [13]. This broadband model is accurate in the frequency range from 100 MHz up to 6 GHz.

Table 1 Parameters for a 2-pole Debye model for wide-band time-domain simulation for different human tissues

Parameter	Brain	Muscles	Bones
$\varepsilon_{r,\infty}$	20.20	21.02	5.48
σ_0 in S/m	0.355	0.699	0.104
$\Delta\varepsilon_{r,1}$	30.89	12.08	3.19
$\Delta\varepsilon_{r,2}$	24.44	34.20	6.20
f_1 in MHz	164.8	212.8	239.6
f_2 in GHz	9.086	11.737	6.707

Recently, specific emphasis has been put on ultra wide-band (UWB) imaging techniques for medical diagnosis and biological detection, which are based, in contrast to X-ray imaging, on low power and non-ionizing radiation while still providing high resolution. These techniques require not only exact dispersive models but also broadband analysis capability. Moreover, the complexity of the materials to investigate is such that an explicit time domain technique with no matrix inversion is the method of choice. In this context, time-domain methods such as the proposed EC FDTD technique may play a crucial role, since they allow almost unlimited description of the geometrical and dispersive properties of these complex materials.

6 Conclusion

The EC FDTD method has been derived for the simplest case of non-dispersive materials and for the case of dispersive materials, including Drude, Lorentz and Debye dispersions.

Within the framework of the EC FDTD a discretized and circuital version of the Poynting theorem both for the time-continuous and time-discrete cases was established. In the former case the theorem has exactly the same form as the well-known one in the space-and-time-continuous world. Whereas the time-discrete theorem with its additional power flow terms could be utilized to derive novel stability criteria based on a Lyapunov energy function. With the help of this generalized energy function stability criteria have been derived for the non-dispersive, the ε-Drude-dispersive and the double-Drude-dispersive case. At least for these two dispersive materials the maximal possible time iteration step was decreased compared to a non-dispersive media due to additional reactances in the 3D equivalent circuit.

Finally several application examples of metamaterial and natural material dispersive structures and devices have been presented. The diversity of the examples indicates the versatility of the proposed EC FDTD method.

By incorporating the divergence information of the electrodynamic field, or in the circuital world of the EC FDTD the conservation of charges, an even improved stability criterion is expected. This will be investigated in the near future.

References

1. A. Taflove and S. Hagness. *Computational Electrodynamics: The Finite-Difference Time-Domain Method*, Third Edition, Artech House, 2005.
2. A. Rennings, S. Otto, A. Lauer, C. Caloz, and P. Waldow. "An extended equivalent circuit based FDTD scheme for the efficient simulation of composite right/left-handed metamaterials," *Proc. of the Euro. Microw. Assoc.*, vol. 2, no. 1, pp. 71–82, March 2006.
3. A. Rennings, S. Otto, C. Caloz, A. Lauer, W. Bilgic, and P. Waldow. "Composite right/left-handed extended equivalent circuit (CRLH-EEC) FDTD: stability, dispersion analysis with examples," *Int. J. Numer. Modell.*, vol. 19, no. 2, pp. 141–172, March 2006.
4. C. Caloz and T. Itoh. *Electromagnetic Metamaterials, Transmission Line Theory and Microwave Applications*, Wiley and IEEE Press, 2005.
5. E. M. Lifshitz, L. D. Landau, and L. P. Pitaevskii. *Electrodynamics of Continuous Media: Volume 8*, Second Edition, Butterworth-Heinemann, 1984.
6. A. Rennings, T. Liebig, C. Caloz, and I. Wolff. "Double Lorentz transmission line metamaterials and their applications to triband devices," *IEEE MTT-S Int. Microwave Symp.*, Honolulu, USA, June 2007, pp. 1427–1430.
7. N. Rouche, P. Habets, and M. Laloy. *Stability Theory by Liapunov's Direct Method*, Springer, 1977.
8. J. B. Pendry. "Negative refraction makes a perfect lens," *Phys. Rev. Lett.*, vol. 85, no. 18, pp. 3966–3969, Oct. 2000.
9. R. A. Shelby, D. R. Smith, and S. Schultz. "Experimental verification of a negative index of refraction," *Science*, vol. 292, pp. 77–79, April 2001.
10. A. Rennings, J. Mosig, S. Gupta, C. Caloz, R. Kashyap, D. Erni, and P. Waldow, "Ultra-compact power splitter based on coupled surface plasmons," *International Symposium on Signals, Systems and Electronics (ISSSE)*, Montréal, Canada, July–Aug. 2007, pp. 471–474.
11. Y. Wang, R. Islam, and G. V. Eleftheriades, "An ultra-short contra-directional coupler utilizing surface plasmon-polaritons at optical frequencies," *Opt. Express*, vol. 14, no. 16, pp. 7279–7290, 2006.
12. P. B. Johnson and R. W. Christy, "Optical constants of the noble metals," *Phy. Rev. B 6*, pp. 4370–4379, 1972.
13. T. Wuren, T. Takai, M. Fujii, and I. Sakagami, "Effective 2-Debye-pole FDTD model of electromagnetic interaction between whole human body and UWB radiation," *Microw. Wirel. Compon. Lett.*, vol. 17, no. 7, pp. 483–485, 2007.

A 3D Isotropic Left-Handed Metamaterial Based on the Rotated TLM Scheme

M. Zedler, C. Caloz and P. Russer

1 Introduction

In this contribution a three-dimensional left-handed metamaterial is presented which is shown to be a physical implementation of the rotated TLM scheme. The key characteristics of the metamaterial in terms of dispersion relation and Bloch impedance as well as composite right/left-handed response are analysed. Simplified models for one-dimensional propagation within the structure are derived. Numerical, circuit-based simulations modelling plane wave incidence at a RH/LH interface are performed. Two realisations of the metamaterial are proposed, one full 3D structure and two planarised structures that maintain the required network topology. Measurements of the full 3D structure and field-simulations of the planarised structure demonstrate the left-handed behaviour of the structures.

Metamaterials are composite materials exhibiting electromagnetic properties not found in nature such as close to zero or negative refractive index [1, 2] first investigated by Veselago [3]. A multitude of fields of application have been proposed that rely on these properties, examples being novel antenna concepts [4, 5], improved imaging systems [6] and approaches reducing the radar cross-section [7, 8].

Metamaterials need to exhibit internal structure [9]. In the early 2000s the synthesis of one- and two-dimensional metamaterials was of primary interest to the scientific community of physics and electrical engineering. Recently, approaches to three-dimensional metamaterials have gathered much interest.

So far five different topologies for 3D left-handed isotropic metamaterials have been proposed: A FDTD-derived structure independently proposed in [10, 11], a structure derived from the rotated TLM scheme [12, 13], a structure consisting of dielectric spheres [14], a 3D extension of the wire/split-ring approach [15, 16],

M. Zedler
Technische Universität München, e-mail: michael.zedler@tum.de

C. Caloz
École Polytechnique, Montréal, e-mail: christophe.caloz@polymtl.ca

P. Russer
Technische Universität München, e-mail: russer@tum.de

and a straightforward extension of the 1D/2D composite right-left-handed (CRLH) structure [17, 18] supporting scalar waves that requires a common ground across all unit cells. The latter was the first 3D left-handed metamaterial demonstrated experimentally.

This contribution is organised as follows: Sects. 2 and 3.1 recall the derivation of the rotated TLM metamaterial [12, 13]. Section 3.2 discusses a simplified schematic model describing waves propagating along a principal axis, allowing easier comparison with other metamaterials based on lumped elements. Section 3.3 shows that rotated TLM metamaterials only support the desired differential mode and that a parasitic common mode does not exist. Section 3.4 recalls the behaviour of the rotated TLM metamaterial in the CRLH case. Section 4 recalls the fabrication and measurement conducted in [13]. Section 5 discusses a planar implementation that preserves the network topology and that permits highly eased fabrication. Parasitic effects resulting from planarisation are analysed and design guidelines are given to mitigate them. Full-wave simulations verify the left-handed behaviour of the metamaterial.

2 Discrete Electrodynamics

In the symmetric condensed node TLM representation of discrete electrodynamics a 12-port scattering matrix, representing the TLM cell, contains all the information of the discretised Maxwell's equations [19, 20, 21, 22]. The field is discretised into cubic cells, the tangential electric and magnetic field components are sampled at the centre of each cube surface. Hence each face has two electrical and two magnetical field components which can be formulated as two incident and two scattered waves per cubic face [20, 22, 23]. The scattering can therefore be described as a 12-port depicted in Fig. 1. It can be represented by a scattering matrix $\mathbf{S} \in \boldsymbol{C}^{\langle 12\times 12\rangle}$, its structure is well known and was, e.g., derived by the method of moments [22]. Here we show that the scattering can be derived by only taking into considering symmetry, matching and losslessness: Assuming that the 12-port adheres to rotational and mirror symmetries with respect to the x-, y-, and z-axes, $\mathbf{S}$ has the form

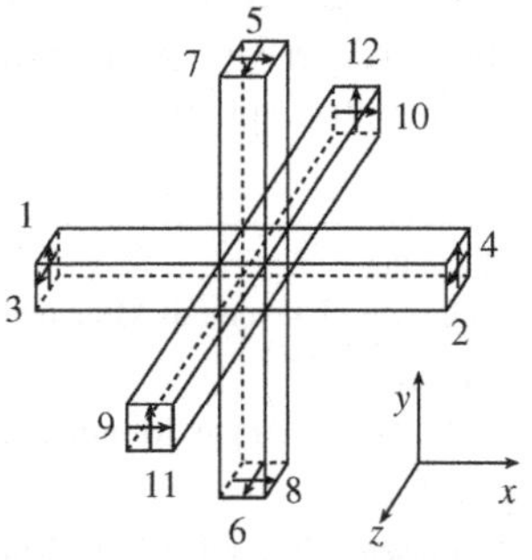

Fig. 1 General space-discretising TLM 12-port [13]

$$\mathbf{S} = \mathrm{e}^{-\mathrm{j}\phi} \begin{bmatrix} \mathbf{S}_A & \mathbf{S}_B & \mathbf{S}_B^T \\ \mathbf{S}_B^T & \mathbf{S}_A & \mathbf{S}_B \\ \mathbf{S}_B & \mathbf{S}_B^T & \mathbf{S}_A \end{bmatrix} \tag{1}$$

with

$$\mathbf{S}_A = \begin{bmatrix} \rho & \tau_1 & 0 & 0 \\ \tau_1 & \rho & 0 & 0 \\ 0 & 0 & \rho & \tau_1 \\ 0 & 0 & \tau_1 & \rho \end{bmatrix} \qquad \mathbf{S}_B = \begin{bmatrix} 0 & 0 & \tau_2 & -\tau_2 \\ 0 & 0 & -\tau_2 & \tau_2 \\ \tau_3 & \tau_3 & 0 & 0 \\ \tau_3 & \tau_3 & 0 & 0 \end{bmatrix} \tag{2}$$

Assuming matched ports ($\rho = 0$), a lossless node ($\mathbf{S}^\dagger = \mathbf{S}^{-1}$), and equal phase delay in all branches ($\arg \tau_1 = \arg \tau_2 = \arg \tau_3, \mathbf{S}_A, \mathbf{S}_B \in \boldsymbol{R}^{\langle 4\times 4\rangle}$) one obtains the two solutions

$$\tau_1 = 0 \qquad \tau_2 = \tau_3 = \frac{1}{2} \tag{3a}$$

$$\tau_1 = 1 \qquad \tau_2 = \tau_3 = 0 \tag{3b}$$

Equation (3b) describes a trivial node not connected to off-axis adjacent nodes and is thus not of further interest. Equation (3a) describes a general space-discretising cell, which is the sought symmetric condensed node TLM scattering matrix. The phase $\exp(-\mathrm{j}\phi)$ represents the phase delay experienced by a wave travelling through the TLM cell.

In order to be able to synthesise a metamaterial, i.e., a physical realisation of the computing scheme, a transformation of the symmetric condensed node TLM scheme needs to be applied. The 12-port cell can be decomposed into two independent six-ports $\tilde{\mathbf{S}}$ and $-\tilde{\mathbf{S}}$ by the coordinate transformation [23, 24]

$$\mathbf{P}^T \cdot \mathbf{S} \cdot \mathbf{P} = \begin{bmatrix} \tilde{\mathbf{S}} & 0 \\ 0 & -\tilde{\mathbf{S}} \end{bmatrix} \tag{4a}$$

$$\mathbf{P} = \begin{bmatrix} \mathbf{P}_{0a} & \mathbf{0} & \mathbf{0} & \mathbf{P}_{0b} & \mathbf{0} & \mathbf{0} \\ \mathbf{0} & \mathbf{P}_{0a} & \mathbf{0} & \mathbf{0} & \mathbf{P}_{0b} & \mathbf{0} \\ \mathbf{0} & \mathbf{0} & \mathbf{P}_{0a} & \mathbf{0} & \mathbf{0} & \mathbf{P}_{0b} \end{bmatrix} \tag{4b}$$

$$\mathbf{P}_{0a}^T = \begin{bmatrix} 1 & 0 & -1 & 0 \\ 0 & 1 & 0 & 1 \end{bmatrix} \qquad \mathbf{P}_{0b}^T = \begin{bmatrix} 1 & 0 & 1 & 0 \\ 0 & 1 & 0 & -1 \end{bmatrix} \tag{4c}$$

This corresponds to a rotation of the polarisations by 45°, as shown in Fig. 2. The transformed scattering matrix is given by

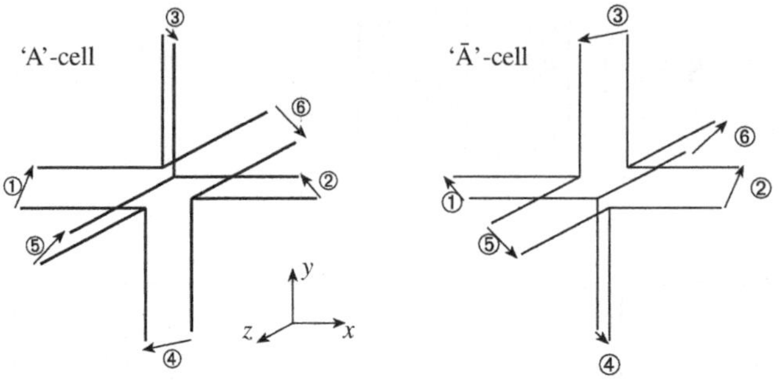

Fig. 2 Rotated TLM half unit cells implementing $\tilde{\mathbf{S}}$. The complete unit cell is constituted by the merging of both half cells [12]

$$\tilde{\mathbf{S}} = \mathrm{e}^{-\mathrm{j}\phi} \begin{bmatrix} \mathbf{0} & \tilde{\mathbf{S}}_0 & \tilde{\mathbf{S}}_0^T \\ \tilde{\mathbf{S}}_0^T & \mathbf{0} & \tilde{\mathbf{S}}_0 \\ \tilde{\mathbf{S}}_0 & \tilde{\mathbf{S}}_0^T & \mathbf{0} \end{bmatrix} \qquad \tilde{\mathbf{S}}_0 = \frac{1}{2} \begin{bmatrix} -1 & -1 \\ 1 & 1 \end{bmatrix} \tag{5}$$

In this contribution the two independent half cell six-ports described by $\tilde{\mathbf{S}}$ and $-\tilde{\mathbf{S}}$ are called 'A' and 'Ā' cells, respectively. A TLM cell that completely samples the electromagnetic field can be established by either nesting the six-port structures of the 'A' and 'Ā' half cells or by a cluster of eight half cells with alternating 'A' and 'Ā' cells [23].

As an aside it shall be noted that the rotation of polarisation by 45° used in (4a) of each face of the space discretising cube can be understood also from a network point of view using ideal transformers, as shown in Fig. 3: Two unrotated ports with polarisations/port voltages V', V'' are transformed into the superposition $(V' \pm V'')/\sqrt{2}$ which are the rotated port polarisations/voltages. Thus from a circuit theoretical point of view the symmetric condensed node TLM can be transformed into rotated TLM by means of transformer networks. In the next section it will be shown that a rotated TLM half cell can be implemented by a lumped element network, therefore together with the aforementioned transformer network also the symmetric condensed node TLM node can be represented using network circuits.

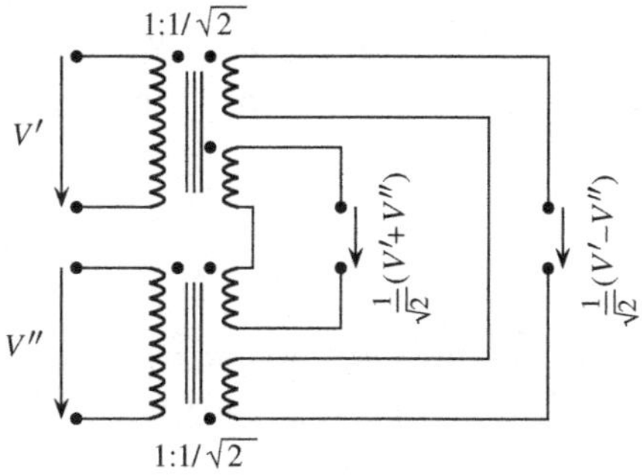

Fig. 3 Representation of a 45° rotation of polarisation by a transformer network

3 Analysis of the Rotated TLM Metamaterial

3.1 Dispersion Relation and Bloch Impedance

An elementary metamaterial cell may be conceived on the basis of rotated TLM cells, by inserting reactances in series to the six cell ports and four admittances connecting the series reactances at a central node forming a virtual ground [12] as shown in Fig. 4. Both half cells can be connected at the virtual ground. Considering that the rotation of polarisation in (4a) can be denoted by an ideal transformer network as of Fig. 3, the symmetrical condensed node TLM computational method can be fully described from a network point of view. If the transformer networks are omitted the computational method – and the metamaterial principle based on this scheme – are called rotated TLM.

Following (4a) the 'A'-cell scattering is described by $\tilde{\mathbf{S}}$ whereas the '$\bar{\text{A}}$'-cell is scattered by $-\tilde{\mathbf{S}}$ (see (4a)). This sign inversion may be interpreted as a frequency-independent phase shifter of 90° attached to each port, which is impractical for any physical realisation. For this reason we consider a metamaterial structure containing solely 'A'-cells. In this case, referring to Fig. 1, a wave incident at port 1 is scattered into ports 5, 6, 9, 10 instead of 7, 8, 11, 12. This means the polarisation is changed from vertical to horizontal after traversing the cell. Further consequences are discussed in Sect. 3.4.

The dispersion relation of a rotated TLM metamaterial consisting of only 'A'-cells is [12]

$$(1+Z_1Y_2)^2 = \frac{1}{4}(3-\cos\chi\cos\eta-\cos\chi\cos\xi-\cos\eta\cos\xi) \tag{6}$$

with χ, η, ξ being the phase shifts observed by a wave traversing the cell along the x-, y-, and z-axis respectively. The Bloch impedance of the rotated TLM cell equals that of the 1D CRLH cell [1] and may shown as [12]

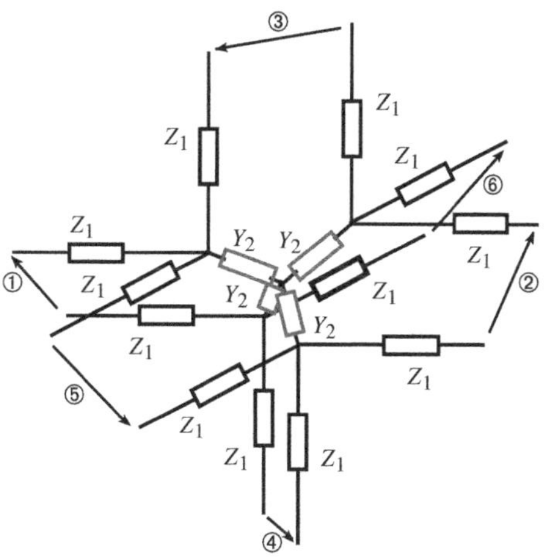

Fig. 4 Lumped element model of a rotated TLM half unit cell implementing the structure shown in Fig. 2. Shunt elements Y_2 shown in red for clarity, series elements denoted by Z_1

$$Z_{\text{Bloch}} = \sqrt{Z_1 Y_2^{-1}} \sqrt{2 + Z_1 Y_2} \tag{7}$$

This shows that there is no angular dispersion, as (7) is independent on χ, η, ξ. This property is unique to three-dimensional TLM and FDTD schemes and is not achievable with two-dimensional discretisations.

3.2 Propagation Along Principal Axes

In the case of propagation along a principal axis the lumped element model of the rotated TLM scheme can be simplified elucidating the relation to one-dimensional ladder network representations of transmission lines: By referring to Fig. 5(a) we seek the solution of propagation along the x-axis;

$$a_{③} = b_{④} \qquad a_{④} = b_{③} \tag{8a}$$
$$a_{⑤} = b_{⑥} \qquad a_{⑥} = b_{⑤} \tag{8b}$$

when a_i and b_i are the incident and reflected wave amplitudes at port i, this yields

$$i_{③} = -i_{④} = 0 \tag{9a}$$
$$v_{⑤} = v_{⑥} = 0 \tag{9b}$$
$$i_{13} + i_{16} = 0 \qquad i_{14} + i_{15} = 0 \tag{9c}$$

where $i_{13} \dots i_{16}$ denote the currents through the shunt elements Y_2, the current flow in each is directed towards the symmetry centre. Equation (9a) enforces ports ③ and ④ to be open circuit, (9b) enforces ports ⑤ and ⑥ to be shorted, and the shunt elements Y_2 decouple due to (9c). Hence the equivalent circuit can be particularised to Fig. 5(b) that can then be simplified into Fig. 5(c). Noting that the impedance matrix of a T-circuit and a flipped T-circuit are equal, a further simplification leads to Fig. 5(d). Therefore in case of 1D propagation the circuit resembles that of the standard 1D CRLH line, except that there are two T-circuits per unit cell and a flip of sign at port ②.

3.3 Parasitic Mode Analysis

While the ports in the network topology of Fig. 5 are fundamentally differential, this Section examines the possible existence of a parasitic common mode.

For the common mode the potentials at node n and $(n+6)$ with $1 \leq n \leq 6$ are equal, and all voltages are defined with respect to the potential at infinity. The circuit simplifies as follows: all shunt elements Y_2 are shorted and the structure simplifies to that shown in Fig. 6. The network in Fig. 6 only contains one type of element, denoted with $Z_1/2$. A network with only one type of element does not

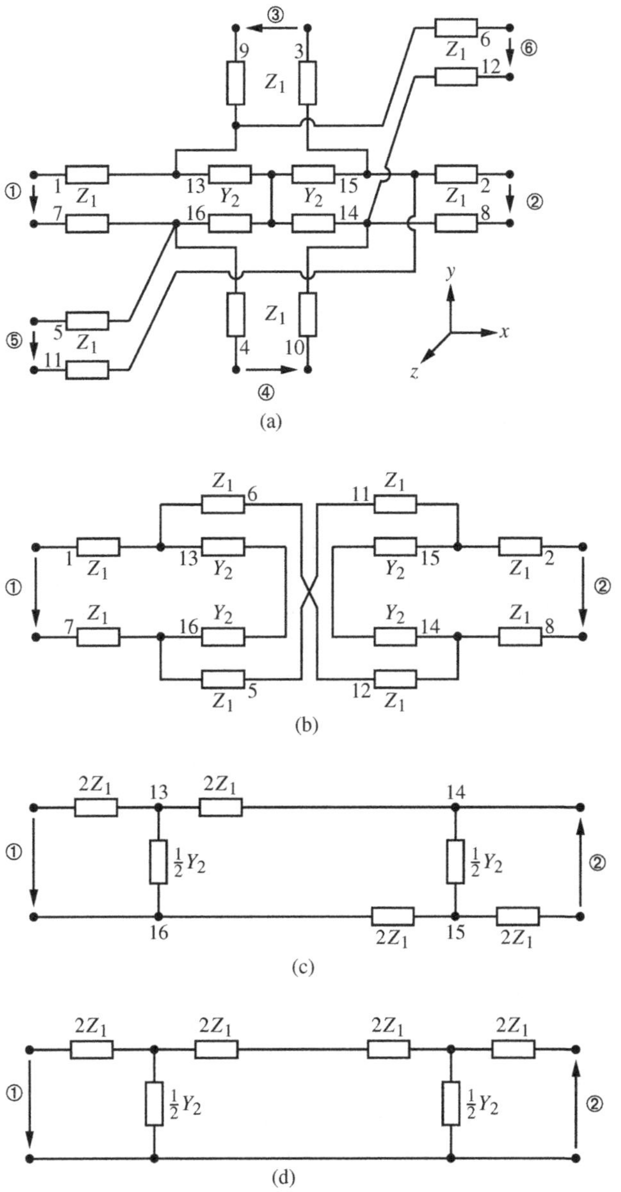

Fig. 5 Equivalent circuits for the rotated TLM 3D metamaterial half unit cell. (**a**) implementation of 'A'-cell. (**b**) particularised circuit for the case of x-axis propagation following Eqs. (8a)–(9c). (**c**) and (**d**) simplified equivalent circuit of (b). Note that in figures (c), (d) port ② is flipped

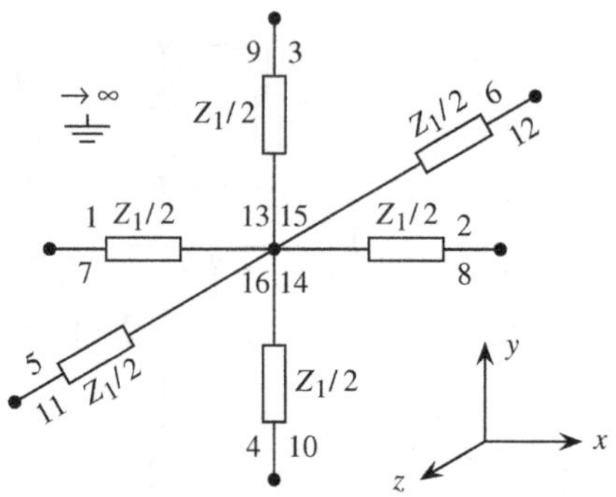

Fig. 6 Topology of the rotated TLM metamaterial half unit cell for the common mode. The Node labels $1\ldots16$ correspond to the labelling scheme in Fig. 5(a)

support wave propagation: Assuming that Z_1 consists of only inductors, then the lack of capacitors forbids propagating waves at all frequencies. The non-existence of a common mode for other network elements representing Z_1 can be shown by frequency-transformation reasoning.

In addition to a common mode one may think of a parasitic quasi-TEM mode to be able to propagate that is not bound to the lumped elements of the RTLM metamaterial cell. This mode would effectively represent an 'artificial dielectric' mode [25]. This mode is unproblematic as the operational frequency of a metamaterial around the Γ-point is different from zero.

3.4 CRLH Behaviour

In a CRLH unit cell the impedance Z_1 is a series resonator $(L_R,\ C_L)$ whereas Y_2 is a parallel resonator $(L_L,\ C_R)$. The right-handed components L_R, C_R account for unavoidable parasitics [1]. The corresponding unit cell for the rotated TLM metamaterial is shown in Fig. 7.

Comparing the dispersion relation of (6) with that of a 1D CRLH medium $(1+Z_1Y_2=\cos\chi)$, one notes that the left hand side in the rotated TLM case is simply the square of that in the 1D CRLH case. This square yields *two* left-handed and *two* right-handed bands, while the 1D case has only one band of each type. Apart from this fact the properties of 1D CRLH structures remain, including the balanced-$\left(\frac{L_RC_L}{L_LC_R}=1\right)$/unbalanced-$\left(\frac{L_RC_L}{L_LC_R}\neq 1\right)$ resonance conditions and their impact on the dispersion relation (gapless/gap transitions between bands) and Bloch impedance (frequency independency/dependency).

The four bands, obtained by inserting the CRLH reactances into (6), are given by

$$\omega^2/\omega_R^2=\left(t+\frac{1}{2}\sqrt{d}\right)\pm\sqrt{\left(t+\frac{1}{2}\sqrt{d}\right)^2-\omega_L^2/\omega_R^2} \tag{10a}$$

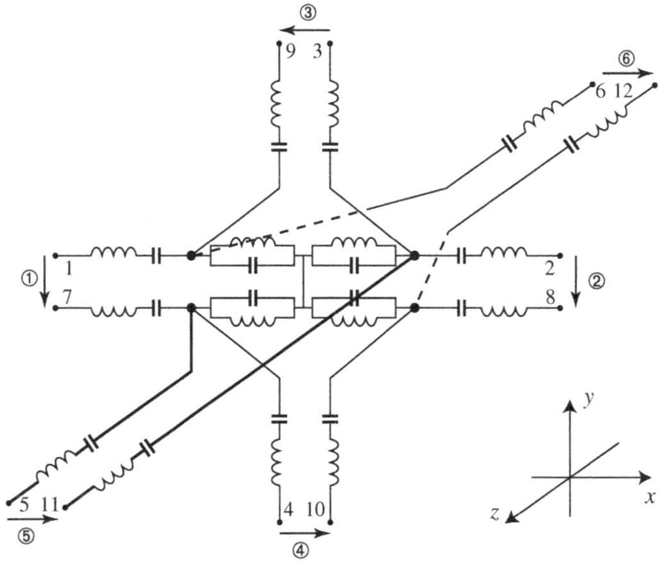

Fig. 7 CRLH rotated TLM metamaterial half unit cell [12]

$$\omega^2/\omega_R^2 = \left(t - \frac{1}{2}\sqrt{d}\right) \pm \sqrt{\left(t - \frac{1}{2}\sqrt{d}\right)^2 - \omega_L^2/\omega_R^2} \tag{10b}$$

where $\omega_R^2 C_R L_R = 1$, $\omega_L^2 C_L L_L = 1$, $2t = \omega_L^2(L_R C_L + L_L C_R) + 1$, $d = \frac{1}{4}(3 - \cos\chi \cos\eta - \cos\chi\cos\xi - \cos\eta\cos\xi)$. The dispersion diagram is shown in Fig. 8 for both the balanced- and unbalanced-resonance cases.

The consequence of the suppressed negative sign in the scattering matrix of the '$\bar{\text{A}}$'-cell (Sect. 3.1) are as follows: (i) the unbalanced-resonance gap of the dispersion diagram appears not at the Γ-point, as in the 1D CRLH metamaterial, but between the X and M points; (ii) similarly the frequency-independency of the Bloch impedance may be shown to exist not near the Γ-point but near the X/M-region. This

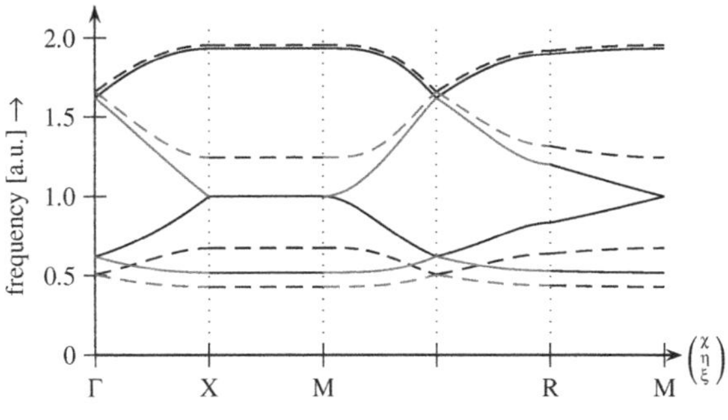

Fig. 8 Dispersion diagram for the 3D CRLH rotated TLM metamaterial for the balanced-resonance case (*solid line*) and unbalanced-resonance (*dashed line*) case. Red lines represent metamaterial operation behaviour

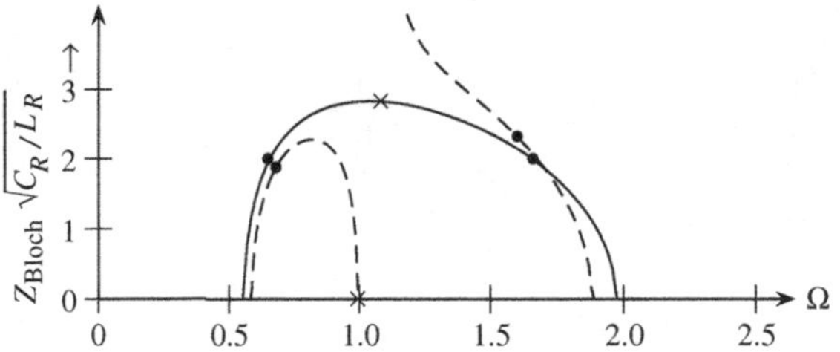

Fig. 9 Bloch impedance of a CRLH rotated TLM metamaterial cell. *Solid line*: balanced-resonance case ($L_LC_R = L_RC_L$), *dashed line*: unbalanced-resonance case ($L_LC_R = L_RC_L$). *Crosses* denote Γ-point operation for a 1D CRLH structure, *dots* denote Γ-point operation of the CRLH rotated TLM metamaterial

second consequence means that the Bloch impedance cannot be constant to first order in the metamaterial frequency range (i.e., around the Γ-point) and that therefore the bandwidth of this 3D rotated TLM metamaterial is restricted compared to 1D CRLH metamaterials. This is the price to pay for the simplification of the structure avoiding phase shifters in the 'Ā' cell. This behaviour of the Bloch impedance is shown in Fig. 9: While for a balanced-resonance 1D CRLH structure the Γ-point corresponds to the frequency $\Omega = \omega\sqrt{L_LC_R} = 1$ and there the Bloch impedance is flat, for the 3D CRLH rotated TLM metamaterial this is not the case.

4 Realisation of the Rotated TLM Metamaterial

4.1 Description of the Structure

The proposed realisation of the CRLH rotated TLM metamaterial, corresponding to the lumped element network of Fig. 7, is depicted in Fig. 10. Shunt inductors are implemented by wires connected to a common centre point, and series capacitors are implemented by metal-insulator-metal plates located between adjacent unit cells. Figure 10(a) shows the two half unit cells in the proper orientation to be then nested. Figure 10(b) shows a cluster of $2 \times 2 \times 2$ nested unit cells. Figure 10(c) shows the complete unit cell prototype. The plate capacitors are realised in PCB with patches on both sides of the substrate which ensures accurate C_L values. The inductors are realised by rigid wires. The unit cell length is 2 cm, the substrate is Rogers 4350B 10 mil, the left-handed values are $L_L \approx 17.3\,\text{nH}$, $C_L \approx 5.4\,\text{pF}$.

4.2 Intuitive Explanation of the Structure

The unit cell is a balanced structure: It has a virtual ground, that is a zero voltage point due to the symmetry of the structure rather than due to physical connection to

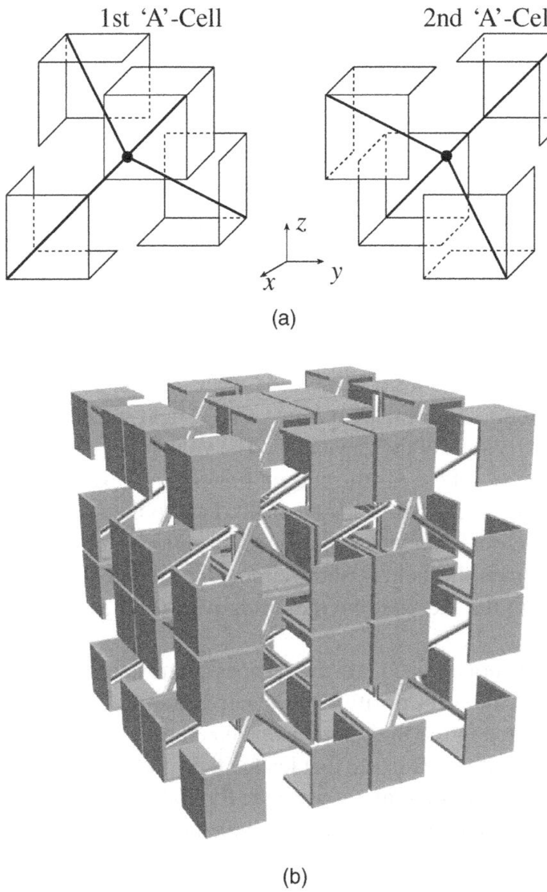

Fig. 10 3D CRLH rotated TLM metamaterial realisation. (**a**) 'A'-type half unit cells, (**b**) a complete $2 \times 2 \times 2$ structure [13]

a physical ground. It consists of two nested cells, the two cells shown in Fig. 10(a). These two cells are electromagnetically decoupled in the sense that they support electromagnetic waves of independent orthogonal polarizations in each of the directions of space, x, y and z.

To gain better insight into the propagation phenomenon along the structure, let us consider in some details the example of a $-xz$-polarized (E-field along the $z = -x$ direction) plane (transverse) wave propagating along the y direction and incident on the structure at the level of a unit cell.

Consider the first 'A' half cell (left-hand side of Fig. 10(a)), displayed in Fig. 11. In this cell, the incident plane wave produces a symmetric voltage difference ($+V$ and $-V$) between the two patches at the input face of the half cell (left-hand side of the structure in Fig. 10(a)). These two patches form capacitors with the patches printed on the opposite faces of the thin substrate slabs, which store electric energy and provide the required series capacitance C_L corresponding to negative

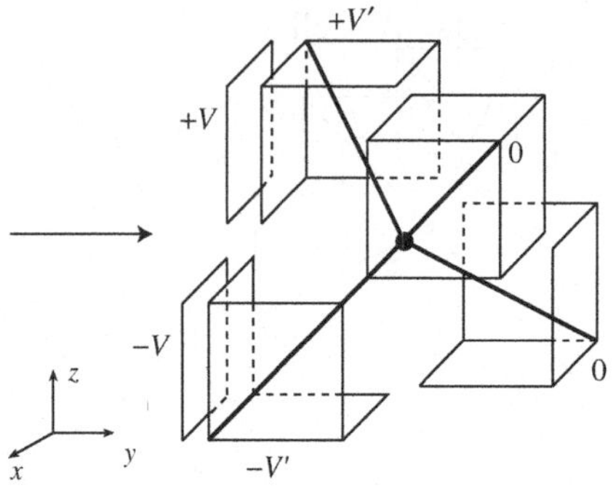

Fig. 11 Intuitive explanation of the structure

permeability. Due to these capacitors, the incident transverse E-field becomes locally longitudinal between the two plates of the capacitors. The voltages at the plates inside the structure are $V' = V - Z_1 I$, where I is the current flowing into the incident port. From this point, the wave 'sees' the wires environment. Due to the symmetry of the structure and to the symmetrical incident voltages, propagation is prohibited along the straight (y) direction since the field is short-circuited at the center node (virtual ground), consistently with the fact that the scattering parameter $\tilde{s}_{21}$ is zero in (5).

While rectilinear propagation is not allowed, off-axis propagation occurs through the four lateral faces of the unit cell, along the positive and negative x- and z-directions. Figure 11 shows how this is realized in the three-dimensional cell as a result of the differential voltages and symmetry of the structure that lead to the voltage difference $\pm V'$ at the off-axis ports. While the E-field rotates toward the four lateral faces in the unit cell, the magnetic flux circulates around the two wire branches extending from the corner voltage points to the virtual ground point, which corresponds to magnetic energy storage and generates the required shunt inductance L_L corresponding to negative permittivity. Note that the directions of the fields indicated in Fig. 11 correspond to the other four independent scattering parameters of (5).

Consider next the second 'A' half cell (right-hand side of Fig. 10(a)). In this cell, the incident $-xz$-polarized E-field does not encounter any metallization at the input plane of the cell, which is therefore transparent to it. The plates with the same polarization at the output plane belong to the next 'A' half cell.

When many cells nested 'A-A' are cascaded along the three directions of space, the plane wave macroscopically travels along the structure along the y-direction in a meandrous fashion from the off-axis cells, which constitutes the essence of the scattering-type propagation along the proposed rotated TLM structure.

4.3 Experimental Verification

Figure 12 shows the unit cell prototype of the CRLH rotated TLM metamaterial. This prototype was measured with a two-port vector network analyser through

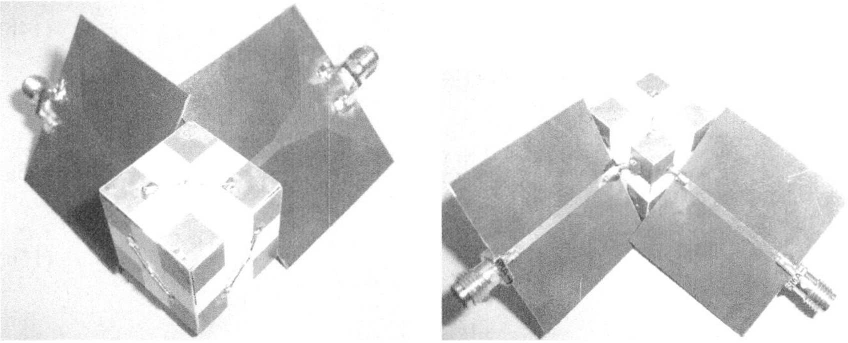

Fig. 12 3D CRLH rotated TLM unit cell with its input and output baluns required for the differential excitation of the measurement setup

baluns (microstrip to parallel-strip transitions) connected at two arbitrary non-aligned ports, while the remaining ports are terminated with $Z_L = 50\,\Omega$ resistors. Note that this excitation appropriately models wave propagation through the structure because the rotated TLM structure is a network with well-defined ports. Upon the basis of the rotated TLM metamaterial it suffices to verify experimentally that the metamaterial cell indeed acts like the lumped circuit of Fig. 5(a). Under this assumption the behaviour of the entire structure can be inferred from the response of the single unit cell.

The extraction procedure of the lumped elements Z_1, Y_2 is derived as follows: Taking the six-port impedance matrix of the ideal rotated TLM cell [12]

$$Z^{\langle 6\times 6\rangle} = \frac{2}{Y_2}(1+Z_1Y_2) + \frac{1}{Y_2}\begin{bmatrix} 0 & 0 & -1 & -1 & -1 & 1 \\ 0 & 0 & 1 & 1 & -1 & 1 \\ -1 & 1 & 0 & 0 & -1 & -1 \\ -1 & 1 & 0 & 0 & 1 & 1 \\ -1 & -1 & -1 & 1 & 0 & 0 \\ 1 & 1 & -1 & 1 & 0 & 0 \end{bmatrix} \tag{11}$$

and terminating the four unconnected ports with Z_L

$$V_2 = -Z_L I_2 \qquad V_3 = -Z_L I_3 \tag{12}$$

$$V_5 = -Z_L I_5 \qquad V_6 = -Z_L I_6 \tag{13}$$

yields a symmetric two-port impedance matrix. This can be converted to a scattering matrix normalised to Z_L with the elements

$$\hat{s}_{1,1}^{\langle 2\times 2\rangle} = \frac{4Z_1^2Y_2 + 8Z_1 - Z_L^2Y_2}{(2Z_1+Z_L)(2Z_1Y_2+Z_LY_2+4)} \tag{14a}$$

$$\hat{s}_{2,1}^{\langle 2\times 2\rangle} = \frac{-2Z_L}{(2Z_1+Z_L)(2Z_1Y_2+Z_LY_2+4)} \tag{14b}$$

Solving for Z_1, Y_2 yields

$$Z_1 = \frac{Z_L}{2}\cdot\frac{1+\hat{s}_{11}+2\hat{s}_{21}}{1-\hat{s}_{11}-2\hat{s}_{21}} \tag{15a}$$

$$Y_2 = -\frac{(\hat{s}_{11}+2\hat{s}_{12}-1)(\hat{s}_{11}-2\hat{s}_{21}-1)}{2Z_L\hat{s}_{21}} \tag{15b}$$

$$1+Z_1Y_2 = \frac{\hat{s}_{11}^2-4\hat{s}_{21}^2-1}{4\hat{s}_{21}} \tag{15c}$$

The scattering parameters $\hat{s}$ are those obtained in the experiment.

Combining (15c) with (6) and assuming propagation along a principal axis, i.e., $\eta=\xi=0$, yields the dispersion relation

$$\frac{1}{2}(1-\cos\chi) = (1+Z_1Y_2)^2 = \left(\frac{\hat{s}_{1,1}^2-4\hat{s}_{2,1}^2-1}{4\hat{s}_{2,1}}\right)^2 \tag{16}$$

The extracted frequency dependency of Z_1, Y_2 is depicted in Fig. 13, showing fairly good agreement with lumped element simulations and thus verifying the assumption that the metamaterial cell acts like the intended lumped element circuit. A comparison of simulation and experimental data for the term $(1+Z_1Y_2)$ is shown in Fig. 14 showing excellent agreement and further validating the lumped element assumption for frequencies up to 2.2 GHz. The reason for the better agreement of Fig. 14 compared to Fig. 13 is that (15c) is more robust compared to (15a)/(b). As the term $(1+Z_1Y_2)$ is the frequency determining part of the rotated TLM metamaterial (see (6)), the dispersion diagram depicted in Fig. 15 shows also good agreement with circuit simulation results up to 2.2 GHz. The expected two left-handed and two right-handed frequency bands are clearly visible, therefore verifying the behaviour of the rotated TLM metamaterial.

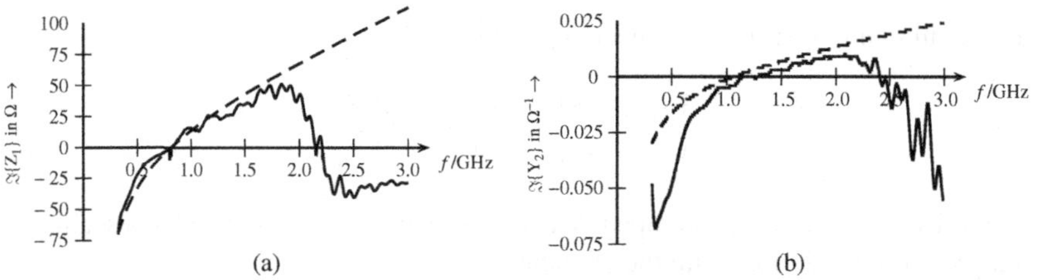

Fig. 13 Lumped element frequency dependency extracted from a measurement of the setup shown in Fig. 12. *Solid line*: measurements, *dashed line*: circuit simulator results using lumped elements only. (**a**) extraction of Z_1, (**b**) extraction of Y_2

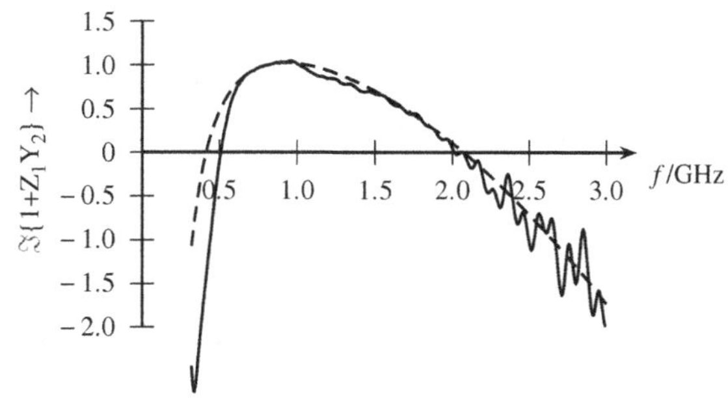

Fig. 14 Comparison of measured (*solid line*) and simulated (*dashed line*) values for $(1 + Z_1 Y_2)$. This term is the frequency determining part of the dispersion relation of the rotated TLM metamaterial.

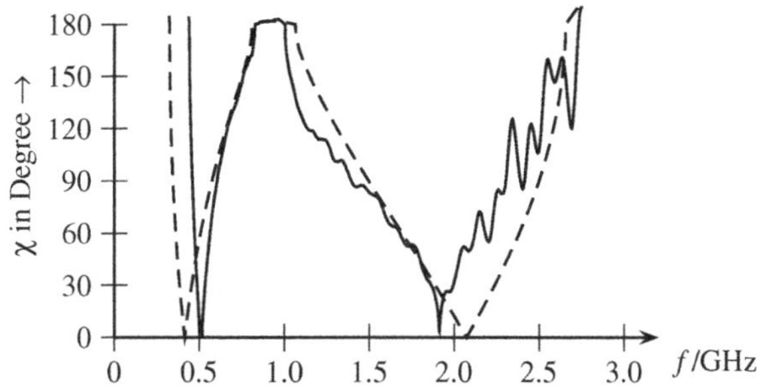

Fig. 15 Dispersion diagram for propagation along a principal axis extracted from a measurement of the setup shown in Fig. 12. *Solid line*: measurements, *dashed line*: circuit simulator results using lumped elements only

5 Planarisation of the Rotated TLM Metamaterial

5.1 Description of the Structure

Fabrication of the 3D rotated TLM metamaterial presented in the previous section is very challenging. The structure is not connective and hence would not hold together without dielectric support. Moreover, the selective metallisation requirement is pending to be solved with 3D rapid prototyping techniques. To overcome these difficulties we propose a planar realisation of the structure preserving its network topology.

Figure 16 shows the cross-sectional view of the planarised rotated TLM metamaterial cell. Figure 17(a) shows the corresponding metal layers. The layers m_1 and m_5 at the bottom and top correspond to the patches in the cell corners in Fig. 10. Layer m_5 is identical with layer m_1 of the overlying cell. Each of the four patches of the layers m_1 and m_5, respectively, is continued into the four neighbouring cells at every corner. These patches produce the capacitive coupling with the neighbouring cells via the patches of layers m_1 and m_4. In the layers m_2 and m_4 the strips are connected to the patches with insets which increase the inductance. Together with the through-connections through layers d_2 and d_3 these strips produce the required inductive coupling. The vertical capacitive coupling is achieved through two series capacitances $m_4 \rightarrow m_5 \rightarrow m_2'$ (where the prime denotes the next unit cell). In-plane capacitive coupling is achieved through two series capacitances $m_4 \rightarrow m_5 \rightarrow m_4'$.

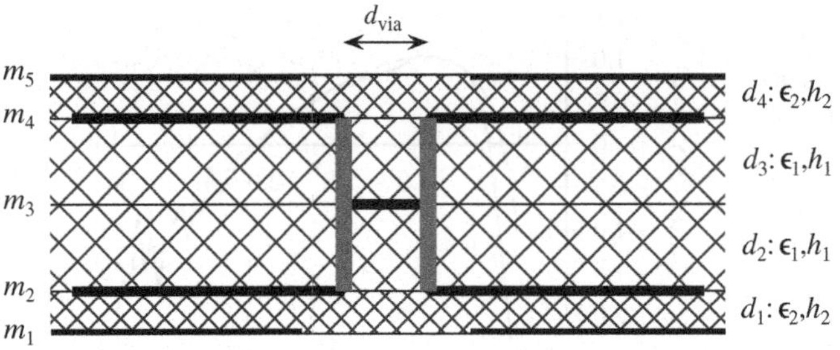

Fig. 16 Cross-sectional view of the planarises CRLH rotated TLM metamaterial. ϵ_1: low permittivity; ϵ_2: high permittivity; h_2 as thin as possible. *Vertical thick lines*: buried via. Via distance d_{via} as small as possible. m_i denote the metallisation layers in the unit cell. If multiple cells are stacked then the adjacent top and bottom layers m_1 and m_5 can be merged into one layer

An alternative configuration requiring no metal-insulator-metal patches is depicted in Fig. 17(b): Here the vertical capacitive coupling is $m_4 \longrightarrow m_2'$. The in-plane capacitive coupling is achieved with interdigital capacitors $m_4 \longrightarrow m_4'$. In addition to the advantage of requiring less layers per unit cell, this configuration also alleviates fabrication tolerances with respect to dielectric layer thicknesses, as the two layers d_1/d_4 are merged into one layer. This advantage comes at a cost; both C_L and L_R of the vertical plate capacitors and the in-plane interdigital capacitors need to be carefully matched.

5.2 *Parasitic Coupling Due to Planarisation*

While ideally the network topology of Figs. 16, 17, 18, 19, 20 should equal that of Fig. 5(a), unwanted capacitive coupling $m_2 \longrightarrow m_4$ needs to be considered. Such parasitic effects couple the otherwise independent two half cells, as shown in Fig. 21.

In order to obtain the dispersion relation we first calculate the impedance matrix of the 12-port by setting up Kirchhoff's equations. One obtains

$$\mathbf{Z} = \begin{bmatrix} \mathbf{Z}_A & \mathbf{Z}_B \\ \mathbf{Z}_B & \mathbf{Z}_A \end{bmatrix} \tag{17}$$

with

$$\begin{aligned}(Y_1 + 2Y_c) \cdot \mathbf{Z}_A &= 2(1 + Z_1(Y_1 + 2Y_c) + \tfrac{Y_c}{Y_1}) \cdot \mathbf{E} \\ &+ \left(1 + \frac{Y_c}{Y_1}\right) \cdot \begin{bmatrix} 0 & 0 & -1 & -1 & -1 & 1 \\ 0 & 0 & 1 & 1 & -1 & 1 \\ -1 & 1 & 0 & 0 & -1 & -1 \\ -1 & 1 & 0 & 0 & 1 & 1 \\ -1 & -1 & -1 & 1 & 0 & 0 \\ 1 & 1 & -1 & 1 & 0 & 0 \end{bmatrix}\end{aligned} \tag{18}$$

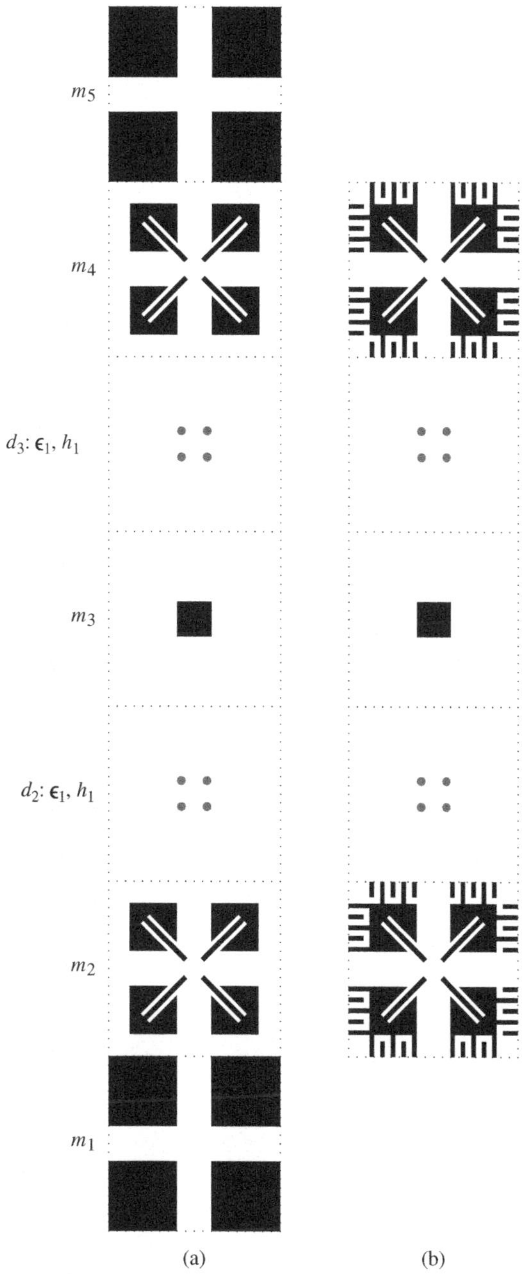

Fig. 17 Exploded top view on the different metal layers for the structure of Fig. 16. (**a**) MIM capacitor implementation (**b**) interdigital capacitor implementation

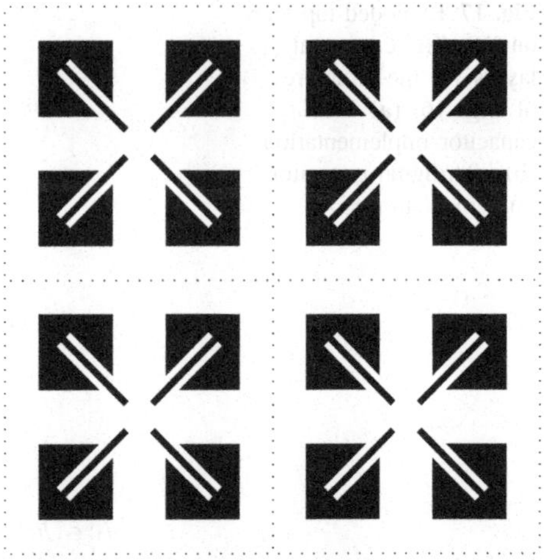

Fig. 18 Top view of layer m_2/m_4 of 2×2 cells of the structure depicted in Fig. 17(a)

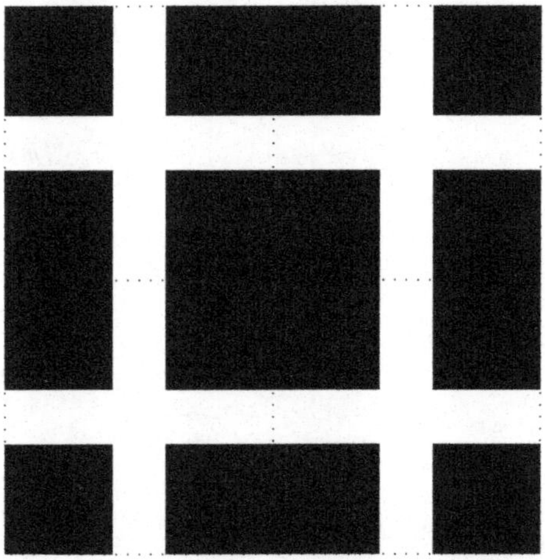

Fig. 19 Top view of layer m_1/m_5 of 2×2 cells of the structure depicted in Fig. 17(a). The coupling patches provide the series capacitive coupling between in-plane adjacent cells

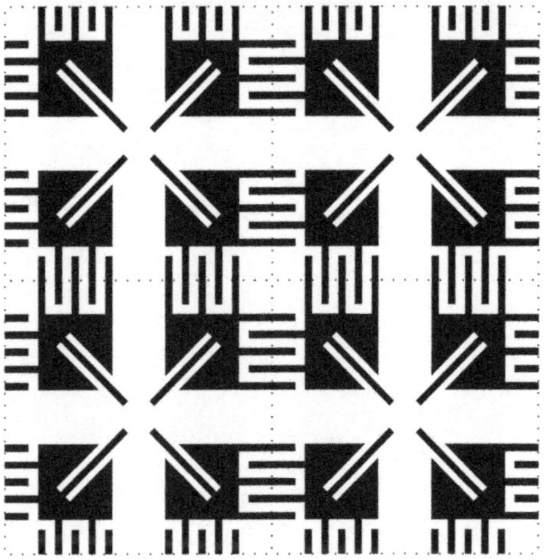

Fig. 20 Top view of layer m_2/m_4 of 2×2 cells of the structure depicted in Fig. 17(b)

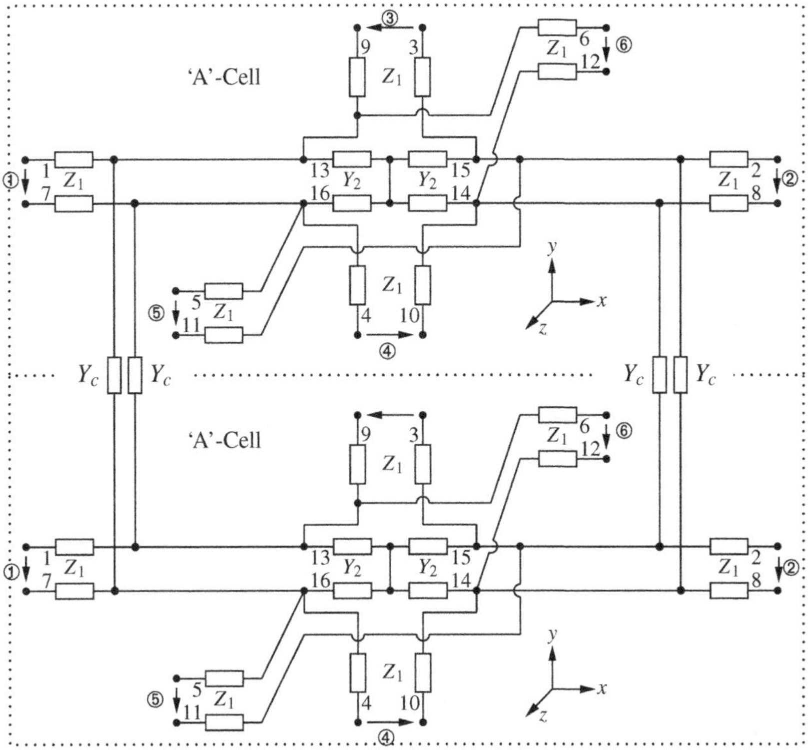

Fig. 21 Planarised unit cell consisting of two nested 'A' cells with parasitic coupling Y_c

$$(Y_1 + 2Y_c) \cdot \mathbf{Z}_B = \frac{Y_c}{Y_1} \begin{bmatrix} -2 & 0 & 1 & 1 & 1 & -1 \\ 0 & -2 & -1 & -1 & 1 & -1 \\ 1 & -1 & 0 & -2 & -1 & -1 \\ 1 & -1 & -2 & 0 & 1 & 1 \\ 1 & 1 & -1 & 1 & 0 & 2 \\ -1 & -1 & -1 & 1 & 2 & 0 \end{bmatrix} \tag{19}$$

Applying to the coupled half cells the Floquet ansatz [20]

$$V = \begin{bmatrix} V_{xA}\exp(-\mathrm{j}\chi/2) \\ V_{xA}\exp(+\mathrm{j}\chi/2) \\ V_{yA}\exp(-\mathrm{j}\eta/2) \\ V_{yA}\exp(+\mathrm{j}\eta/2) \\ V_{zA}\exp(-\mathrm{j}\xi/2) \\ V_{zA}\exp(+\mathrm{j}\xi/2) \\ V_{xB}\exp(-\mathrm{j}\chi/2) \\ V_{xB}\exp(+\mathrm{j}\chi/2) \\ V_{yB}\exp(-\mathrm{j}\eta/2) \\ V_{yB}\exp(+\mathrm{j}\eta/2) \\ V_{zB}\exp(-\mathrm{j}\xi/2) \\ V_{zB}\exp(+\mathrm{j}\xi/2) \end{bmatrix} \qquad I = \begin{bmatrix} I_{xA}\exp(-\mathrm{j}\chi/2) \\ -I_{xA}\exp(+\mathrm{j}\chi/2) \\ I_{yA}\exp(-\mathrm{j}\eta/2) \\ -I_{yA}\exp(+\mathrm{j}\eta/2) \\ I_{zA}\exp(-\mathrm{j}\xi/2) \\ -I_{zA}\exp(+\mathrm{j}\xi/2) \\ I_{xB}\exp(-\mathrm{j}\chi/2) \\ -I_{xB}\exp(+\mathrm{j}\chi/2) \\ I_{yB}\exp(-\mathrm{j}\eta/2) \\ -I_{yB}\exp(+\mathrm{j}\eta/2) \\ I_{zB}\exp(-\mathrm{j}\xi/2) \\ -I_{zB}\exp(+\mathrm{j}\xi/2) \end{bmatrix} \tag{20}$$

the non-trivial solution of $V = \mathbf{Z} \cdot I$ in V and I gives the dispersion relation

$$\begin{aligned}\frac{1}{4}&(3-\cos\chi\cos\eta-\cos\chi\cos\xi-\cos\eta\cos\xi)-(1+Y_2Z_1)^2\\ &=-\frac{Y_cZ_1}{1+Y_2Z_1}\left[\cos\eta\cos\xi+(1+Z_1(Y_2+2Y_c))(\cos\eta-\cos\xi)\right]\\ &\quad-\frac{Y_cZ_1}{1+Y_2Z_1}\left[4Y_2^2Z_1^2+8Y_2Z_1+3+4Y_cZ_1(1+Y_2Z_1)\right]\end{aligned} \tag{21}$$

In the limiting case $Y_c \to 0$ the dispersion relation of the planarised rotated TLM metamaterial reduces to that of (6) as expected.

The dispersion diagram corresponding to (21) for a balanced-resonance CRLH unit with $\frac{L_L}{L_R} = 1$ and $Z_1 Y_c = \frac{1}{5}$ is plotted in Fig. 22. The isotropic behaviour around the Γ-point is approximately maintained. As both Y_c and Z_1 are produced by plate capacitors (Fig. 16), a reasonable corresponding design guideline is

$$h_1 \geq 5\frac{\epsilon_1}{\epsilon_2}h_2 \tag{22}$$

with respect to Fig. 16. In order to satisfy (22) several interdependent conditions need to be met:

1. unit cell size $\ll \frac{1}{4}\lambda$ [1],
2. the via distance d_{via} plus via diameter need to be small compared to the unit cell size,
3. the via diameter is related to h_1 through the achievable aspect ratio, but (22) needs to be fulfilled,
4. the thickness of the thin dielectric layers d_1 and d_4 needs to have very small tolerances in order to assure good control over C_L.

While (4) can be overcome by using interdigital capacitors as shown in Fig. 17(b) instead of coupling plates, the design of such a balanced-resonance CRLH unit cell

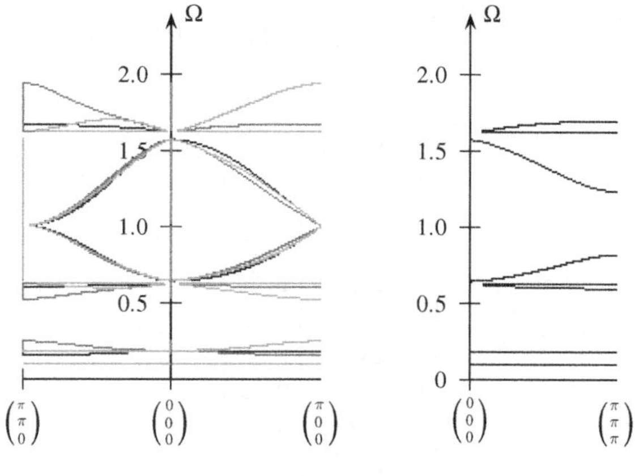

Fig. 22 Dispersion diagram to (21) for a balanced-resonance CRLH unit cell with $\frac{L_L}{L_R} = 1$ and $Y_c Z_1 = \frac{1}{5}$. For $\Gamma \to$ X black represents propagation along χ, red represents propagation along η, green represents propagation along ξ. For $\Gamma \to$ M black represents propagation along the $\chi = \eta$, red represents propagation along $\chi = \xi$, green represents propagation along $\eta = \xi$

is more challenging (see Sect. 5.1): Both C_L and L_R of the vertical plate capacitors and the in-plane interdigital capacitors need to be matched.

5.3 Numerical Calculation of the Dispersion Relation

Numerical results of the planarised 3D rotated TLM structure were calculated for a planarised structure with coupling patches in m_1/m_5 using CST MWS's Jacobi-Davidson eigenmode solve and Ansoft HFSS's eigenmode solver. The unit cell edge length is $a = 20\,\text{mm}$, and the dielectrics are $h_1 = 5\,\text{mm}$, $h_2 = 1\,\text{mm}$, $\epsilon_1 = 1$, $\epsilon_2 = 10$. The structure's dimensions are taken as

Parameter	Value in units of $a/64$
via distance	$4\sqrt{2}$
via diameter	1
m_2/m_4 strip inset width	6
m_2/m_4 strip width	1
m_2/m_4 strip length	28
m_2/m_4 patch to cell boundary distance	4
m_2/m_4 patch edge length	20
m_1/m_5 patch edge length	24

The dispersion diagram for $\Gamma \rightarrow$ X along the x-axis is shown in Fig. 23, displaying good agreement between CST MWS and Ansoft HFSS. The lower right-handed mode is a perturbed plane-wave mode not described by the network model. It corresponds to artificial dielectrics as described in [25]. The left-handed mode, corresponding to the lower left-handed mode of Fig. 8, is clearly visible. The electric field distribution of the left-handed mode for $\chi = \frac{1}{6}\pi$, $\eta = \xi = 0$ is shown in

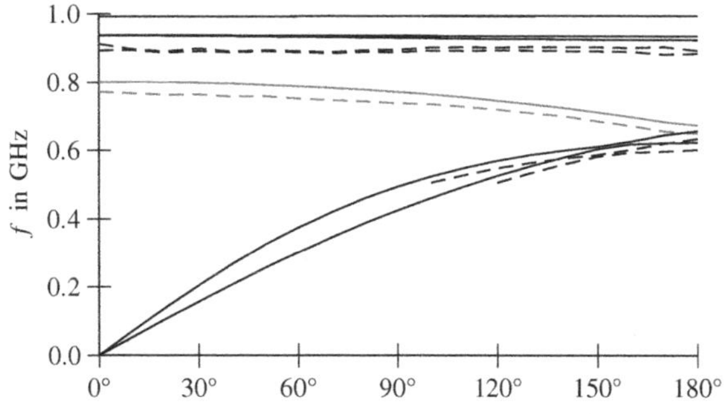

Fig. 23 Numerical calculation of the dispersion diagram of the planarised unit cell for $\Gamma \rightarrow X$ along the x-axis. *Solid lines*: CST MWS's Jacobi-Davidson Eigenmode solver, *dashed lines*: HFSS Eigenmode solver. *Red lines*: left-handed band. Every 10° sampled

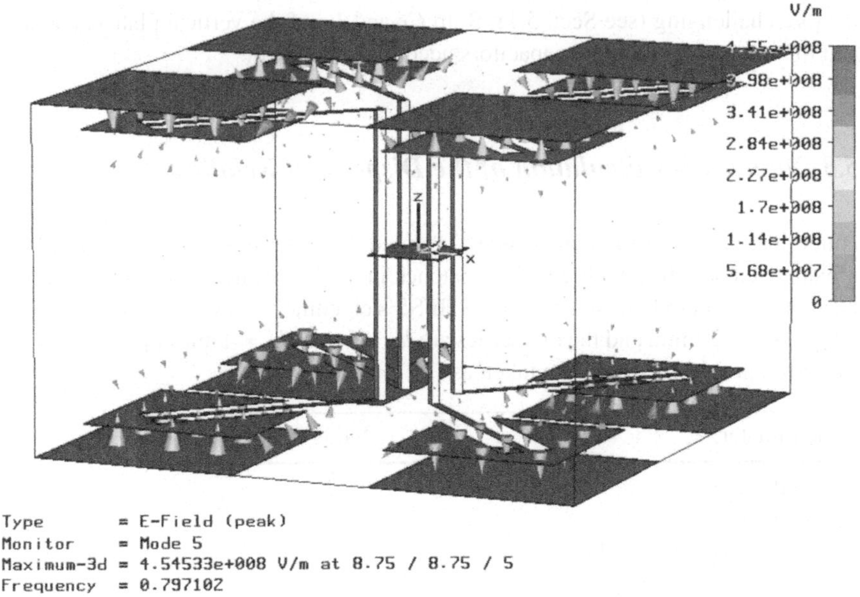

Fig. 24 Electric Field distribution for $\chi = \frac{1}{6}\pi,\ \eta = \xi = 0$ for the left-handed mode (see Fig. 23)

Fig. 24. The high confinement of the electric field in the plate capacitors proves the assumption of lumped element behaviour and accordingly the homogeneity requirement of metamaterials as well as the low impact of the parasitic capacitance Y_c for the simulated configuration.

6 Conclusion and Outlook

A 3D isotropic left-handed metamaterial based on the rotated transmission line matrix (TLM) scheme was presented. Firstly, the discrete electrodynamics theory of the rotated TLM scheme was briefly revisited. The corresponding analysis of the 3D metamaterial was recalled in terms of its dispersion relation and composite right/left-handed (CRLH) response. For the first time the impact of unbalanced-resonance on the Bloch impedance, differential/common mode excitation, as well as principal axes propagation were discussed, the latter yielding a 1D ladder network analogy.

A practical realisation of this 3D metamaterial, based on metal-insulator-metal capacitors and thin wires, was proposed and demonstrated experimentally. A simplified planarised implementation, preserving the same network topology, was proposed with two alternative configurations. Parasitic effects resulting from planarisation are analysed and design guidelines are given to mitigate them. Finally, this structure was demonstrated by a full-wave simulation to exhibit the expected 3D-isotropic left-handed behaviour.

While current fabrication of 3D metamaterials remains a challenge, the planarised structure presented in this work may provide a solution for practical 3D metamaterials in the near future.

Acknowledgements The authors would like to thank Ning Yang for assistance in the measurements.

References

1. C. Caloz and T. Itoh, *Electromagnetic Metamaterials*. Wiley, 2006.
2. N. Engheta and R. Ziolkowski, *Electromagnetic Metamaterials*. Wiley, 2006.
3. V. G. Veselago, "The electrodynamics of substances with simultaneously negative values of ϵ and μ," *Sov. Phys. Usp.*, p. 509, 1968.
4. S. Lim, C. Caloz, and T. Itoh, "Metamaterial-based electronically-controlled transmission line structure as a novel leaky-wave antenna with tunable radiation angle and beamwidth," *IEEE Trans. Microw. Theory Tech.*, vol. 53, no. 1, p. 161, 2005.
5. M. Schüßler, J. Frese, and R. Jakoby, "Design of compact planar antennas using LH-transmission lines," *IEEE MTT-S Int. Microwave Symp Dig*, p. 209, 2004.
6. N. Fang, H. Lee, C. Sun, and X. Zhang, "Sub-diffraction-limited optical imaging with a silver superlens," *Science Magazine*, vol. 308, p. 534, 2005.
7. A. Alù and N. Engheta, "Achieving transparency with plasmonic coatings," *Phys. Rev. B*, vol. 72, p. 016623, 2005.
8. J. Pendry, D. Schurig, and D. Smith, "Controlling electromagnetic fields," *Science*, vol. 312, p. 1780, 2006.
9. S. Tretyakov, *Analytical Modeling in Applied Electromagnetics*. Artech House, 2003.
10. W. Hoefer, P. So, D. Thompson, and M. Tenzeris, "Topology and design of wide-band 3d metamaterials made of periodically loaded transmission lines," *2005 Int. Microwave Symposium Digest, Long Beach*, p. 1443, 2005.
11. A. Grbic and G. Eleftheriades, "An isotropic three-dimensional negative-refractive-index transmission-line metamaterial," *J. Appl. Phys.*, vol. 98, 2005.
12. M. Zedler and P. Russer, "Investigation on the dispersion relation of a 3D LC-based metamaterial with an omnidirectional left-handed frequency band," *2006 Int. Microwave Symposium Digest, San Francisco*, p. 1477, 2006.
13. M. Zedler, C. Caloz, and P. Russer, "Circuital and experimental demonstration of a 3D isotropic lh metamaterial based on the rotated TLM scheme," *2007 Int. Microwave Symposium Digest, Honolulu, Hawaii*, p. accepted for publication, 2007.
14. I. Vendik, O. Vendik, and M. Odit, "Isotropic artificial media with simultaneously negative permittivity and permeability," *Microwave and Optical Technol. Letts.*, vol. 48, p. 2553, 2006.
15. P. K. Mercure, R. P. Haley, A. Bogle, and L. Kempel, "Three-dimensional isotropic metamaterials," *2005 Int. Antennas and Propagation Society Symposium Digest*, p. 623, 2005.
16. T. Koschny, L. Zhang, and C. Soukoulis, "Isotropic three-dimensional left-handed metamaterials," *Physical Review B*, vol. 71, p. 121103, 2005.
17. P. Alitalo, S. Maslovski, and S. Tretyakov, "Three-dimensional isotropic perfect lens based on LC-loaded transmission lines," *J. Appl. Phys.*, vol. 99, 2006.
18. P. Alitalo, S. Maslovski, and S. Tretyakov, "Experimental verification of the key properties of a three-dimensional isotropic transmission-line superlens," *J. Appl. Phys.*, vol. 99, 2006.
19. W. J. R. Hoefer, "The transmission line matrix (TLM) method," in *Numerical Techniques for Microwave and Millimeter Wave Passive Structures*, T. Itoh, Ed. New York: John Wiley & Sons, 1989, pp. 496–591.

20. P. Russer, *Electromagnetics, Microwave Circuit and Antenna Design for Communications Engineering*. Artech House, 2006.
21. G. Kron, "Equivalent circuits to represent the electromagnetic field equations," *Physical Rev.*, vol. 64, no. 3, p. 126, 1943.
22. M. Krumpholz and P. Russer, "On the dispersion in TLM and FDTD," *IEEE Trans. Microwave Theory Techn.*, vol. 42, no. 7, pp. 1275–1279, 1994.
23. P. Russer, "The alternating rotated transmission line matrix (ARTLM) scheme," *Electromagnetics*, vol. 16, no. 5, p. 537, 1996.
24. A. Wlodarczyk, "Representation of symmetrical condensed tlm node," *Electronics Letters*, vol. 28, no. 18, p. 1686, 1992.
25. R. E. Collin, *Field Theory of Guided Waves*, 2nd ed. IEEE Press, 1990.

Connection Subnetworks for the Transmission Line Matrix (TLM) Method

Petr Lorenz and Peter Russer

We present different kinds of connection subnetworks for the Transmission Line Matrix (TLM) method [1, 2]. The connection subnetworks can be used for the modeling of excitation regions, for the connection between different simulation regions and for the hybridization of the TLM method with other numerical methods.

Connection subnetworks are connected to the TLM simulation domain either at the outer boundary of the simulation region or inside the simulation region. They are represented in terms of scattering matrices. It is shown that the properties of the scattering matrix of a connection subnetwork are similar to the properties of the scattering matrix of a TLM cell. Consequently, connection subnetworks can be easily implemented in the TLM algorithm.

First, connection subnetworks based on parallel and series adaptors of Wave Digital Filters (WDFs) [3] are presented. They are suitable for the modeling of discrete source regions. The scattering matrices of connection subnetworks can be computed automatically during pre-processing stage of an electromagnetic simulation. The properties of the WDF-based subnetworks are described and compared to the properties of the scattering matrix of a TLM cell.

Second, connection subnetworks based on ideal transformers are shown [4]. They can be used for the modeling of modal excitations and for hybridization. The scattering matrix of these subnetworks is given and its properties are discussed.

We show an example of impulsive excitation of antenna structures using the WDF-based connection subnetworks. As an example of hybridization the Transmission Line Matrix – Multipole Expansion (TLMME) method is presented. In the TLMME method the Transmission Line Matrix (TLM) method is combined with the Multipole Expansion (ME) method to yield a potentially exact radiating boundary condition. The connection of the TLM method to the ME method is done on an outer

Petr Lorenz
Rohde & Schwarz GmbH & Co. KG, Mühldorfstr. 15, 81671 München, Germany,
e-mail: petr.lorenz@ieee.org

Peter Russer
Institute for High Frequency Engineering, Technische Universität München,
Arcisstr. 21, 80333 München, Germany, e-mail: russer@tum.de

boundary of a spherical simulation domain. The network model of the connection is represented by means of a connection subnetwork consisting of ideal transformers.

1 Connection Subnetworks Based on Wave Digital Filter Networks

The connection subnetworks based on parallel and series adaptors of *Wave Digital Filters* (WDFs) [3] can be used for the modeling of discrete sources. This type of sources are useful, for instance, for the modeling of excitation of antenna structures. In this Section the connection subnetworks based on wave digital filters are described and it will be shown how they can be applied to model discrete sources distributed over multiple TLM cells.

Referring to Fig. 1 it can be shown [5] that the 2D TLM method and, in particular, the scattering matrix of the TLM cell center are related to parallel and series adaptors of wave digital filters. The scattering matrices of the 2D TLM shunt and series nodes are equivalent to the scattering matrices of parallel and series adaptors, respectively. Both the 2D and 3D scattering matrices $\mathbf{S}_{TLM}$ of the TLM cell center have the following properties:

- real,
- frequency independent,

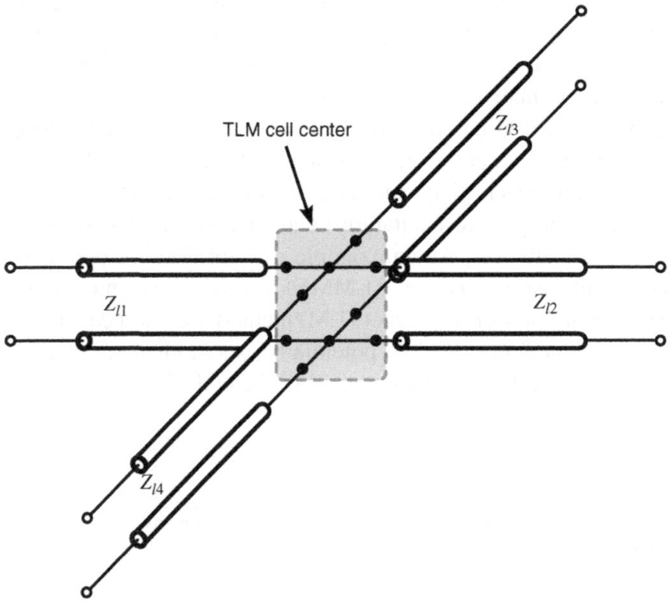

Fig. 1 2D TLM shunt node and its TLM cell center; the length of the link lines is half the spatial discretization step, i.e., $\Delta l/2$

- energy conservative ($|\det(\mathbf{S}_{TLM})| = 1$),
- time-reversible ($\mathbf{S}_{TLM} \cdot \mathbf{S}_{TLM} = \mathbf{I}$; $\mathbf{I}$ is the identity matrix) and
- the eigenvalues equal ± 1. If all the TLM port reference impedances are equal then the scattering matrix is also symmetric and unitary.

Consequently, it is natural to establish a class of connection subnetworks for the TLM method which are based on WDF networks.

The connection subnetwork based on WDFs is an $N+1$ port network with N *TLM ports* connected to the TLM mesh and one *source port*. The source port is the input port to the TLM model. The TLM ports of the connection subnetwork are connected to the TLM link lines at the interfaces between adjacent TLM cells. Consequently, the mapping type between network and field quantities is the *boundary oriented field mapping* [6].

Schematic pictures of canonic parallel and series connection subnetworks exciting one field polarization are shown in Figs. 2 and 3. The link line reference impedances are denoted as Z_{l1} and Z_{l2} and the source port reference impedance as Z_p.

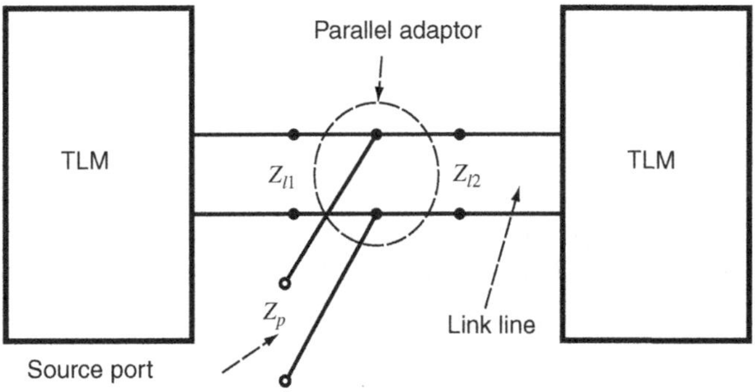

Fig. 2 Canonic parallel connection subnetwork; the reference impedances of the TLM link lines are Z_{l1} and Z_{l2} and the reference impedance of the source port is Z_p

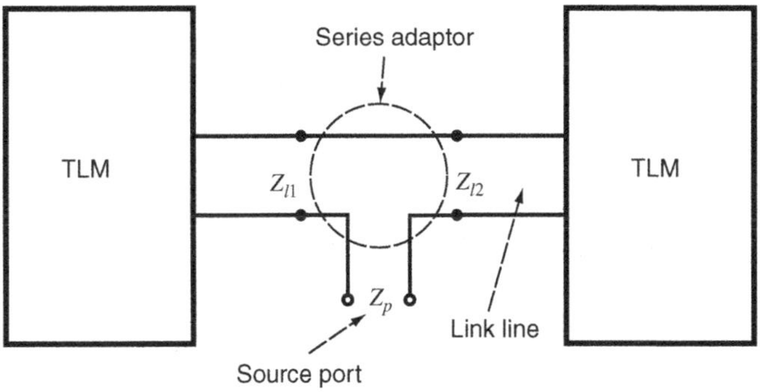

Fig. 3 Canonic series connection subnetwork; the reference impedances of the TLM link lines are Z_{l1} and Z_{l2} and the reference impedance of the source port is Z_p

We can see that the canonic parallel connection subnetwork can be seen as a parallel adaptor placed at the interface of two TLM link lines. The parallel connection subnetwork has the following properties:

1. A common voltage (i.e., electric field intensity) is impressed at the interface between adjacent TLM cells.
2. Different line impedances of the TLM link lines and the source port are taken into account.
3. The parallel connection subnetwork is represented by a scattering matrix $\mathbf{S}_p$. In particular, the canonic parallel connection subnetwork in Fig. 2 is characterized by a (3×3) scattering matrix. Examples of scattering matrices $\mathbf{S}_p$ are given at the end of this Section.

The scattering matrix $\mathbf{S}_p$ of the connection subnetwork is

- real,
- frequency independent,
- energy conservative $(|\det(\mathbf{S}_p)| = 1)$,
- time-reversible $(\mathbf{S}_p \cdot \mathbf{S}_p = \mathbf{I})$ and
- the eigenvalues equal ± 1. If all the TLM port impedances and the source port impedance are equal then the scattering matrix is also symmetric and unitary.

A series connection subnetwork is dual to the parallel connection subnetwork. The properties of the series connection subnetwork are the same as those of the parallel one except that a common current (i.e., magnetic field intensity) is impressed at the interface between adjacent TLM cells. The properties of the series connection subnetwork are

1. A common current (i.e., magnetic field intensity) is impressed at the interface.
2. Different line impedances of the TLM link lines and the source port are taken into account.
3. The series connection subnetwork is characterized by a scattering matrix $\mathbf{S}_s$ with the same properties as those of the parallel connection subnetwork discussed earlier.

To the source port of the connection subnetwork three different source types can be connected. These are

1. ideal voltage source,
2. ideal current source,
3. power source with defined internal impedance Z_s (admittance $G_s = \frac{1}{Z_s}$).

The different source types are shown in Fig. 4.

The WDF realizations of these sources can be connected directly to the source port of the connection subnetwork. The WDF realizations of the different source types with wave digital filters are shown in Fig. 5. We use here the same convention for symbols of WDF adaptors and wave flow diagrams as in [3].

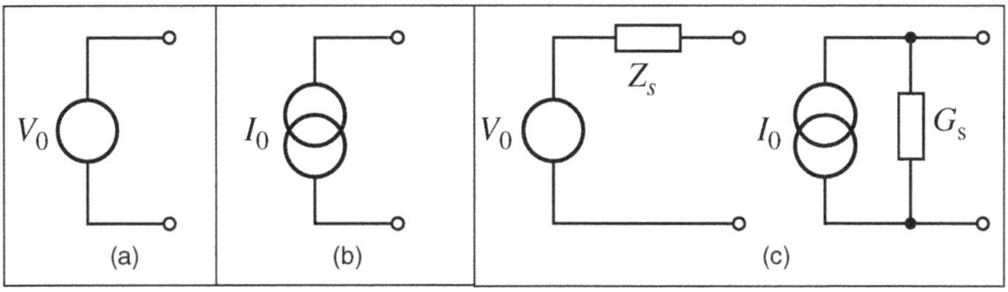

Fig. 4 Lumped sources which may be connected to the source port of the connection subnetwork

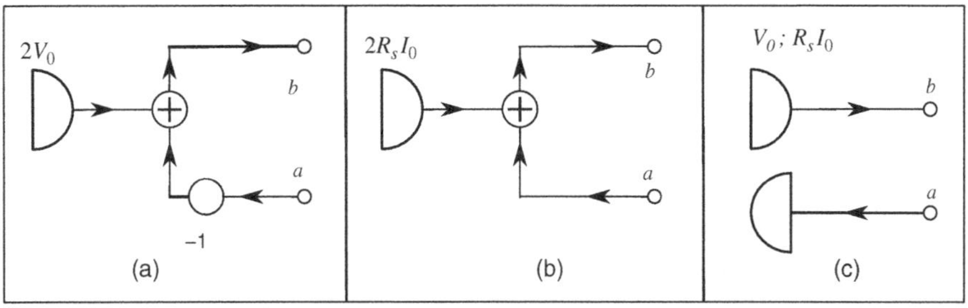

Fig. 5 WDF representations of the sources

The power source can also be seen (using Thévenin theorem) as an ideal voltage source connected at infinity to one end of a transmission line with characteristic impedance Z_S. The other end of the transmission line is connected to the source port.

The connection subnetworks just described implement following features into the TLM algorithm:

- Well-defined excitation of field amplitudes and energy.
- Impulsive characterization of electromagnetic structures with linear time-invariant properties. This is possible since only one source port is available for the excitation and the TLM model is a discrete linear time-invariant (LTI) system.
- Direct calculation of scattering parameters from the incoming and reflected wave amplitudes.

An example of an impulsive excitation of bowtie antenna is given in Sect. 3.1.

1.1 Example – Canonic Parallel Connection Subnetwork

The scattering matrix $\mathbf{S}_p$ of a canonic parallel connection subnetwork referred to the source port impedance $Z_p = \frac{377}{2}\ \Omega$ with $Z_{l1} = Z_{l2} = 377\ \Omega$ reads

$$\mathbf{S}_p = \begin{bmatrix} 0 & \frac{1}{2} & \frac{1}{2} \\ 1 & -\frac{1}{2} & \frac{1}{2} \\ 1 & \frac{1}{2} & -\frac{1}{2} \end{bmatrix}. \tag{1}$$

The scattering matrix referred to the source port impedance $Z_p = 50\ \Omega$ with $Z_{l1} = Z_{l2} = 377\ \Omega$ reads

$$\mathbf{S}_p = \begin{bmatrix} 0.58071 & 0.20964 & 0.20964 \\ 1.58071 & -0.79036 & 0.20964 \\ 1.58071 & 0.20964 & -0.79036 \end{bmatrix}. \tag{2}$$

It is easy to check that $\mathbf{S}_p\mathbf{S}_p = \mathbf{I}$ and $|\det(\mathbf{S}_p)| = 1$. The matrices have thus the same properties as the scattering matrix of the TLM method. Consequently, they can be directly implemented into the TLM algorithm.

2 Connection Subnetworks Based on Ideal Transformer Networks

For the modeling of modal sources with spatially varying field distribution the connection subnetworks based on ideal transformer networks can be used. This type of connection subnetwork can also be used for hybridization of the TLM method with other methods as will be shown in Sect. 3.2.

One possible realization of such connection subnetwork is shown in Fig. 6. Here we consider to have N TLM ports and M modal ports. The incoming and reflected wave quantities of the TLM ports are denoted as $a_1, a_2, \ldots, a_N$ and $b_1, b_2, \ldots, b_N$, respectively. In a similar way, the incoming and reflected wave quantities of the modal ports are denoted as $a_{N+1}, a_{N+2}, \ldots, a_{N+M}$ and $b_{N+1}, b_{N+2}, \ldots, b_{N+M}$, respectively. The transformer ratios are denoted by n_{ij} with $i = 1 \ldots M$ and $j = 1 \ldots N$.

2.1 Scattering Matrix of Ideal Transformer Network

The equations of a transformer network as shown in Fig. 6, with equal reference resistances at the ports, can be summarized in the equation

$$\begin{bmatrix} \mathbf{I} & -\mathbf{N} \\ \mathbf{N}^T & \mathbf{I} \end{bmatrix} \tilde{\mathbf{b}} = \begin{bmatrix} -\mathbf{I} & \mathbf{N} \\ \mathbf{N}^T & \mathbf{I} \end{bmatrix} \tilde{\mathbf{a}} \tag{3}$$

with

$$\mathbf{N} = \begin{bmatrix} n_{11} & n_{12} & \ldots & n_{1N} \\ n_{21} & n_{22} & \ldots & n_{2N} \\ \ldots & \ldots & \ldots & \ldots \\ n_{M1} & n_{M2} & \ldots & n_{MN} \end{bmatrix} \tag{4}$$

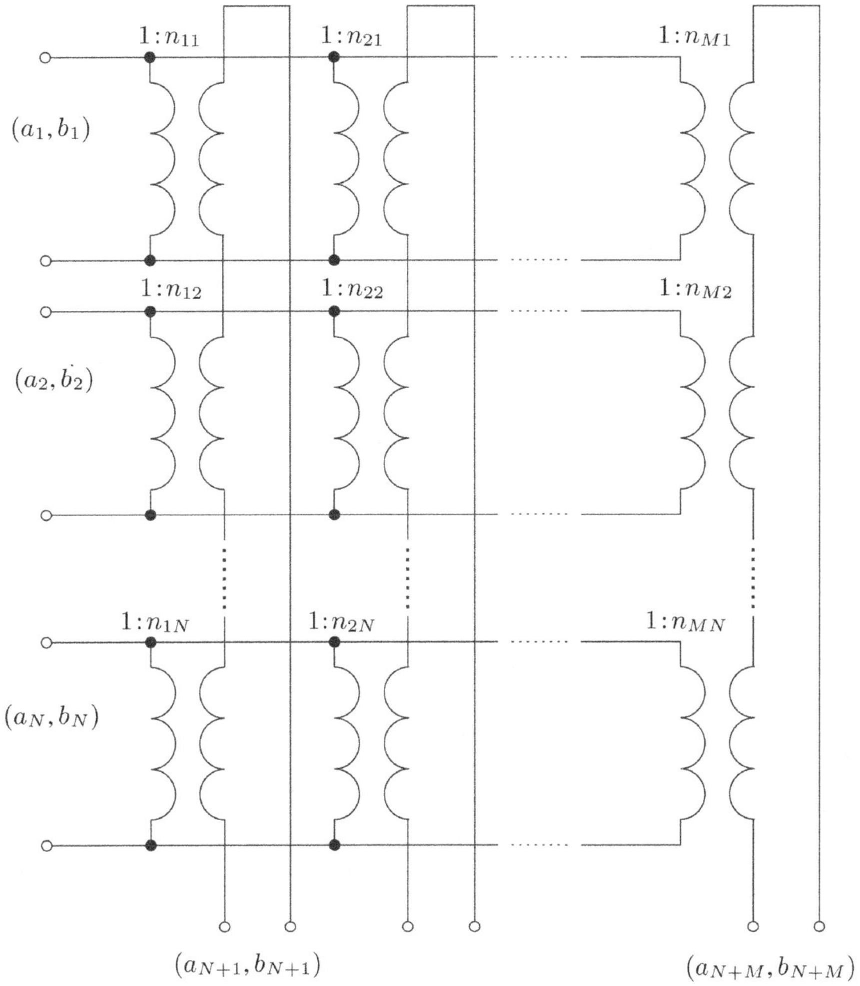

Fig. 6 Connection subnetwork consisting of ideal transformers

summarizing the transformer ratios and

$$\tilde{\mathbf{a}} = \left[a_{N+1}\ a_{N+2}\ \dots\ a_{N+M}\ a_1\ a_2\ \dots\ a_N\right]^T, \tag{5a}$$

$$\tilde{\mathbf{b}} = \left[b_{N+1}\ b_{N+2}\ \dots\ b_{N+M}\ b_1\ b_2\ \dots\ b_N\right]^T \tag{5b}$$

being the reorganized vectors of incoming and reflected waves. **I** denotes the identity matrix.

The reorganized scattering matrix $\tilde{\mathbf{S}}$ is computed as

$$\tilde{\mathbf{S}} = \begin{bmatrix} \mathbf{I} & -\mathbf{N} \\ \mathbf{N}^T & \mathbf{I} \end{bmatrix}^{-1} \begin{bmatrix} -\mathbf{I} & \mathbf{N} \\ \mathbf{N}^T & \mathbf{I} \end{bmatrix} = \begin{bmatrix} \tilde{\mathbf{S}}_4 & \tilde{\mathbf{S}}_3 \\ \tilde{\mathbf{S}}_2 & \tilde{\mathbf{S}}_1 \end{bmatrix}, \tag{6}$$

where the inverse matrix can be obtained using the technique of *Schur complement* (see Appendix A).

The scattering matrix $\mathbf{S}$ is related to the reorganized scattering matrix $\tilde{\mathbf{S}}$ as follows

$$\mathbf{S} = \begin{bmatrix} \tilde{\mathbf{S}}_1 & \tilde{\mathbf{S}}_2 \\ \tilde{\mathbf{S}}_3 & \tilde{\mathbf{S}}_4 \end{bmatrix}. \tag{7}$$

The resulting scattering matrix of the connection subnetwork reads

$$\mathbf{S} = \begin{bmatrix} \mathbf{I} - 2\mathbf{N}^T\mathbf{S}_A^{-1}\mathbf{N} & \mathbf{N}^T\mathbf{S}_A^{-1} + \mathbf{N}^T - \mathbf{N}^T\mathbf{S}_A^{-1}\mathbf{N}\mathbf{N}^T \\ 2\mathbf{S}_A^{-1}\mathbf{N} & -\mathbf{S}_A^{-1} + \mathbf{S}_A^{-1}\mathbf{N}\mathbf{N}^T \end{bmatrix}, \tag{8}$$

with

$$\mathbf{S}_A = \mathbf{I} + \mathbf{N}\mathbf{N}^T, \tag{9}$$

where $\mathbf{S}_A$ is the Schur complement.

2.2 Examples

In the following let us examine some special cases of the scattering matrix (8).

2.2.1 Orthonormal Square N

In this case the matrix $\mathbf{N}$ satisfies

$$\mathbf{N}\mathbf{N}^T = \mathbf{N}^T\mathbf{N} = \mathbf{I}. \tag{10}$$

Consequently, the Schur complement (9) and its inverse read

$$\mathbf{S}_A = 2\mathbf{I}; \qquad \mathbf{S}_A^{-1} = \frac{1}{2}\mathbf{I}. \tag{11}$$

The scattering matrix (8) of the connection subnetwork is simplified to

$$\mathbf{S} = \begin{bmatrix} \mathbf{0} & \mathbf{N}^T \\ \mathbf{N} & \mathbf{0} \end{bmatrix}. \tag{12}$$

For this particular case all incoming waves of one representation are transformed to reflected waves of another representation.

2.2.2 $\mathbf{N}^T$ is a Right Inverse of N

In this case the matrix $\mathbf{N}$ satisfies

$$\mathbf{N}\mathbf{N}^T = \mathbf{I} \tag{13}$$

but

$$\mathbf{N}^T\mathbf{N} \neq \mathbf{I}. \tag{14}$$

The Schur complement reads

$$\mathbf{S}_A = 2\mathbf{I}; \qquad \mathbf{S}_A^{-1} = \frac{1}{2}\mathbf{I}. \tag{15}$$

The scattering matrix (8) is reduced to

$$\mathbf{S} = \begin{bmatrix} \mathbf{I} - \mathbf{N}^T\mathbf{N} & \mathbf{N}^T \\ \mathbf{N} & \mathbf{0} \end{bmatrix}. \tag{16}$$

2.2.3 Ideal Transformer: N-Ports at the Primary Side, One-Port at the Secondary Side

In this example the transformer ratios are denoted by n_i with $i = 1, \ldots, N$. The matrix $\mathbf{N}$ is reduced to a vector

$$\mathbf{N} = \begin{bmatrix} n_1 \; n_2 \; \ldots \; n_N \end{bmatrix}. \tag{17}$$

The $(N+1) \times (N+1)$ S-matrix is given by

$$\mathbf{S} = D \begin{bmatrix} \frac{1}{D} + n_1^2 & n_1 n_2 & n_1 n_3 & \ldots & n_1 n_N & -n_1 \\ n_2 n_1 & \frac{1}{D} + n_2^2 & n_2 n_3 & \ldots & n_2 n_N & -n_2 \\ \vdots & \vdots & \vdots & \vdots & \vdots & \vdots \\ n_N n_1 & n_N n_2 & n_N n_3 & \ldots & \frac{1}{D} + n_N^2 & -n_N \\ -n_1 & -n_2 & -n_3 & \ldots & -n_N & \frac{1}{D} + 1 \end{bmatrix} \tag{18a}$$

with

$$D = -\frac{2}{1 + n_1^2 + \ldots + n_N^2}. \tag{18b}$$

Note that if $n_1^2 + \ldots + n_N^2 = 1$, i.e., if $\mathbf{N} \cdot \mathbf{N}^T = 1$, then $D = -1$ and the scattering matrix has a zero at the last element. This is the same result as the one we have obtained in (16).

For the special case of $N = 1$, the scattering matrix computed from (18a) is reduced to

$$\mathbf{S} = \frac{1}{1 + n_1^2} \begin{bmatrix} 1 - n_1^2 & 2n_1 \\ 2n_1 & n_1^2 - 1 \end{bmatrix}, \tag{19}$$

which may be confirmed with the results given in the literature (e.g., see [7] p. 41).

3 Numerical Examples

3.1 Bowtie Antenna

In this example the input impedance of a broadband planar bowtie antenna is computed. The purpose of this example is to demonstrate the performance of the developed excitation techniques by means of connection subnetworks.

The dimensions of the antenna structure are shown in Fig. 7. The input impedance of the bowtie antenna is the main antenna characteristics of interest.

The bowtie antenna was discretized with a spatial discretization step $\Delta l = 0.25\,\text{mm}$ in a computational box with dimensions $50\,\text{mm} \times 50\,\text{mm} \times 50\,\text{mm}$ resulting in a total number of 8×10^6 TLM cells. At the outer boundary of the simulation domain absorbing boundary conditions were applied.

For the excitation a lumped power source (see Fig. 4(c) and Fig. 5(c)) with a connection subnetwork distributed across 2×2 TLM cell interfaces was used. The antenna was excited with a discrete Dirac impulse. The WDF representation of the connection subnetwork is shown in Fig. 8. The port impedances of the ports were chosen so that realizability is ensured (see [3], pp. 271–272).

The 9×9 scattering matrix of the connection subnetwork reads

$$\mathbf{S} = -\frac{1}{4}\begin{bmatrix} 0 & \mathbf{V}_1{}^T & \mathbf{V}_1{}^T \\ 2\mathbf{V}_1 & \mathbf{M}_1 & \mathbf{M}_2 \\ 2\mathbf{V}_1 & \mathbf{M}_2 & \mathbf{M}_1 \end{bmatrix}, \tag{20a}$$

where

$$\mathbf{V}_1 = \begin{bmatrix} 1 & 1 & 1 & 1 \end{bmatrix}^T, \tag{20b}$$

$$\mathbf{M}_1 = \frac{1}{2}\begin{bmatrix} 3 & -5 & 3 & 3 \\ -5 & 3 & 3 & 3 \\ 3 & 3 & 3 & -5 \\ 3 & 3 & -5 & 3 \end{bmatrix} \tag{20c}$$

and

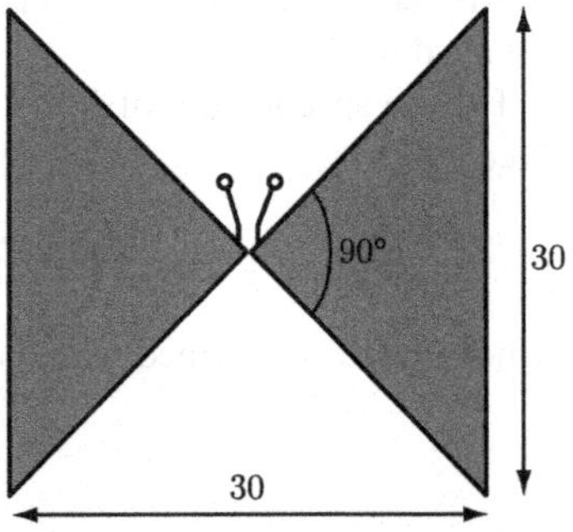

Fig. 7 The bowtie antenna; all dimensions are given in mm

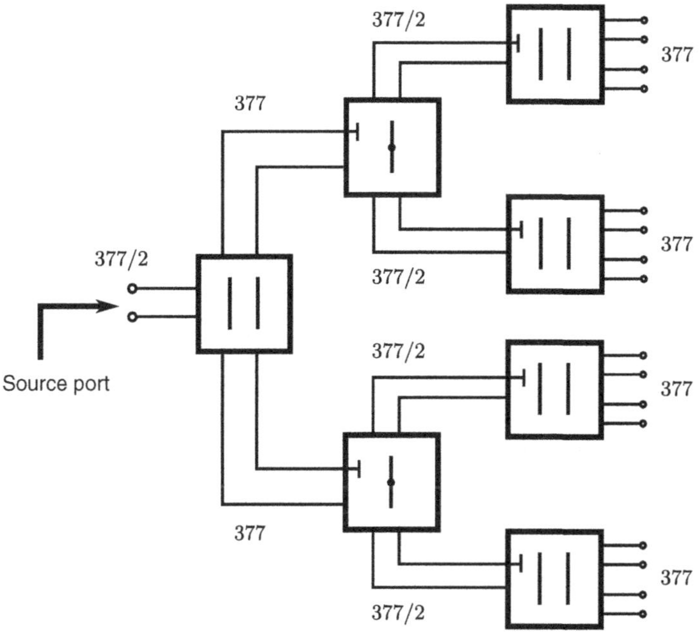

Fig. 8 WDF representation of a connection subnetwork distributed over a region of 2×2 TLM cells; the port impedances are also shown, with the source port impedances equal to $377/2\ \Omega$ and the TLM link line impedances equal to $377\ \Omega$

$$\mathbf{M}_2 = -\frac{1}{2}\begin{bmatrix} 1 & 1 & 1 & 1 \\ 1 & 1 & 1 & 1 \\ 1 & 1 & 1 & 1 \\ 1 & 1 & 1 & 1 \end{bmatrix}. \tag{20d}$$

It can be checked by direct computation that $\mathbf{S} \cdot \mathbf{S} = \mathbf{I}$ and $|\det(\mathbf{S})| = 1$.

The result of the computed input impedance using the connection subnetwork is shown in Fig. 9. We can see a good agreement with method of moments (MoM) result presented in [8] and measurement result presented in [9].

In Fig. 10 the first 150 time steps of the reflected wave at the source port are shown. From the returns to zero of the signal can be clearly seen that non-physical modes are present in the simulation. However, since the TLM model is linear and time-invariant and since there is only one source port, the results for low frequency components are correct.

3.2 Transmission Line Matrix – Multipole Expansion (TLMME) Method

The hybrid *Transmision Line Matrix – Multipole Expansion* (TLMME) method, which is given as an example of hybridization using connection subnetworks based

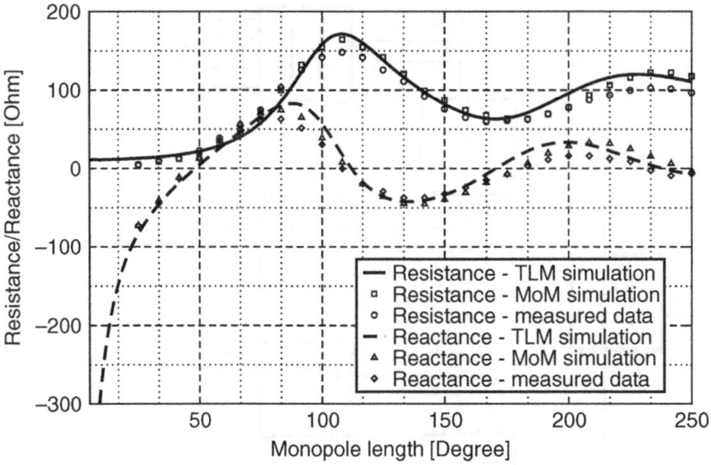

Fig. 9 Input impedance of the bowtie antenna. MoM simulation results are taken from [8] and measurement results from [9]

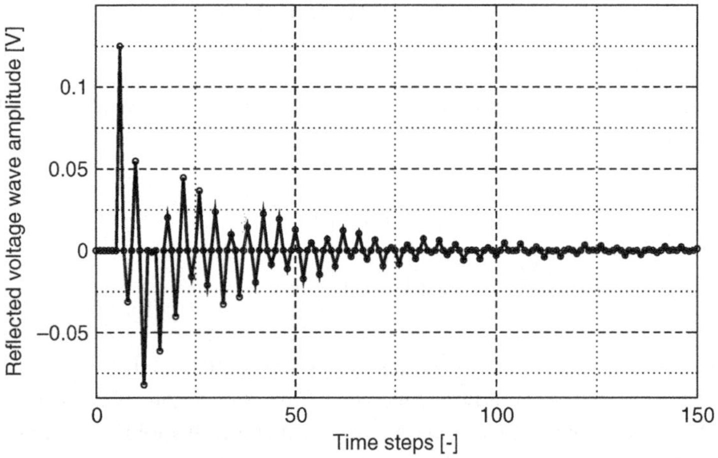

Fig. 10 Time domain signal for 150 time steps of the reflected wave at the source port

on ideal transformers, yields a potentially exact solution for general radiating electromagnetic structures in homogeneous media. In contrast to the *Absorbing Boundary Condition* (ABC) approach (see, e.g., [10, 11, 12, 13, 14, 15]), the TLMME models the complete radiated field including the near-field.

In this hybrid method the problem space is divided in two subspaces $\mathscr{R}_1$ and $\mathscr{R}_2$ connected at a common boundary $\mathscr{S}$ – see also Fig. 11. Subspace $\mathscr{R}_1$ represents the TLM region where the discretized structure and all the sources are modeled and subspace $\mathscr{R}_2$ represents the homogeneous background space where the radiating electromagnetic field is represented in terms of transverse electric (TE) and

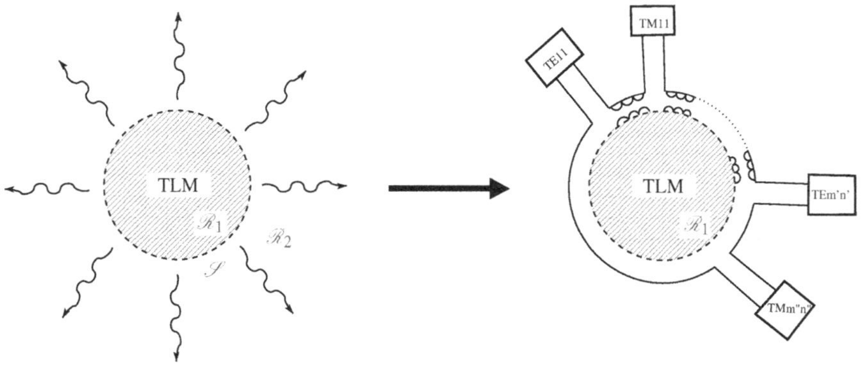

Fig. 11 *Left*: schematic representation of the division of the unbounded space in subspaces $\mathscr{R}_1$ and $\mathscr{R}_2$ with the common boundary $\mathscr{S}$. *Right*: schematic representation of the global network model of the TLMME simulation

transverse magnetic (TM) radiation modes. The expansion of the radiated field into the radiation modes is referred to as the *Multipole Expansion* (ME).

The connection between subspaces $\mathscr{R}_1$ and $\mathscr{R}_2$ is established from its network representation. The connection subnetwork, which connects subspaces $\mathscr{R}_1$ and $\mathscr{R}_2$ contains only ideal transformers [6, 14], and is connected to lumped element equivalent circuits representing the impedances of the radiation modes.

The basis one-forms [17] of the ME are orthogonal and form a complete set. A compact lumped element LC ladder network representing the coupling between the electric and magnetic field intensities may be established [16].

In the TLM region the spherical boundary is approximated by cubical TLM cells. The connection of the outer faces of the TLM cells on the spherical boundary to the spherical modes is modeled by a connection subnetwork based on ideal transformers as described in Sect. 2.

In TLMME the tangential electric field on the interfacing surface (i.e., on the sphere $\mathscr{S}$) is expanded into two sets of vector basis functions $\mathbf{e}_n^{(1)}$ and $\mathbf{e}_m^{(2)}$ with indices $n = 1,2,\ldots,N$ and $m = 1,2,\ldots,M$. The electric fields can be written as

$$\mathbf{E}_t^{(1)} = \sum_n^N V_n^{(1)} \mathbf{e}_n^{(1)}, \tag{21a}$$

$$\mathbf{E}_t^{(2)} = \sum_m^M V_m^{(2)} \mathbf{e}_m^{(2)}, \tag{21b}$$

with superscripts $^{(1)}$ and $^{(2)}$ denoting the quantities corresponding to spaces $\mathscr{R}_1$ and $\mathscr{R}_2$, respectively. We call the modal amplitudes $V_n^{(1)}$ and $V_m^{(2)}$ *equivalent voltages*. The TLM basis functions $\mathbf{e}_n^{(1)}$ and the ME basis functions $\mathbf{e}_m^{(2)}$ [18] are orthonormal, i.e., they satisfy

$$\int_{\mathscr{S}} \mathbf{e}_i^{(1)} \cdot \mathbf{e}_j^{(1)} = \begin{cases} 1 & \text{if } i = j \\ 0 & \text{if } i \neq j \end{cases}, \tag{22a}$$

$$\int_{\mathscr{S}} \mathbf{e}_i^{(2)} \cdot \mathbf{e}_j^{(2)} = \begin{cases} 1 & \text{if } i = j \\ 0 & \text{if } i \neq j \end{cases}. \tag{22b}$$

Next, we define the coefficient n_{mn} by

$$n_{mn} = \int_{\mathscr{S}} \mathbf{e}_m^{(2)} \cdot \mathbf{e}_n^{(1)} \, \mathrm{d}\mathscr{S} \tag{23}$$

and the matrix $\mathbf{N}$ as

$$\mathbf{N} = \begin{bmatrix} n_{11} & n_{12} & \dots & n_{1N} \\ n_{21} & n_{22} & \dots & n_{2N} \\ \dots & \dots & \dots & \dots \\ n_{M1} & n_{M2} & \dots & n_{MN} \end{bmatrix}. \tag{24}$$

It is shown in [5] that the connection between the TLM region and the ME region is represented by a connection subnetwork consisting of ideal transformers with the scattering matrix $\mathbf{S}$ according to (8) and the transformer ratio matrix $\mathbf{N}$ according to (24) with (23).

The equivalent voltages $V_m^{(2)}$ in (21b) represent the input of the corresponding radiation mode. Each TLM port of the surface element s^c contributes to the total radiation. The lumped element equivalent circuit model of the connection circuit connecting the TLM network representing the simulation domain with the ladder networks representing the radiation modes is shown in Fig. 12. In [4] it has been shown, that the connection subnetwork contains only ideal transformers.

3.2.1 TLMME Simulation of a Dipole Antenna

To examine the performance of the TLMME method, the input impedance of a flat dipole antenna is computed. It is well known and generally valid that the input impedance is much more sensitive to the accuracy of the modeling method than other antenna parameters like the radiation pattern (cf., e.g., [19]).

For the absorbing boundary condition to work properly, there must be a certain distance between the geometric object and the computational boundary. In this example a dipole antenna of 6 mm length is placed directly at the boundary of the simulation region and its input impedance is computed.

For the comparison the following simulation setups are used:

1. Reference simulation with ABC,
2. TLM with ABC on the boundary of a cubical simulation region,
3. TLM with ABC on the boundary of a spherical TLM region,
4. TLMME boundary condition on the boundary of a spherical TLM region.

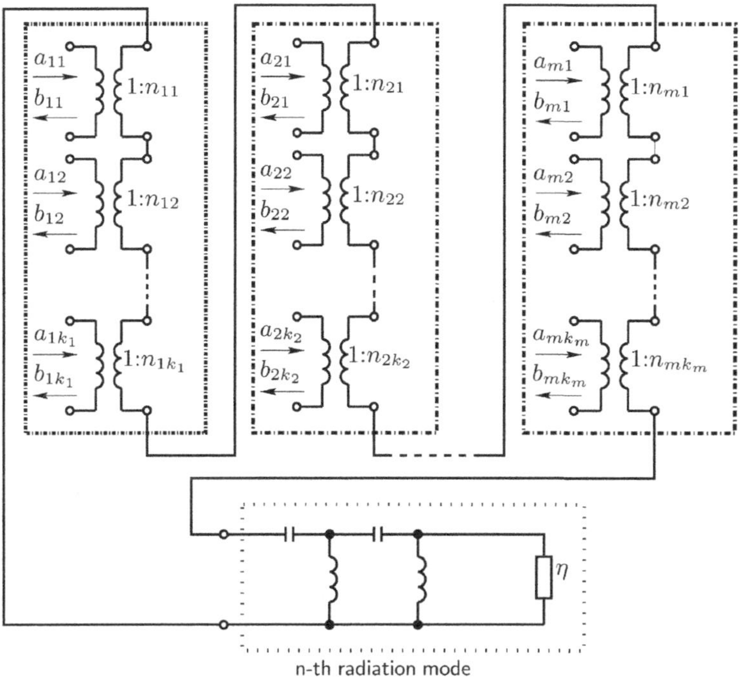

Fig. 12 Network model of the connection of one radiation mode to the TLM simulation domain

The discretized dipole antenna inside the spherical TLM region is shown in Fig. 13. Except in the case of the reference simulation, the dipole antenna is touching directly the boundary of the simulation region.

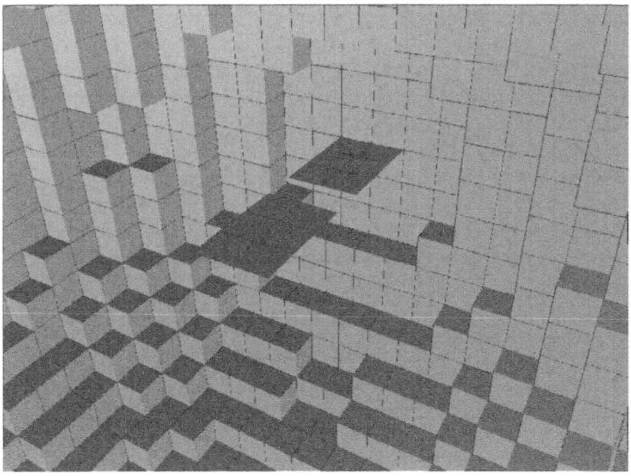

Fig. 13 Dipole antenna inside spherical TLM region. The antenna is touching the boundary of the spherical simulation region

For the reference simulation a simulation domain of the size of 50 mm × 50 mm × 50 mm was used. The other parameters were chosen the same as those described below for the other simulation setups.

The setup for the ABC and TLMME simulations was:

- Simulation domain of 20 mm × 20 mm × 20 mm,
- Uniform spatial discretization with $\Delta x = \Delta y = \Delta z = 1$ mm,
- Number of TLM iterations $n = 500$ time steps,
- Radius of the spherical TLM region $r = 10$ mm,
- Excitation by a discrete Dirac impulse at the source port of a connection subnetwork distributed over a region of 2×1 TLM cells with reference port resistance equal to $377/2\ \Omega$.

For the TLMME simulation only TM_{n0} radiation modes were considered and the number of modes was $M = 25$.

The results of the computed input impedance for the different setups are shown in Fig. 14. We can see that the results of ABC simulations when the dipole is touching the boundary are not correct in both cases, the cubical simulation region and the spherical simulation region, respectively. The reactive part of the input impedance does not properly reflect the resonant behaviour of the antenna.

On the other hand, we can see a good agreement between the TLMME simulation and the reference simulation, even if the antenna is touching the boundary of the simulation region. The imaginary part reflects correctly the resonant behaviour of the antenna and also the real part agrees well with the reference simulation.

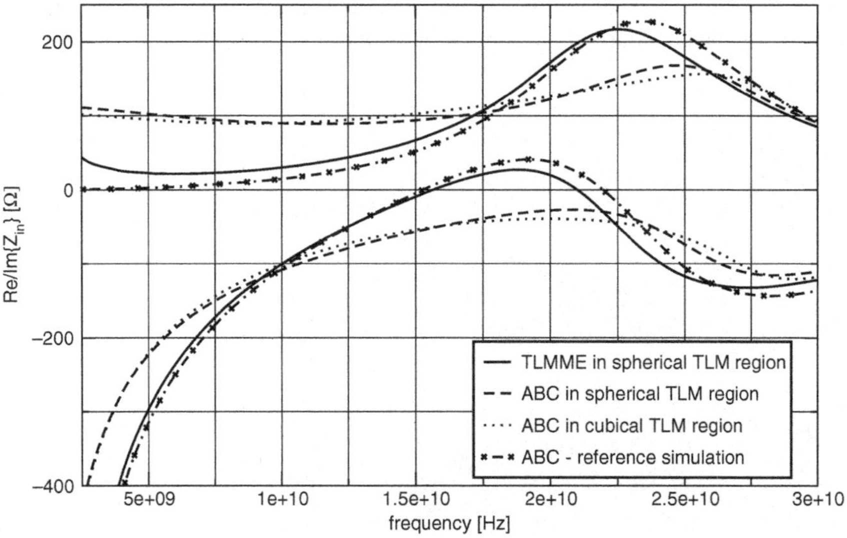

Fig. 14 Real and imaginary parts of the input impedance of the 6 mm dipole for different simulation setups

4 Conclusions

We have presented connection subnetworks for the TLM method suitable for the modeling of excitation regions, for the connection between different simulation regions and for the hybridization of the TLM method with other methods. We have shown that the connection subnetworks have similar properties as the scattering matrices of TLM cells and can be thus easily implemented into the TLM method.

For the modeling of lumped sources connection subnetworks based on wave digital filter networks were introduced. In a numerical example we have outlined how parallel and series adaptors can be interconnected to model a source distributed across multiple TLM cell interfaces. For modal sources and hybridization purposes the connection subnetwork based on ideal transformer network was introduced and its scattering matrix was derived.

We have shown simulation results for the calculation of the input impedance of a bowtie antenna excited with a discrete Dirac impulse. It was shown on the time-domain signal of the reflected wave at the source port that non-physical modes were present. However, since the TLM model is a linear time-invariant system and since only one source port is excited, the result of the input impedance in the frequency range of interest was not affected. In this way the shortest possible time for the simulation was achieved.

As an example of a hybrid method using the connection subnetwork based on ideal transformers the Transmission Line Matrix – Multipole Expansion method was given. The results of the calculation of the input impedance of a dipole antenna using the TLMME method were shown.

In the TLMME example we have compared the computed input impedance of a dipole antenna located at the boundary of the simulation domain. It was shown that the TLMME method gives better result of the input impedance of the antenna than the TLM method with ABC. In contrast to the TLM method with ABC the input impedance computed by the TLMME method models correctly the resonant behaviour of the dipole antenna even if the antenna structure was touching directly the boundary of the simulation domain.

Appendix A Block matrix and Schur complement

In the following consider the matrices $[\mathbf{A}]_{p\times p}$, $[\mathbf{B}]_{p\times q}$, $[\mathbf{C}]_{q\times p}$, $[\mathbf{D}]_{q\times q}$, where $\mathbf{D}$ is nonsingular. Let $[\mathbf{M}]_{(p+q)\times(p+q)}$ be the block matrix (partitioned matrix)

$$\mathbf{M} = \begin{bmatrix} \mathbf{A} & \mathbf{B} \\ \mathbf{C} & \mathbf{D} \end{bmatrix}. \tag{25}$$

The transpose of $\mathbf{M}$ is given by

$$\mathbf{M}^T = \begin{bmatrix} \mathbf{A} & \mathbf{B} \\ \mathbf{C} & \mathbf{D} \end{bmatrix}^T = \begin{bmatrix} \mathbf{A}^T & \mathbf{C}^T \\ \mathbf{B}^T & \mathbf{D}^T \end{bmatrix}. \tag{26}$$

The Schur complement $\mathbf{S}_A$ is defined as

$$\mathbf{S}_A = \mathbf{A} - \mathbf{B}\mathbf{D}^{-1}\mathbf{C}. \tag{27}$$

Using the Schur complement the inverse matrix $\mathbf{M}^{-1}$ can be calculated as

$$\begin{bmatrix} \mathbf{A} & \mathbf{B} \\ \mathbf{C} & \mathbf{D} \end{bmatrix}^{-1} = \begin{bmatrix} \mathbf{I} & \mathbf{0} \\ -\mathbf{D}^{-1}\mathbf{C} & \mathbf{I} \end{bmatrix} \begin{bmatrix} \mathbf{S}_A^{-1} & \mathbf{0} \\ \mathbf{0} & \mathbf{D}^{-1} \end{bmatrix} \begin{bmatrix} \mathbf{I} & -\mathbf{B}\mathbf{D}^{-1} \\ \mathbf{0} & \mathbf{I} \end{bmatrix}. \tag{28}$$

References

1. W. J. R. Hoefer. The transmission line matrix (TLM) method. In T. Itoh, editor, *Numerical Techniques for Microwave and Millimeter Wave Passive Structures*, pages 496–591. J. Wiley, New York, 1989.
2. C. Christopoulos. *The Transmission–Line Modeling Method TLM*. IEEE Press, New York, 1995.
3. A. Fettweis. Wave digital filters: Theory and practice. *Proc. IEEE*, 74(2):270–327, Feb. 1986.
4. P. Russer, M. Mongiardo, and L. B. Felsen. Electromagnetic field representations and computations in complex structures III: network representations of the connection and subdomain circuits. *Int. J. Numerical Modelling*, 15(1):127–145, Jan. 2002.
5. P. Lorenz. *Transmission Line Matrix – Multipole Expansion Method*. Ph.d dissertation, Technische Universität München, Germany, July 2007.
6. M. Krumpholz and P. Russer. A field theoretical derivation TLM. *IEEE Trans. Microwave Theory Techn.*, 42(9):1660–1668, Sep. 1994.
7. K. C. Gupta, R. Garg, and R. Chadha. *Computer-Aided Design of Microwave Circuits*. Artech House, 1981.
8. C. J. Leat, N. V. Shuley, and G. F. Stickley. Triangular-patch model of bowtie antennas: validation against Brown and Woodward. *IEE Proc.-Microw. Antennas Propag.*, 145(6):465–470, Dec. 1998.
9. G. H. Brown and O. M. Woodward. Experimentally determined radiation characteristics of conical and triangular antennas. *RCA Rev.*, 13:425–452, Dec. 1952.
10. B. Engquist and A. Majda. Absorbing boundary conditions for numerical simulation of waves. *PNAS*, 74(5):1765–1766, May 1977.
11. J. P. Berenger. A perfectly matched layer for the absorption of electromagnetic waves. *J. Comp. Phy.*, pages 185–200, Oct. 1994.
12. J. A. Morente, J. A. Porti, and M. Khalladi. Absorbing boundary conditions for the TLM method. *IEEE Trans. Microwave Theory Tech.*, pages 2095–2099, Nov. 1992.
13. C. Eswarappa and W. J. R. Hoefer. Implementation of Berenger absorbing boundary conditions in TLM by interfacing FDTD perfectly matched layers. *Electronics Letters*, 31: 1264–1266, July 1995.
14. N. Peña and M. M. Ney. Absorbing-boundary conditions using perfectly matched-layer (PML) technique for three-dimensional TLM simulations. *IEEE Trans. Microwave Theory Tech.*, Pages 1749–1755, Oct. 1997.

15. J. L. Dubard and D. Pompei. Simulation of Berenger's perfectly matched layer with a modified TLM node. *IEE Proc. Microwaves, Antennas, Propagat.*, 144:205–207, June 1997.
16. P. Russer. Network-oriented modeling of radiating electromagnetic structures. *Turk. J. Elec. Engin.*, 10(2):147–162, 2002.
17. P. Russer. *Electromagnetics, Microwave Circuit and Antenna Design for Communications Engineering*. Artech House, Inc., Boston, Second edition, 2006.
18. P. Lorenz and P. Russer. Hybrid transmission line matrix – multipole expansion (TLMME) method. In *Fields, Networks, Computational Methods, and Systems in Modern Electromagnetics, A tribute to Leopold B. Felsen*, volume 97 of *Springer Proceedings in Physics*, pages 157–168, Berlin, Heidelberg, 2004. Springer.
19. C. A. Balanis. *Antenna Theory*. John Wiley & Sons, Inc., Second edition, 1997.

RFID

Li Yang, Amin Rida, Anya Traille and Manos M. Tentzeris

1 RFID Introduction

Radio Frequency Identification (RFID) is an emerging technology for the identification of objects and/or personnel. RFID is recognized as one of the technologies capable of realizing a complete ubiquitous computing network due to its strong benefits and advantages over traditional means of identification such as the optical bar code systems. It has recently found a tremendous demand due to emerging as well as already existing applications requiring more and more automatic identification techniques that facilitate management, increase security levels, enhance access control and tracking, and reduce labor force. A brief listing of RFID applications that find use on a daily basis is:

- Warehouse Management Systems
- Retail Inventory Management
- Toll Roads
- Automatic Payment Transactions
- High Value Asset Tracking and Management
- Public Transportation
- Automotive Industry
- Livestock Ranching
- Healthcare and Hospitals

Li Yang
Georgia Electronic Design Center, School of Electrical and Computer Engineering, Georgia Institute of Technology, Atlanta, GA 30332-0250, USA

Amin Rida
Georgia Electronic Design Center, School of Electrical and Computer Engineering, Georgia Institute of Technology, Atlanta, GA 30332-0250, USA

Anya Traille
Georgia Electronic Design Center, School of Electrical and Computer Engineering, Georgia Institute of Technology, Atlanta, GA 30332-0250, USA

Manos M. Tentzeris
Georgia Electronic Design Center, School of Electrical and Computer Engineering, Georgia Institute of Technology, Atlanta, GA 30332-0250, USA

P. Russer, U. Siart (eds.), *Time Domain Methods in Electrodynamics*,

- Pharmaceutical Management Systems
- Military
- Marine Terminal Operation
- Manufacturing
- Anti-counterfeit

As an illustration to how an RFID system works a block diagram of a Passive RFID tag is illustrated in the Fig. 1. Passive RFID tags use the power induced across its antenna terminals received by an RFID reader antenna for powering itself and communication between the tag and the reader by using backscatter technique. RFID reader antennas typically use loop or array antenna configurations depending on the frequency of operation. Active tags on the other hand use a battery as a power source for turning on the IC and for communicating with the reader. At the heart of the passive tag to which the antenna sends and receives signals from is an IC; shown as part of Fig. 1. The IC contains a passive power converter which rectifies and filters the AC power induced across the antenna into a continuous DC mode signal, which powers a signal modulator/demodulator and the logic control within the IC. In addition, the AC voltage across the tag antenna is also demodulated into its corresponding bit scheme, which is fed into the logic controller of the IC. The logic controller decodes the instructions received from the reader, and also sends out information about the tag namely the unique identification number of the tag also known as the GUID number, which is modulated and fed back to the tag antenna to be transmitted to the reader.

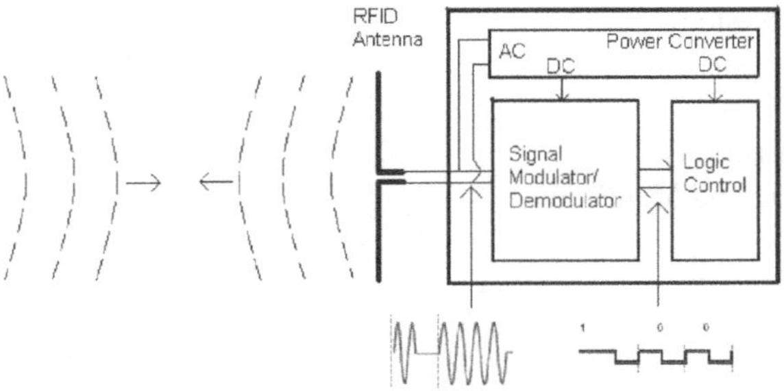

Fig. 1 Block Diagram of Passive RFID Tag

2 RFID Tag Antenna Design

RFID faces several challenges in the design and development phases of its RF hardware, mainly since modern RFID systems operate at high frequencies such as UHF Frequency bands. UHF RFID bands are preferred in most applications for its added benefit over low frequency systems especially in increased read range and higher data rate transfers as well as size reduction of the RFID system. The main challenge is the existence of various different UHF frequency bands ranging from 866 to 964 MHz for Europe (866–868 MHz), Asia (864–964 MHz) and US (902–928 MHz). The universal operation of the RFID necessitates the use of wideband antennas covering one or more of these three bands. Another is the dimensions of a tag design. In limited assigned area, the antenna should meet requirements to optimize the RFID system power-performance, especially for passive or semi-active configurations, where the only energy source is the incoming reader energy. The third obstacle is from the harsh environment in which dual polarization property of a tag antenna is preferred.

In this section, various novel approaches in wideband adaptive impedance matching, compact design method, and polarization improvement are proposed. These approaches are supplemented with modeling, simulation, and measurement results.

2.1 Wideband Adaptive Impedance Matching

2.1.1 Design

A major challenge in RFID antenna design is the impedance matching of the antenna (Z_{ANT}) to that of the IC (Z_{IC}). For years, antennas have been designed primarily to match either 50 Ω or 75 Ω loads. However, RFID chips primarily exhibit complex input impedance, making matching extremely challenging. A block diagram of a passive RFID tag is shown in the Fig. 2. The antenna matching network must provide the maximum power delivered to the IC which is used to store the data that is transmitted to/received from the reader.

In this design approach, a very compact configuration (in an area less than $3\,\text{in} \times 3\,\text{in}$) of a $(\lambda/2)$ dipole antenna was developed, where λ is the free space wavelength. The $(\lambda/2)$ antenna design is favorable for its quasi- omnidirectional radiation pattern; a fundamental requirement for RFID's to allow for the communication between the RFID tags and the RFID reader, independent of the orientation of the tags. The step by step design is illustrated in the figures below. Figure 3a shows the main radiating element which if stretched from one end to the other corresponds to a length of $\sim 16\,\text{cm}$ (which is $\lambda/2$ around the center frequency 935 MHz in air). The tapering of the antenna was chosen for maximum current flow (hence

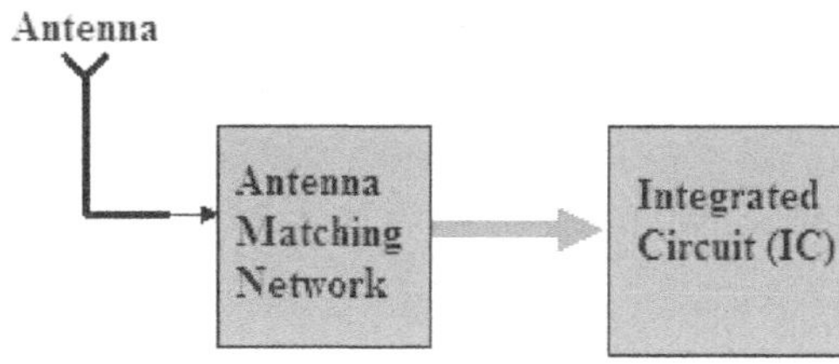

Fig. 2 Block diagram of a passive RFID tag

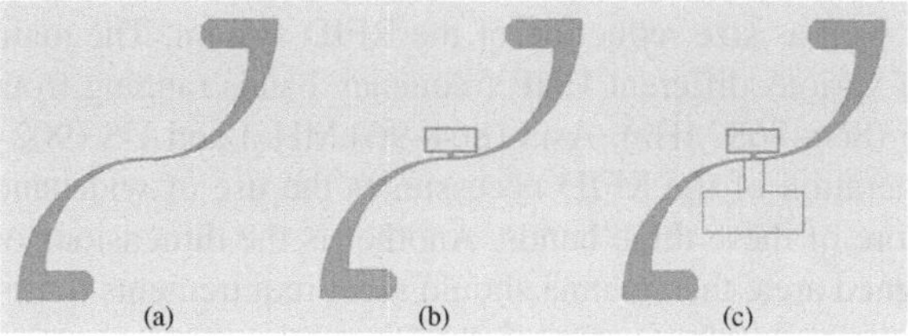

Fig. 3 Step by step antenna design showing (**a**) Radiating body (**b**) Radiating body plus double inductive stub and (**c**) final antenna structure with the resistive stub

optimum efficiency) and to achieve a high bandwidth. The $\lambda/2$ antenna was folded as shown in Fig. 3 at a distance ($\sim$0.16 λ) not to cause any significant current perturbation (keeping maximum radiation efficiency), while making the design more compact. Without loss of generality, in this design the overall matching network is designed to conjugately match an RFID chip with a high capacitive impedance of $Z_{IC} = 73\text{–}j113\ \Omega$. Figure 3 also shows the step by step procedure used in the design. To satisfy the conformality RFID requirements, the proposed antenna was fabricated on flexible 4-mil Liquid Crystal Polymer (LCP) that is an organic material which has a dielectric constant ε_r of ~ 3.10 and a loss tangent $(\tan\delta)$ of 0.0019.

The IC used for this design has four ports; two input ports namely RF1 and RF2 which may be connected to a single or dual antenna configuration, one ground port (V_{ss}) to connect to the second arm of the antenna (for example in the dipole-based design of this paper) and an open port (V_{dd}) to measure or drive the bias voltage of the IC when conducting measurements. The two input ports are identical and can be used either for single or dual antenna topologies. For single antenna structures RF2 and Vss ports are shorted. The resistive shorting stub and the double inductive stub as illustrated in Fig. 4 constitute the overall matching network. The resistive stub

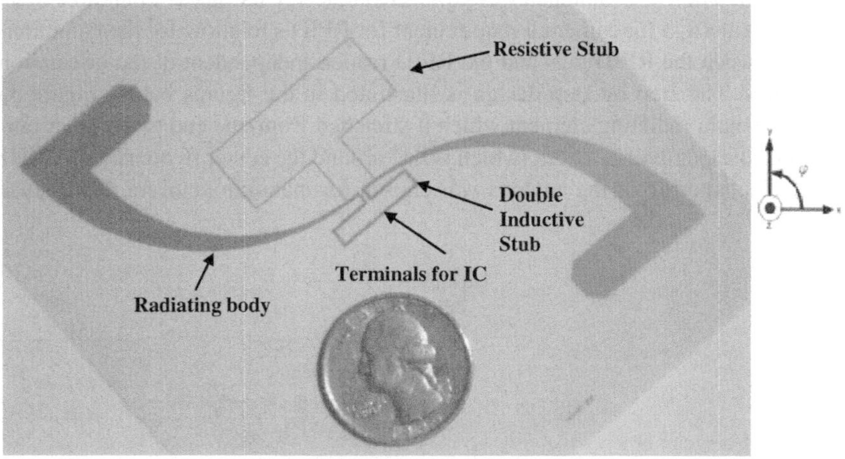

Fig. 4 RFID antenna structure showing stubs

is used to tune the resistance of the antenna to match that of the IC. In this design the size and shape (thin long loop shaped line) of the resistive stub were designed to have an optimum match to $Z_{IC} = 73\text{–}j113\ \Omega$. The double inductive stub is composed of two inductive stubs to provide symmetry on both sides of the antenna. The double inductive stub also serves as the reactive tuning element of the antenna.

The feeding point of the antenna is at the bottom part of the double inductive stub where an IC would be surface mounted. Figure 4 illustrates the final structure.

The stubs were designed to have a center frequency f_0 at 895 MHz with a bandwidth of 70 MHz operating from 860 to 930 MHz (European and U.S. frequencies). A wide frequency sweep has also been performed up to 5 GHz where no parasitic radiation has been observed for this antenna. Those variables can be fine tuned to optimize the antenna characteristics on the RFID tag at any frequency and matched to any IC impedance.

The structure was simulated and optimized. The input impedance of the simulated antenna design is shown in Fig. 5. As it can be observed the RFID UHF band (860–930 MHz) is outside the antenna self-resonance peak, resulting in a more flat impedance response against frequency. This yields to a bandwidth of $\sim 8\%$ which is predominantly realized by the finite slope of the reactance of the antenna in the frequency of interest.

The simulated impedance at the center frequency $f_0 = 895\,\text{MHz}$ is $57.46 + j112.1\ \Omega$ which results in a return loss $\text{RL} < -18\,\text{dB}$. This antenna has a bandwidth of $\sim 8\%$ (70 MHz) where the bandwidth is defined by a Voltage Standing Wave Ratio (VSWR) of 2 (alternatively a RL of $-9.6\,\text{dB}$). The tapering of this S-shaped antenna along with the matching techniques (resistive and inductive stubs) allow for the first-ever $3\,\text{in} \times 3\,\text{in}$ RFID antenna with such a high bandwidth ($\sim 8\%$).

The return loss of this antenna was calculated based on the power reflection coefficient which takes into account the capacitance of the IC:

$$\left|s^2\right| = \left|\frac{Z_{IC} - Z_{ANT}{}^{*}}{Z_{IC} - Z_{ANT}}\right|^2$$

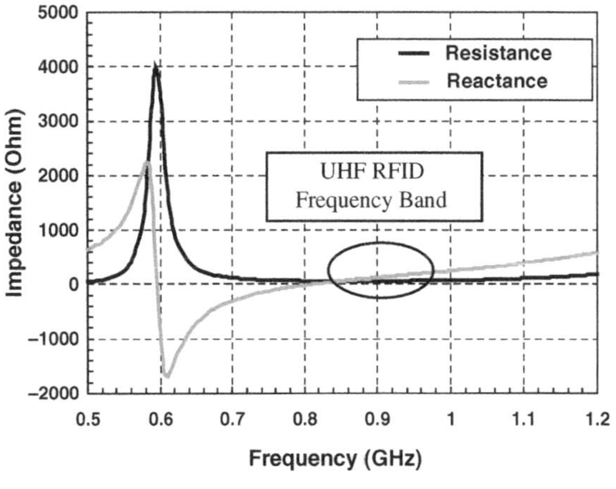

Fig. 5 Simulated input impedance of the S-shaped antenna

Z_{IC} represents the impedance of the IC and Z_{ANT} represents the impedance of the antenna with Z_{ANT}^* being its conjugate.

The simulated radiation pattern and radiation efficiency were numerically computed by introducing an RLC boundary along with the port impedance that simulates the behavior of the IC (with its complex impedance feed). In Fig. 6 the 2-D radiation plot is shown for the phi = 0° and phi = 90° where an omnidirectional pattern is realized. The radiation pattern throughout the bandwidth of the antenna has also shown to have an omnidirectional pattern similar to that of a classic ($\lambda/2$) dipole antenna.

A directivity of 2.10 dBi is achieved with a radiation efficiency of 97%. The omnidirectional radiation is one of the most fundamental requirements for RFID's to allow for their reading/writing operation independent of the orientation of their antenna with respect to the reader. Preliminary experimental results have verified the shape of Fig. 6 theoretical ones.

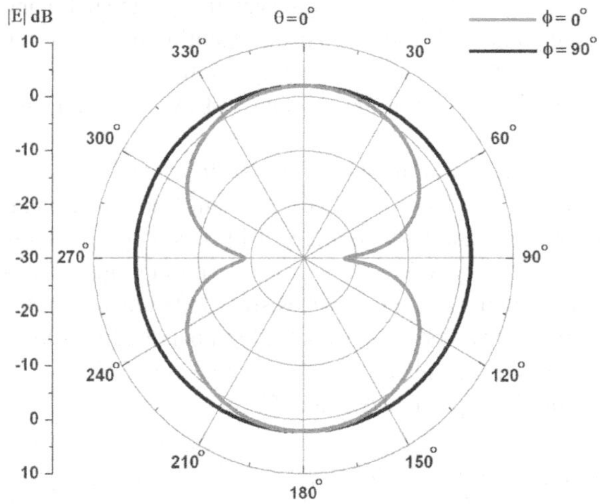

Fig. 6 2D far-field radiation plot

2.1.2 UHF RFID Antenna Modeling

In order to obtain a thorough understanding of the power reflection caused by any mismatch at the terminals of the feed structure of the antenna, a wideband equivalent circuit model has been developed. This model serves as a benchmark for the design of an RFID antenna to theoretically match any Z_{IC} for maximum power flow resulting in optimum antenna efficiency and an excellent read range.

Based on a physical approach, an equivalent lumped element circuit model was derived. The system level design and simulation tool Advanced Design System (ADS) was used to simulate the behavior of the circuit model (S_{11} parameter) and resulted in a negligible error function ($< 10^{-5}$). Figure 7 shows the agreement between the lumped element circuit S parameters with that of the structure (from the full wave simulator).

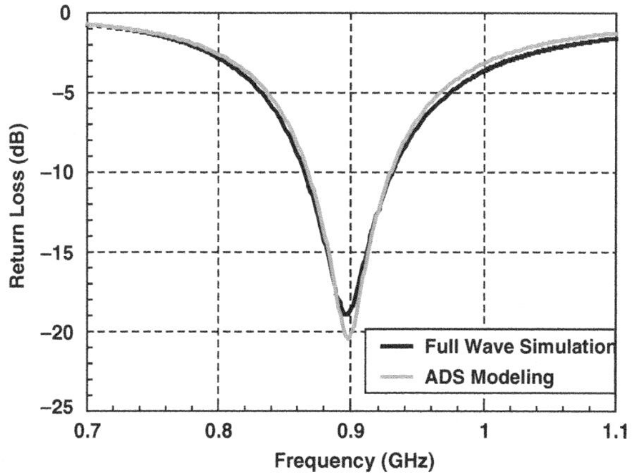

Fig. 7 S_{11} for the exact structure and for the equivalent circuit model

Figure 8 shows the detailed equivalent circuit of the radiating body only. Each arm of the radiating body consists of a resistor in series with an inductor, the combination of which models the metal effects. A capacitor in series with a resistor, which are located in parallel with the previous combination of Ls and Rs model the substrate effects. Finally, the capacitive coupling or E-Field coupling between the two arms of the S-shaped antenna is modeled by the top and bottom capacitors (for air and dielectric capacitive coupling respectively). This lumped element model covers a frequency range 700 → 1100 MHz as shown in Fig. 7.

The circuit configuration in Fig. 8 can be simplified to the one shown in Fig. 9 by using symmetry and direct circuit analysis.

Since the double inductive stub is connected in series with the radiating body (the antenna is now fed in the center of the double inductive stub), the equivalent circuit model of the second stage design (radiating body plus inductive stub) as shown in Fig. 3 has the same circuit elements configuration as the one in Fig. 9, with change in values only. The final stage of the design has the circuit model configuration shown in Fig. 10. Due to the configuration of the resistive shorting stub (connected in parallel with the radiating body plus inductive stub), the components: Rs2, Ls2, Rp2, Cp2 are introduced as shown in Fig. 9 and model the same effects as those discussed previously for radiating body circuit model (Fig. 9).

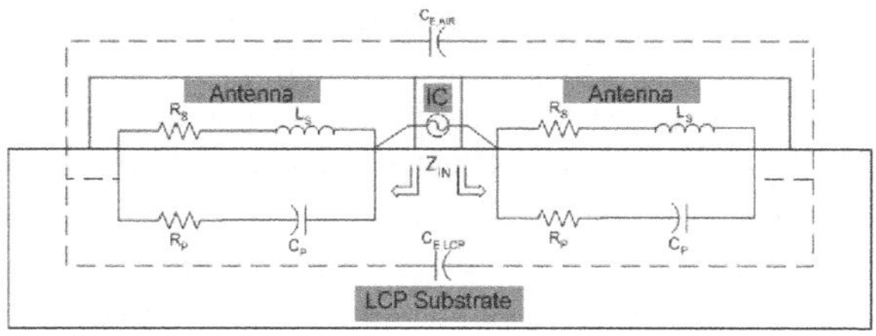

Fig. 8 Cross-sectional detail showing equivalent lumped element model of RFID antenna shown in Fig. 2a and b

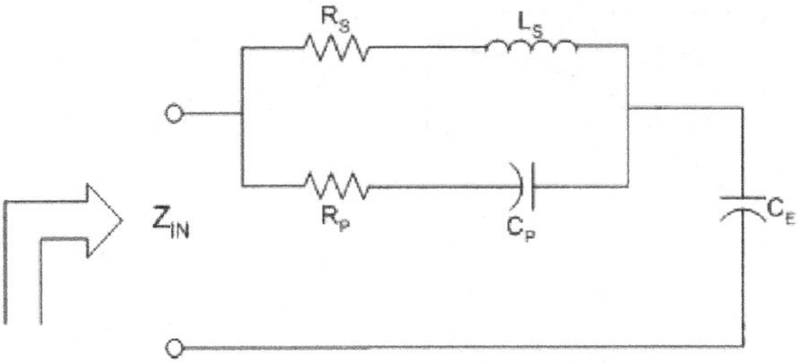

Fig. 9 Equivalent circuit model of RFID antenna shown in Fig. 2a and b

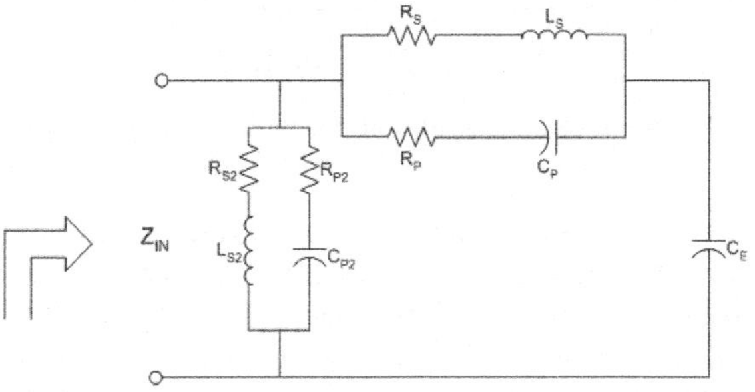

Fig. 10 Equivalent circuit for antenna structure shown in Fig. 2c

The equivalent circuit shows how stubs can be used to tune the impedance in order to match to any IC. Parametric sweeps can be used along different stubs structures (for example loops structures can be used for adding series inductance or parallel capacitance). The resistance of the antenna is mainly determined by the radiating body and can be tuned by the two stubs as shown above. This model also helps to determine the amount of loss (as parallel resistance and capacitance) due to the substrate loss which helps in understanding radiation efficiency as a function of the substrate.

2.1.3 Measurements

In order to accurately measure the input impedance of the RFID antennas, numerous problems should be taken into consideration. First of all, a traditional probe station was not suitable for our tests due to the undesired shorting effect of the metallic chuck, which was behaving as a spurious ground plane for the dipole antennas. To tackle the problem, a custom-made probe station using wood and high density polystyrene foam was built. This type of foam was selected due to its low dielectric constant (1.06) resembling that of the free space. A $5 \cdot \lambda/2$-thick foam station was designed in order to ensure minimum backside reflections of the antenna.

It was also taken into account the fact that the antennas were balanced structures and a typical GS probe connected to a regular coaxial cable would provide an unbalanced signal. To prevent a current difference between the dipole arms, a $\lambda/4$ balun

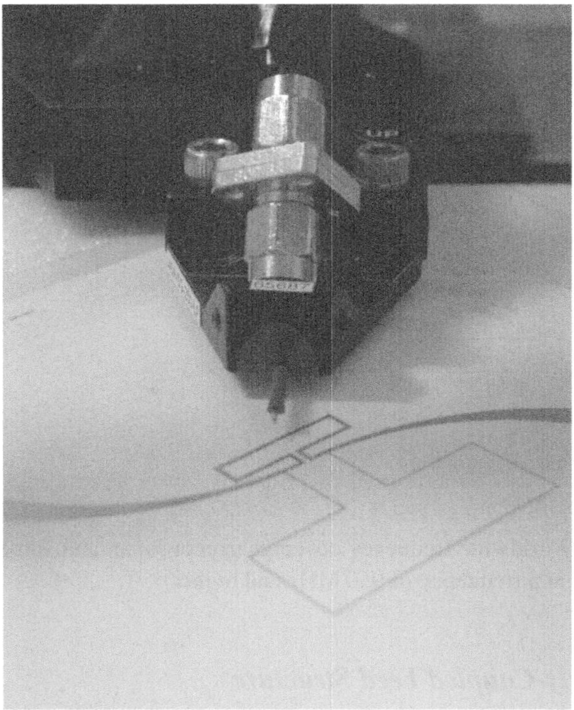

Fig. 11 Photograph of the probe plus S-shaped antenna

with an operational bandwidth of 840–930 MHz (which covers the band of interest for this design) was used. After all the above mentioned precautions were taken and minding about the calibration process, S parameters were measured with an Agilent 8510C S parameter receiver (Fig. 11) and transformed to Z_{IN} or Z_{ANT}.

Figures 12 and 13 show a very good agreement between the simulated results and the measurements for the antenna input impedance and S_{11} parameter respectively. The demonstrated antenna bandwidth allows for a universal operation of

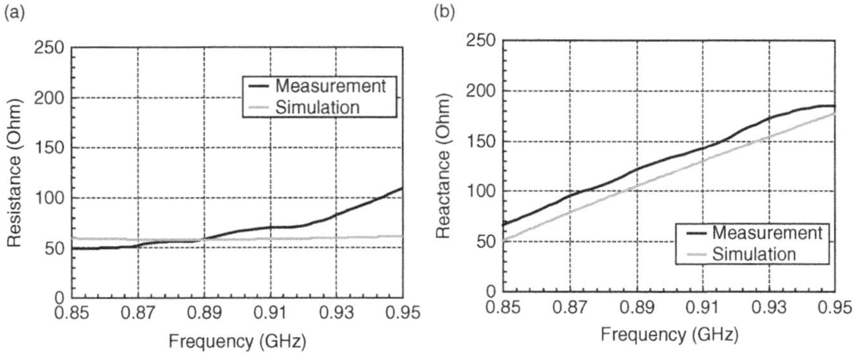

Fig. 12 Measured and simulated data of input impedance: (a) Resistance (b) Reactance

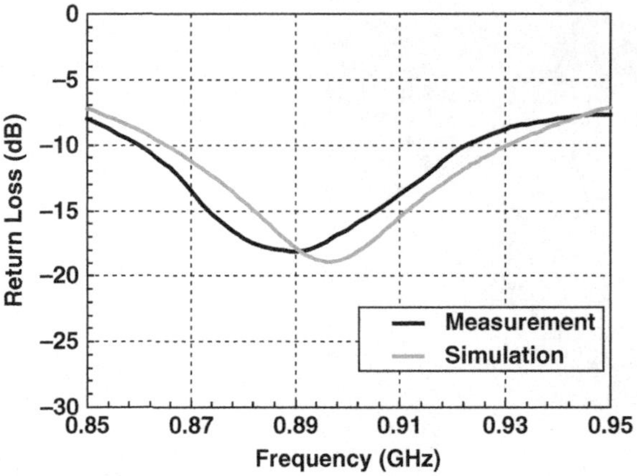

Fig. 13 Measured and simulated data of return loss

the proposed UHF RFIDs (Worldwide frequency coverage except Japan and some Asian countries that operate at a frequency of 950 MHz and higher).

2.2 Compact Inductively-Coupled Feed Structure

In the serial stub feed structure, the dimensions of the resistive shorting stub and the double inductive stub are comparable with the main radiating body. These stubs limit the miniaturization of the antenna size as a tradeoff for the enhanced bandwidth. For an ultra-compact RFID antenna design, a matching technique without any stubs is highly preferred. An inductively-coupled feed structure, similar to the one shown in Fig. 14, is such an effective way for impedance matching. The antenna consists of a feed loop with two terminals and a radiating body. The two terminals

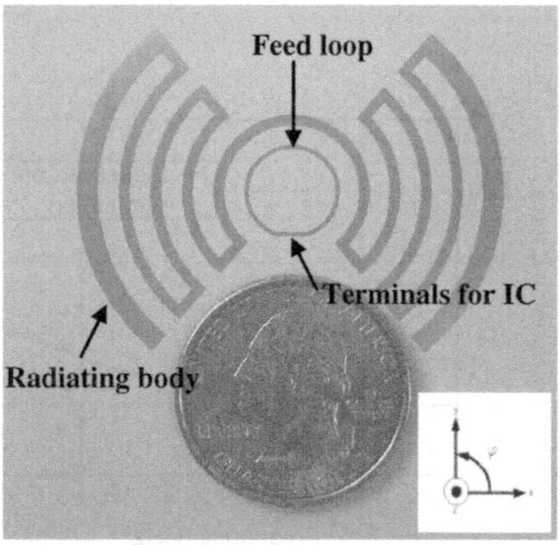

Fig. 14 Inductively coupled feed antenna fabricated on LCP substrate

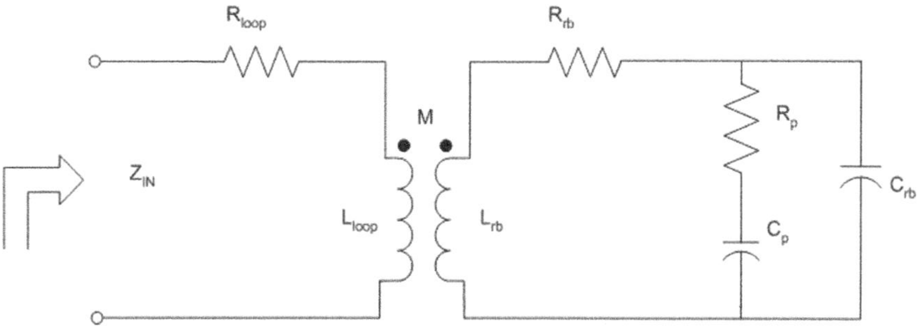

Fig. 15 Lumped element model for inductively coupled feeding RFID

of the loop are connected to the IC, and the feed "communicates" with the antenna body through mutual coupling.

The inductively-coupled structure can be modeled as a transformer. Figure 15 depicts the equivalent lumped element model, where R_{rb} and R_{loop} are the individual resistances of the radiating body and the feed loop. M is the mutual inductance and L_{loop} is the self-inductance of the feed loop. R_p and C_p are representing the substrate effects.

Compared with the serial stub feed structure in which off-resonance operation is preferred for better resistance/reactance stability, the inductively-coupled structure is designed at the resonant frequency. This is because at the resonant frequency f_0, assuming that the substrate effect is minimal, the components of antenna input resistance R_{zin} and reactance X_{zin} can be predicted as

$$R_{zin} = \frac{(2\pi f_0 M)^2}{R_{rb}} + R_{loop} \quad (1)$$

$$X_{zin} = 2\pi f_0 L_{loop} \quad (2)$$

Thus, at the resonant frequency, the resistance is mainly controlled by M and R_{rb}, and the reactance is dependent only upon L_{loop}. In this way, the antenna input resistance and input reactance can be adjusted independently. Therefore, inductively coupled feeding structures present one of the theoretical optimum solutions to effectively match an antenna to arbitrary chip impedances, especially for RFID IC's with significant imaginary part.

An ultra-compact inductively-coupled feed antenna was designed in an area less than 1.7 in × 1.4 in, as illustrated in Fig. 14. The substrate was a flexible 4mil LCP that has a dielectric constant ε_r of ∼3.10 and a loss tangent (tan δ) of 0.0019. Since the inductively-coupled feed structure is designed at the resonant frequency, the impedance response is sharper than the one in the serial stub structure, resulting in a narrower bandwidth. Therefore, the inductively-coupled antenna is more applicable to operate at specific bands. In this case, the antenna was designed to cover the European RFID band, ranging from 865 MHz to 868 MHz. The target RFID chip impedance was 73–j113 Ω. It has to be noted that the performance of this design is quite insensitive to fabrication tolerances and variations.

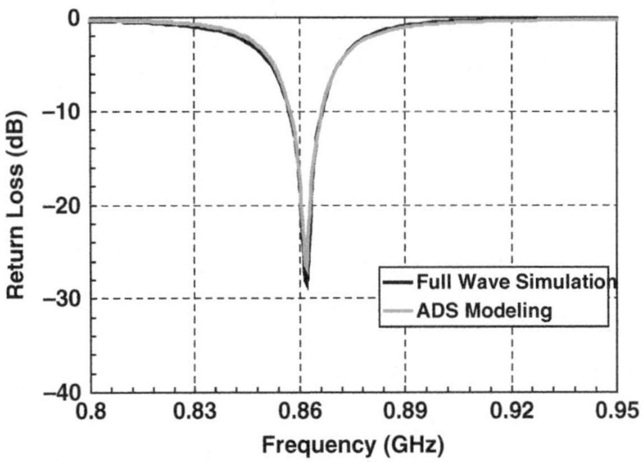

Fig. 16 Comparison of return loss full wave simulation results with ADS modeling

The lumped element values of the equivalent circuit were optimized by ADS simulation software, including the substrate effects [8]. It can be seen that the model data and the full wave simulated impedance data demonstrate a very good agreement, as shown in Fig. 16.

The measured and simulated input impedances of the antenna in the interested band are illustrated in Figs. 17 and 18 with a very good agreement. At 866 MHz, a measured input impedance of $60.8 + j102.5\,\Omega$ is achieved, resulting in a -30 dB return loss at that frequency. The return loss below -10 dB extends from 858 MHz to 869 MHz, covering the whole European UHF band which extends from 865 MHz to 868 MHz and accounting for most fabrication variations, as shown in Fig. 17.

Since the radiating body is basically a $\lambda/2$ dipole, the radiation pattern looks similar to the one of a dipole, as shown in Fig. 19. A directivity of 1.99dBi is achieved with 90% radiation efficiency.

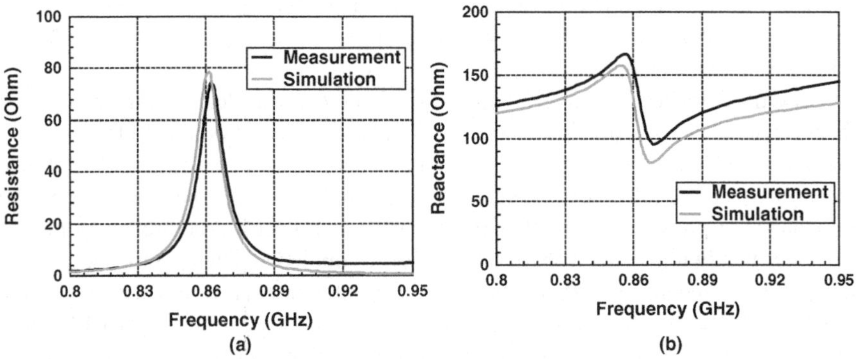

Fig. 17 Measured and simulated data of input impedance: (**a**) Resistance (**b**) Reactance

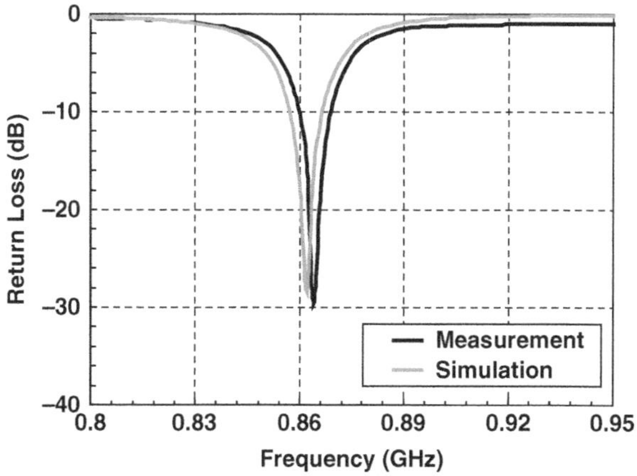

Fig. 18 Measured and simulated data of return loss

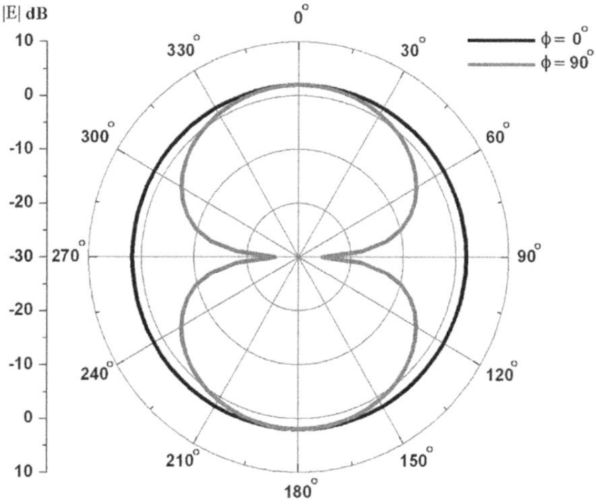

Fig. 19 2D far-field radiation plot

2.3 Polarization-Diversity Solutions for Harsh Environments

Most antennas in UHF RFID tags, similar to the one shown in Fig. 3, are linearly (vertical or horizontal) polarized. In the presence of harsh environments (e.g. containers, industrial machinery) with multiple reflections (i.e. metal) which cause multi-path effects, the transmitted/received plane waves undergo polarization changes (depolarization). For instance, a vertically polarized transmitted wave can reach a tag at its blind spot in the radiation pattern. This causes the RFID tag not to be read. In order to prevent this, in this paper polarization diversity is proposed for a better readability of UHF RFID tags, including the use of both vertical and linear polarized antennas. For dipole antennas, the two orthogonal-polarization antennas

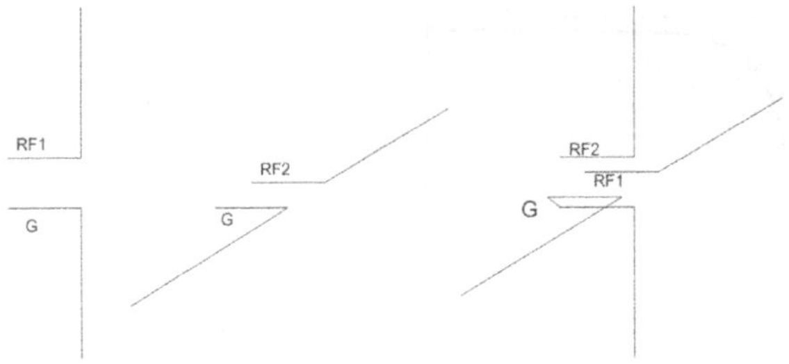

(a) vertical polarization (b) horizontal polarization (c) dual polarization

Fig. 20 Polarization definition in dual antenna design

can be oriented orthogonal to each other as shown in Fig. 20. These two antennas are identical in dimensions and shape, so that the identical signals arriving at these two different branches are in-phase and uncorrelated. The two antennas are also connected with a shorting stub to guarantee an in-phase arrival of both polarization signals to the demodulator in the IC, when the data needs to be retrieved from the combined reception of the two antennas.

Figure 21 shows a 7.5 cm × 5 cm dual polarized antenna. The shorting stub that connects the "top left" legs of the design both provides the resistive matching and is also used to "dc-short" the two orthogonal dipole antennas. The dc-shorting is also utilized to receive in-phase signals at the two identical input ports. The "bottom right" legs are connected to the ground to achieve the necessary signal – ground excitation. The connecting stubs are used to change the resistance and inductance of the dual antenna structure to conjugately match to the chip impedance. The bandwidth of this antenna is 7.78% (70 MHz). The radiation efficiency is 93% since the current flow adds up constructively for the far-field electromagnetic radiation, something that results in an optimized performance in terms of the read range and environment versatility. The radiation pattern of the dual polarized antenna is omnidirectional with a maximum directivity of 2.25 dBi. Figure 22 shows the

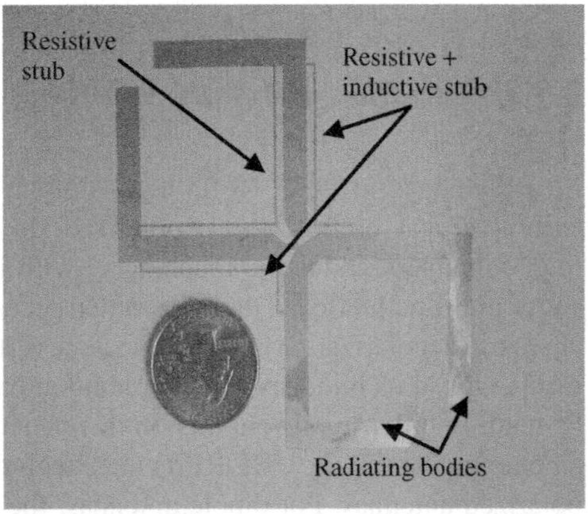

Fig. 21 Dual antenna structure

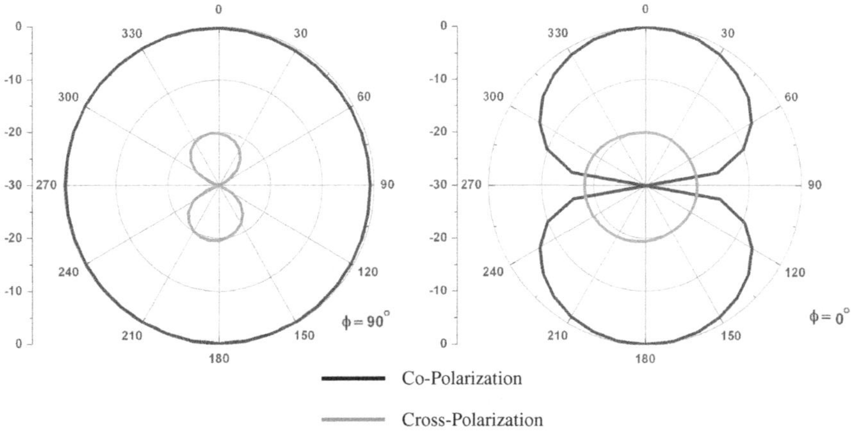

Fig. 22 Simulated co-polarization and cross-polarization radiation patterns

co-polarization and cross-polarization characteristics of the antenna. A cross-polarization suppression larger than 18 dBi is obtained.

3 RFID Reader Design

As RFID's find an ever increasing range of applications, designing and implementing multistandard/universal RFID systems becomes a critical issue for their large-scale implementation. Despite the fact that most of the research effort has been dedicated to the (active, semiactive or passive) RFID's, the readers play a very important role in the range and the coverage. To alleviate problems from harsh environments and provide omni coverage, the RFID readers require compact antennas with high efficiency, broadband performance and circular-polarization with low cross-polarization levels. In addition, the two UHF RFID bands (active: 432–435 MHz, passive: 866–954 MHz) that have been recently approved, impose the need for dual-band antennas, which is extremely challenging due to the large physical size of the UHF wavelengths. This section proposes a novel circular-polarization antenna that relies on the combination of a dual-square loop and a cross-dipole, which allows for the efficient coverage of BOTH bandwidths for universal operation (US, Asia, Europe). This innovative approach could be extended to 4 G, WiMax and wireless broadband multimedia in millimeter-wave range and could set the foundation for ubiquitous sensing/cognitive radio nodes.

3.1 Introduction

The recent advances in cost-effective low-power electronics and packaging have enabled the RFID tag as a likely substitute for barcodes in industries such as access control, parcel and document tracking, distribution logistics, automotive systems, and livestock or pet tracking. In these applications data are contact-free transferred to a local querying system (reader or interrogator) from a remote transponder (tag) including an antenna and a microchip transmitter. A suitable antenna for these tags must have low cost, low profile and especially small size. The RFID tags also present challenges in behavioral modeling and simulation of the antenna and module/package integration in parameters such as the pad capacitance, the estimation of the parasitics due to the proximity of IC and antenna, and the identification of a low-cost low-loss light material. On the other side, antennas for RFID readers require circular polarization and wideband/multiband performance in order to enable operation in different environments and standards (US, Europe, Asia). This paper presents for the first time the design, modeling and optimization of a dual-band circular-polarized antenna for universal UHF RFID readers in the two most common bands for UHF RFID applications (432–435 MHz-active RFIDs/866–954 MHz-passive RFIDs).

3.2 RFID Readers Specifications

To satisfy the frequency requirements for a universal supply chain tracking across Europe, USA and Asia, a global UHF RFID reader must have an antenna that can operate with a least a 9.6% bandwidth from 866 MHz to 954 MHz, assuming mostly passive RFID tags. In addition, this reader has to be "flexible" enabling its functionality in different environments that potentially de-polarize linearly polarized waves (water, food, wood, plants, and liquids). Finally, numerous RFID readers have to be portable or quasi-portable, limiting their size to only 30 cm $\times$ 30 cm.

Currently, there are only a limited number of commercial RFID readers with circular-polarization antennas covering the whole universal (866–954 MHz) band. Still, they could provide the benchmarking design specifications for reader antennas. A gain greater than 5dBi is required for a read range of at least 10 m (30–40 ft) for a beamwidth larger than $65^\circ \times 65^\circ$, while maintaining an axial ratio of less than 4 dB is required to guarantee low cross-polarization. In addition, a VSWR less than 2:1 throughout the entire frequency range is also necessary to provide adequate matching with the reader system, and to satisfy power efficiency requirements. The transverse (face) dimensions should be smaller than 30 cm $\times$ 30 cm. It has to be noted that the maximum regulated output power is currently 4 Watts globally. In addition to the above features, future-generation RFID reader antennas are anticipated to demonstrate the following improvements:

- Mechanical stability to provide portability and resistance against harsh environments.
- A VSWR decreased down to 1.5:1 (return loss < -14 dB) for even higher power efficiency and range.
- Multiband capabilities including lower UHF bands approved for active RFID's (e.g. 432–435 MHz).

Especially the large requirement is very challenging since no dual-band broadband circular-polarization antennas have been published in the past, and the antenna shown in this paper will be the first ever reported.

3.3 RFID Reader Antenna Modeling and Design

About seven years ago, it was found that a dual-rhombic loop can also serve as a broadband CP antenna. A bandwidth of 20% for AR < 2 dB was obtained for the dual-rhombic loop antenna. The bandwidth of the dual-rhombic loop can be further improved by adding a pair of parasitic rhombic loops inside the original loop. The impedance matching of the broadband CP antenna is achieved by incorporating a broadband balun into the feeding structure. The gain of the dual-rhombic loop antenna is enhanced by employing it as the excitation element of a short backfire antenna (SBA). The configuration of a CP dual-rhombic loop antenna is shown in Fig. 23. The square (W1 × W1) dual-rhombic loop is printed on a thin square (Ls × Ls) dielectric substrate (0.254 mm) with a low dielectric constant (RT/Duroid with $\varepsilon_r = 2.2$). To radiate a CP wave, a pair of small gaps is introduced on the dual-rhombic loop. A pair of square (W2 × W2) parasitic rhombic loops (also with a small gap) is added inside the dual-rhombic loop for bandwidth enhancement. An important feature of this loop antenna is that the sense of circular polarization may be adjusted by changing the positions of the small gaps on the loops using RF switches such as microelectromechanical systems or p-i-n diodes. The CP rhombic loop is driven at its center by a broadband balun, which was also fabricated on the same type of RT/duroid 5880 substrate. This kind of broadband balun has been widely used to excite a dipole antenna. The broadband balun can excite the balanced mode by making use of the coupling between the microstrip line printed on one side of the substrate to the slot which is etched on the other side (it also serves as the ground plane for the microstrip line) of the substrate. A good impedance matching can be achieved by adjusting the length (ls) of the slot, the height (hm) and the length (lm) of the microstrip line. The printed CP dual-rhombic loop is backed by a square (Lg × Lg) copper plate (i.e., a ground plane) at a height h for unidirectional radiation and fed through an SMA connector.

For simplicity of experimental verification of our simulation results, we designed the above antenna in the frequency range 4–8 GHz. The plan is to scale down this design to 865–965 MHz possibly including the upper DVB band for cellular phone data communication. The modeling and simulation were performed using initially NEC1.1 and then with the full-wave TLM technique (Microstripes 6.5 CAD tool) and the results are shown below.

It is found (Fig. 24(a)) that the bandwidth for AR < 2dB of the dual-rhombic loop antenna with parasitic loops is about 46%. For comparison, the simulated AR for the dual-rhombic loop antenna without the parasitic loops is also plotted in this figure, which shows a 2-dB AR bandwidth of only 15%. Obviously, the parasitic loops play an important role for the bandwidth enhancement. The reason for this

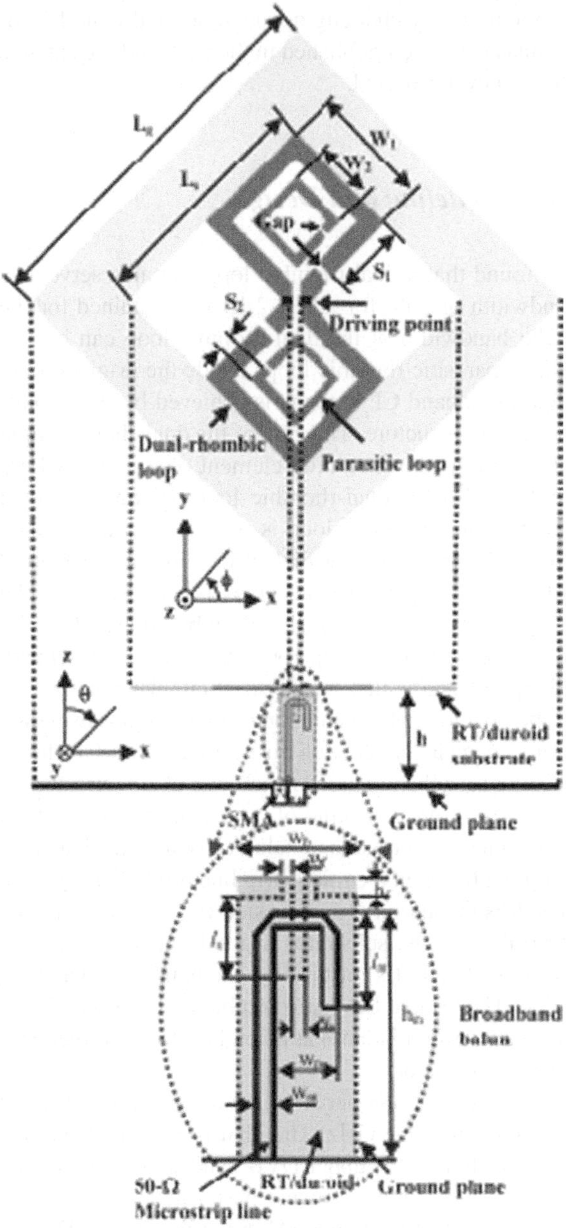

Fig. 23 Dual-rhombic loop CP antenna

effect is that the dual-rhombic loop can only create one minimum AR point while the parasitic loops can produce an additional minimum AR point. An appropriate combination of the two minimum AR points leads to a significant enhancement for the CP bandwidth. The gain of the antenna maintains a value around 8 dBi over the

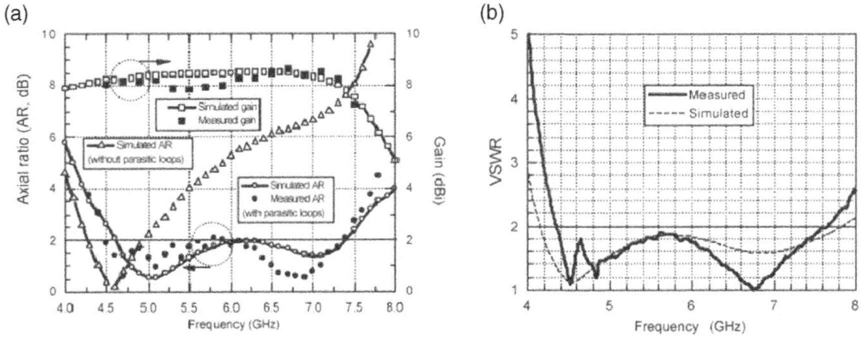

Fig. 24 ((**a**)-*left*) AR, ((**b**)-*right*) VSWR of the dual-rhombic antenna

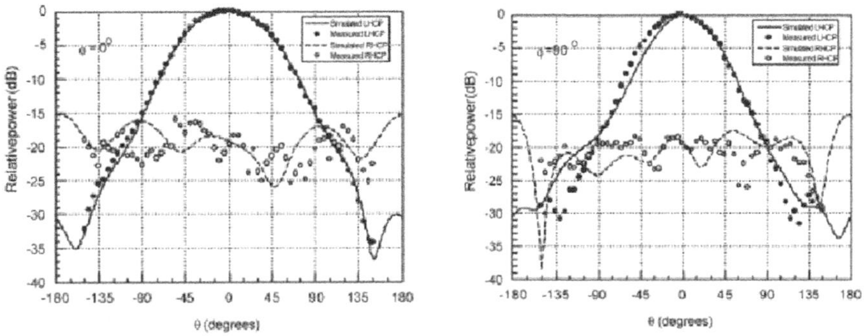

Fig. 25 Radiation patterns at 6 GHz for two principal planes (LHCP: Left-hand-circular-polarization, RHCP: Right-hand-circular-polarization)

2-dB AR bandwidth. The simulated and measured results for the voltage standing-wave ratio (VSWR) are presented in Fig. 24(b). The bandwidth for it about 50%, entirely covering the bandwidth. The radiation patterns (Fig. 25) at 6 GHz at $= 0^\circ$ and 90° verify the effective performance of this antenna.

Numerical Modeling of Car Antennas

Stefan Lindenmeier and Joachim Brose

1 Introduction

Modern cars are usually equipped with a quite impressive number of wireless services like audio and TV broadcasting, cellular telephone, navigation systems, traffic information or remote control as seen exemplarily in Fig. 1. While in former times a single rod antenna was completely sufficient, nowadays the demand for additional services like radar, short range communication and WiFi e.g. enforces a complete new philosophy for the integration of this extreme number of different antennas into the car body. Car manufacturers and designers will never accept many antenna locations spread over the car. So, the only solution to overcome this is to combine as many antennas as possible into a few 'multifunctional antennas'. Examples are multifunction window antennas or miniaturized multifunction roof antennas as well as hidden multifunctional structures underneath plastic parts of the car's skin. The vicinity of different antennas where even some of them are transmitting while others are receiving and the integration into the complicated structured car body demands for an accurate simulation of the near field and far field coupling of the antennas and the resulting pattern changes. Furthermore with an increasing portion of electronic steering and controlling equipment the coupling between the antennas and these units should be investigated for defining optimum positions for the antennas, the control units and the cabling.

In a mobile environment EM reception is characterized by interference and shadowing. Modern receivers can cope with such narrow localized fading effects using antenna diversity. So beside field and coupling calculation also the improvements achievable with antenna diversity in a 'real' mobile scenario should be simulated in time domain in order to obtain relevant statistics of the overall behaviour of diversity antennas in real traffic scenarios.

Stefan Lindenmeier
Institute for High Frequency Technology, Faculty for Electronics and Information Theory Bundeswehr University, Munich, 85577 Neubiberg, Germany

Joachim Brose
Institute for High Frequency Technology, Faculty for Electronics and Information Theory Bundeswehr University, Munich, 85577 Neubiberg, Germany

P. Russer, U. Siart (eds.), *Time Domain Methods in Electrodynamics*,

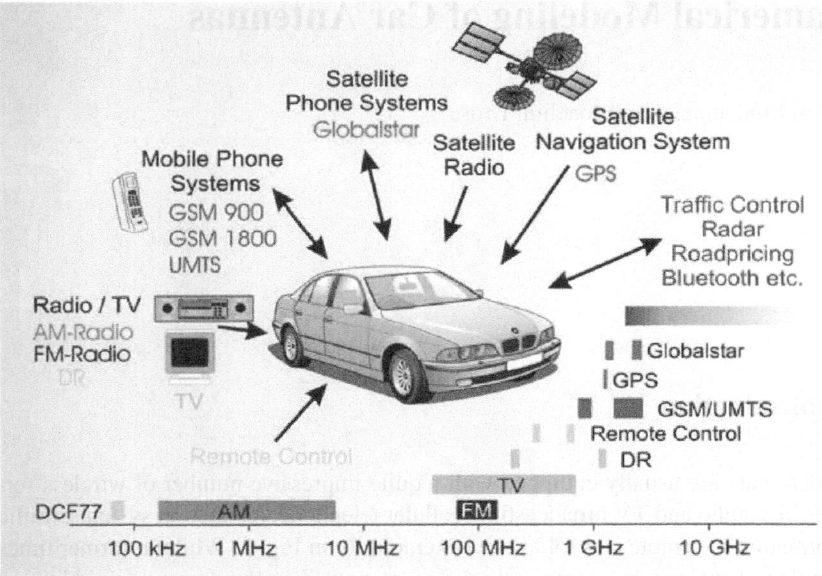

Fig. 1 Required antennas with an actual car

2 Widely Used Methods of Modeling

In the following common methods for antenna modeling are introduced and compared with respect to their capability in different fields of applications.

2.1 Volume Discretization

Methods with volume discretization are widely used in electromagnetics. For car applications the whole volume of the car and its surrounding as far as reactive EM fields are to be expected must be discretized as volume cells as e.g. cubes (see Fig. 2) which have to be small compared to the wavelength. Along that 'nearfield border' absorbing elements [2, 29] have to be introduced. For each volume cell Maxwell's laws have to be fulfilled with to the differential quotients replaced by the ratio of the corresponding differences. The main advantage of these methods is that it solves the fields easily for compound dielectric or magnetic materials and that as a consequence of the step-by-step technique no matrix decomposition is needed. Main disadvantage is the need to calculate the complete discretized air-filled volume of car interior and the surrounding as well. The solvers use preferably the time domain (Finite Differences in the Time Domain, FDTD [12, 34, 35, 36] and Transmission Line Matrix (TLM-) Method [6, 10, 30]) but also the frequency domain (Finite Element Method, FEM).

The number of volume cells can be reduced using a non-orthogonal [4, 13, 27] mesh or local mesh refinement [11, 37]. Examples for antenna simulations are [18, 19, 31, 32].

Fig. 2 Principle of volume discretization used for Finite Differences, Transmission Line or Finite Elements solvers

2.2 Surface Discretization

Mobiles are mainly built of thin sheet metal. So, solving Maxwell's equations along the metal car surface offers the great advantage that only the surface data have to be entered into the field solver (see example in Fig. 3). This data are usually available from the car manufacturer as e.g. NASTRAN files. For each discretized surface element the currents induced on the structure by sources or incident fields are calculated with the numerical solution of integral equations derived from Maxwell's law. A special sort of surface descretising methods are wiregrid methods, where structures are described by a mesh of wire segments.

Usually not only ideally conducting surfaces but also lossy materials, lumped elements or dielectric coated elements can be considered. With this so-called Method of Moments (MoM, [5, 6, 28]) we simulate window antennas or even antennas on carbon fibre bodies e.g. [3].

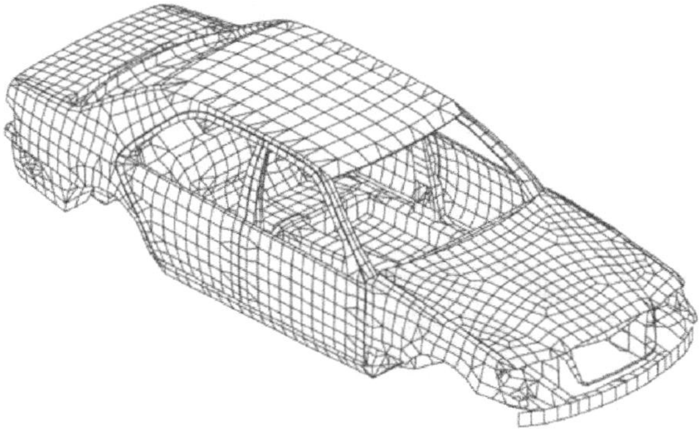

Fig. 3 Principle of surface discretization used for the Method of Moments

The disadvantage of the MoM is the need of solving a matrix equation the size of which increases dramatically when car antennas for frequencies above 1 GHz have to be calculated.

The Method of Moments solves Maxwell's laws in the frequency domain. Examples for MoM antenna calculations can be found in [7, 8, 9, 14, 16, 22, 23]

2.3 Quasi-Optical Methods

As for frequencies above 1 GHz an extremely high number of subdivided volume or surface elements is necessary, quasi-optical methods might offer faster solutions for antenna problems. The conditions to use the geometrical optics (GO), physical optics (PO) or unified theory of diffraction (UTD) approach is that the concerned elements are large in terms of wavelength and that in general no radiating sources are mounted directly onto these elements. Further on only a limited choice of canonical element shapes, like plates, cylinders, ellipsoids or cones is available.

Figure 4 shows a sketch for a typical optical ray scenario with a single ray which is reflected and scattered.

Under unfavorable conditions the ray tracing might lead to extreme high numbers of re-reflections. So, these quasi-optical approaches usually set limits for the number of pursued reflections in order to reduce the simulation effort.

An example of the multiple effects with optical methods in a mobile situation is shown in Fig. 5, where a GSM on-glass antenna at the upper part at a rear window of a car is investigated taking into account the direct wave (line-of-sight propagation), the scattered wave, the creeping wave (refraction), the reflection and the resulting interference caused by this multi path propagation. With these effects the antenna radiation pattern is not omni-directional in the equatorial plane as desired but radiates preferably to the rear side (Fig. 6).

Which one of the described solvers should be preferably used with a given problem is not only based on the required data input but also on the processing time. The frequency reliance of the discretization leads to a frequency dependency in the required CPU time t_{CPU}. With FE (volume discretization $n \sim f^3$ and solver $t_{\text{CPU}} \sim n^2$)

Fig. 4 Typical Ray-Tracing with incident EM-wave and car

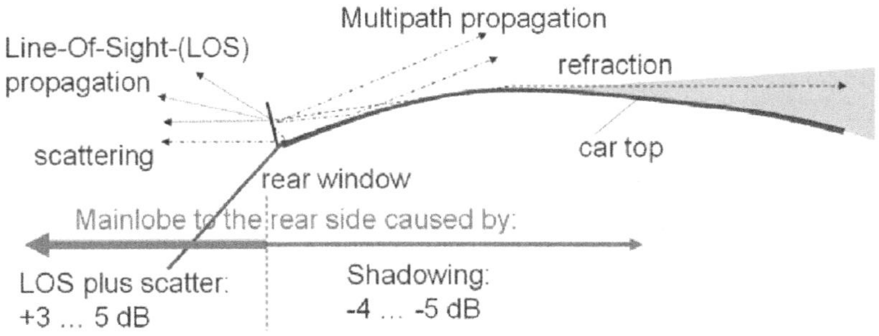

Fig. 5 Using quasi-optical methods with a GSM on-glass antenna

being n the number of elements and MoM (surface discretization $n \sim f^2$ and decomposition $t_{\text{CPU}} \sim n^3$) t_{CPU} grows roughly with f^5 to f^6 while the time which ray tracing needs depends on the number of elements n and number of regarded reflections k as $t_{\text{CPU}} \sim n^{k+1}$.

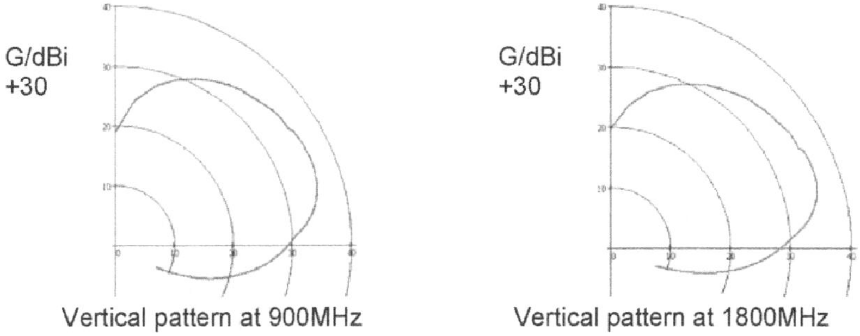

Fig. 6 First approach estimation of losses from defraction along a car roof using the UTD

2.4 Hybrid Methods

Since many years intensive efforts have been undertaken towards the combining of field solvers [1, 15, 17, 20, 21, 31, 32, 33] using an exact physical-mathematical interface between the different formulations. Actual hybrid solvers now include several or all of the mentioned methods described before (MoM, FE and ray tracing). Figure 7 shows the application of a hybrid solver with a rear window antenna equipped car. For dielectric elements volume discretization is the best suited method, for edges, corners and elements close to the antennas MoM fits best and for smooth parts in a sufficient distance from the antennas ray tracing delivers acceptable results.

We often have to deal with electromagnetic coupling among antennas or car parts of arbitrary shape which are separated by wide free space regions. While the field simulation of very general structures exhibiting complex geometry and lossy

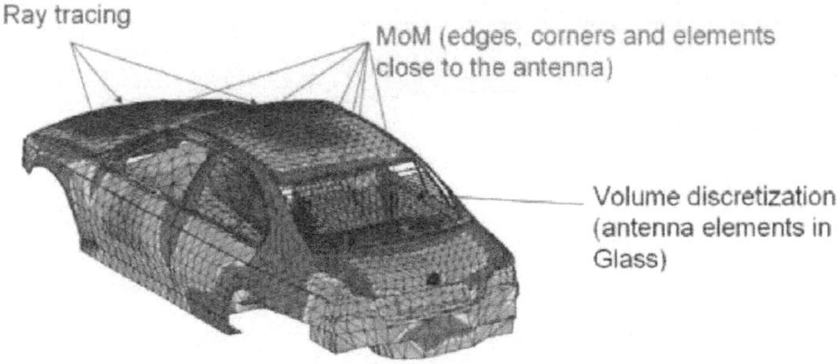

Fig. 7 Hybrid solver with FM rear window antenna and car

materials may be performed by volume discretizing methods the numerical modeling of wide free space regions requires high computational effort. Surface discretizing Integral Equation (IE) methods are very efficient for the electromagnetic field analysis of radiating structures which are embedded in free space. However, in the case of complex dielectric structures and with consideration of metallization thickness, these methods require a high effort for analytical processing. In order to combine the advantages of both methods and to overcome their limits the Transmission Line Matrix Integral Equation (TLM-IE) method has been introduced in [17, 19, 20, 21, 31]. In the TLM-IE Method the volume discretizing TLM Method for the numerical modeling of the near field is combined with the integral equation method for the description of the electromagnetic field propagation in wide free space regions. In the TLM-IE method each of the interacting objects is embedded into a closed spatial subdomain with a boundary surface S_B. The material containing regions are enclosed by further equivalent source surfaces S_S inside the discretized subdomains where the electromagnetic field is sampled in each time step. In Fig. 8, three interacting objects (gray) are shown with these surfaces. Inside the boundaries S_B a volume discretizing method like the TLM method is applied in time domain. Following the surface equivalence theorem of Schelkunoff-Huygens, we use free space Green's functions G in time domain for calculation of the field on the S_B surfaces on base of the field on the S_S surface which represents an equivalent source of radiation along these boundaries at any time step. That way, radiating open boundary conditions are achieved at the boundaries S_B of the discretized regions. In comparison to pure volume discretizing methods the computational effort of the hybrid methods is reduced by more than one order in magnitude.

Fig. 8 Hybrid TLM-IE scheme and EM coupling among discretized subdomains

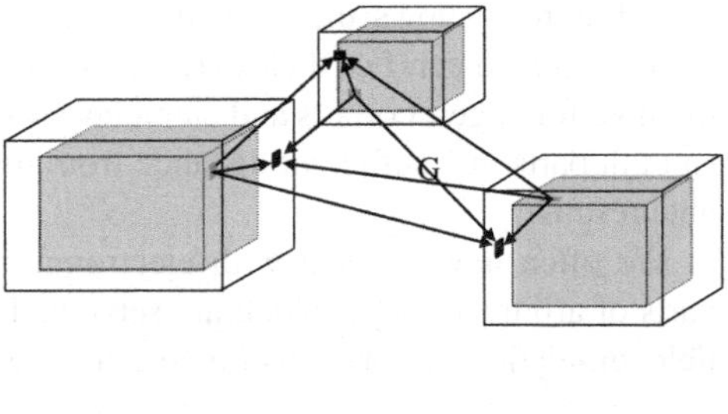

3 Comparison of Calculation and Measurement

Usually the quality of calculation is judged by comparison of calculation results with appropriate measurements. With antennas the differences are mostly graded as imperfection of the theoretical approach. From our experience over many years this estimate is true in most cases – but not generally. So, judging a solver by comparison with measurements has to be done very carefully. The well known limitations of the field solvers are of course at first the error-free implementation of the physical laws into a numerical representation. Especially for the decomposition of the matrix a high numerical precision is necessary. From the change from continuous to stepped distributions the solver results are only valid for defined points.

Furthermore solvers show incomplete separation of currents on both sides of thin metal sheets which are in reality caused by the skin effect. Last not least, the stability of the solution might be low. Generally these limitations are well known if wide spread solvers are used and systematic errors shouldn't occur any more. So, the vehicle and the antenna data input remains the crucial uncertainty.

As examples for comparisons of calculation and measurement. Figure 9 shows the results of an innovative TLM solver applied to a microwave antenna array for an automotive sensor [32] while Fig. 10 gives the comparison of radiation patterns achievable with GSM mobile antennas like the depicted 'low profile antenna' and a digitized car surface.

A further example concerning the simulation of a resistive layer embedded in dielectrics is given in Fig. 11. It demonstrates that screen antennas can be reliably simulated even with a MoM solver. The resistive layer was simulated with a grid of resistive wires and the lossy glass with a lossy dielectric coating. Diameters, loss factor and conductivity where matched to result in the very same frequency dependent impedance like the real window antenna.

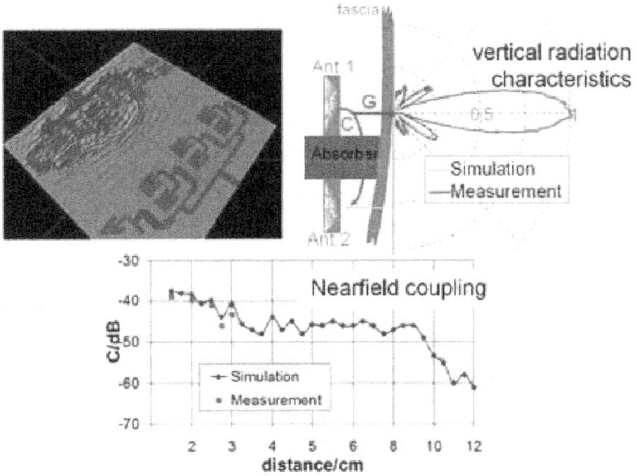

Fig. 9 Comparison of simulation and measurement with the sensors

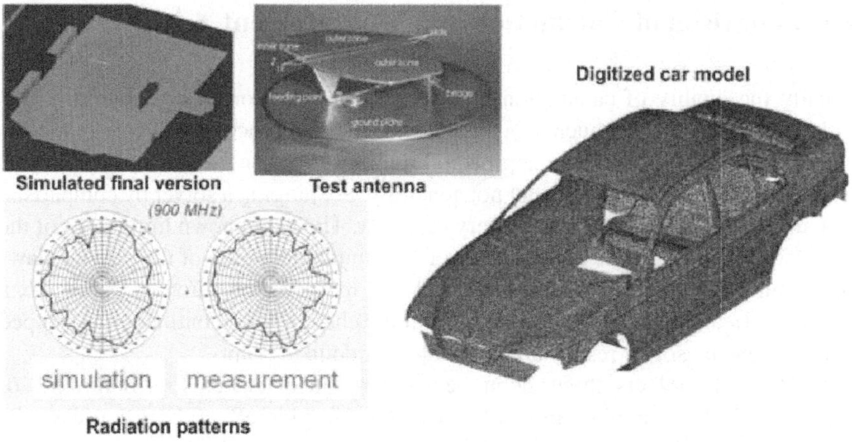

Fig. 10 GSM antenna, digitized car model with simulated and measured radiation patterns

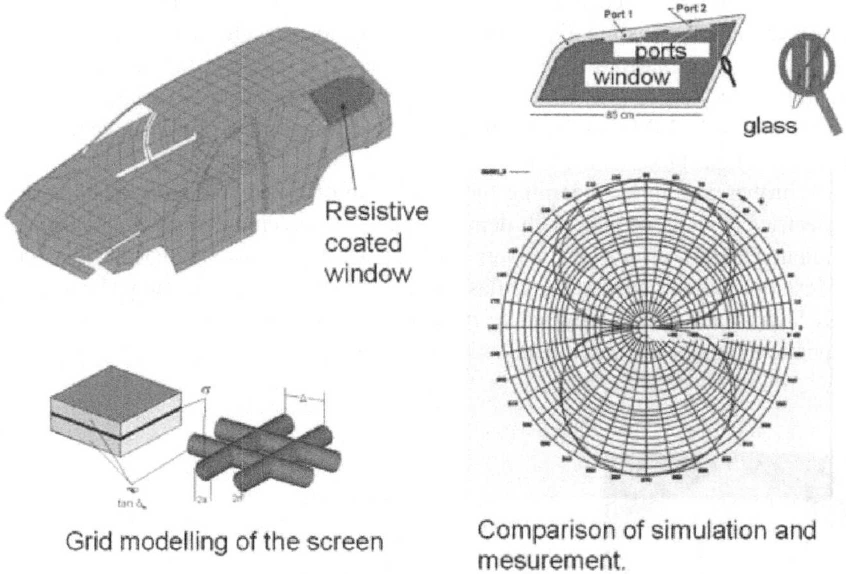

Fig. 11 MoM simulation of a window antenna with a resistive layer

As shown in these examples antenna simulations in the automobile field offer quite acceptable results. The main reasons for greater deviations are from our experience mainly caused by inaccurate, faulty or incomplete data input. Typical errors include inaccurate geometry data, wrong intersections, double input of elements, unknown rf data especially from glass, plastic, weather-strips, spot welds or materials close to the antennas or the unknown conductivity and inductance of hinges and ground straps.

4 Numerical Simulation of Antenna Diversity in Time Domain

The measurement and the assessment of the antenna behavior in a space- depending disturbed EM field is extremely time consuming. Therefore the simulation of an antenna equipped car moving through such fields offers great advantages for rapid system evaluation.

4.1 Simulation of Rice or Rayleigh Distributed Fields

The basic multi path situation [26] is sketched in Fig. 12. Assume a vehicle driving in x-direction and receiving the desired signal plus a co-channel interferer plus noise. Signal and interferer are scattered, reflected or diffracted from a number of close or far away obstacles and arrive at different times at the vehicle's antennas.

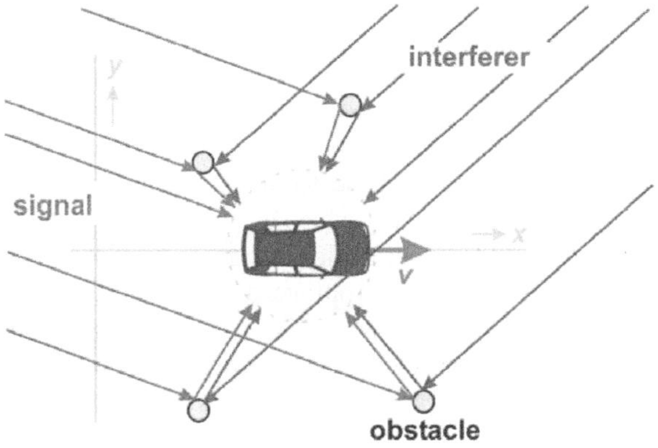

Fig. 12 Simulation set-up with multiple scattering objects

The superposition of all of them gives the resulting field strength at each locus. The complex output voltage at each car antenna depends on the directions of the waves, on their phase angle and on the phase shifts from the antenna. The obstacles are spread several times statistically and represent each time a certain receiving situation.

4.2 Theory of Diversity Operation

Because of the interference from the described multi path propagation the field strength of the EM waves shows frequency depending very deep narrow localized fading and broad flat maximum values where the distance between minimum and maximum is close to a quarter wave length (see Fig. 13).

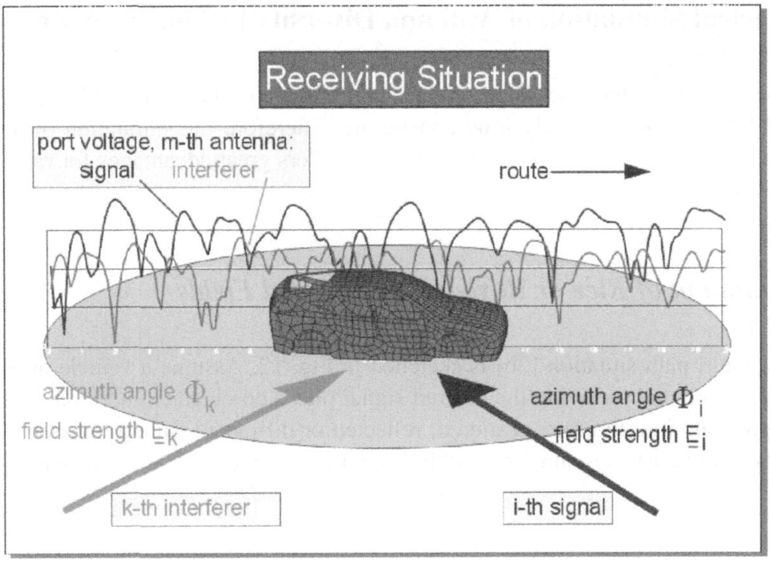

Fig. 13 Signal and interferer voltage during a short part of the test drive

If the receiving antenna is at a certain moment located at such a minimum the reception is probably disturbed. A second antenna – which even can be located relatively close to the first one – might be able to get sufficient signal strength. An intelligent receiver choosing in every situation the appropriate antenna could increase the signal quality. Beside different diversity approaches like space, frequency and time diversity there now exists also a number of different strategies for optimization of the achievable improvement and hardware expense like switching diversity, maximum ratio diversity and more. Figure 14 shows a combined 3-antenna system with adaptive pattern tracking (maximum gain) plus switching diversity. As long as the antennas are 'independent' enough or 'de-correlated' the processor in such an intelligent receiver could choose the antenna with the actually highest SNR or steer the overall antenna pattern by phase adjustment. In the case of a mobile phone system with transmit and receive frequencies not spaced too far away from each other and with constant driving direction the adaptive pattern control offers an additional signal improvement while transmitting.

Diversity systems are usable for digital systems [24, 25] as well if the processor is synchronized with the serial data frames. Figure 15 shows exemplarily a short part of a test drive with two antennas. Whenever the signal drops below the noise floor bit errors occur. The ratio of faulty bits over transmitted bits is the bit error rate (BER). Usually antennas are used, the BER of which is – more or less – similar.

If the diversity processor chooses always the better one of these two antennas the number of remaining bit errors will decrease considerably. Our investigations proved that with n independent antennas each showing the same $BER_{\text{single antenna}}$ the diversity bit error rate $BER_{\text{diversity}}$ will be:

$$BER_{\text{diversity}} \approx (BER_{\text{single antenna}})^n$$

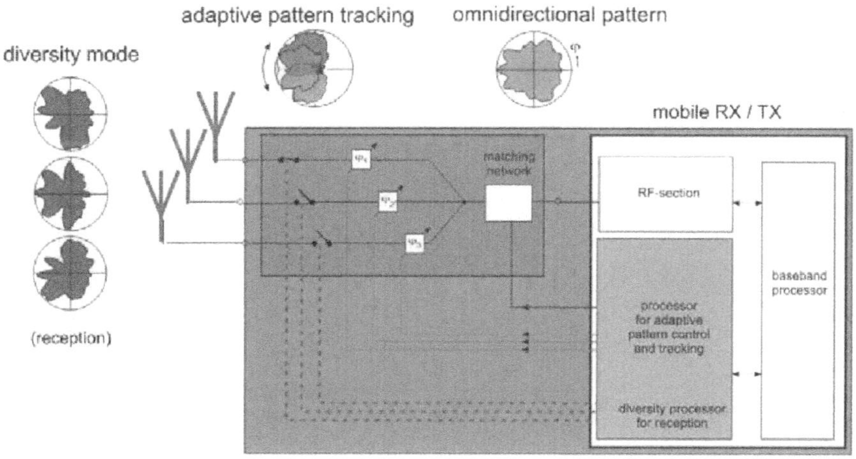

Fig. 14 Diversity and adaptive pattern tracking system

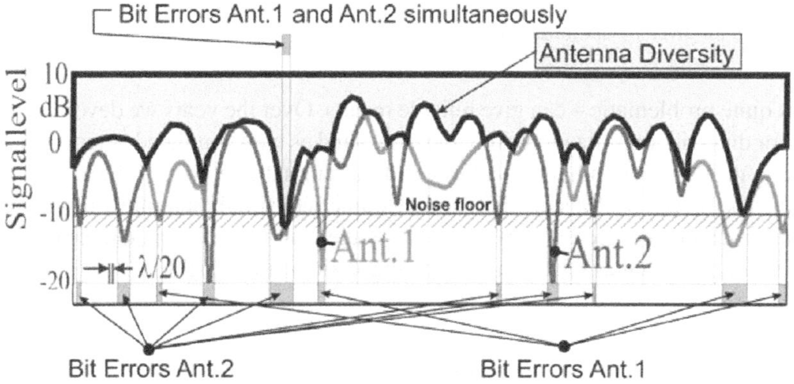

Fig. 15 Measured signal levels in a 2-antenna System

Just to give an impression of the (measured) extent of this dependency, imagine a 4-antenna diversity system in a scenario delivering a BER of 0.14. If the antennas are partially correlated here leading to merely 3.2 independent antennas the resulting BER is only 0.0017. That means that this 4-antenna diversity system's BER is about two orders in magnitude smaller than that of a single antenna system!

4.3 Mixed Diversity Optimization

As mentioned, the optimization of the diversity efficiency of automobile antennas is extremely time-consuming as only a sufficient number of test drives with the actual antenna configuration through a representative environment – the definition of

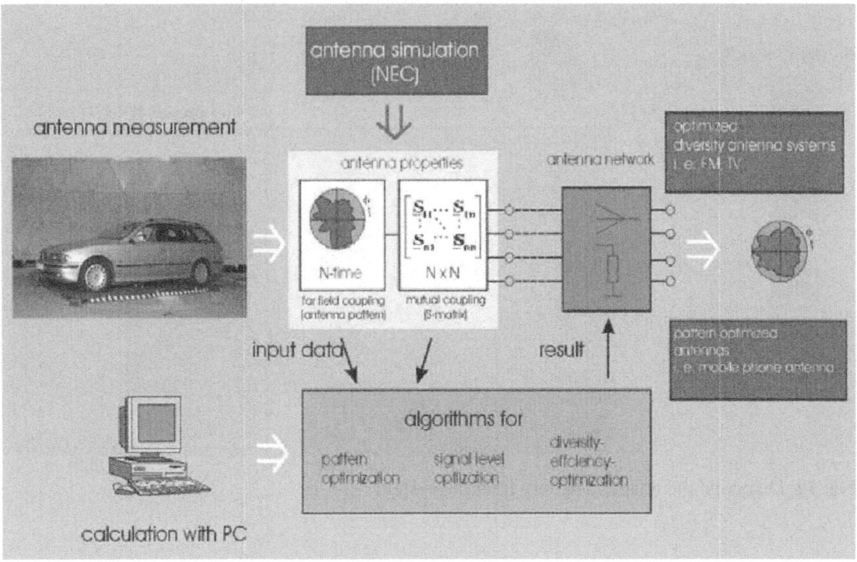

Fig. 16 Set-up for diversity optimization

which is quite problematic – can give reliable results. Over the years we developed a system for diversity optimization using measured and as well simulated antenna data containing the complex radiation patterns and the mutual coupling of the antennas (Fig. 16). The diversity efficiency is found by numerical simulation of the test drive where miles can be covered within seconds. As the results are obtainable nearly at once, determined changes of the antenna configurations are possible.

4.4 Drive Simulation in the Time Domain

With several independent antennas at the vehicle, the signal and the interferer output voltages are usually different at each antenna port.

The separated voltages at two antennas during a short part of a simulated test drive are shown in Fig. 17. Only some times the signal voltage at both antennas drops below the interference voltage at the same time even for a very unfavorable signal over interferer ratio as assumed in the simulation for Fig. 17. But very often only one of the two antenna signals is disturbed. The simulated diversity efficiency was compared extensively with the results of real test drives in order to figure out the most suitable parameters for the simulated obstacles, the signal over interferer ratio S/I, the drive length or the number of different scenarios. It shows for example, that the 'number of independent antennas' is not changed by the S/I ratio or that the simulation (and the measurement) become inaccurate if the diversity efficiency is very high (as drop-outs then become very seldom). A few results of this investigation are shown in Fig. 18.

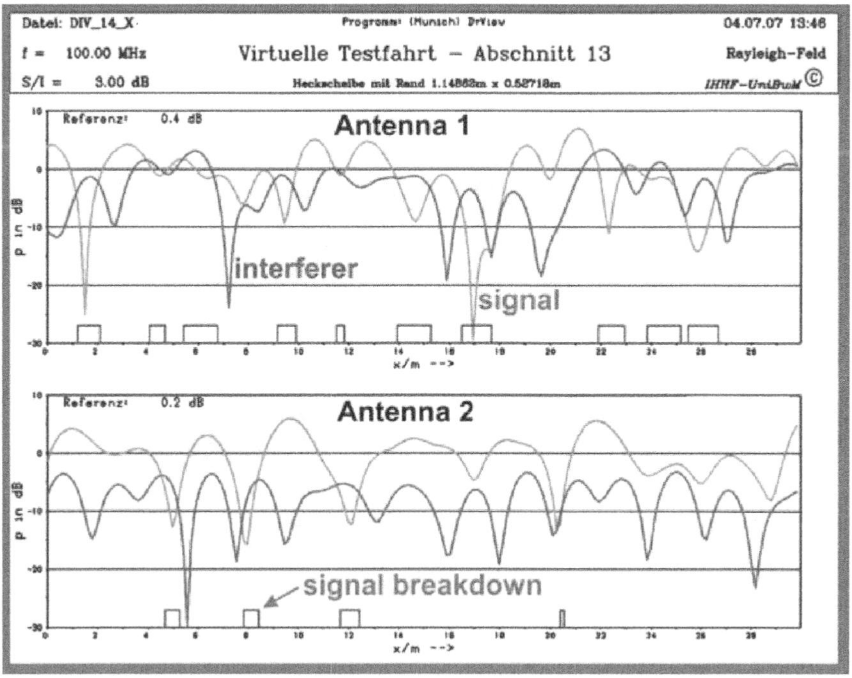

Fig. 17 Output voltages of 2 antennas

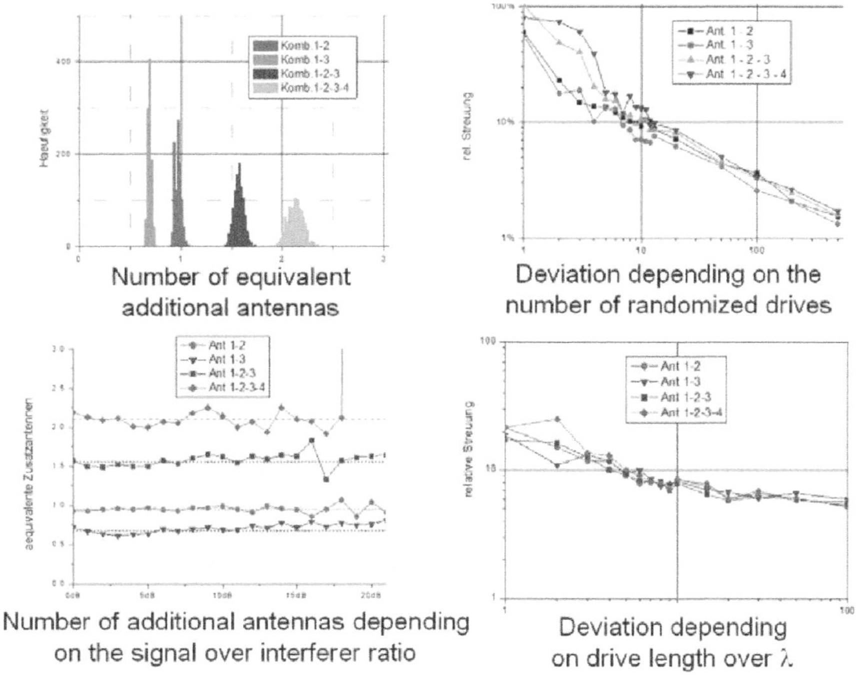

Fig. 18 Statistical testing of the 'virtual drive' characteristics

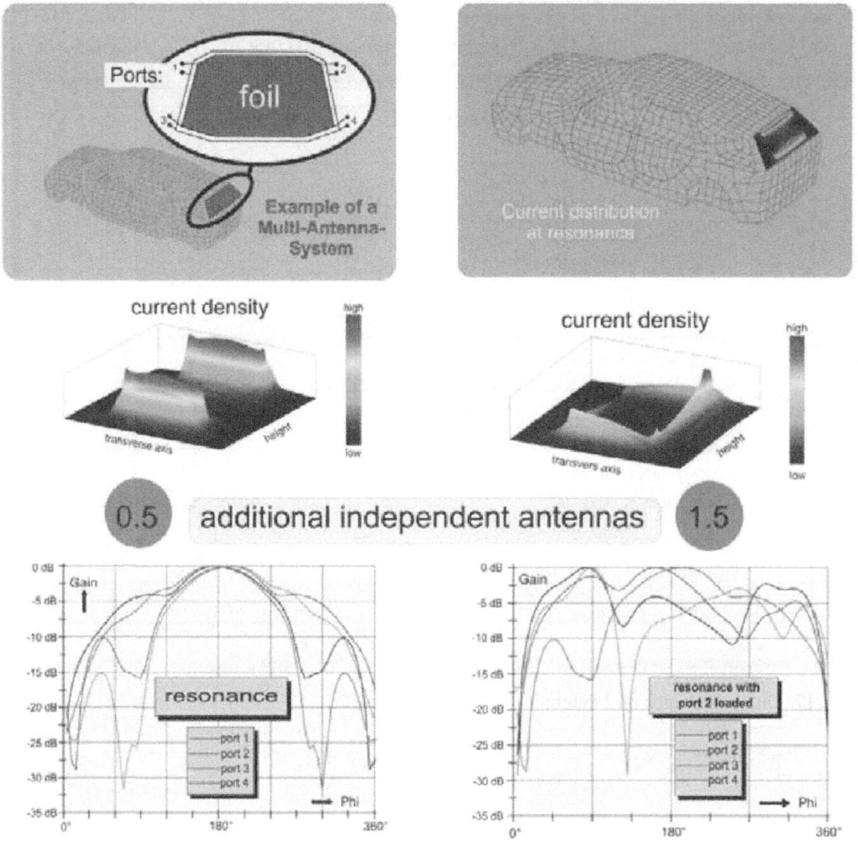

Fig. 19 Multiport antenna close to resonance and diversity enhancement by appropriate reactive port loading

One of the most interesting advantages in the field of antenna simulation with vehicles is the visualization of currents and fields from which quite often strange antenna behavior becomes comprehensive. From very sensitive neighboring antennas for example it is known that the diversity efficiency is quite poor. This behavior was tested with a rear window antenna as shown in Fig. 19. Again a resistive layer was used as antenna with a strong coupling between the antenna ports. At resonance the induced currents at the upper and lower edge are strongly dominating, resulting in similar radiation patterns at all ports with an extremely poor number of only 0.5 additional antennas. Because of the distinct port coupling simple reactive port loadings at the other ports can change the patterns now leading to one more additional independent antenna without changing any antenna geometry or location.

5 Conclusion

Meanwhile numerous well suited field solvers for simulations of antennas on cars are available. The methods used are volume discretising methods, the method of moments and ray tracing. Very few programs combine these methods into a hybrid solver giving the user the opportunity to choose the most qualified solver for every portion in mixed problems. Especially with wide-spread solvers only few systematic errors keep undiscovered and therefore their reliability is in most cases more than acceptable for an estimation of the complex field scenario around antenna and car. The main problem with car antennas and therefore the main error source is from our experience the provision of the correct RF-relevant data of the car and the antenna elements. Usually a greater part of these data is not even available and this makes the data input act as the crucial bottleneck for the correct field solution. Beside this, differences between calculated and measured data may also be caused by the finite EM purity of the antenna test range especially if the antennas or the car show resonances or if the antennas have poor cross polarization attenuation (vertical whip antenna with horizontal signal polarization e.g.). The main advantages of this numerical modeling are the multiple results which lead to a better understanding of the antenna behavior, the possibility to simulate cars which do not yet exist and the massive time save with the optimization of antenna diversity during virtual test drives in time domain.

References

1. Altintas A, Lindenmeier S, Russer P (1999)Time-Domain Equivalent Edge Currents. Progress in Applied Computational Electromagnetics, ACES Symposium Digest, pp 622–627
2. Berenger J P (1994) A Perfectly Matched Layer for the Absorption of Electromagnetic Waves. J Comp Phys, vol 114, no 2, pp 185–200
3. Brose J, Lindenmeier H (2003) Optimized Reactive Port-Loads of a Multi-Antenna-System for Enhanced Diversity-Effectiveness. International ITG-Conference on Antennas, Berlin
4. Hao Y, Railton C J (1998) Analyzing Electromagnetic Structures with Curved Boundaries on Cartesian FDTD Meshes. IEEE Transactions MTT, vol 46, no 1, pp 82–88
5. Harrington R F (1993) Field Computation by Moment Methods. IEEE Press Series on EM Waves
6. Hoefer W R (1989) The Transmission Line Matrix (TLM) Method. Techniques for Microwave and Millimeter Wave Passive Structures. ed Itoh T, J Wiley, New York
7. Hopf J, Lindenmeier H, Reiter L (2003) Mobile Multimedia Antenna Systems for Station Wagons and the Achievable Diversity Effectiveness Obtained by Analysis of Virtual Test Drives. Radio Science, vol 38, no 2
8. Hopf J, Reiter L, Lindenmeier S (2007) Compact Multi-Antenna System for Cars with Electrically Invisible Phone Antennas for SDARS Frequencies. International ITG-Conference on Antennas
9. Hopf J, Lindenmeier S, Reiter L (2006) Crossed Frame Antenna for Automotive Application for the New US Satellite Digital Audio Radio Service (SDARS). IEEE-MTTS 36th European Microwave Conference, Manchester, UK
10. Johns P B (1987) A Symmetrical Condensed Node for the TLM Method. IEEE Transactions MTT, vol 35, pp 370–377

11. Kim I S, Hoefer W J R (1990) A Local Mesh Refinement Algorithm for the Time Domain Finite Difference Method Using Maxwell's Curl Equations. IEEE Transactions MTT, vol 38, pp 812–815
12. Kim I S, Hoefer W J R (1990) The Dispersion Characteristics and the Stability Factor for the TD-FD Method. Electronic Letters, vol 26, no 7, pp 485–487
13. Lee J-F, Palandech R, Mittra R (1992) Modelling Threedimensional Discontinuis in Waveguides Using Non-orthogonal FDTD Algorithm. IEEE Transactions MTT, vol 40, pp 346–352
14. Lindenmeier S, Reiter L (2005) Multiband antenna systems for terrestrial and satellite radio services. IEEE-MTTS International Microwave Symposium, Long Beach, USA
15. Lindenmeier S, Russer P (1997) Design of Planar Circuit Structures with an efficient Magneto-Static Field Solver. IEEE MTT-S Digest, vol 3, pp 1807–1810
16. Lindenmeier S, Russer P (2000) Automatic Optimization of High Gain Antenna Arrays. VDE World Microtechnologies Congress, Proceedings, vol 2, pp 121–124
17. Lindenmeier S, Russer P (2001) The ARB Method. International Journal of Numerical Modeling, Special Issue Electronic Networks, Devices and Fields, vol 14, pp 115–128
18. Lindenmeier S, Bader B, Russer P (1997) CAD-design of viahole fed planar antennas using space discretizing field solvers. Europ. Microwave Symposium Digest, vol 2, pp 848–853
19. Lindenmeier S, Bader B, Russer P (1997) Investigation of Various H-Shaped Antennas with an ATLM Field-Solver. IEEE MTT-S Digest, vol 3, pp 1365–1368
20. Lindenmeier S, Pierantoni L, Russer P (1999) Numerical Modeling of Transient Radiated Interferences in Time Domain by the Hybrid ARB Method. International Journal of Numerical Modeling: Electronic Networks, Devices and Fields, pp 295–310
21. Lindenmeier S, Christopoulos C, Russer P (2000) Methods for the Modeling of Thin Wire Structures with the TLM Method. IEEE-MTT-S International Microwave Symposium, Boston, USA
22. Lindenmeier S, Hopf J, Reiter L (2005) Integrated Microwave Antenna Systems in Mobile Applications. XXVIIIth General Assembly of International Union of Radio Science, New Delhi, India
23. Lindenmeier S, Hopf J, Reiter L, Brose J (2005) Multiband antenna systems for terrestrial and satellite radio services. IEEE-MTT-S International Microwave Symposium Workshop, Long Beach
24. Lindenmeier S, Reiter L, Barié D, Hopf J (2007) Antenna Diversity for Improving the BER in Mobile Digital Radio Reception Especially in Areas with Dense Foliage. International ITG-Conference on Antennas, Munich
25. Lindenmeier S, Reiter L, Hopf J, Barié D, (2007) Antenna Diversity for the Improvement of Satellite Radio Reception in Fading Scenario. Invited paper and inaugural lecture, Conference on Wave Propagation in Communication, Microwave Systems and Navigation, Chemnitz
26. Meier C, Terzis A, Lindenmeier S (2007) Investigation and Suppression of Multipath Influence on Indoor Radio Location in the Millimetre Wave Range. Conference on Wave Propagation in Communication, Microwave Systems and Navigation, Chemnitz
27. Mezzanotte P, Roselli L, Sorrentino R (1995) A Simple Way to Model Curved Metal Boundaries in FDTD Algorithm Avoiding Staircase Approximation. IEEE Microwave and Guided Waves Letters, Band 5, no 8, pp 267–270
28. Morita N, Kumagai N, Mautz J R (1990) IE Methods for Electromagnetics. Artech House, Boston, London
29. Mur G (1981) Absorbing Boundary Conditions for the Finite Difference Approximation of the Time Domain Electromagnetic Field Equations Meshes. IEEE Transactions EMC, vol 23, pp 377–382
30. Nielsen J S, Hoefer W J R (1993) Generalized Dispersion Analysis and Spurious Modes of 2-D and 3-D TLM Formulations. IEEE Transactions MTT, vol 11, pp 1375–1384
31. Pierantoni L, Lindenmeier S, Russer P (1999) Efficient Analysis and Modeling of the Radiation of Microstrip Lines and Patch Antennas by the TLM-Integral Equation (TLM-IE) Method. International Journal of Numerical Modeling: Electronic Networks, Devices and Fields, pp 329–340

32. Pierantoni L, Lindenmeier S, Rozzi T (2003) Efficient Modeling of the Near Field Coupling between Phased Array Antennas. European Microwave Conference, Munich
33. Pierantoni L, Lindenmeier S, Massaro A, Rozzi T (2004) Analysis of the EM coupling between satellite antennas. Europ Microwave Conference, Amsterdam, Netherlands
34. Pogatzki P, Kulke R, Sporkmann T, Köther D, Tempel R, Wolff I (1994) A Comprehensive Evaluation of Quasi-static 3D-FD Calculations for more than 14 CPW Structures – Lines, Discontinuities and Lumped Elements. IEEE MTT-S Digest, pp 1289–1292
35. Wolff I (1992) Finite Difference Time Domain Simulation of Electromagnetic Fields and Microwave Circuits. International Journal Num Model, vol 5, pp 163–182
36. Yee K S (1966) Numerical Solution of Initial Boundary Value Problems Involving Maxwell's Equations in Isotropic Media. IEEE Transactions AP, vol 14, pp 302–307
37. Zivanovic S S, Yee K S, Mei K K (1991) A Subgridding Method for the Time Domain Finite Difference Method to Solve Maxwell's Equations. IEEE Transactions MTT, vol 39, no 3, pp 471–479

Time-Domain Modelling of Group-Delay and Amplitude Characteristics in Ultra-Wideband Printed-Circuit Antennas

Hung-Jui Lam, Yinying Lu, Huilian Du, Poman P.M. So and Jens Bornemann

1 Introduction

With the release of the 3.1–10.6 GHz band for ultra-wideband (UWB) operation, a variety of typical UWB applications evolved; examples are indoor/outdoor communication systems, ground-penetrating and vehicular radars, wall and through-wall imaging, medical imaging and surveillance, e.g. [1, 2]. Many future systems will utilize handheld devices for such short-range and high bandwidth applications. Therefore, the realization of UWB antennas in printed-circuit technologies within relatively small substrate areas is of primary importance. And a number of such antennas with either microstrip, e.g. [3, 4, 5, 6, 7, 8, 9, 10] or coplanar waveguide feeds, e.g. [11, 12, 13, 14, 15, 16, 17, 18, 19, 20, 21, 22, 23], and in combined technologies, e.g. [24, 25], have been presented recently, mostly for the 3.1–10.6 GHz band, but also for higher frequency ranges, e.g. [26].

Since UWB systems involve the transmission and reception of short pulses, the variations of radiated amplitudes and phases over frequency contribute to the distortion of the pulse. While the amplitude variation is usually indicated by changes in the peak gain or radiation patterns, the frequency-dependent phase variation is often

Hung-Jui Lam
Department of Electrical and Computer Engineering, University of Victoria, P.O. Box 3055 STN CSC, Victoria, B.C., V8W 3P6, Canada

Yinying Lu
Department of Electrical and Computer Engineering, University of Victoria, P.O. Box 3055 STN CSC, Victoria, B.C., V8W 3P6, Canada

Huilian Du
Department of Electrical and Computer Engineering, University of Victoria, P.O. Box 3055 STN CSC, Victoria, B.C., V8W 3P6, Canada

Poman P.M. So
Department of Electrical and Computer Engineering, University of Victoria, P.O. Box 3055 STN CSC, Victoria, B.C., V8W 3P6, Canada, e-mail: Poman.So@ECE.UVic.CA

Jens Bornemann
Department of Electrical and Computer Engineering, University of Victoria, P.O. Box 3055 STN CSC, Victoria, B.C., V8W 3P6, Canada

P. Russer, U. Siart (eds.), *Time Domain Methods in Electrodynamics*,

omitted, and related data is published only sporadically, e.g., [5, 7, 17, 26]. In order to quantify this behavior, one of two methods is usually applied.

First, in the frequency domain, the spherical wave front in the far field is detected for each frequency, from which the apparent phase center along the antenna surface or axis can be calculated. Alternatively, the phase variation in the near field over the main beam is computed for different phase center points moved from a reference point on the surface of the antenna. Then a valid phase center location is detected if the phase variation over the main beam is within a few degrees. These methods are complicated and time-consuming [26].

Secondly, in the time domain, a transient analysis is performed which leads to the group delay. A pulse, whose frequency spectrum covers the bandwidth of the antenna, is generated, applied at the antenna input and its radiated pulse detected. Both pulses are Fourier transformed and their phase response recorded. The group delay is obtained from the derivative of the phase variation with respect to angular frequency [7].

In this paper, the Transmission-Line Matrix (TLM) method in the time domain is utilized to determine the group delay of two printed circuit UWB antennas. The first one is a recently developed, new coplanar-waveguide antenna [27], the second a published microstrip antenna [9, 10] with so far no information about phase variations.

2 Coplanar UWB Antenna

Figure 1 shows the layout and the superimposed coordinate system of the UWB antenna in coplanar technology. It uses an FR4 substrate of 1mm thickness, an area of 30 mm × 40 mm (W × L), a permittivity of $\varepsilon_r = 4.7$ and a loss tangent of

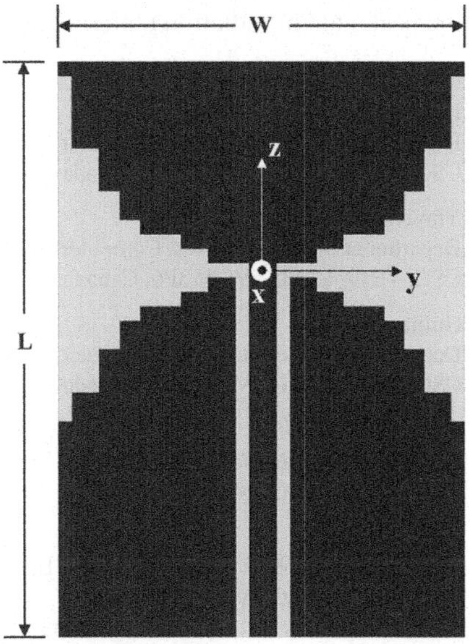

Fig. 1 Layout and coordinate system of UWB antenna in coplanar technology

$\tan\delta = 0.018$. It appears to be a stepped version of a similar antenna presented in [20]. However, there are two fundamental differences. First of all, the antenna in [20] is a slot radiator, which maintains metallic strips at the left and right edges of the substrate. Such metallic strips are missing in Fig. 1 and thus result in a somewhat conical shape of the radiating profile – similar to a tapered slot antenna. Secondly, the stepping is chosen such that the smallest dimension is 0.5 mm. This contributes to low manufacturing sensitivity. However, it also influences the characteristic impedance of the feeding coplanar waveguide, which is significantly higher than the 50 Ω coaxial line to be connected at the input. (Note that the coaxial line is also used to physically connect the two ground planes.) As we will show later, this mismatch is not to the detriment of the antenna performance.

The coplanar UWB antenna was designed using the finite-element software HFSS®. For the evaluation of the group-delay characteristics, the antenna was also analyzed by the TLM time-domain field solver MEFiSTo-3D®. Figure 2 shows a comparison between the input reflection coefficients obtained with both methods. Note that the connection of the input of the antenna to a coaxial cable is included in both methods. Good agreement is observed, thus verifying the antenna's performance at its input terminal. The input return loss as computer by HFSS between 3.1 GHz and 10.6 GHz is better than 9.4 dB.

The peak gain, computed using HFSS at the dots and spline interpolated, is shown in Fig. 3. Its variation versus frequency is comparable to other UWB printed-circuit antennas found in the literature. Note that the direction of the peak gain varies with frequency and, therefore, is not an indication of the amplitude variation in a specific direction.

Such a variation is presented by the normalized radiation pattern. The E-field variation with angle and frequency in the yz-plane (c.f. Fig. 1) is demonstrated in Fig. 4. (For E-plane and H-plane radiation patterns in other planes, the reader is referred to [27].) As we will calculate the amplitude variation using a time-domain technique in the next chapter, it is important to note that in the direction of $\theta = \phi = \pi/2$, the variation versus frequency in Fig. 4 is in the order of 8 to 9 dB.

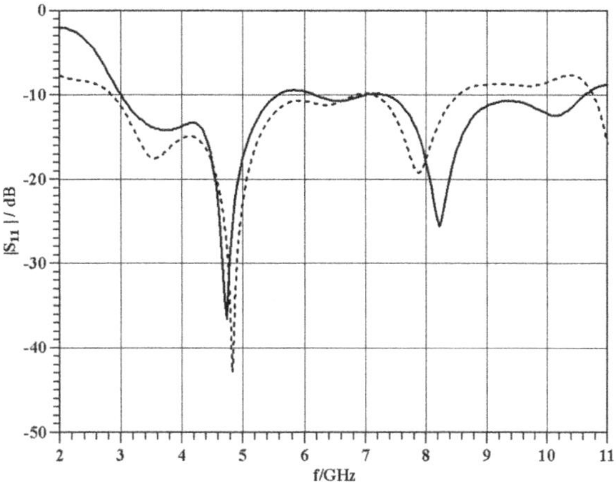

Fig. 2 Comparison of input reflection performance between HFSS (*solid line*) and MEFiSTo-3D (*dashed line*)

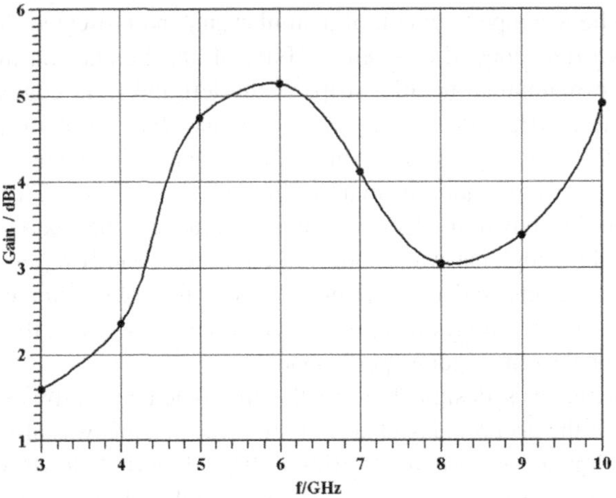

Fig. 3 Peak gain of the UWB antenna in CPW technology computed by HFSS (*dots*) and spline interpolated (*solid line*)

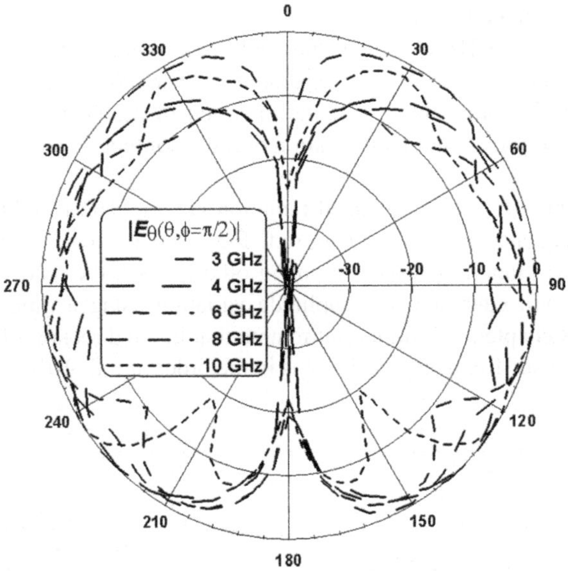

Fig. 4 Normalized E-plane radiation pattern (computed with HFSS) in the yz-plane (c.f. Fig. 1) at various frequencies between 3 GHz and 10 GHz

3 Group Delay and Amplitude Variation

In the first part of this chapter, we will demonstrate the time-domain calculation of the group delay and amplitude variation at the example of the coplanar UWB antenna presented in Chap. 2. The second part applies the same technique to the microstrip antenna presented in [9, 10].

3.1 Coplanar Antenna

Figure 5 shows the setup in MEFiSTo-3D. Since the problem is symmetric with respect to a magnetic wall in the xz-plane (all other walls are absorbing boundaries), only half of the computational space is required. The input of the antenna is excited with a pulse covering the entire frequency spectrum of application. At a point in the far field, probes detect the vertical polarization E_θ and the horizontal polarization E_ϕ. Their orientation with respect to Fig. 1 are depicted in Fig. 6. Note that the coaxial input port and a reference port are included.

Input and detected signals are Fourier transformed to obtain amplitude and phase responses. The group delay is obtained from the derivative of the phase response. Figure 7 shows the input time-domain signal together with its corresponding amplitude (in dB) and phase spectra. Note that the duration of the pulse is about 0.4 ns and the phase variation is in the order of hundreds of degrees.

The radiated signals E_θ (solid lines) and E_ϕ (dashed lines) as detected by the probes in Fig. 5 and their amplitude and phase spectra are shown in Fig. 8. Figure 8a and b confirm that the main polarization is vertical (E_θ) since the detected signal in horizontal polarization (E_ϕ) is at least more than 20 dB below the vertical component. Figure 8c shows the phase variation now in thousands of degrees, which is a result of the ringing of the detected time signal in Fig. 8a. Moreover, notice that the main part of the received pulse in Fig. 8a looks similar to a negative derivative of the input pulse rather than the original input signal in Fig. 7a. Such behaviour is common in antennas that radiate pulses covering a significant frequency spectrum, e.g. [28].

Figure 9a and b show the amplitude and group-delay responses, respectively, of the coplanar UWB antenna fed by a coaxial cable. The amplitude response in the main polarization (solid line) is between −40 and −50 dB which is due to the small

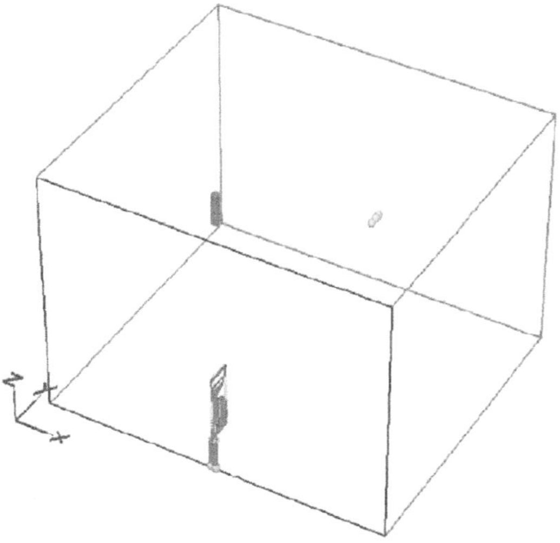

Fig. 5 Setup of one half of the coplanar UWB antenna in MEFiSTo-3D including coaxial input port, probes and coaxial reference port

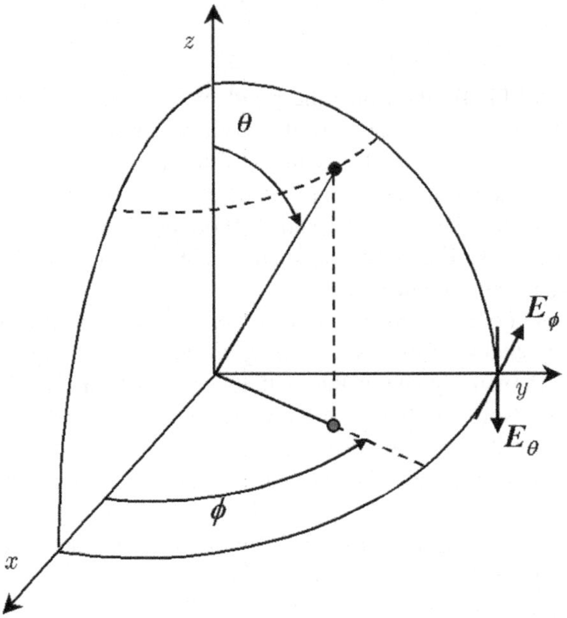

Fig. 6 Orientation of field components received by probes in Fig. 5 with respect to Fig. 1

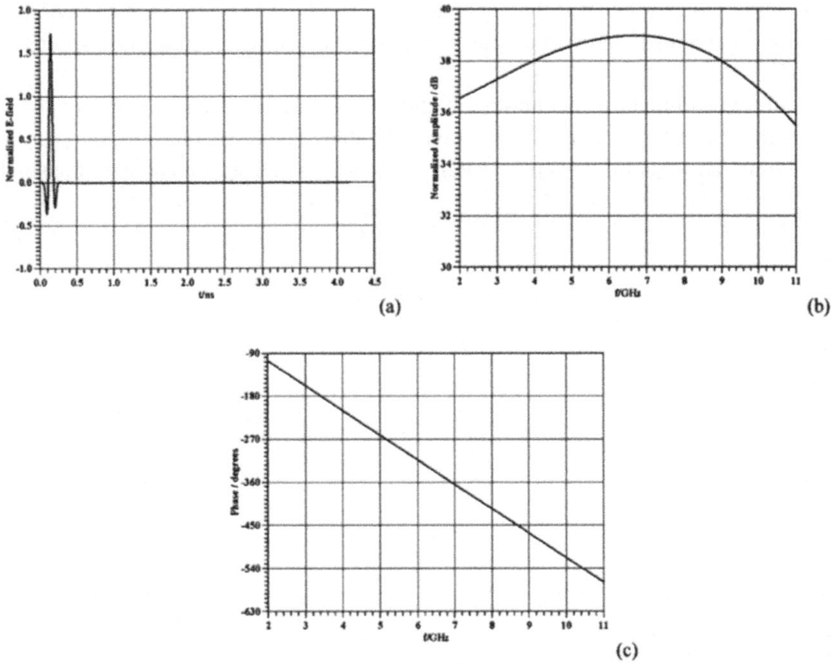

Fig. 7 Time-domain signal (**a**), amplitude spectrum (**b**) and phase spectrum (**c**) at the input of the coaxial cable feeding the coplanar antenna (c.f. Fig. 5)

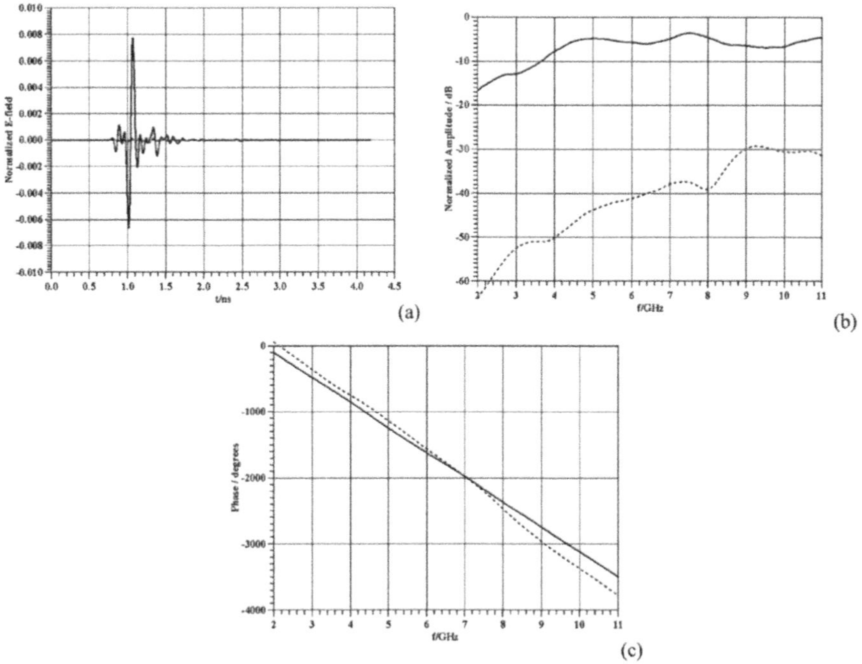

Fig. 8 Radiated time-domain signal (**a**), amplitude spectrum (**b**) and phase spectrum (**c**) detected by the probes; E_θ (*solid lines*) and E_ϕ (*dashed lines*)

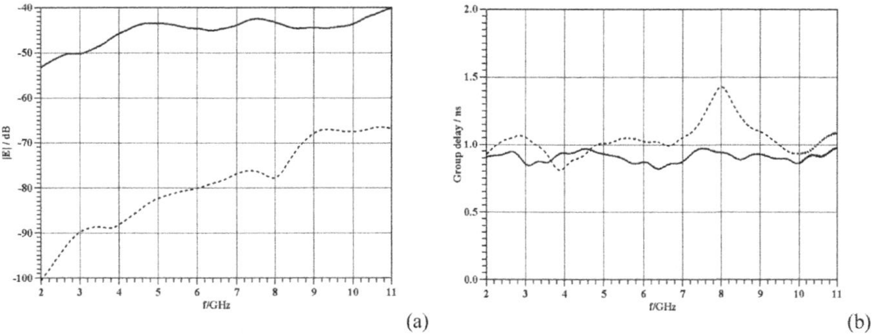

Fig. 9 Amplitude response (**a**) and group-delay characteristic (**b**) of coplanar UWB antenna; vertical polarization E_θ (*solid lines*), and horizontal polarization E_ϕ (*dashed lines*)

effective area of the receiving probes. Since the variations in amplitude and phase (group delay) determine the distortion of the pulse transmitted by the antenna, the respective values – as read from the data plotted in Fig. 9 – are summarized below for both vertical (VP) and horizontal (HP) polarizations.

Frequency range:	3.1 GHz–10.6 GHz
Amplitude variation:	< 8.7 db (VP); < 23 dB (HP)
Group-delay variation:	< 163 ps (VP); < 620 ps (HP)

Note that the amplitude variation of 8.7 dB in vertical polarization (E_θ) is in very good agreement with the radiation patterns displayed in Fig. 4 for individual frequencies between 3 GHz and 10 GHz. Since Fig. 9 was obtained from data computed by the time-domain solver MEFiSTo-3D and Fig. 4 from that of the frequency-domain package HFSS, this agreement (together with Fig. 2) verifies the design and performance of the coplanar UWB antenna.

3.2 Microstrip Antenna

In order to compare the results obtained for the coplanar UWB antenna with those of a different antenna, we apply the above time-domain method to the microstrip UWB antenna presented in [9, 10].

As a verification of the model, Fig. 10 shows the input reflection coefficient (in dB). The VSWR measurements in [9, 10] have been converted to reflection coefficients and are shown as dash-dotted lines. The data from HFSS are shown as dashed lines and are in reasonable agreement with measurements. Note that the HFSS model includes the connection to a coaxial cable. In order to reduce the computational domain, i.e., shorten the long microstrip feed line shown in [9], the coaxial connector could not be modelled in MEFiSTo-3D. Therefore, and especially in the higher frequency range, the agreement between measurements and MEFiSTo-3D is not as good as that with HFSS. However, the basic shape and the reasonably small discrepancies validate the numerical computations.

After exciting the microstrip antenna with a pulse shown in Fig. 7, detecting the radiated signal and calculating amplitude and phase responses, the data presented in Fig. 11 is obtained. Between 3 GHz and 10 GHz, the amplitude variation in vertical polarization is similar to that of the coplanar UWB antenna in Fig. 9a. The signal level difference between horizontal and vertical polarizations in Fig. 11a is smaller than that in Fig. 9a. This is due to the fact that the x-component of the electric field represents the main polarization in a microstrip line if the antenna is oriented in the same way as the coplanar one in Fig. 1.

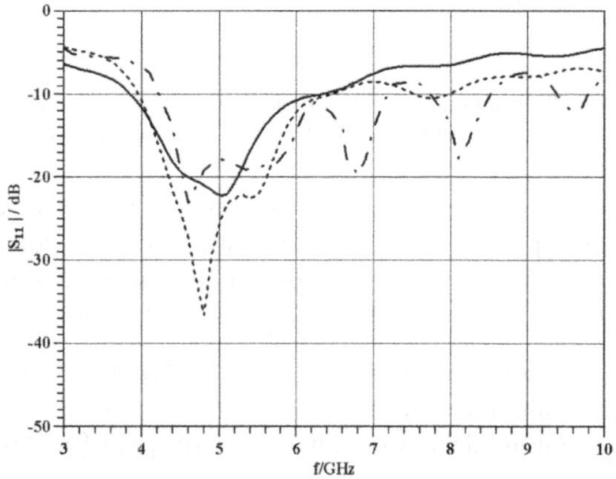

Fig. 10 Input reflection coefficient in dB of the microstrip UWB antenna of [9, 10]; calculated values from VSWR measurements in [9, 10] (*dash-dotted line*), HFSS (*dashed line*) and MEFiSTo-3D (*solid line*)

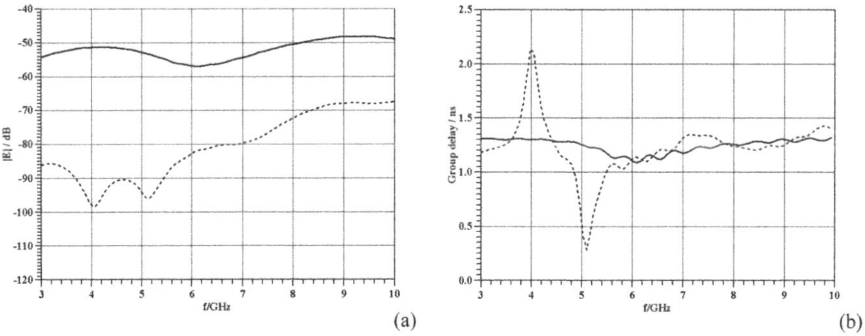

Fig. 11 Amplitude response (**a**) and group-delay characteristic (**b**) of the microstrip UWB antenna in [9, 10]; vertical polarization E_θ (*solid lines*), and horizontal polarization E_ϕ (*dashed lines*)

The group delay performances of the microstrip antenna are inferior to those of the coplanar antenna in both polarizations. The following values are obtained:

Frequency range:	3.0 GHz–10.0 GHz
Amplitude variation:	$<$ 8.8 db (VP); $<$ 31 dB (HP)
Group-delay variation:	$<$ 239 ps (VP); $<$ 1.9 ns (HP)

3.3 Comparison

Both the coplanar and the microstrip antenna display nearly omnidirectional radiation patterns with characteristics slightly distorting towards 10 GHz (c.f. [9, 10] and [27] for details). Over the 3.1–10.6 GHz range, the input reflection coefficient of the coplanar antenna is superior to that of the microstrip antenna. The amplitude variations in vertical polarization are comparable; in horizontal polarization, however, it is 8 dB in favour of the coplanar antenna. The group-delay variations of the coplanar antenna are much smaller than those of the microstrip antenna and, therefore, the coplanar structure of Fig. 1 is better suited for UWB applications.

It is noted that a smaller group-delay variation ($<$ 100 ps) is reported in [7] for a microstrip UWB antenna with two slots in the radiating patch. However, the gain of that antenna is lower that the one reported in Fig. 3 and even drops below 0 dB above 9.8 GHz [7].

4 Conclusion

Time-domain techniques, applied here in form of the TLM solver MEFiSTo-3D, present a viable option for the analysis and modelling of UWB printed-circuit antennas. Amplitude characteristics extracted from the time-domain solution agree well with frequency-main methods, which are used for the design of UWB antennas. The computation of group-delay data in an actual application of pulsed transmission is one of the clear advantages of time-domain over frequency-domain techniques.

The time-domain modelling procedure presented here is applied to two different printed-circuit UWB antennas, and agreement with frequency-domain computations and measurements is demonstrated.

Acknowledgement This work was supported by the TELUS Grant in Wireless Communications.

References

1. L. Yang and G.B. Giannakis, "Ultra-wideband communications: an idea whose time has come", IEEE Signal Proc. Mag., Vol. 21, pp. 26–54, Nov. 2004.
2. International Telecommunication Union, Radiocommunication Study Groups, "Framework for the introduction of devices using ultra-wideband technology", Document 1/85(Rev.1)-E, 09 Nov. 2005.
3. K. Kiminami, A. Hirata and T. Shiozawa, "Double-sided printed bow-tie antenna for UWB communications", IEEE Antennas Wireless Propagat. Lett., Vol. 3, pp. 152–153, 2004.
4. J. Liang, C.C. Chiau, X. Chen and C.G. Parini, "Printed circular disc monopole antenna for ultra-wideband applications", IEE Electronics Lett., Vol. 40, No. 20, pp. 1246–1247, Sep. 2004.
5. S.H. Choi, J.K. Park, S.K. Kim and J.Y. Park, "A new ultra-wideband antenna for UWB applications", Microwave Opt. Technol. Lett., Vol. 40, No. 5, pp. 399–401, Mar. 2004.
6. J. Liang, C.C. Chiau, X. Chen and C.G. Parini, "Study of a printed circular disc monopole antenna for UWB systems", IEEE Trans Antennas Propagat., Vol. 53, pp. 3500–3504, Nov. 2005.
7. Z.N. Low, J.H. Cheong and C.L. Law, "Low-cost PCB antenna for UWB applications", IEEE Antennas Wireless Propagat. Lett., Vol. 4, pp. 237–239, 2005.
8. J. Liang, C.C. Chiau, X. Chen, and C.G. Parini, "Printed circular ring monopole antennas", Microwave Opt. Technol. Lett., Vol. 45, No. 5, pp. 372–375, June 2005.
9. C.-C. Lin, Y.-C. Kan, L.-C. Kuo and H.-R. Chuang, "A planar triangular monopole antenna for UWB communication", IEEE Microwave Wireless Comp. Lett., Vol. 15, pp. 624–626, Oct 2005.
10. H.R Chuang, C.C. Lin and Y.C. Kan, "A printed UWB triangular monopole antenna", Microwave J., Vol. 49, pp. 108–120, Jan. 2006.
11. N. Fortino, G. Kossiavas, J.Y. Dauvignac, and R. Staraj, "Novel antennas for ultrawideband communications", Microwave Opt. Technol. Lett., Vol. 41, No. 3, pp. 166–169, May 2004.
12. W. Wang, S. S. Zhong, and S.-B. Chen, "A novel wideband coplanar-fed monopole antenna", Microwave Opt. Technol. Lett., Vol 43, No. 1, pp. 50–52, Oct. 2004.
13. A.M. Abbosh, M.E. Bialkowski, M.V. Jacob and J. Mazierska, "Investigations into an LTCC based ultra wideband antenna", Proc. Asia-Pacific Microwave Conf., 4p., Suzhou, China, Dec. 2005.
14. C.T.H. Lim, "A GCPW-fed printed antenna for UWB applications", Proc. Asia-Pacific Microwave Conf., 3p., Suzhou, China, Dec. 2005.
15. X. Chen, J. Liang, P. Li, L. Guo, C. C. Chiau and C.G. Parini, "Planar UWB monopole antennas", Proc. Asia-Pacific Microwave Conf., 4p., Suzhou, China, Dec. 2005.
16. H.K. Lee, J.K. Park and J.N. Lee, "Design of a planar half-circle shaped UWB notch antenna", Microwave Opt. Technol. Lett., Vol 47, No. 1, pp. 9–11, Oct. 2005.
17. T.-G.Ma and C.-H. Tseng, "An ultrawideband coplanar waveguide-fed tapered ring slot antenna", IEEE Trans. Antennas Propagat., Vol. 54, pp. 1105–1110, Apr. 2006.

18. Y.-C. Lee, S.-C. Lin and J.-S. Sun, "CPW-fed UWB slot antenna", Proc. Asia-Pacific Microwave Conf., 4p., Yokohama, Japan, Dec. 2006.
19. S. Nikolaou, D.E. Anagnostou, G.E. Ponchak, M.M. Tentzeris and J. Papapolymerou, "Compact Ultra Wide-Band (UWB) CPW-fed elliptical monopole on Liquid Crystal Polymer (LCP)", IEEE AP-S Int. Symp. Dig., pp. 4657–4660, Albuquerque, USA, July 2006.
20. E.S. Angelopoulos, A.Z. Anastopoulos and D.I. Kaklamani, "Ultra-wideband bow-tie slot antenna fed by a cpw-to-cpw transition loaded with inductively coupled slots", Microwave Opt. Technol. Lett., Vol. 48, No. 9, pp. 1816–1820, Sep. 2006.
21. X.-L. Liang, S.-S. Zhong and W. Wang, "UWB printed circular monopole antenna", Microwave Opt. Technol. Lett., Vol. 48, No. 8, pp. 1532–1534, Aug. 2006.
22. J.-S. Sun, Y.-C. Lee and S.-C. Lin, "New design of a CPW-fed ultrawideband slot antenna", Microwave Opt. Technol. Lett., Vol. 49, No. 3, pp. 561–564, Mar. 2007.
23. D.-B. Lin, I.-T. Tang and M.-Y. Tsou, "A compact UWB antenna with CPW-feed", Microwave Opt. Technol. Lett., Vol. 49, No. 3, pp. 564–567, Mar. 2007.
24. Z.N. Chen and X. Qing, "Research and development of planar UWB antennas", Suzhou, China, Dec. 2005.
25. B.L. Ooi, G. Zhao, M.S. Leong, K.M. Chua and C.W.L. Albert, "Wideband LTCC CPW-fed two-layered monopole antenna", IEE Electronics Lett., Vol. 41, No. 16, pp. 9–10, Aug. 2005.
26. K. Rambabu, H.A. Thiart, J. Bornemann and S.Y. Yu, "Ultrawideband printed-circuit antenna", IEEE Trans. Antennas Propagat., Vol. 54, pp. 3908–3911, Dec. 2006.
27. H.-J. Lam and J. Bornemann, "Ultra-wideband printed-circuit antenna in coplanar technology", 2007 IEEE EMC-S Int. Symp. Dig., TU-PM-1-7, 4p, Honolulu, USA, July 2007.
28. D. Ghosh, A. De, M.C. Taylor, T.K. Sarkar, M.C. Wicks and E.L. Mokole, "Transmission and reception by ultra-wideband (UWB) antennas", IEEE Trans. Antennas Propagat. Mag., Vol. 48, pp. 67–99, Oct. 2006.

On the Modeling of Ultra Wide Band (UWB) Radiating Structures

Bruno Biscontini, Uwe Siart and Peter Russer

Due to a number of unique features such as high data rate, multipath immunity, fine range resolution, Ultra Wide Band (UWB) radio is an emerging attractive technology for a number of applications in wireless communications, networking, radar, imaging, and positioning systems. A UWB system operates using low-power ultra-short information carrying pulses achieving a very large frequency bandwidth. This makes the performance of the UWB receiver significantly depending on the frequency transfer function of the ultra wideband radio channel. Indeed, ideally the received signal should match to the transmitted signal in bandwidth, signal shape and time. Any mismatch results in loss of performance of the UWB radio link. The shape of the received time-domain pulse is determinated by several effects among other things by radio channel and antenna signal transfer characteristics. In this context, more then in narrowband systems, the antenna has a crucial impact on the performance of the whole communication system and works together with the other system components to meet the design goals [1, 2, 3]. Depending on the application, the antenna may be required to be non-dispersive and having unidirectional radiation pattern over a large impedance bandwidth, small size, high gain etc. This scenario requires often novel antenna structures having at the same time properties common to frequency-independent antennas, small-element antennas and reflector antennas [4].

To address those challenges, in this work we exploit advanced numerical modeling techniques to design a novel UWB planar multilayered cylindrical reflector antenna array. We first describe the design of UWB radiating elements and then their circular cylindrical reflector array configuration. To achieve small size, the

B. Biscontini
Munch University of Technology, Institute for High Frequency Engineering, Arcisstr. 21 80333, Munich Germany, e-mail: biscontini@mytum.de

U. Siart,
Munch University of Technology, Institute for High Frequency Engineering, Arcisstr. 21 80333, Munich Germany, e-mail: uwe.siart@tum.de

P. Russer
Munch University of Technology, Institute for High Frequency Engineering, Arcisstr. 21 80333, Munich Germany, e-mail: russer@tum.de

P. Russer, U. Siart (eds.), *Time Domain Methods in Electrodynamics*,

geometry of the UWB radiating element is chosen to be planar. The synthesis and the optimization of the radiating element is done using genetic algorithm (GA) in connection with method of moments (MoM). To achieve the required unidirectional characteristics the radiating element is embedded into a substrate-superstrates layered structure. The obtained geometry of the substrate-superstrates structure, similar to a double ellipsoid, is optimized to achieve the best radiation characteristics over the frequency band from 3 to 21 GHz. The design and the optimization of the UWB cylindrical reflector array is performed on the basis of a simplified model based on cylindrical Green's functions. In Sect. 1 the synthesis of the UWB radiating element is described. In Sect. 2 the numerical results with the cylindrical reflector are given. In Sect. 3 the simplified modeling of the UWB array will be discussed. Eventually in Sect. 4 the conclusions will be given.

1 UWB Antenna Element Synthesis

Although there are several different ways to analyze UWB antennas [2, 5] there is no systematic method to design the geometry of a UWB antenna. Due to their flexibility and global convergence capabilities to synthesize and optimize the radiating element geometry, we use genetic algorithms(GA) [6, 7].

In the synthesis of the radiating element the electromagnetic problem is solved by means of the electric field integral equation (EFIE). To solve the integral equation we use MoM expanding the current into a series of Rao Glisson Wilson vector basis functions (RWG) [8]. Since the RWG basis functions are supported by a triangular mesh, the surfaces S' of the radiating elements will be meshed by triangles. The RWG basis function are described by a pair of triangles that form one rectangular cell. In order to determine a relation between the geometry and the genetic algorithm process we map the meshing into bits. The bit combinations defining the meshing cells of the antenna geometry element play the role of the "genes". The combinations of genes form chromosomes which give the complete bits representation of the antenna geometry.

The genetic algorithm optimizes the antenna geometry to reach the required antenna specifications. The genetic process starts with the initialization of random chromosomes. In the first step we impose general optimization limits to the antenna geometry. Considering that we intend to design a small-element antenna we impose the geometry to be symmetric with respect to the excitation point. We also impose the maximum and the minimum dimensions of the geometry. Considering the symmetry of the structure, the number of parameters to be optimized is reduced by a factor of two. We also wish the antenna to be frequency-independent and therefore to exhibit a gradually varying of the geometry. This is a considerable simplification to the GA process since only the structure profile can be changed. It is interesting to observe that just some general properties of the structure geometry have been specified.

Before running the genetic algorithm the population number and the number of iterations must be chosen. The population number (N_{pop}), i.e., the maximum number of structures that may be considered, is set to $N_{pop} = 16$. We note that a pure random search would lead to $12^{21} = 4.6 \times 10^{22}$ possible structures. The number of iteration is 40. The mutation per iteration is set to 10% of the total bits in all chromosomes (population). Random single point crossover is used to mate the selected parents. The selection strategy for parents is based on proportionate selection (roulette-wheel selection). In this way the probability of each chromosome to be selected is proportional to its score value. The scoring scheme is chosen to maximize the antenna return loss. As score the negative logarithmic return loss weighted and averaged over the considered frequency interval is used. To higher scores a low probability to be chosen as a parent is assigned. According to this criterion, the solution by the GA is guided to converge towards structures that have return loss below -12 dB or -15 dB or -20 dB in the frequency range of 3–21 GHz. Fig. 1 shows the geometry of the optimum generated structure and the computed return loss.

Considering that the antenna radiates into a pattern governed by a current supported by the radiator geometry, in case of UWB antennas, it can be expected that the current density distribution keeps as uniform as possible on the radiator surface over the bandwidth. This effect, typical for the frequency-independent antennas, can be observed in Fig. 2(a), (b), (c) and (d) where the absolute value of the current density distribution is plotted over the antenna geometry at 3, 7, 11 and 21 GHz.

Figure 3 shows the output pulse obtained by a Gaussian pulse excitation. This shows the non-dispersive nature of the radiating element. However a small wave front distortion occurs when the transmitted pulse has traveled through the antenna, see Fig. 3(a) and (b).

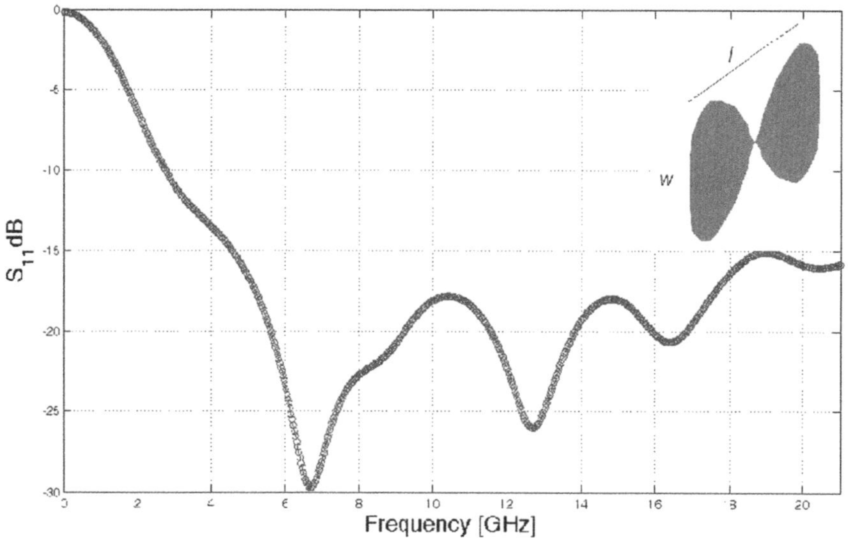

Fig. 1 Best generated structure and its return loss

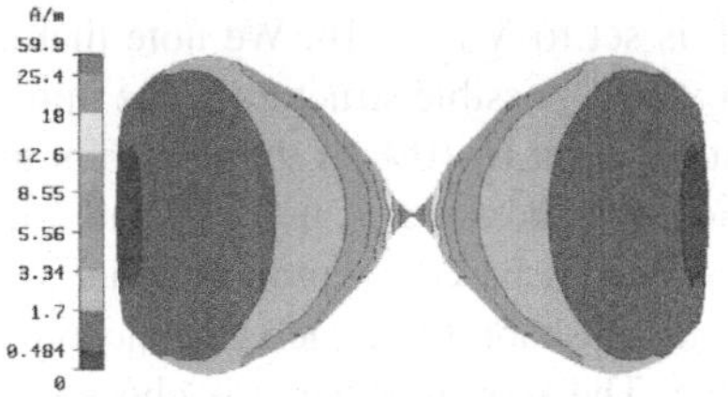

(a) Amplitude plot of the current density distribution at 3 GHz.

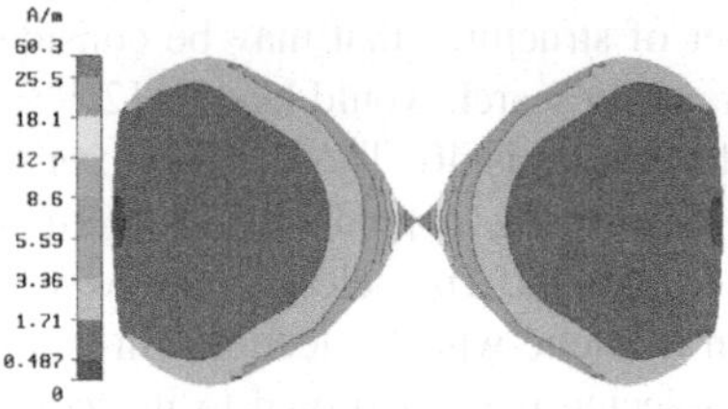

(b) Amplitude plot of the current density distribution at 7 GHz.

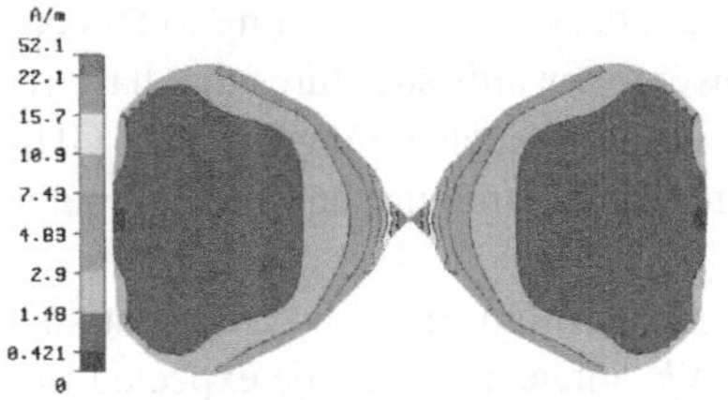

(c) Amplitude plot of the current density distribution at 11 GHz.

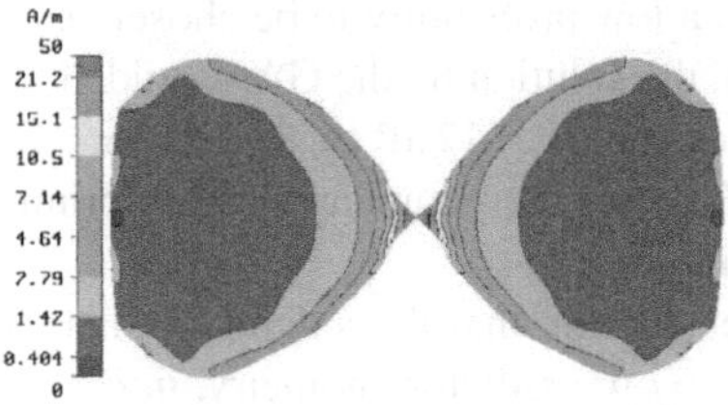

(d) Amplitude plot of the current density distribution at 21 GHz.

Fig. 2 Current density distribution behavior over the frequency

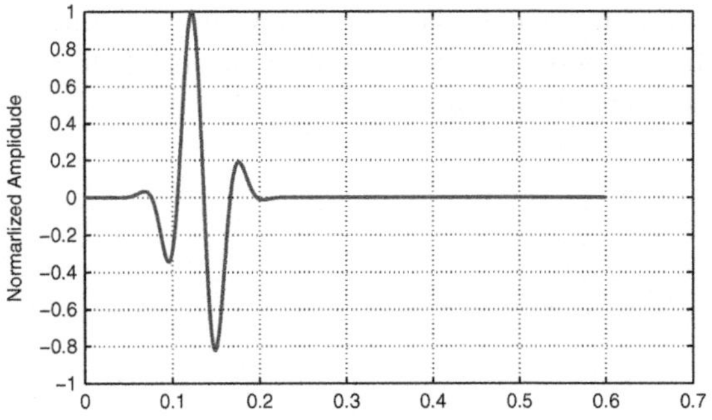

(a) Input Gaussian Pulse.

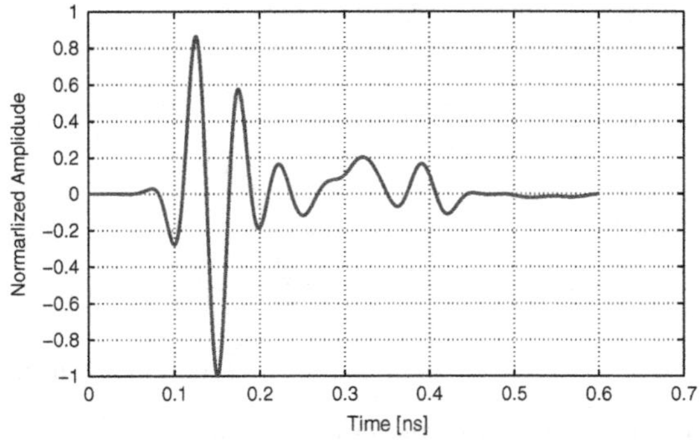

(b) Output Gaussian Pulse.

Fig. 3 Time domain behavior with Gaussian excitation

1.1 Substrate-Superstrates Configuration

A reflector improves the radiation of the energy in one direction. However its vicinity yields to severe constrains to the performance of planar Ultra Wide Band (UWB) antennas. Indeed for a very broadband frequency range as 3–21 GHz, the reflector at low frequency tends to be electrically near to the antenna degrading the antenna input impedance whereas at the higher frequencies acts as scattering body deteriorating the radiation pattern. For this reason, to mitigate the influence of the reflector the antenna elements should exhibit a unidirectional radiation pattern at higher frequencies. To achieve a unidirectional radiation pattern, we place the radiating element in the middle of a substrate of dielectric constant $\varepsilon_1 = 2.2$. Additional superstrates with dielectric constants $\varepsilon_2 = 3.37$, $\varepsilon_3 = 6$ and $\varepsilon_4 = 9.2$, respectively are used to maximize the radiation efficiency on the upper half-plane, Fig. 4. The thicknesses of the layers are chosen in a way to minimize the propagation of surface waves [7, 9]. The overall behavior of the surface waves is similar that in the single layer case. The lower dominant mode is the TM_1 which above cut-off is independent from the substrate thickness. The next excited modes are the TE_1 and TM_2 [7, 9]. The above considerations, are in general valid for the case of infinite extension of the dielectric layers. In realty the substrate layers exhibit finite extension. In order minimize the radiation from the dielectric, the edges are rounded, Fig. 6(a). To optimize the radiation pattern a double elliptical configuration of the dielectric layers is used, Fig. 6(a). In Fig. 5 a

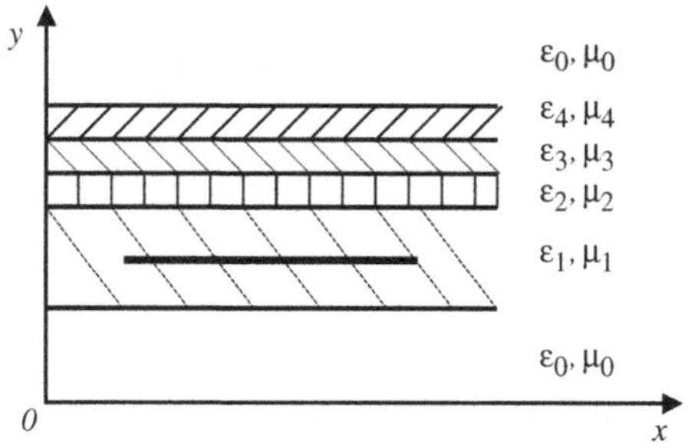

Fig. 4 The substrate-superstrates configuration

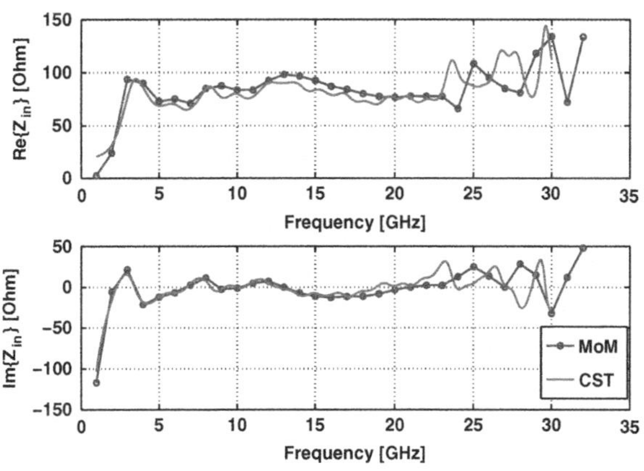

Fig. 5 Input impedance for the double elliptical configuration computed with CST and the infinite extension case computed with MoM

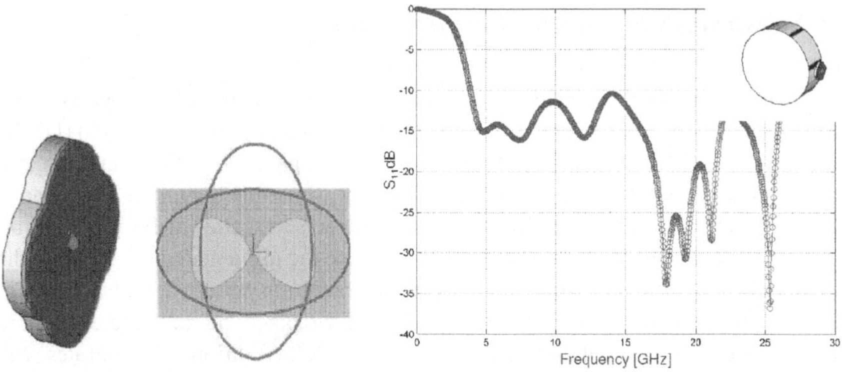

(a) (b)

Fig. 6 Optimized radiating element (**a**). Return loss of the UWB placed near the cylindrical reflector (**b**)

comparison between the input impendence computed with our MoM (infinite extension of the layers) and with CST Microwave Studio for the case of the double elliptical configuration is given. It is interesting to note that there is basically no difference between the double elliptical configuration and the case when the layers are infinite extended [7].

2 Numerical Results with the Cylindrical Reflector

In this section the numerical results are shown for the case in which the antenna is placed near a cylindrical reflector. The cylindrical reflector has a radius of 6 cm and height of 5.5 cm. The radiating element is place at the distance of $\lambda/4$ at the frequency of 5.5 GHz. In Fig. 6(b) the return loss (S_{11}dB) is reported and good matching from 3 to 21 GHz and above can be observed. In Fig. 7 the gain plot in the elevation and azimuth cuts are plotted. We can observe that the antenna exhibits good performance at up to 21 GHz in nearly all the cases for either the elevation and the azimuth.

3 UWB Circular Cylindrical Reflector Antenna Array

We consider a circular antenna array composed by 8 UWB radiating elements mounded around the cylindrical reflector, Fig 8. Due to the very large frequency spam and the high complexity of the electromagnetic problem, the design and the optimization exceeds the computation capabilities of a standard desktop computer.

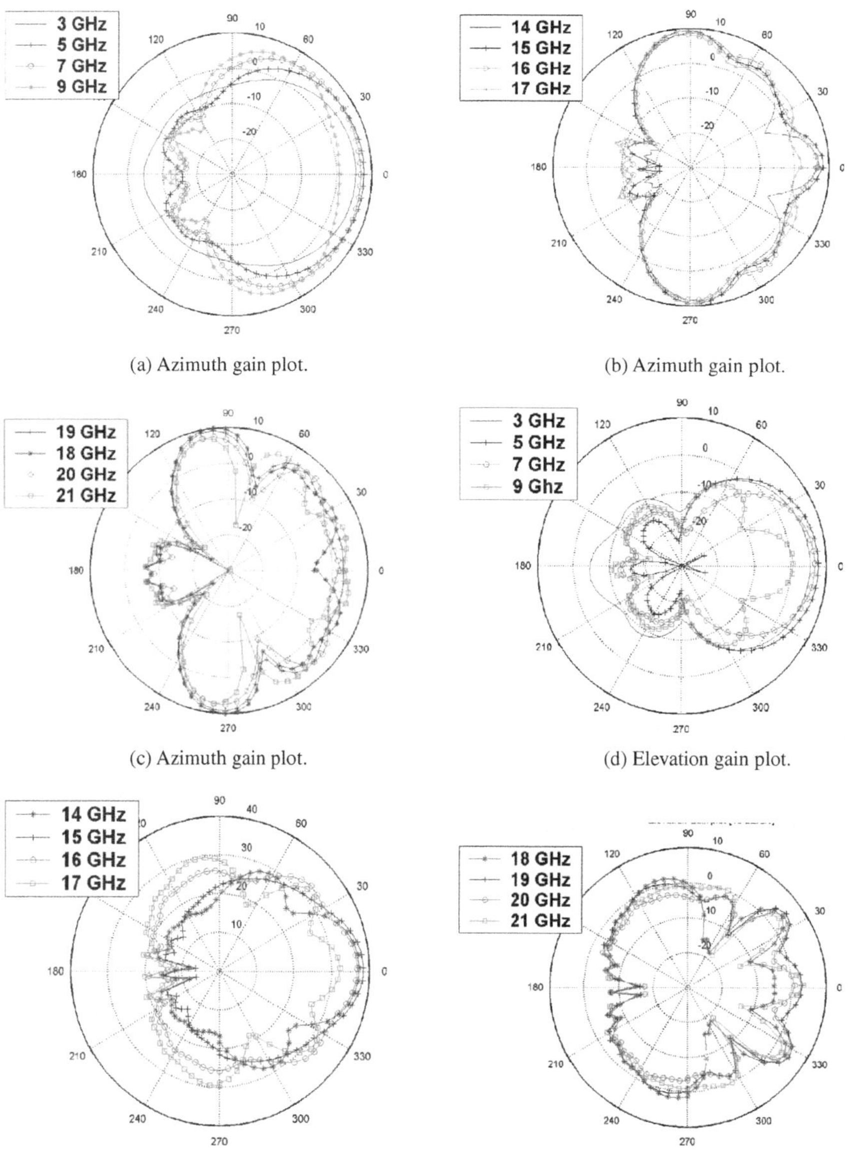

(a) Azimuth gain plot.

(b) Azimuth gain plot.

(c) Azimuth gain plot.

(d) Elevation gain plot.

(e) Elevation gain plot.

(f) Elevation gain plot.

Fig. 7 Gain azimuth and elevation plots

Therefore it is often better to model the electromagnetic problem using a simplified physical model. In particular the mutual coupling between the UWB antenna elements has to be understood to achieve an efficient optimization of the diameter of the array and cylindrical reflector.

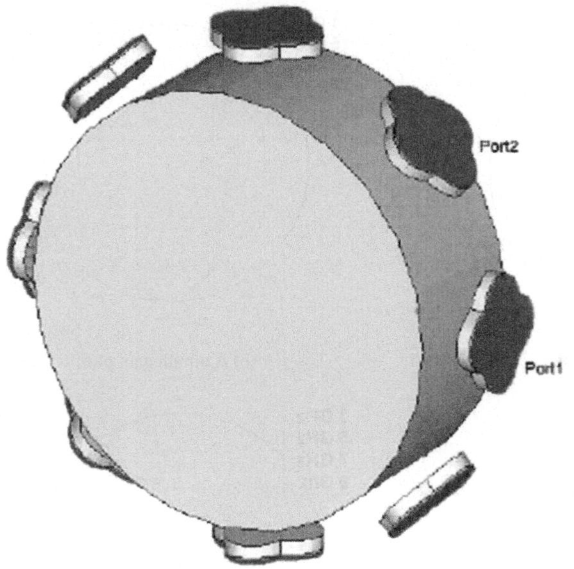

Fig. 8 Array geometry

In the case of short pulse excited antenna array the mutual coupling between the radiators can be efficiently described as superposition of multiples echoes scattered by the antennas and superimposed to the direct radiated signal [10]. Since the echoes appears in different times the overall summation becomes negligible and less relevant that in the case of narrowband arrays excited with CW signals [10].

Neglecting the mutual coupling, we propose to exploit a simplified model based on cylindrical Green's functions to find the optimal reflector radius and array [11, 12].

The dyadic Green's functions for the cylindrical reflector (for ϕ- and z-orientated Hertzian dipoles) are computed using Generalized Transmission Line (GTL) method [11, 12]. The optimal reflector radius and array diameter are found by optimizing the farfield radiation patterns of the ϕ- and z-orientated Hertzian dipoles as function of the GTL parameter. The structure is optimized 4–13 GHz since above 13 GHz the effect of the substrate is expected to be stronger as it is was previously shown. To demonstrate the potentiality of the model, the optimized reflector radius and array diameter are used in a fullwave simulation and the numerical results are shown in Sect. 3.1.

3.1 Numerical Results for the UWB array

The simulated array structure is shown in Fig. 8. The circular array is composed of 8 radiating elements. The scattering parameters obtained exciting the port 1 are shown Fig. 9(a) and (b). The port numbering is shown in Fig. 8. Due the complexity

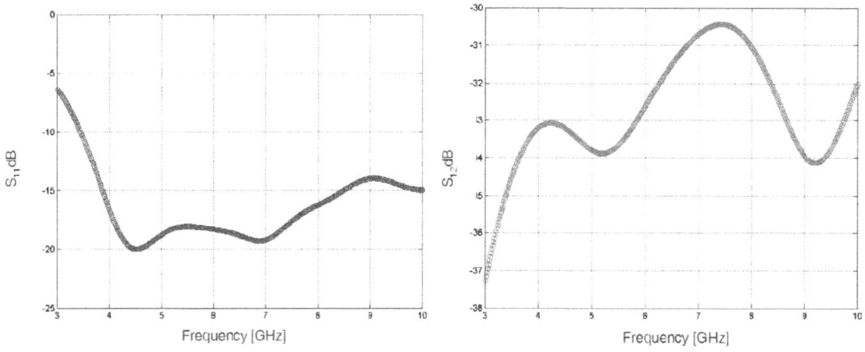

(a)|S11|dB (b)|S12|dB

Fig. 9 $|S_{11}|$dB (**a**), $|S_{11}|$dB (**b**) computed from 0 to 10 GHz

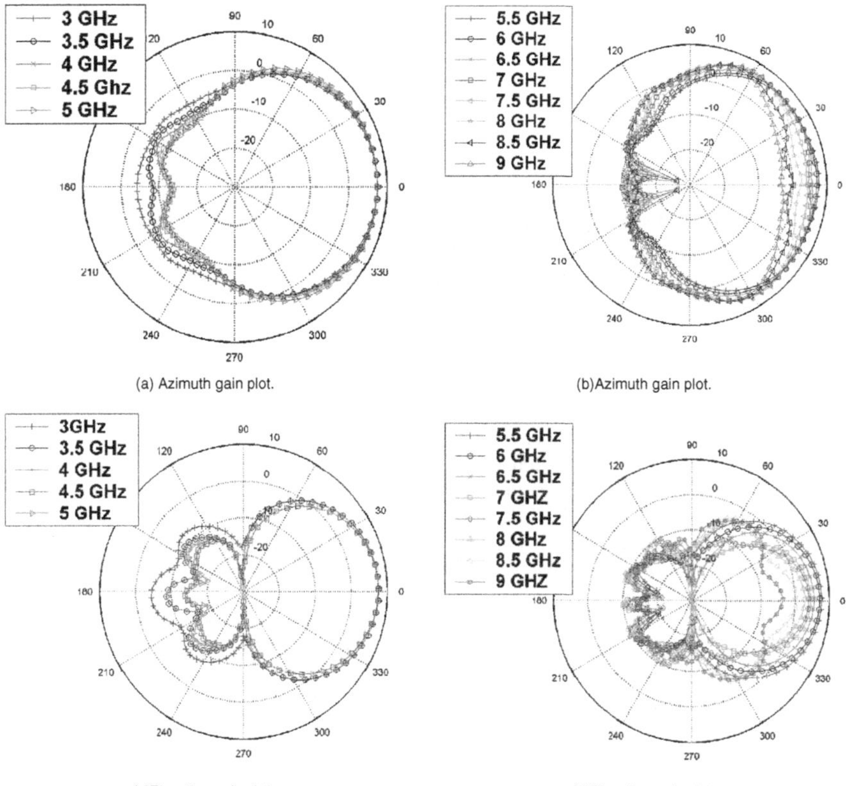

(a) Azimuth gain plot. (b)Azimuth gain plot.

(c)Elevation gain plot. (d)Elevation gain plot.

Fig. 10 Gain azimuth and elevation plots

of the electromagnetic structure the array could be simulated only until 10 GHz. From Fig. 9(b) the low coupling between the elements and the UWB behavior of the array can been seen. In Fig. 10 the elevation and azimuthal radiation patterns are shown and a similar behavior to the case of one dipole only can be observed.

4 Conclusions

In this work we have presented several methods to achieve a design of a novel UWB cylindrical reflector antenna array. A novel planar multilayered cylindrical reflector antenna for UWB applications has been presented. The planar UWB radiating element has been designed using a combination of genetic algorithm and integral equation method. We have discussed how genetic algorithms can be used for antenna geometry optimization considering general geometrical properties of the radiating elements. We have shown that a UWB antenna element embedded in a dielectric substrate-superstrates structure can exhibit good performance near a cylindrical reflector up to 21 GHz. A simplified model for the design and the optimization of UWB array has been discussed and verified by means of fullwave electromagnetic simulation.

References

1. R. A. Scholtz, D. M. Pozar and W. Namgoong,"Ultra-wideband radio," *EURASIP Journal on Applied Signal Processing*, vol. 3, pp. 252–272, 2005.
2. H. G. Schantz, "Introduction to ultra wideband antennas," *IEEE UWBST Conference Proceedings*, 2003.
3. T. Do-Hong and P. Russer, " Wideband Smart Antenna Array Applications," *IEEE Microwave Magazine*, pp. 57–67, March 2004.
4. H. Schantz, *The Art and the Science of Ultrawideband Antennas*, Artech House, Boston, 2005.
5. P. Russer, *Electromagnetics, Microwave Circuit and Antenna Design for Communications Engineering*, Artech House, Boston, 2003.
6. F. J. Villegas, T. Cwik, Y. Rahmat-Samii and M. Manteghi, "A parallel electromagnetic genetic-algorithm (EGO) application for patch antenna design," *IEEE Trans. Antennas Propagat.*, vol. 52, no. 9, pp. 2424–2435, Sept. 2004.
7. B. Biscontini et al., "A Novel Antenna for Ultra Wide Band (UWB) Intelligent Antenna Systems Biscontini," *Microwave Symposium Digest, 2006. IEEE MTT-S International*, June 2006, San Francisco, USA.
8. S. M. Rao, D. R. Wilton, and A. W. Glisson, "Electromagnetic scattering by surfaces of arbitrary shape," *IEEE Trans. Antennas Propagat.*, vol. AP-30, pp. 409–418, May 1982.
9. N. G. Alexópoulos and D. R. Jackson, "Fundamental superstrate (cover) effects on printed circuit," *IEEE Trans. Antennas and Propagat.*, vol. AP-32, pp. 807–816, Aug. 1984.
10. M. Ciattaglia and G. Marrocco,"Investigation on Antenna Coupling in Pulsed Arrays," *IEEE Trans. Antennas Propagat.*, vol. 54, no. 3, pp. 835–843, March 2006.
11. B. Biscontini, M. Burger and P. Russer, *Fields, Networks, Computational Methods and Systems in Modern Electromagnetics: A tribute to Leopold B. Felsen*, Springer Proceedings in Physics, Berlin, Heidelberg Springer, vol. 97, pp. 129–142, 2004.

12. B. Biscontini and P. Russer, "A Novel Approach for Designing Circular Antenna Arrays for Ultra Wide Band (UWB) Applications," *In Proceedings of the 22nd Annual Review of Progress in Applied Computational Electromagnetics ACES*, March 12–16, 2006, Miami, FL, USA, March 2006, pp. 198–204.
13. W. Sorgel and W. Wiesbeck, "Influence of the antennas on the ultra-wideband transmission," *EURASIP Journal on Applied Signal Processing*, pp. 296–305, 2005.
14. Z. N. Chen, X. H. Wu, H. F. Li, and M. Y. W. Chia, "Consideration for source pulses and antennas in UWB radio systems," *IEEE Trans. Antennas Propagat.*, vol. 52, no. 7, pp. 1739–1748, Jul. 2004.
15. D. R. Jackson and N.G. Alexópoulos, "Gain enhancement methods for printed circuit antennas," *IEEE Trans. Antennas and Propagat.*, vol. AP-33, no. 9, pp. 976–987, Sept. 1985.

An Efficient Electromagnetically Optimized Design and Realization of Pseudo-Elliptic All-Metal Cavities Filters

Dr. Savvas Kosmopoulos and Nikolaos Sidiropoulos

1 Introduction

In the actual design of pseudo-elliptic cross-coupled resonator filters, the bypass or cross-couplings are most often represented by irises, coupling screws, or other physical elements [1, 2]. Another mechanism that has been reported in literature [3], is the use of higher order modes to provide additional paths for the signal between different points along the in-line waveguide structure. For this new class of pseudo-elliptic function filters, which is based on compact in-line TM_{110} mode cavity configurations, no structural folding is required, since, the bypass couplings are implemented through the non-resonating TE_{10}/TE_{01} modes. In this way arbitrarily allocation of transmission zeros can be implemented. A limitation of such an approach, is that "spurious" responses and/or transmission zeros might appear above or below the pass-band of the designed filter. This is usually caused by aperture resonances or by fundamental (e.g., TE_{101}) or higher order mode resonances in the cavities. In order to shift in frequency and/or eliminate such spurious responses, an alternative design procedure, based on the proper and accurate dimensioning of the coupling irises by means of a full-wave electromagnetic simulator, is proposed in [3]. This approach, however, involves excessive design run-time scenarios, directly related to the lengthy optimization process required (repeated irises mechanical dimensioning versus corresponding electromagnetic simulations), in order to obtain the requested filter performance. This is particularly true when both imaginary and real transmission zeros are involved contemporary in the design, in order to guarantee the requested GD vs. frequency behavior stability for various ambient temperatures [5]. Indeed, for such filter design procedures, the usually involved numerical optimization technique should be accompanied by adequate pure electromagnetic optimizations (see Fig. 1).

The design procedure proposed here, uses the adequately modified main design constrains of the procedure for folded waveguide filters with cross-couplings

Dr. Savvas Kosmopoulos
Space Engineering S.p.A., via dei Berio No.91, 00155, Rome, Italy

Nikolaos Sidiropoulos
Space Engineering S.p.A., via dei Berio No.91, 00155, Rome, Italy

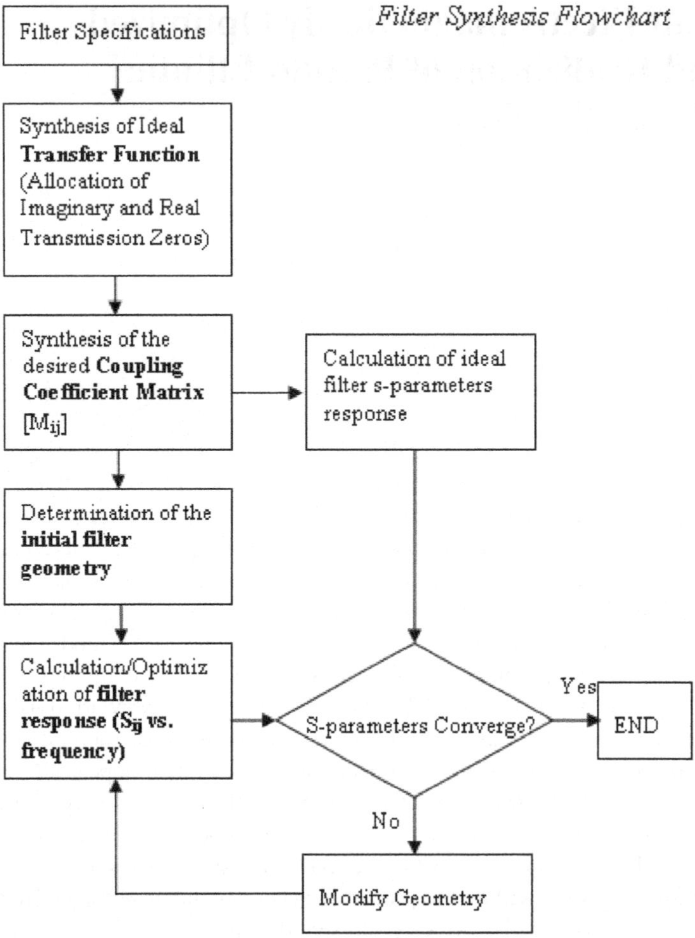

Fig. 1 The Proposed Design Flowchart

presented in [4], as combined to the fully-canonical or "folded coupling matrix" ($[M_{ij}]$) given in [5]. The aim for such design strategy is to improve the efficiency (accuracy vs. run-time) in evaluating the theoretical $|S_{ij}|$ values for the input, output and inter-stage resonators. Indeed, using an in-house built fast converging full-wave electromagnetic simulator, based on the Mode Matching method, the initial overall filter's mechanical dimensions are obtained, using as optimization target the overall theoretical $|S_{ij}|$ matrix directly derived from $[M_{ij}]$. Afterwards, the design proceeds with the fine-tuning of the proposed filter configuration by means of μ Wave Wizard^{TM}, a commercially available by Mician full-wave electromagnetic simulator.

The above described design approach is applied for the design, simulation and realization of a 6th order X-Band pseudo-elliptic function waveguide filter, to be compliant to the very stringent requirements, in terms of out-of-band rejection and

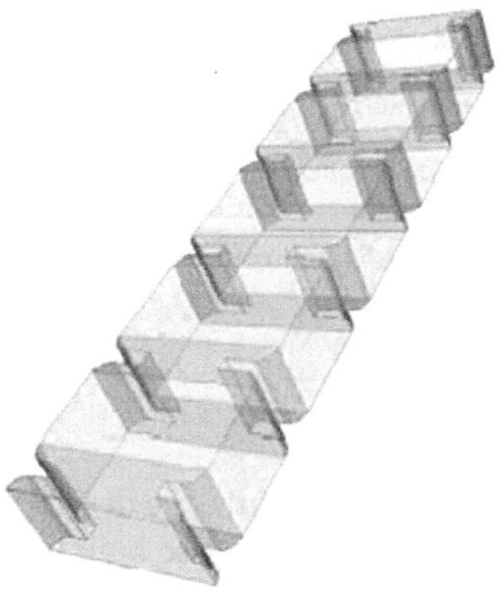

Fig. 2 Layout of the initial 5th order X-Band BPF design

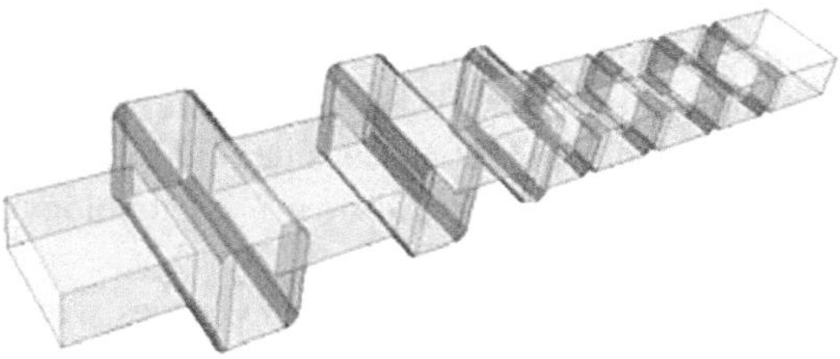

Fig. 3 The Layout of the final simulated 6th order X-Band BPF

stability versus temperature variations, of the COSMO/SKYMED Calibrator [7]. This design based on a compact in-line TM_{110} mode tuneless cavity configuration, has initially been started proposing a simple Chebyshev 5th order topology as shown in Fig. 2. Furthermore, after various fast-converging optimisation itirations, the final non-conformable filter topology shown in Fig. 3 has been obtained.

Once the final filter design is obtained, it's immunity to multipaction and ionization effects is investigated, in order to specify the filter's power handling capabilities. To do so, the voltages that appear in the critical areas and frequencies inside the filter are evaluated by means of the full wave electromagnetic simulator of Ansoft, $HFSS^{TM}$. These voltages are afterwards confronted to the multipaction threshold voltages established by ESA [8]. Hence, the input power that yields to RF breakdown is evaluated.

2 The Design Procedure

In order to underline the most important features of the herewith proposed design procedure, one may start from the requirements of the COSMO/SKYMED Calibrator input BPF, as shown in Table 1.

In order to realize these requirements, a theoretical synthesis and design procedure is followed here, initially considering the general filter topology shown in

Table 1 The COSMO/SKYMED calibrator input BPF requirements

Parameter	Value	Unit
Center Frequency	9.6	GHz
Bandwidth	420	MHz
Return Loss (all ports)	≥ 20	dB
Insertion Loss	≤ 0.2	dB
Group Delay Variation	0.6	nsec

Fig. 2. Afterwards, the design is developed toward the topology shown in Fig. 3, by means of an in-house built CAD tool, based on the theoretical approach presented in [5]. The final outcome of this initial design procedure is the derivation of the coupling matrix shown in Table 2.

To do so, a precise verification that the required number of transmission zeros can be achieved with the selected topology is performed, following the simple algorithm outlined in [5]. For instance, a six-pole quasi-elliptic function filter with two attenuation poles on each side of the pass-band zone (see the imaginary axis in Fig. 4), can be represented by a 7×7 topology matrix, since it includes the input and output waveguides.

In general, in such matrices the entries with zero value indicate the absence of coupling, while unity entries underline all possible direct and/or cross couplings. The next step in design such in-line BPFs, involves an optimization algorithm which varies the entries (non-zero elements) of the coupling topology matrix until the response of the matrix produces the desired filter characteristics. Note that the optimization procedure involved here, being extremely fast as it is based on analytically evaluated Gradients, includes constraints on the coupling elements such as the sign and/or the order of magnitude. This is especially important in in-line filter configurations since the coupling of propagating (non-resonating) modes used for bypass coupling decrease by approximately one order of magnitude per bypassed cavity. Once a coupling matrix satisfying the specifications is obtained, an initial filter topological configuration is produced, by respecting the following coupling mechanism:

Table 2 The theoretically obtained Coupling Coefficients Matrix

	Source	1	2	3	4	5	6	Load
S	0.0	1.0988	0.0	0.0	0.0	0.0	0.0	0.0008
1	1.0988	0.0	0.9355	0.0	0.0	0.0	−0.059	0.0
2	0.0	0.9355	0.0	−0.645	0.0	0.0298	0.0	0.0
3	0.0	0.0	−0.645	0.0	−0.608	0.0	0.0	0.0
4	0.0	0.0	0.0	−0.608	0.0	0.644	0.0	0.0
5	0.0	0.0	0.0298	0.0	0.644	0.0	0.9378	0.0
6	0.0	−0.059	0.0	0.0	0.0	0.9378	0.0	1.103
L	0.0008	0.0	0.0	0.0	0.0	0.0	1.103	0.0

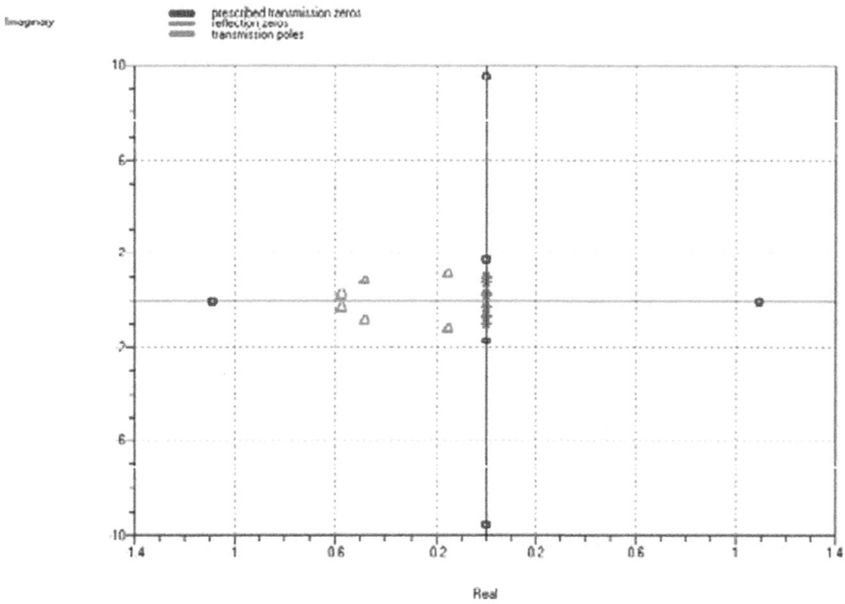

Fig. 4 Transmission zeros allocation of the 6th order X-Band BPF

1. The coupling of the TM_{110} mode in adjacent cavities (see Fig. 2), is generated by the field components propagating through the irises, in the proximate broad wall having an offset from the center.
2. Input and output waveguides are directly offset-connected to the respective cavity walls to provide the coupling of the field components of the perpendicular polarized TE_{10}/TE_{01} waveguide modes and the TM_{110} cavity mode. Since the TE_{10}/TE_{01} modes are accessible (but not resonating) within the cavities, it can be exploited to transfer portions of the signal energy in parallel to the TM_{110} resonance mode of the cavity. Therefore, the modes of the inter-cavities irises (or input/output) waveguides simultaneously couple the TM_{11} resonance mode and the TE_{10}/TE_{01} non-resonating modes.
3. The main inter-cavity couplings as well as those between the input/output waveguides and, hence, the corresponding TM_{110} resonances are controlled by the iris/port center offsets and geometry. Furthermore, bypass couplings are also depended on these geometrical parameters as well (particularly on their lengths). Independent control of the main and bypass couplings is chiefly obtained by the port/iris offset locations at one cavity. If we consider for example no offset, the TM_{110} mode is not excited, and energy would be transferred solely into the non-resonating TE_{10}/TE_{01} cavity modes (with reflections at the cavity entry and opposite walls).

Now, once an initial design for the cavities, irises and offsets is obtained, the entire filter structure may be optimized using as aim the theoretically obtained S_{ij} parameters extrapolated from the ideal (theoretical) coupling coefficients matrix

(M_{ij}). However, such an optimization procedure may lead into a non-conformable filter topology, quite different from the initially considered geometrical topology. Indeed, Fig. 3 illustrates such a device, being composed by two cascaded filter structures; i.e., a three cavities pseudo elliptic BPF along with a three cavities irises coupled K-inverted Chebychev BPF), being, however, fully compliant to filter design requirements (see for the performance results incorporated within the following section).

3 Performance Evaluation Results

The example chosen to demonstrate the efficiency of the BPF design procedure followed here, involves a six-poles design with two imaginary attenuation poles on each side of the pass-band zone and two real transmission zeros (see Fig. 4). The fourth attenuation pole, being used in the synthesis here, was placed at the cutoff frequency of the input/output waveguide. Therefore, input-to-output bypass coupling is required and, hence, all possible entries of the Coupling Coefficients Matrix, being derived according to the procedure proposed in [5], are present. Furthermore, the S_{21}, S_{11} and GD vs. frequency behaviour of the theoretical 6th order pseudo-elliptic BPF, directly derived from the M_{ij} Coupling Coefficients Matrix, is shown in Figs. 5 and 6 respectively.

Now, the simulated by means of μ Wave Wizard^{TM} behaviour vs. frequency of the same BPF shown in Figs. 7 and 8, approximates very tight those shown in Figs. 5 and 6 respectively. This is mostly due to the fact that finite length has been

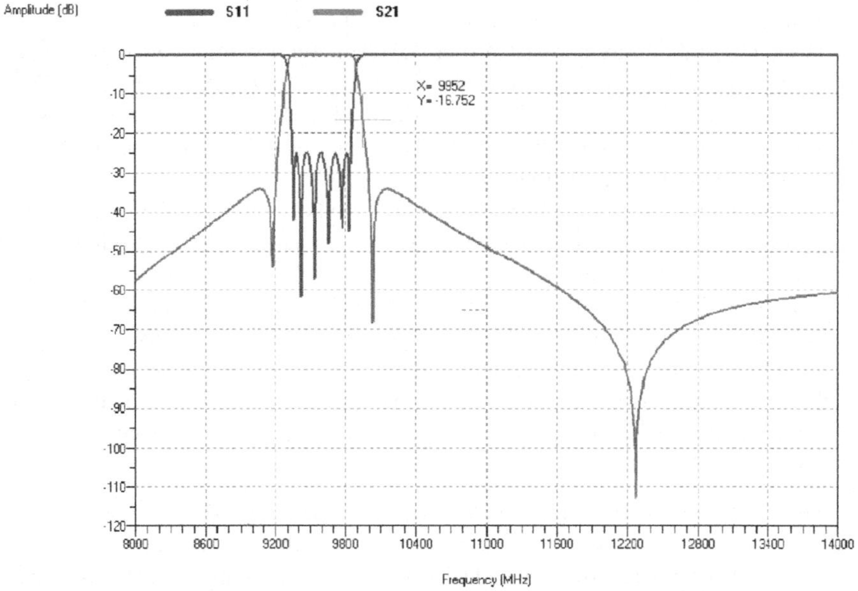

Fig. 5 S_{21}/S_{11} vs. Frequency behavior

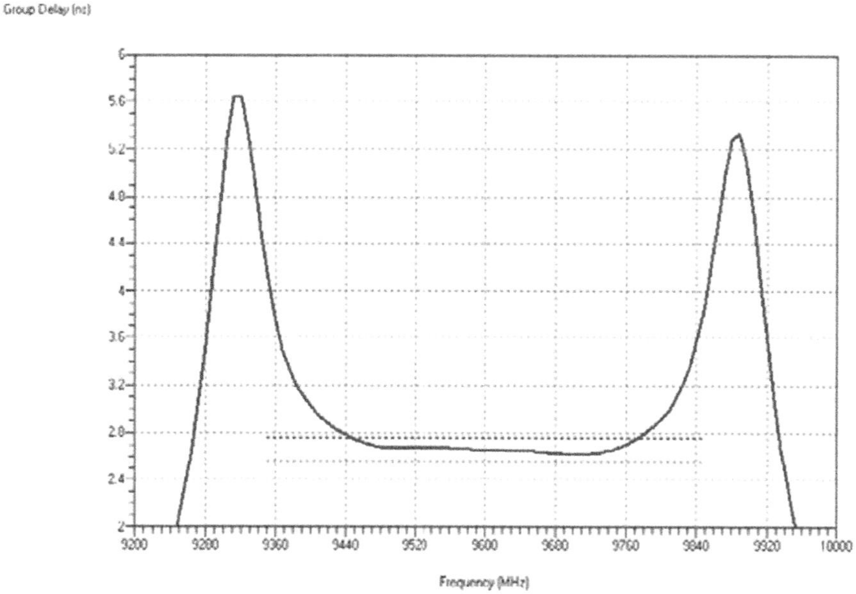

Fig. 6 The optimized GD vs. Frequency behavior

introduced to the inter-cavity irises, as clearly stated in the theoretical design procedure description given in the design procedure section.

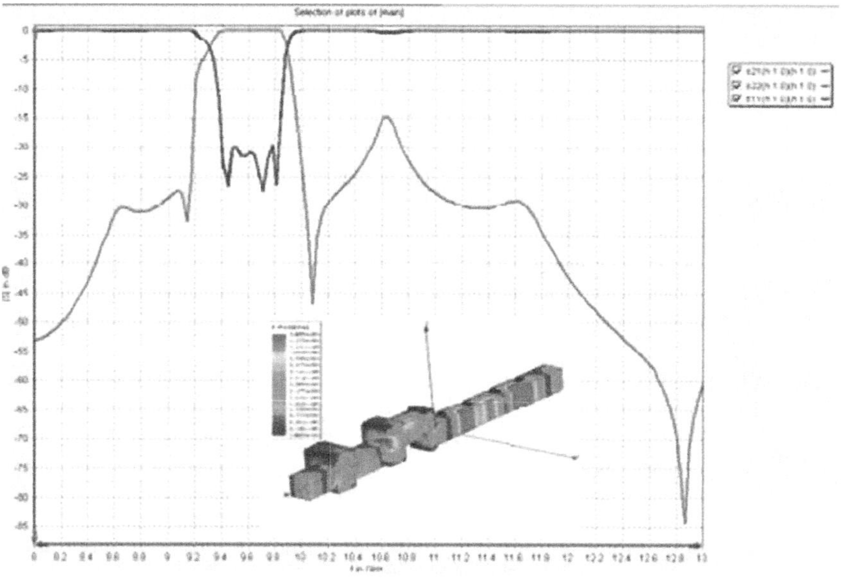

Fig. 7 S_{21}/S_{11} vs. frequency behavior (μ Wave Wizard^{TM})

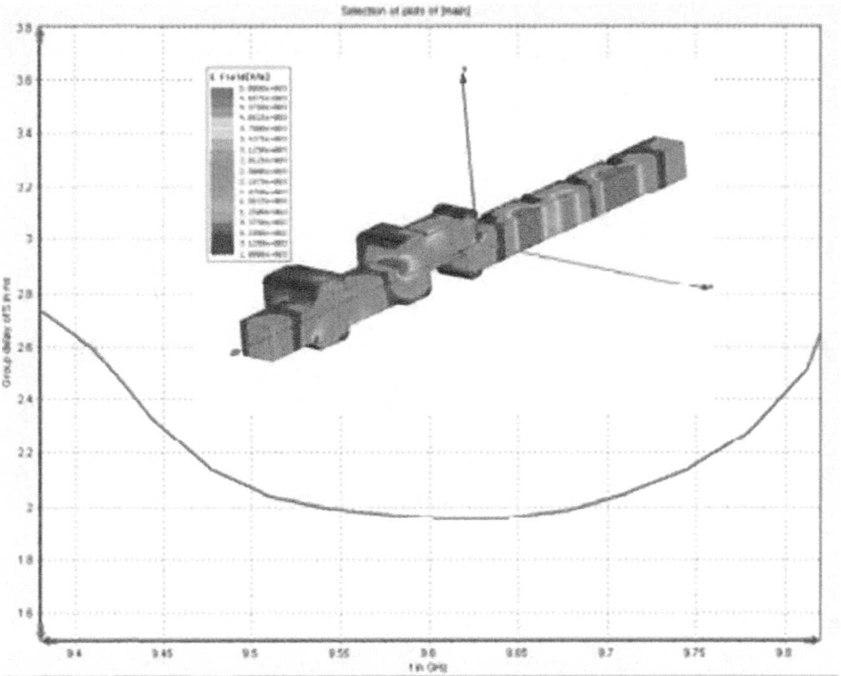

Fig. 8 The optimized GD vs. frequency behavior (μ Wave Wizard^{TM})

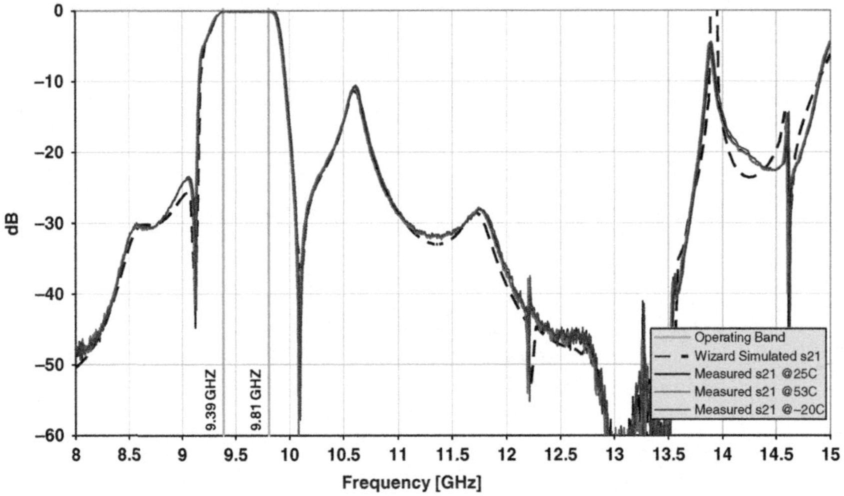

Fig. 9 Simulated vs Measured s21 Results

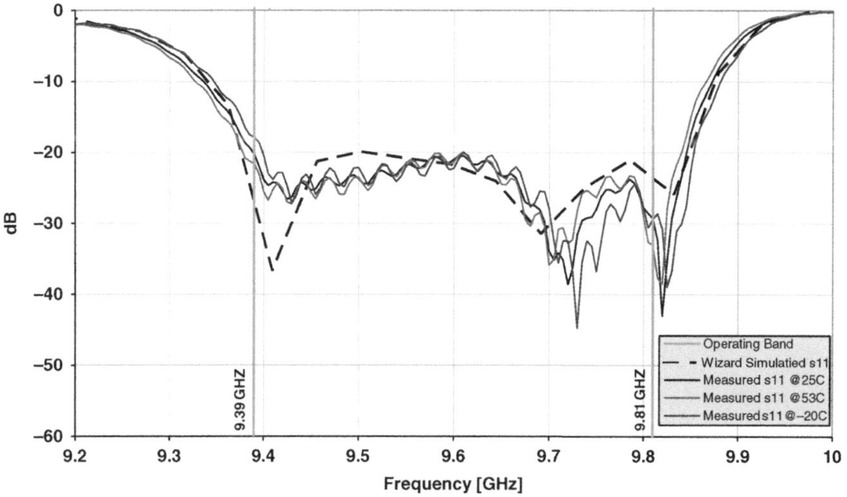

Fig. 10 Simulated vs Measured s11 Results

Indeed, it is the optimization of these irises finite lengths that leaded both into a good agreement between the synthesized and simulated out-of-band zeros allocation and the corresponding S_{11} vs. frequency filter's requested behaviour.

Finally, Figs. 9–11 illustrate the measured S21, S11 and Group Delay vs. frequency behaviour of the realized 6th order X-band pseudo-elliptic BPF (see Fig. 12),

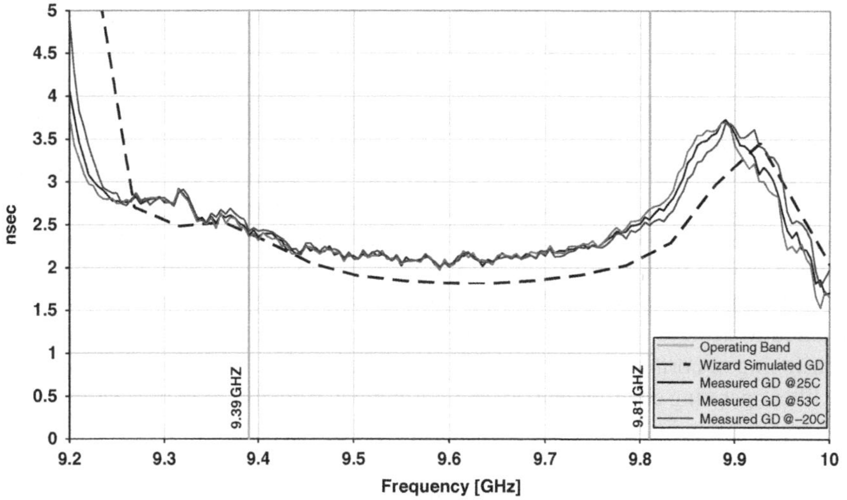

Fig. 11 Simulated vs. Measured Group Delay Results

Fig. 12 The realized 6th order X-Band pseudo-elliptic BPF

in an environment where temperature varies from – 20° C up to +53° C. Indeed, from these figures it is obvious the very close agreement between simulated and measured results (e.g., one-shot design/realization approach).

4 Power Handling Capabilities and Multipaction Analysis

The power handling capabilities of the designed filter are investigated into this section. The phenomena that determine and limit these capabilities are possible discharges that may occur between the internal metal surfaces of the filter, when the input power exceeds a certain level. These discharges can damage the internal surfaces of the filter by permanent corrosion, resulting in the degradation of the RF signal in the output of the component, or even in total component failure. It is thus necessary to take into account these effects in order to determine the maximum RF input power level for safe operation.

There are two different physical phenomena that can both result in RF breakdown inside the filter: Ionization and Multipaction. Ionization, that produces corona discharge, is correlated to the gas pressure inside the component and the dielectric strength that is used to fill it. In normal temperature and pressure conditions the dielectric strength of air is 3 KV/mm. Thus, extremely high input power levels are required to achieve internally these breakdown voltage values, considering the dimensioning of normal waveguide filters, like irises filters. For this reason, and when the filter operating environment doesn't differ much from that of normal temperature and pressure conditions, no analysis is carried out for corona discharge phenomena.

Multipaction, on the other hand, is explained as the phenomenon of vacuum discharge between two surfaces, caused by the resonant growth of free electrons driven by the strong RF fields applied inside the structure. It is vital to conduct multipaction analyses and/or tests mainly for space mission components only, as very low pressure conditions (below 10^{-2} Torr) are needed for this phenomenon to take place. At these low pressure conditions, the surfaces gap distance is much lower than the mean free path of the electrons [10], thus the free electrons are accelerated towards the internal surfaces without impacting on each other. Nevertheless, also in the case of in orbit high power components, venting, in case it can be applied, is sufficient to prevent multipaction. In this study however, perfect vacuum conditions are supposed.

The multipaction analysis is based on the confrontation between the maximum voltages that appear inside the component, and the multipaction threshold voltages that are provided by the Hatch and Williams susceptibility diagrams [8]. In order to evaluate the voltages between the metal surfaces of the filter, a full-wave electromagnetic simulation software capable of representing the field density inside the filter with accuracy is required. In the present analysis, HFSS™ by Ansoft Corporation has been chosen. This software allows for easy calculations of the exact value of the voltage between two internal points. This is essential for conducting an accurate multipaction analysis.

To begin with, the most critical areas for multipaction inside the filter have to be individuated. These are small gap regions and/or regions of high electric field density, where high voltages are expected. Figure 13 shows the electric field vector inside the filter structure, at the YZ plane in the middle of the filter. It is known, from electromagnetic theory, that the electric field density takes maximum values at this plane. This is more clearly shown in Fig. 14.

As shown in Figs. 13 and 14, the field strength inside the cavities of the 3rd order X-Band Chebyshev partial BPF is much higher than in any other part of the filter. Subsequently, the analysis can be focused in these cavities, and more specifically in their center. In fact, the integration lines, used to calculate the conducted voltages by means of HFSS™, are shown in the center of these cavities in Fig. 14. Using

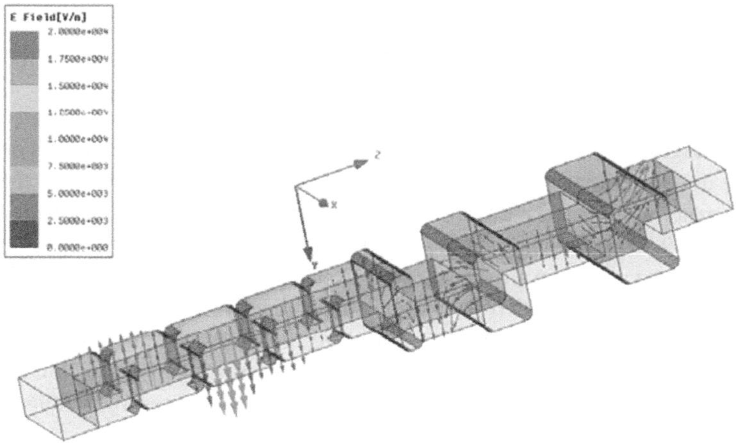

Fig. 13 E-field Vector inside the Filter

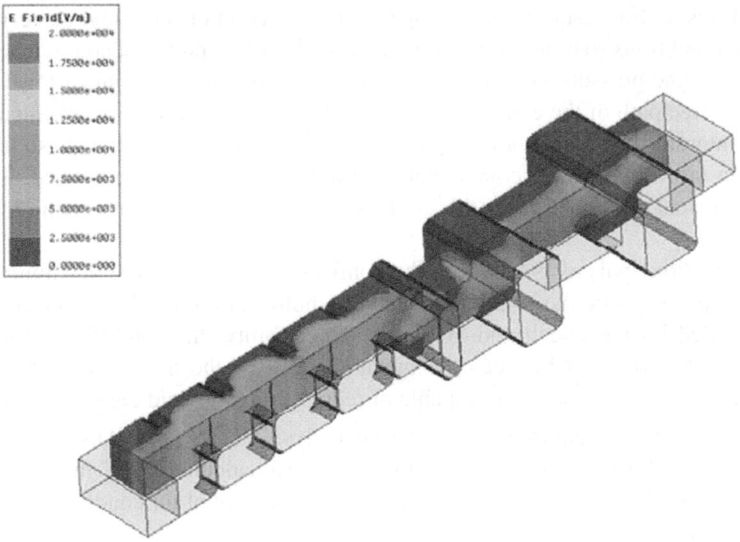

Fig. 14 E-field Magnitude inside the Filter

1 Watt input power and the integration lines, the normalized voltages that appear between the top and bottom walls of these three cavities are calculated.

As mentioned, the multipaction analysis is focused to the critical areas of the filter only. Similarly, not all frequencies are equally susceptible to multipaction. For example, the rejection band of the filter can be totally excluded from the analysis, as very low power levels pass through the filter at these frequencies. On the other hand, it is known [9] that the most susceptible frequencies to multipaction, i.e. the frequencies at which the higher voltages inside most filter structures appear, are the two extremes of the filtering band. This can be considered a general rule of thump. Thus, the analysis can be focused at these frequency ranges only, though the herewith presented analysis considers the whole filtering band for completeness reasons. Both field representations of Figs. 13 and 14 are obtained at the frequency of 9.5 GHz, i.e., inside the operating band of the filter.

The next step is to extract the multipaction threshold voltage from the Hatch and Williams diagrams [8], presented in Fig. 15. The threshold voltage depends on the following three parameters: the analysis frequency f, the gap distance d and the inner surface material properties. Regarding the frequency, the analysis is conducted from 9.15 GHz to 9.95 GHz, thus covering the filtering band and part of the rejection band. The gap distance d is in this case always constant and equal to 10.16 mm, equal to the height of the WR90 type of waveguide. Finally, the internal surfaces of the filter are considered silver plated, in order to reduce the ohmic losses, and subsequently the curves for Ag/Au of Fig. 15 are taken into account. The following formula can thus be utilized in order to calculate the multipaction threshold voltages, beyond which multipaction is expected to occur according to [8]:

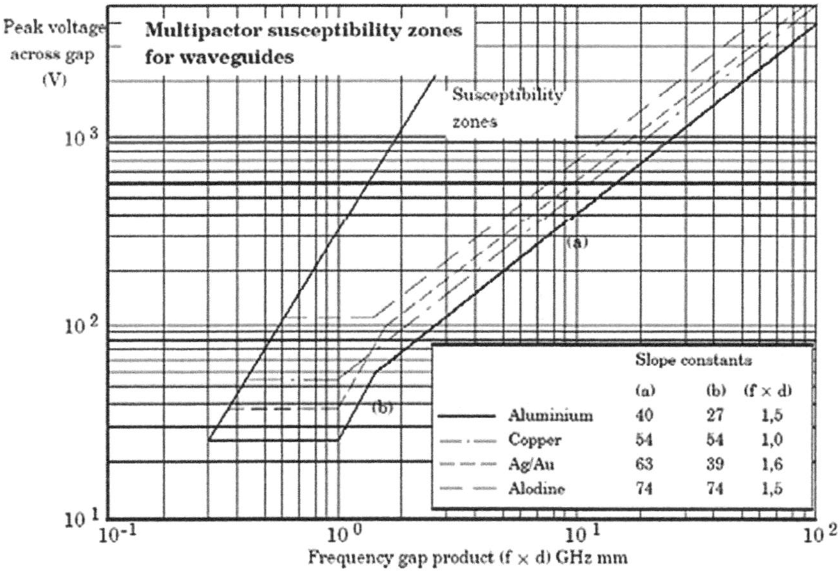

Fig. 15 Multipaction Susceptibility Masks

$$V_{\text{th}} = 63 \times f \times d \tag{1}$$

where

V_{th} the multipaction threshold voltage in [Volt],
f the analysis frequency in [GHz],
d the internal surfaces gap distance in [mm].

The threshold power P_{sc} i.e. the power of a CW at the filter's input port that generates the threshold voltage of eq. (1) inside the filter is calculated by:

$$P_{sc} = \frac{V_{\text{th}}^2 V_{\text{in}}^2}{2 Z_{\text{in}} V_{\text{peak}}^2} \tag{2}$$

where

V_{th} the threshold voltage calculated by means of (1),
Z_{in} the input port impedance, calculated by means of HFSS^{TM},
V_{in} the input incident peak voltage,
V_{peak} the peak voltage inside the filter, calculated inside HFSS^{TM}.

Regarding the input voltage V_{in}, it can be easily calculated by the formula:

$$V_{\text{in}} = \sqrt{2 \times Z_{\text{in}} \times P_{\text{in}}} \tag{3}$$

where P_{in} is set equal to 1 Watt, as already mentioned.

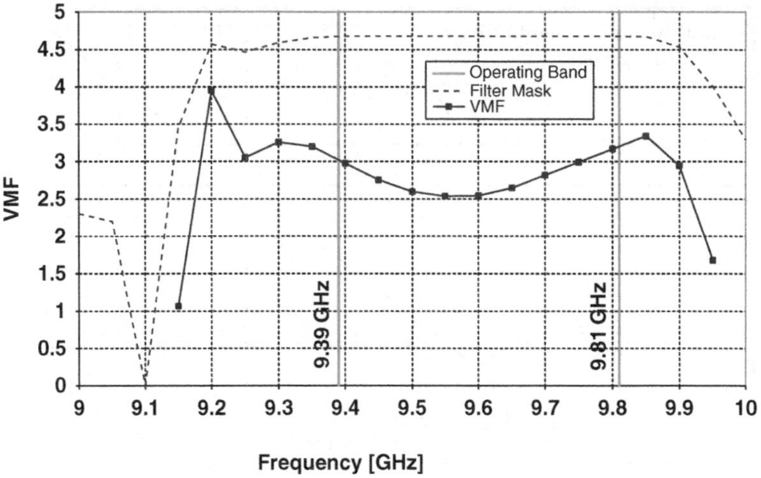

Fig. 16 VMF vs Frequency plotting

It is often used in literature to express the ratio of the peak voltage by the input voltage as the Voltage Magnification Factor:

$$\mathrm{VMF} = {}^{V_{\mathrm{peak}}}\!/\!_{V_{\mathrm{in}}} \tag{4}$$

The voltage magnification factor represents the amplification of the voltage inside the filter in respect to the input incident voltage and is a characteristic of the filter. A plotting of the calculated VMF versus frequency is shown in Fig. 16. As expected, it takes maximum values at the edges of the filtering band and very low values at the rejection band.

Equation (2) can be rewritten including VMF as:

$$P_{sc} = \frac{V_{\mathrm{th}}^2}{2Z_{\mathrm{in}}\mathrm{VMF}^2} \tag{5}$$

Looking at eq. (5), it is shown that the higher the VMF is, i.e., the higher the amplification of the voltage inside the filter, the lower the required power at the input of the filter for multipaction. High Q filters present high VMF.

Figure 17 shows P_{sc}, i.e., the calculated single carrier power level at the input of the filter that is expected to result in multipaction phenomena inside the filter in various frequency points.

As shown in Fig. 17, the minimum value of P_{sc} is calculated to be almost 4350 Watt inside the operating band. Above this input power level multipaction is likely to happen, if perfect vacuum conditions are applied. On the other hand, the amplification stages, which are placed before the transmission filter, are by nature wide-band, in respect to the operating and the filter band. In other words, high power levels are injected to the filter in a frequency band that is much wider than the operating one.

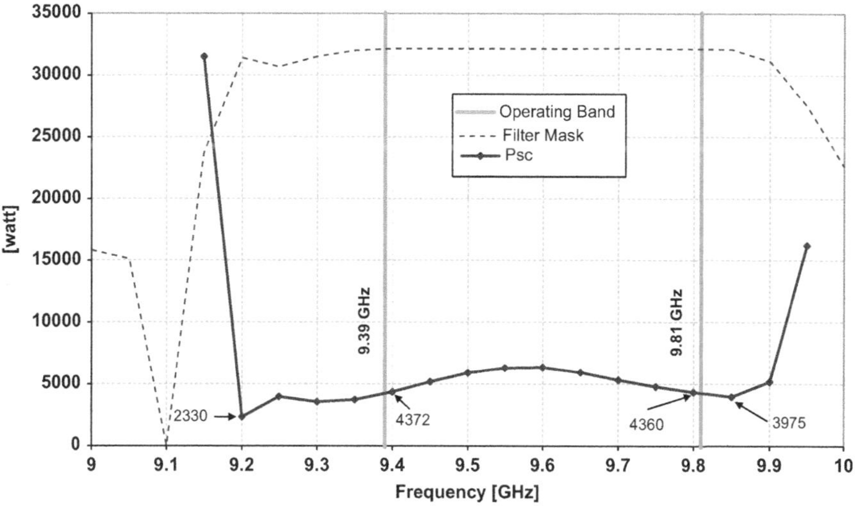

Fig. 17 P_{sc} level vs. Frequency plotting

Thus, it is safer to consider the absolute minimum value of P_{sc} (at 9.2 GHz), in order to determine the multipaction power threshold, and not the local minimum inside the operating band. Subsequently, P_{sc} is calculated equal to 2330 Watt.

Additionally, according to the ESA analysis procedure found in [8], a multipaction margin of 8 dB for metallic RF paths at analysis level is sufficient to exclude the necessity for further hardware testing. Following this approach, the filter can be considered space qualified, from the multipaction point of view, when a single carrier power level of less than 370 Watt is injected. This input power limitation can be furthermore extended by hardware testing, as the 8 dB margin is considered too much conservative. Hardware multipaction testing though is difficult and costly to perform. A more immediate way to increase the maximum allowed input power level is to use Alodine for the internal surfaces, in the place of silver. The use of Alodine, as shown in Fig. 15, results in higher multipaction threshold voltages, allowing for more injected input power. On the other hand, the ohmic losses of the filter rise by the use of Alodine, in respect to silver plating.

Conclusions

An in-line TM_{110} and TE_{10}/TE_{01} modes filters design procedure is presented here, as a complementation to the state-of-the-art waveguide filter design. The exact allocation of the contemporary presented imaginary and real transmission zeros at arbitrary frequencies offers, indeed, many flexible solutions for a large variety of filters applications both for terrestrial and space communications. The validity of the design procedure followed here is clearly demonstrated by the performance evaluation of the X-Band COSMO/SKYMED Calibrator input BPF, being simulated by means of a fast converging in-house build Mode Matching-base CAD too and the commercially available by Mician µ Wave Wizard^{TM}. The very good agreement between the simulated performances and those obtained by means of actual measurements verify, indeed, the validity of the proposed herewith design/realization procedure as an "one-shot" fast converging designing approach.

A thorough multipaction analysis of this BPF is in addition conducted, as a post process to the optimized design, in order to determine the maximum allowed input power level for multipaction free operation, in the case that the designed BPF is to be used as a transmission high power filter in space missions. In fact, the single carrier power threshold is determined with an accurate way, using the exact electric field distribution, calculated inside the filter by means of a full-wave electromagnetic simulator (HFSS^{TM} of Ansoft Corp.).

References

1. G. L. Matthaei, L. Young, and E. M. T. Jones, "Microwave filters, impedance- matching networks, and coupling structures", New York: McGraw-Hill, 1964 (Reprint Norwood: Artech House, 1980).
2. M. Guglielmi, "Simple CAD procedure for microwave filters and multiplexers", IEEE Trans. Microwave Theory Tech., vol. 42, pp. 1347–1352, July 1994.
3. U. Rosenberg, S. Amari, J. Bornemann, "In-line TM_{110}-Mode Filters With High-Design Flexibility by Utilizing Bypass Couplings of Nonresonating $TE_{10/01}$ Modes", IEEE Trans. Microwave Theory Tech., vol. 51, No. 6, pp. 1735–1742, June 2003.
4. Jan Kocbach, Kjetil Folgerø, "Design Procedure for Waveguide Filters with Cross-Couplings", In the proceedings of 2002 IEEE MTT-S, pp. 1449–1452.
5. R. J. Cameron, "Advanced Coupling Matrix Synthesis Techniques for Microwave Filters", IEEE Trans. Microwave Theory Tech., vol. 51, No. 1, pp. 1–10, Jan. 2003.
6. J. Uher, J. Bornemann and U. Rosenberg, "Waveguide Components for Antenna Feed Systems. Theory and CAD", Artech House, Norwood 1993.
7. "COSMO-SkyMed Active Calibrator: Performance Analyses Document", Space Engineering, Rome, Italy, 25 Nov. 2004.
8. "Multipaction design and test" ECSS-E-20-01A, 5 May 2003.
9. "Comparative Study of the Power Capability of Space Broadband Antenna Filters in Ku-band" 5th International Workshop on Multipactor, Corona and Passive Intermodulation in Space RF Hardware, 12–14 Sep. 2005, ESTEC, Noordwijk, The Netherlands.
10. "RF Multicoupler design techniques to minimize problems of corona, multipaction and stability", NASA CR-61381.

Simulation of Coplanar Devices Accessing Nano Systems

F. Peretti, G. Csaba and P. Lugli

We present a series of electromagnetic simulations of coplanar components, which can be used to probe nanoscale and quantum systems. Frequency domain, finite element simulations are used to model waveguides and resonators with gap sizes in the micrometer range. Equivalent circuit models can be coupled to the electromagnetic simulations in order to understand the behaviour of a nanosystems in an electromagnetic environment.

1 Introduction

The formidable advances in nanotechnology has lead in recent years to the realization of a variety of nanostructures and nanodevices with very interesting and appealing properties. However one problem remains though open: fabricating electrical contacts to a nanosystem is often more difficult than realizing a complex, functional nanosystem itself. The challenge is both technological and fundamental. The technological challenge is straightforward to understand, as nanodevices are often fabricated with technologies, which are difficult to integrate with conventional microelectronics fabrication methods. The more fundamental challenge is that contacts bring the nanosystem into interaction with a much larger environment. The interaction with such "thermal bath" can disrupt or alter the "delicate" quantum dynamics properties on the basis of which the device self operates. It would be much less disruptive to access nanosystems without any "hard" connection provided by metallic leads, as for instance via a wireless interaction through electromagnetic fields.

Microwave coplanar devices are very well-suited for this purpose. These structures can be fabricated with micron-scale gaps, thus confining the electromagnetic

F. Peretti,
Institute for Nanoelectronics Technical University of Munich, D-80333 Munich, Germany

G. Csaba
Institute for Nanoelectronics Technical University of Munich, D-80333 Munich, Germany

P. Lugli
Institute for Nanoelectronics Technical University of Munich, D-80333 Munich, Germany

field to an extremely small volume. This result in a strong coupling to a nanoscale system inserted into such volume. The field of the waveguide/resonator can therefore modulate (switch) the nanosystem. Moreover, the fabrication technology of coplanar devices is highly compatible with micro/nano technologies.

If the coupling between the microwave components and the nanosystem is sufficiently strong, then the state of the quantum system can be deduced from a microwave measurement. In other words, the interaction is "two way" and the reaction on the field of the nanosystem can be used for its read-out. Examples of such scheme are already available. Waveguides made from normal conductors can be used to manipulate and read-out nanomagnetic systems [1]. Such waveguides have also been proposed for the manipulation of cold atoms [2]. Superconducting waveguides can achieve extremely high quality factors and can be used both to manipulate and read-out of delicate quantum circuits [3, 4].

Understanding such hybrid structures under 'week' electromagnetic excitation requires (i) an electromagnetic model of the microwave structure, (ii) a model of the nanosystem behaviour (e.g. from quantum mechanics) and (iii) an integration/coupling of the electromagnetic model with the model of the nanosystem. The present paper is mainly concerned with the electromagnetic model, but an outline is given in the last section on how to handle the coupling. For details the reader is referred to our recent publications [5, 6].

The finite-element simulation of the microwave components is a challenging task. The length of the structures lies in the cm range (that is, the reciprocal of the resonance frequency of GHz–10 GHz structures), while the thickness and the gap of the structures are in the submicron and micron ranges, respectively. Our investigation deals with two types of structures: (1) coplanar superconducting resonators (2) coplanar waveguides made of normal conductors. For each of these structures we perform finite-element, frequency-domain electromagnetic simulations with the commercial software tool HFSS™ [7, 8], and compare the results with the results on simplified equivalent-circuit approaches. We also show the field distributions around the waveguides. We believe that field-controlled access to nanosystems can have a big potential in future nanoelectronic applications. Our chapter serves as a guide to the electromagnetic design of the microwave components for such applications.

2 Methodology: The High Frequency Structure Simulator (HFSS)

HFSS [7] is an electromagnetic frequency domain Finite Element Method Full Wave solver. The numerical algorithms used by of an electromagnetic simulation tool can be based, in principle, on different techniques. In general it is possible to divide the numerical frequency domain solvers into two main types: the integral equations solvers and the differential equations solvers.

The former methods solve Maxwell's equations in the integral form. Such solvers are commonly known as BEM (Boundary Element Method) solvers. The latter group includes Finite Difference Method (FDM), MOM (Method of Moments) and Finite Element Method (FEM) solvers, which differ from the method they use to solve the electromagnetic problem [9]. Such methods perform a sub-division ("meshing" process) of the structure into sub parts. Each of this sub-part is related, through appropriate geometrical functions, to the rest of the structure. Such functions, that depend on the geometrical properties of the structure, integrated with the Maxwell's equations, constitute the group of equations (expressed through matrices) that are numerically computed in order to solve the electromagnetic fields.

A FEM is generally appropriate for electromagnetic problems when the structure under evaluation is not too large to be reasonably solved with the available resources (e.g. memory and CPU speed). Even when sufficient resources are available, the total dimensions of the structure should not exceed a given value (a rough rule of thumb is the following: the dimensions of the structure to be simulated should not exceed 200 times the excited in lengths).

The aspect ratio of a structure is defined as the ratio between the biggest and the smallest dimension of the structure. High aspect ratio structures are extremely difficult to mesh. The rule of thumb tells us that the aspect ratio of the structure should not exceed thousand to one. Examples of structures exceeding such value are long, thin wire antennas and 'planar' structures with stratified dielectrics. For extreme aspect ratios a solver exploiting the Method of Moments [10] could be more accurate than a Finite Element Method solver. Coplanar resonators have very high aspect ratios and it is necessary to guide the "meshing" process. Examples of such process are shown in the next paragraphs.

We choose HFSS for our simulation, because the calculations are quite demanding computationally and this code is relatively robust and well-optimized. The solution process of HFSS consists of subdividing the model into elements (tetrahedrons) and to define the basis functions for each tetrahedron. The group of tetrahedrons constitutes the "mesh" of the structure, while each basis function defines conditions between nodal locations in the overall mesh of tetrahedrons. The fields are discretized by means of basis functions in order to obtain matrices to be solved. The results of such process are electromagnetic field values. Such procedure is performed iteratively and, at each step, the computation of the error is performed. Based on the error analysis the process is repeated and the new solutions are compared to with the solution. The procedure goes on until the user-specified criteria for convergence are met.

Structures having localized regions of strong field that are very small compared to the external dimensions are also critical in HFSS. In order to solve them properly the density of tetrahedrons has to be much bigger in such zones with respect to the rest of the structure. If the regions with large electromagnetic fields are really small compared to the rest of the structure the automatic meshing process cannot be accurate enough. Therefore it becomes necessary to manually help the "meshing" process by defining regions with a larger number of nodes. Examples of structures that require such an approach to be solved are coplanar microwave waveguides. In such case, the region where the electromagnetic field is confined is very small compared to the dimensions of the entire system. As a rule, scattering parameters of a structure converge faster then fields. Therefore accurate field distributions also require additional mesh refinement.

3 Simulation of Coplanar Superconductive Resonators

3.1 Model Assumptions

It is possible, as a first approximation, to consider a superconductor like a normal material showing a perfect conducting behaviour. The limits of validity of such approximation depend on many factors. First of all, in a real superconductor the fields penetrate the material causing a variation of the effective permittivity and possibly also losses. In the perfect conductor approximation, there is no electromagnetic field penetration. How much the penetration of the field affects the behaviour of the system (compared to the ideal situation of perfect conductivity) depends on the thickness of the superconductor and on the characteristics of the superconductive material. In many applications [11] the presence of the superconductor minimally affects the behaviour of the device. This is almost true when the guided wavelength in the system is much bigger than the thickness of the superconductor. A further approximation of our approach consists in neglecting the kinetic inductance of the superconducting electrons (which normally adds to the geometric one). We have estimated a few percent error in the simulations due to the latter approximation.

An alternative approach to superconductivity, different from the perfect conducting approach, is the surface impedance approximation [12]. Such method, in principle, can provide a good approximation of superconductivity. However it is still not exact (because Maxwell's equations are not solved inside the conductor). The limits of validity of such approach depend on the penetration depth as compared to the superconductor thickness.

From now on we will model superconductivity using a perfect conductor approximation.

3.2 Design of the Coplanar Resonator

The resonator is going to be described is very similar to the one fabricated and measured in [3, 4]. The only crucial difference is that, in order to be correctly simulated in HFSS, it is necessary to remove the tapering (linear increase of the inner waveguide dimensions connected to the external ports). Therefore all the performed simulations concern the inner part of the resonator. Another important (but not crucial) difference is the radius of each bend of the waveguide. In the fabricated resonator it is around 150 μm, in the simulated one is more than 200 μm. A larger bend is easier to simulate compared to smaller one, because it requires a lower level of accuracy of the meshing process, therefore helping the convergence of the solving algorithm. We have verified that significant differences can arise with bending radiuses smaller than 40 μm, while, for the simulated structures, the results do not change as long the radius is larger than 150μm.

It is important to mention that in HFSS it is necessary to provide the exact dimensions not just of the device, but also of the boundaries where the electromagnetic equations are solved. Therefore in the coplanar resonator is set within an air box, having the same dimensions of the underlying substrate. The external box surrounding the entire chip has a perfect conductive boundary.

Figures 1 and 2 show the simulated resonator.

The waveguide in the middle of the structure has the following dimensions: the central strip is 10 μm wide, the gap between the central strip and the outer conductor is 5 μm; the outer conductor is in touch with the external box, the thickness of the perfect conductor is 200 nm.

In the resonator waveguide there are two capacitances located at a distance of 23.5 mm one from the other. The shape of each capacitance is described in Figure 2b. Each capacitor extends for 100 μm with a gap dimension of 2 μm. The capacitance has a value of 6 fF [3, 4] and can be calculated also using HFSS. It is possible in HFSS by setting up an equivalent circuit analysis. The chip aspect ratio has a value of 10^5.

Figure 1b shows the different layers of the resonator: a silicon substrate, followed by a silicon dioxide layer and by the superconductive (conductive) layer and vacuum box above it.

The distance between the capacitors is directly related to the resonant frequency of the resonator. Therefore it is expected that a change in such distance will result in a shifting of the resonant frequency. The value of 6 fF of each capacitance

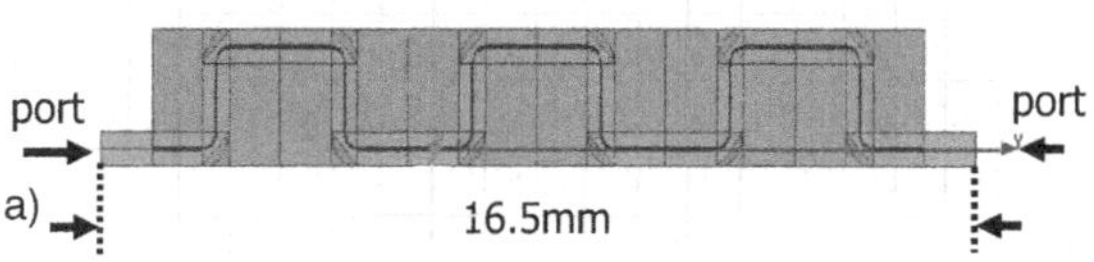

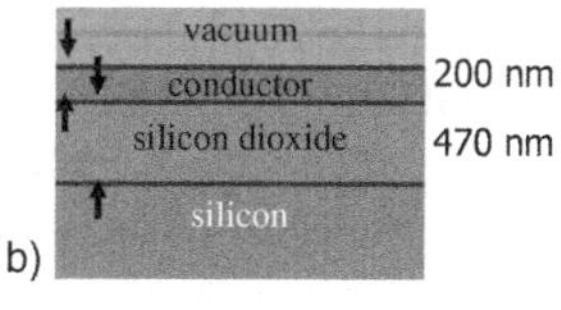

Fig. 1 (**a**) Coplanar superconducting resonator sketched in HFSS. The red part (region of the waveguide) and the purple part (region of the capacitors) are both superconductor (modelled in HFSS as perfect conductor). The position of the two ports is visible in the picture. The length of the entire structure is 16.5 mm. (**b**) Layered structure of the resonator

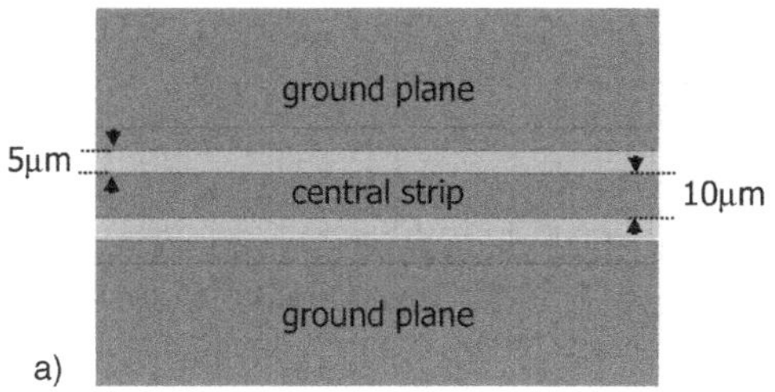

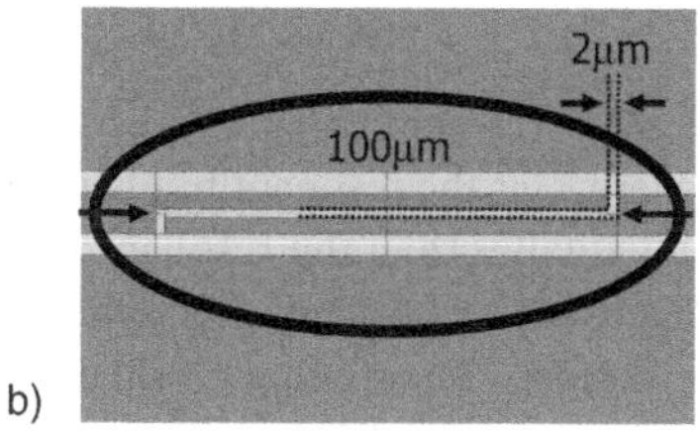

Fig. 2 (**a**) Superconducting coplanar resonator waveguide dimensions. The red part is the superconductor (perfect conductor in HFSS). The gaps between the superconductors confine the coplanar propagating mode. The inner strip is 10μm width, the gap is 5 μm width. (**b**) Capacitive discontinuity dimensions

is also very important because it affects the quality factor of the resonator. It is possible, for special applications such as quantum bit coupling, to use capacitors with a smaller value of the capacitance [3]. The substrate is made of a layer of silicon dioxide located on top of a silicon layer. The height of silicon dioxide is 470 nm. The dimensions of the entire chip are: 15 mm long, 3 mm wide, 1 mm high.

Two ports have been set in the device, for each of them just the first mode is computed. It is straightforward to check that the first computed mode is a coplanar mode with electromagnetic field lines coming from the inner conductor to the outer one or vice versa. Therefore it corresponds to the quasi TEM (Transverse Electromagnetic) mode propagating in a common coplanar device. The ports are perfectly matched with semi-infinite transmission lines having the same section and characteristic of the port. This is also automatically performed in HFSS.

In order to compute the impedance it is also necessary to set a computational line corresponding to the path where the impedance calculation is performed (such calculation is automatically performed by HFSS integrating the voltage along the line). The result of the impedance calculation of the resonator is 48.5Ω. This is almost independent on the excitation frequency. The measured value is 50Ω, in a quite good agreement with the simulated one. The difference found with the simulation comes from two reasons: one is the perfect conductor approximation, the other is the finite distance of the external perfect conducting box to the waveguide. The first factor seems to be more important in our case.

In order to solve the entire chip it is necessary to create a solution setup indicating the central resonant frequency, the number and the characteristic of the frequency sweeps. The central excitation frequency of 2.5 GHz and the frequency sweep from 2.57 GHz to 2.595 GHz have been chosen. The chosen sweep is an interpolating sweep with a step of 5 MHz. The 'interpolating sweep tool' in HFSS is applicable for structures exhibiting sharp resonances, and where fast sweep techniques fail. On a state of the art workstation, with plenty of memory, the simulation time can vary from a few hours up to a few days.

The computed transmission coefficient is reported in Fig. 3a. As we can see, there is a clear resonant frequency at a value slightly higher than 2.58 GHz. The quality factor associated to this curve can be computed using Eq. (1).

$$Q = \frac{\frac{f_0}{f_2 - f_1}}{1 - S_{21}(f_0)} \tag{1}$$

where f_0 is the resonant frequency, f_1 and f_2 are the frequencies respectively greater and smaller than the resonant frequency at which the transmission is −3 dB compared to the transmission at resonance, $S_{21}(f_0)$ is the transmission coefficient at resonance. The computed quality factor is around 10^5, in fair agreement with the measurements of [3].

The measured resonant frequency has values between 2.5 GHz and 3.1 GHz, depending on the thickness of the silicon dioxide layer. The order of magnitude for the measured loaded quality factor is 10^6. Therefore it is clear that the simulated

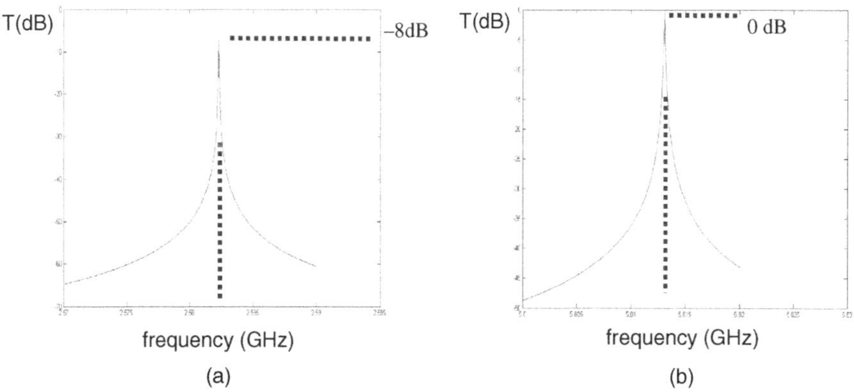

Fig. 3 (**a**) Transmission coefficient of the coplanar superconductive resonator in the range of frequencies from 2.57 to 2.59 GHz. (**b**) Transmission coefficient of the resonator in the range of frequencies from 5.8 to 5.83 GHz

results are in a quite good agreement with the measurements. In particular, concerning the resonance frequencies, the simulated values are in the range of the measured ones. Limits of our simulations are due to the accuracy (already brought at the limit with long computational times) and due to perfect conductor approximations. The latter reason can cause a few-percent shift of the simulated resonant frequency.

The same coplanar resonator has been solved for different setups. In particular a set up with a central frequency of 5.8 GHz and a frequency sweep from 5.8 GHz to 5.83 GHz has been added. The Transmission coefficient is reported in Fig. 3b. Comparing such results with the measurements performed in [3], the resonant frequency occurs in the measured range from 5.5 up to 6.2 GHz. The simulated quality factor is 10^6 in agreement with the measured one. The reason we found a better agreement with respect to the previous simulated frequencies, in terms of quality factor, is probably due to the fact that, at higher frequencies, the dimension of each tetrahedral is smaller, resulting in a finer mesh and therefore in a better accuracy of the entire simulation. We performed a 1 GHz wide frequency sweep around the resonant frequencies, which showed the absence of parasitic resonances.

The resonant frequency of the device is actually related to the distance between the capacitors. The resonant frequency in Fig. 3a corresponds to the situation in which half of the guided wavelength occurs in the middle of the two capacitors. The higher resonant frequencies correspond to the situation in which multiples of half of the guided wavelength occur in the middle of the two capacitors. In order to show such relation it is possible to change the distance between the capacitors and repeating the simulations. It is straightforward to check a shift of the resonant frequency by changing such distance. The quality factor is not affected by the frequency shift.

4 Atomic Trap Waveguides

Magnetic traps using DC currents to generate trapping fields are well established in physics. When microwave excitations are applied on the waveguides, it gives new opportunities for the manipulation of cold atoms [2]. In a joint collaboration with the experimental group of [2], coplanar waveguides for atomic traps applications have been designed and analysed by using HFSS [7]. What makes such waveguides suitable for atomic trap applications is the possibility of exploiting the electromagnetic field generated above the chips at a distance of a few micrometers. An optimal arrangement of many waveguides together creates the proper potential profile capable of generating an atomic trap. We numerically calculate (using HFSS) S-parameters and electromagnetic field distributions generated by coplanar waveguides. Electromagnetic field distributions and losses calculations will be shown. In addition simulated S-parameters will be compared to the measured ones.

5 HFSS Analysis of Coplanar Waveguides

The analysed coplanar waveguide has two tapering parts at both ends in order to be connected to conventional sources (such as coaxial cables or chip probes). In the case of the coplanar superconductive resonators examined before, the tapering had been removed in order to speed up the simulations. The waveguide between the two capacitors was the only part of the structure we analysed. We have verified that the resonant behaviour of the device is not function of the tapering, but depends only on the characteristics of the inner waveguide, of the capacitors and on their distance from each other. In particular we have proven that varying the distance between the capacitors it is possible to tune the resonance frequency.

In conductive coplanar waveguides, the electromagnetic field is going to penetrate the conductor and to produce losses also in the tapering zone. Therefore the entire simulation of the chip has been performed, since tapering of the waveguide may cause undesired resonances and it is important to take it into account in the modelling.

The entire simulated chip is 25 mm long, 3 mm wide and made from gold as illustrated in Fig. 4a. The substrate is made of two layers (Fig. 4b). The bottom layer is silicon (525 μm thick), the layer over the silicon substrate in touch with the golden stripes is polymide (which is 6 μm thick). Buried inside the silicon substrate there is a golden wire having a section of $50\,\mu\text{m} \times 5\,\mu\text{m}$ going from one side of the waveguide to the other one. Such wire extends along the direction perpendicular to the waveguide. Its purpose is to provide a static magnetic field for the localization of the cold atoms. In our simulations, no current is considered in the wire. The golden stripes thickness is 1 μm.

Concerning the horizontal dimensions of the waveguide, a distinction between the inner waveguide (internal part of the tapering) and outer waveguide (external end of the tapering) has to be made. The inner waveguide has an internal strip width of 3.6 μm, the gap between the inner strip and outer stripes is 2 μm, the outer stripes are

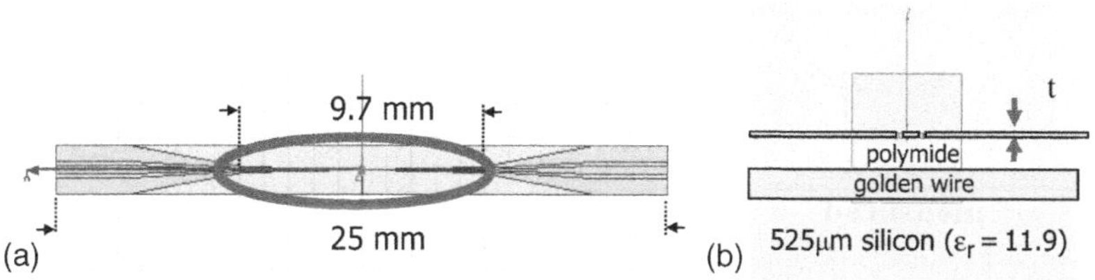

Fig. 4 (**a**) Coplanar conductive waveguide. The length is 25mm. the length of the inner waveguide (without tapering) is 9.7mm. (**b**) Cross section of the coplanar waveguide showing the various layers of. The stripe thickness t is 1 μm

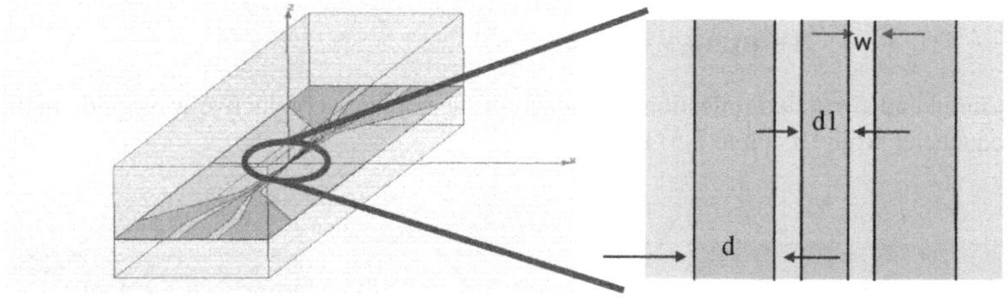

Fig. 5 Coplanar conductive waveguide. The inner waveguide has very small dimensions: the central strip width d1 is 3.6 μm, the ground plane width d is 6 μm, the gap w is 2 μm

6 μm width as illustrated in Fig. 5. The outer waveguide (at each port) has a central strip width of 0.28 mm, a gap of 0.11 mm, 0.5 mm is the width of the outer stripes. The outer stripes represent the ground and they are in touch to the outer box of the chip. As in the case of the coplanar superconductive resonator, an outer air box, on top of the chip, is created. The air box thickness is the same of the substrate. The entire outer box (with the chip inside) has perfect conductor boundary conditions.

Concerning the golden conductor, a complete calculation of the inner fields has been carried out.

We excite a single quasi-TEM (coplanar) mode at the ports, in the same way described previously. A first port-only simulation is performed resulting in a port impedance of 49Ω. This is in good agreement with the measured value of 50Ω.

A full simulation of the device has been performed at a frequency of 6 GHz. A frequency sweep from DC up to 9 GHz has been also performed. The measured and simulated reflection coefficients are shown in Fig. 6. From the figure it is clear that measurements and simulations are in good agreement. Only a little mismatch in the range of frequencies between 6 and 7 GHz is present. The reason is straightforward: the measured chip, had, by accident, a length of 23 mm instead of 25 mm. In addition the electromagnetic field distributions and the losses have been analyzed.

The losses calculations are performed in HFSS by integrating the density of lost power inside each volumes of the chip (such as silicon substrate, golden stripes, polymide layer). From our calculations it turns out that the larger losses occurs in the conductor, especially in the central golden strip. The silicon substrate and polymide losses are very small in comparison.

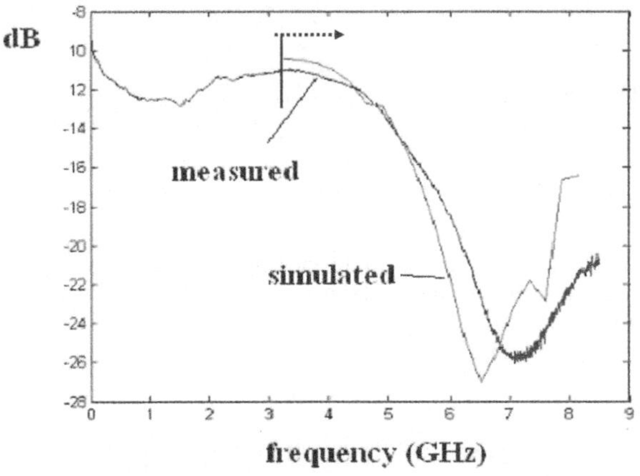

Fig. 6 Measured and simulated reflection coefficient of the coplanar conductive waveguide in the range of frequencies from DC up to 8.5 GHz

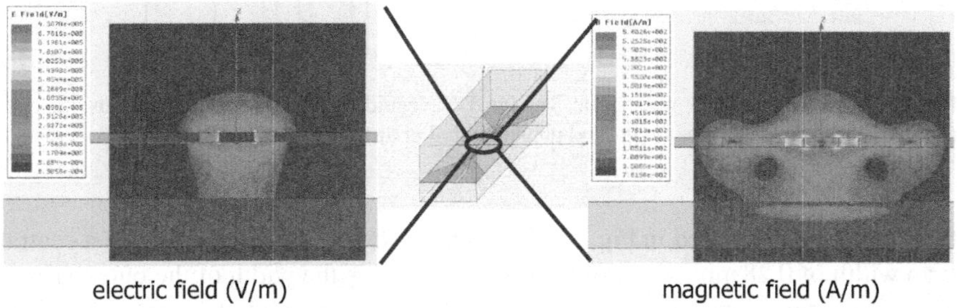

Fig. 7 Electric and magnetic field distributions in planes perpendicular to the chip waveguide. The calculations are performed on a 20 μm × 20 μm squared section. The magnetic field penetrates entirely the conductor

The electric and magnetic field distributions in sections perpendicular to the waveguide, in the very middle of the entire chip, are actually responsible of the atomic trap generation on top of the chip. The cold atom trapping takes place over the chip at a distance of 2–10 μm from it. Figure 7 shows such field distributions. It is worth mentioning that the magnetic field penetrates the entire conductor, causing the largest contribute to the losses. Such result is in agreement with the penetration depth of gold of 900 nm (big almost as the conductor thickness) at such frequencies.

6 Modelling of the Resonator-Nanosystem Interaction

As an example of how to model the interaction between nanosystems and electromagnetic fields, we have investigated the interaction of single-domain magnetic particles and coplanar resonators or waveguides. Even for very small magnetic particles (in the size range of 10–100 nm) the coupling with the resonator is strong, and we can show that resonators made of resistive conductors have sufficiently high quality factor to read-out the state of such nanoparticles. Qualitatively, the magnet/resonator coupling can be understood by considering the magnet as a tiny piece of material with a nonlinear permeability. The nonlinear permeability changes locally the inductance of the resonator, which then can be detected as a change in its transmission, or phase shift.

Our modelling approach is illustrated in Fig. 8. We generate an equivalent circuit model of the resonator, an equivalent circuit of the nanosystem and interconnect the lumped models in a circuit simulator. Commercial electromagnetic field solvers (among them HFSS) are capable to extract lumped circuit parameters from a finite-element electromagnetic field solution. The lumped circuit representation can be analyzed by a circuit simulator, such as PSPICE. Each mode of the electrical field at a particular port corresponds to a port (terminal pair) in the circuit model. If we assume that the resonator supports only one mode at each port and the equivalent circuit is a three port, two of those ports correspond to the 'real' (coaxial) inputs of the resonator and the third one is connected to the current loop in the resonator gap.

Nanoscale magnets can be modeled as tiny oscillators, with their ferromagnetic resonance frequencies lying in the 1 GHz–10 GHz range. This resonance frequency depends on their shape and the applied external magnetic field. We consider an approximately 80 nm by 100 nm size, ellipsoid-shaped permalloy nanomagnet. If there are no strong external fields, its magnetization points along the longer axis (positive or negative x direction in Fig. 9). A weak static bias field is applied in the positive x direction and a very weak, oscillating field along the z axis.

We numerically solved micromagnetic equations for several frequencies of the H_z^{ext} external field [6]. The Fourier transform of the average M_z magnetization component as the function of excitation frequency is shown in Fig. 9.

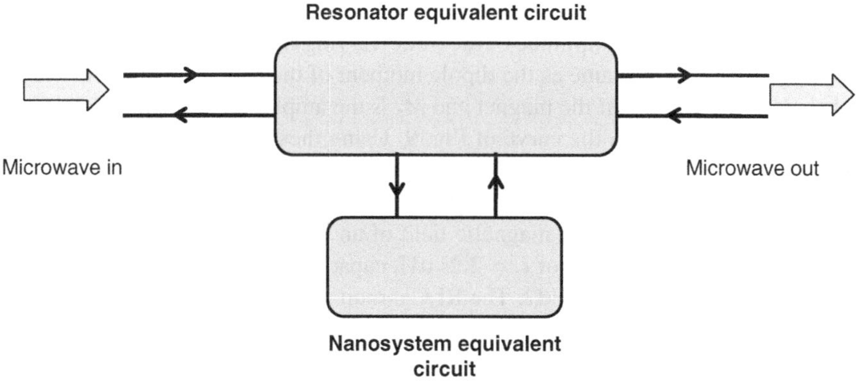

Fig. 8 The electromagnetic part of the problem is modeled as a three-port: two of ports are real microwave connections, while the third one connects to the equivalent circuit of the nanosystem

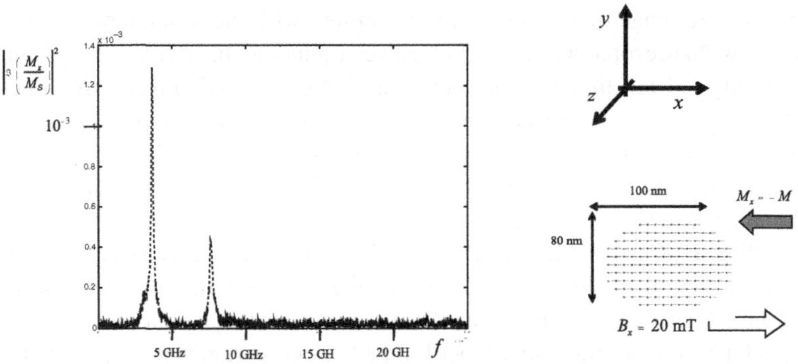

Fig. 9 A near single-domain nanomagnet is pumped with a varying frequency, $B_z = 20.0\,\mathrm{mT}$ electromagnetic field. The Fourier transform of the response is shown The magnet shows resonance. $f = 3.6\,\mathrm{GHz}$ – this resonance frequency strongly depends on the state of the magnet and the external field. The magnetization is normalized to the value of saturation magnetization, M_s

The magnetization precesses around the x axis and the precession frequency depends on the state of the magnet (i.e. whether it points to the negative or positive x direction).

Now we investigate this nanomagnet when it is placed in the center of coplanar resonator. Sufficiently far from the nanomagnet (i.e. a few hundred nanometers distance) its field can be approximated by a point dipole and is indistinguishable from the magnetic field of a current loop. It means that the field-interaction between a nanomagnet placed in the resonator can be modeled by connecting an equivalent circuit of the magnet to the resonator circuit model. In the first approximation, the equivalent circuit can be modeled as a lumped *RLC* circuit.

A time varying magnetic field of amplitude $\hat{B}_z$ induces a $\hat{V} = Aj\omega\hat{B}_z$ voltage in the current loop of the equivalent circuit (A is the surface of the loop). This voltage gives rise to a current of $\hat{I} = \frac{\hat{V}}{Z}$, where $Z = R + j\omega L + \frac{1}{j\omega C}$ is the complex impedance of the *RLC* circuit. The amplitude of the (time-varying) dipole moment generated by the loop should be the same as the dipole moment of the nanomagnet: $\hat{I} \cdot A = \hat{M}_z V$, where V is the volume of the magnet and $\hat{M}_z$ is the amplitude of the magnetization, which can be seen from the curve of Fig. 9. Using these formulas and the relation between the loop surface and its inductance one can 'fit' the resonance curve of the *RLC* circuit. For example the magnetic field described by the curve of Fig. 9 is indistinguishable from the magnetic field of an *RLC* circuit with a $1\,\mu\mathrm{m} \times 1\,\mu\mathrm{m}$ area loop, a loop inductance of $L = 2.21\,\mu\mathrm{H}$, capacitance of $C = 8.84 \cdot 10^{-16}\,\mathrm{F}$ and a serial resistance of $R = 4.2\,\mathrm{k}\Omega$. The RLC circuit representing the magnet couples inductively to a center section of the resonator, as it is sketched in Fig. 10.

After determining the circuit parameters of the nanomagnet, it can be connected to the equivalent circuit of the resonator [6]. Preliminary results show that the nanomagnet pulls the resonant frequency of the resonator by a few megahertzs, which is detectable [1]. The same modeling methodology could be applied to model the

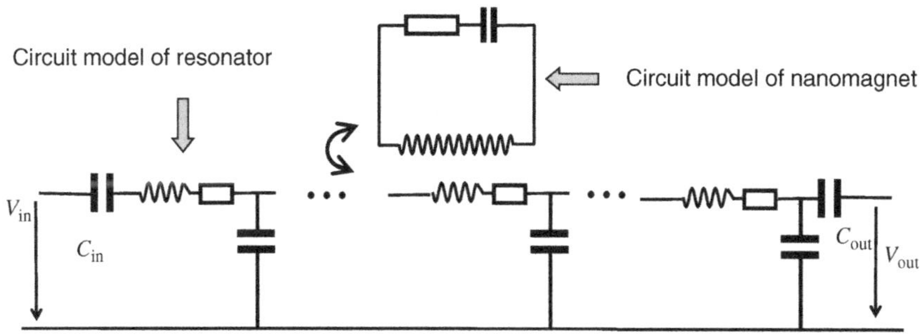

Fig. 10 Illustration of the coupling between the nanomagnet and the resonator

coupling of resonators to other types of nanosystems, such as polar molecules, flux or charge quantum bits [4] or quantum dots.

7 Summary and Conclusions

We have presented elements of a modelling framework to understand the interaction of microwave structures and nanosystems. We performed electromagnetic field simulations to model resonators and waveguides with extreme aspect ratios. Such simulations push finite-element simulation codes to their limit, but the obtained simulation results are in good agreement with measurements. The field simulations can be used to obtain circuit models, which in turn provide a tool to model the interaction of nanosystem with microwave electromagnetic fields.

Acknowledgements We thank to Prof. Gross, Thomas Nyemzcyk, Susanne Hofmann for the collaboration on the superconductive coplanar resonators. We also acknowledge to Prof. Theodore Hänsch, Dr. Philipp Treutlin, Pascal Böhi for the fabrication and measurement of the coplanar waveguides and for the useful work performed together. We are grateful to Prof. Dirk Grundler for bringing into our attention works related to resonator-nanomagnet interaction and to Prof. Árpad Csurgay and Prof. Wolfgang Porod for their advices regarding equivalent circuit construction. We acknowledge the support of the Deutsche Forschungsgemeinschaft (DFG) through the Collaborative Project SFB 631.

References

1. J. Podbielski, F. Giesen, M. Berginski, N. Hoyer, D. Grundler: "Spin configurations in nanostructured magnetic rings: From DC transport to GHz spectroscopy", Superlattices and Microstructures, vol. 37, issue 5, May 2005, pp. 341–348.
2. P. Treutlein, T. W. Hänsch, J. Reichel, A. Negretti, M. A. Cirone, T. Calarco, "Microwave potentials and optimal control for robust quantum gates on an atom chip", Phys. Rev. A 74, 022312 (2006).

3. L. Frunzio, A. Wallraff, D. Schuster, J. Majer, R. Schoelkopf, "Fabrication and characterization of superconducting circuit QED devices for quantum computation", IEEE Transactions on Applied Superconductivity, vol. 15, Issue 2, part 1, June 2005, pp. 860–863.
4. A. Wallraff, D.I. Schuster, A. Blais, L. Frunzio, R-S. Huang, J. Mayer, S. Kumar, S.M. Girvin, R.J. Schoelkopf, "Strong coupling of a single photon to a superconducting qubit using circuit quantum electrodynamics", Nature, vol. 431, pp. 162–167 (2004).
5. G. Csaba, Z. Fahem, F. Peretti, P. Lugli, "Circuit Modeling of flux Qubits interacting with superconducting waveguides", Journal of Computational Electronics, Springer Netherlands, ISSN 1569-8025 15728137, January 19, 2007.
6. G. Csaba, A. Màtyàs, F. Peretti, P. Lugli, "Circuit modelling of coupling between nanosystems and microwave coplanar waveguides", International Journal of Circuit Theory and Applications 35, 3, May/June 2007, pp. 315–324.
7. Z.J. Cendes, " Vector finite elements for electromagnetic field computation", Magnetics IEEE Transactions on, vol. 27, issue 5, pp. 3958–3966, September 1991.
8. http://www.ansoft.com/products/hf/hfss/
9. M. N. O. Sadiku, "Numerical Techniques in Electromagnetics", published 2001, CRC Press, ISBN 0849313953.
10. D. H. Werner, "A Method of Moments Approach for the Efficient and Accurate Modeling of Moderately Thick Cylindrical Antennas", IEEE Transactions on Antennas and Propagation, vol. 46, no. 3, March 1998.
11. J. Kessler, R. Dill, P. Russer: "Field Theory Investigation of High-Tc Superconducting Coplanar Waveguide Transmission Lines and Resonators", IEEE Transactions on Microwave Theory and Techniques, Sep 1991, vol. 4, issue 9, pp. 1566–1574, ISSN 0018-9480.
12. L. H. Lee, W. Gregory Lyons, T. P. Orlando, S. M. Ali, R. S. Withers, "Full-Wave Analysis of Superconducting Microstrip Lines on anisotropic Substrates Using Equivalent Surface Impedance Approach", IEEE Transactions on Microwave Theory and Techniques, vol. 41, no. 12, December 1993.

Time-Domain Measurements of Electromagnetic Interference

Stephan Braun, Arnd Frech and Peter Russer

1 Introduction

Traditionally measurements of electromagnetic interference are performed in frequency domain. Seventy years ago radio noise meters have become available [1] to detect radio noise. Many improvements have been made over the past 70 years. Today EMI receivers and spectrum analyzers are used for emission measurements in frequency domain. EMI receivers perform a band-pass filtering of the input signal by the intermediate frequency (IF)-filter and use a weighting of the instationary emitted signal in the peak, quasipeak, average and rms detector modes. In order to perform a characterization of a device under test for a certain frequency band the filtering and weighting by a detector has to be performed at a certain number of frequency bins sequentially. For a typical emission measurement several thousand frequency bins have to be measured sequentially. Due to such a large number of frequency bins and long time constants of the detector functions as well as the instationary emissions a typical characterization of the emission of a device takes several hours.

2 Time-Domain EMI Measurement System

An efficient way to reduce the measurement time is to perform the measurements in time-domain. In the following the theory and application of a real-time operating time-domain EMI (TDEMI) measurement system are presented. With a time-domain EMI (TDEMI) measurement system the input signal is digitized by an

Stephan Braun
Institute for High-Frequency Engineering, Technische Universität München,
e-mail: stephan.braun@tum.de

Arnd Frech
Institute for High-Frequency Engineering, Technische Universität München, e-mail: frech@tum.de

Peter Russer
Institute for High-Frequency Engineering, Technische Universität München,
e-mail: russer@tum.de

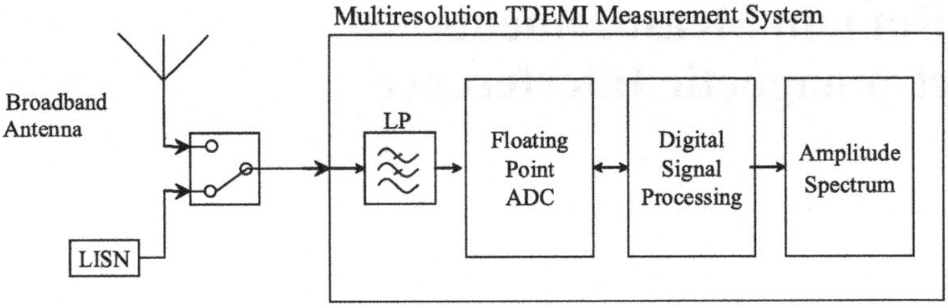

Fig. 1 Multiresolution time-domain EMI measurement system

analog-to-digital converter (ADC) system. Spectral estimation is performed by a fast Fourier transform (FFT). The block diagram of a multiresolution TDEMI measurement system is shown in Fig. 1. The EMI signal is received via a broadband antenna [2]. By a multiresolution analog-to-digital Converter (ADC) system a floating point analog-to-digital conversion is performed [3]. The measured and digitized EMI signal is processed via digital signal processing and the amplitude spectrum is displayed.

An algorithm that allows measurements in the peak, average and rms detector modes to be simulated has been presented in [4]. One of the first systems where based on a digital storage oscilloscope and used the fast Fourier transform as presented by C. Keller [5]. An algorithm for the quasipeak detector mode has been presented in [6]. With both algorithms almost all requirements that are given by CISPR 16-1-1 are fulfilled [7]. The only requirement that could not be fulfilled by these algorithms is the continuous processing of the input signal. However continuous processing is mandatory in order to provide the requested IF-signal and quasipeak-meter signal. Such auxiliary output signals can only be provided by a system that is operating in real-time. Another benefit of the real-time operating system is processing of the EMI-signal without gaps during the dwell time.

Due to the real-time operation the system can emulate the complete behavior of several thousand parallel EMI receivers. A detailed comparison between the analogue signal processing and the digital signal processing of the real-time operating system is shown. The EMI signal is digitized by an analog-to-digital converter system. The frequency band that has to be processed during the characterization of a device under test is subdivided into several sub-bands. The sub-bands are measured sequentially. Each sub-band is down-converted into its baseband by a digital down-conversion (DDC) unit. The down-converted signal is down-sampled and processed by a short-time fast Fourier transform (STFFT). The STFFT generates a spectrogram that shows a discretization in time and frequency [8]. At each calculated discrete frequency the evaluation is performed by a digital detector.

2.1 Fast Fourier Transform

Digital spectral estimation is achieved via the discrete Fourier transform (DFT). Algorithms for DFT computations that exploit symmetry and repetition properties of the DFT are defined as FFT. The DFT formulation considers periodic repetition of the signal in time-domain and is given as follows:

$$X[k] = \sum_{n=0}^{N-1} x[n] e^{\frac{-j2\pi kn}{N}} \tag{1}$$

2.2 Short-Time Fast Fourier Transform

EMI signals however consist of stationary and transient signals as well as noise. In order to model the exact behavior of an EMI receiver the STFFT is used. By the use of a Gaussian window function the IF-Filter is modelled according to its impulse bandwidth, noise-bandwidth and the filter masks as defined in CISPR 16-1-1. By the STFFT a spectrogram is calculated. This spectrogram shows a disctretization in frequency and time. The resolution in frequency is described by the bin-width $\triangle f$. The resolution in time is described by a time step T_{sBB}. The inverse of the time-step is called baseband sampling frequency f_{sbb}. The STFFT calculates as follows:

$$X[t,k] = \sum_{n=0}^{N-1} x[n+t] w[n] e^{\frac{-j2\pi kn}{N}} \tag{2}$$

where $w[n]$ is a Gaussian window function that models the IF-filter of an EMI receiver [6]. The relation between the samples that are processed by a single FFT and the number of samples where shifting is performed is described by the overlap factor O_f. The number of FFT calculations N_O that have to be performed for N samples is described by:

$$N_O = \frac{1}{1-O_f} = \frac{f_{sbb}}{\triangle f} \tag{3}$$

The discretization in the time-domain has to be high enough to fulfill the Nyquist criterion. An overlapping factor of about 75% is sufficient to achieve a sufficiently high discretization for the simulation of the various detector modes.

2.3 Spectrogram

A spectrogram shows a discretization in frequency and time. An example of a spectrogram is shown in Fig. 2. For a certain point in time e.g. t_1 an instantaneous

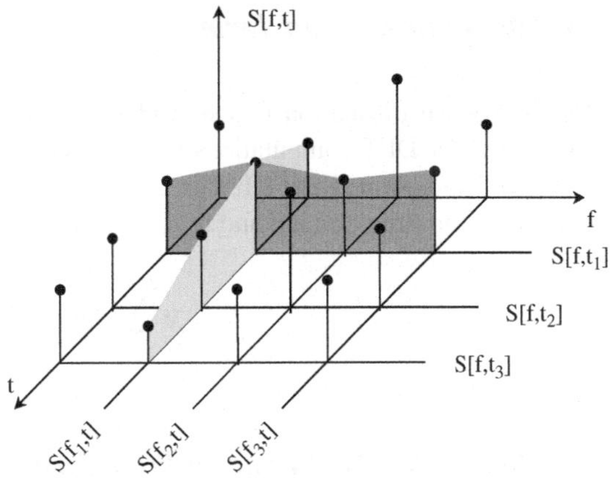

Fig. 2 Spectrogram

spectrum of the emission of a device under test is described. For instationary signals the amplitude of the spectra changes over the time. The signal in time domain of a certain frequency bin f_1 corresponds to the demodulated IF-signal of an EMI receiver which is tuned to the frequency f_1. Thus such an implementation corresponds to several parallel EMI receivers, with a demodulated and sampled IF signal.

2.4 Detector Modes

The signal in time domain extracted from the spectrogram at each frequency bin is evaluated by a detector mode. This corresponds to a parallel evaluation of parallel sampled IF-signals by a set of parallel detectors. In a TDEMI measurement system several thousand frequency bins are evaluated by the peak, average, rms, cispr-average and quasipeak detector modes. The peak detector formulation is obtained by:

$$X_p = MAX\{X[n] | n \in \{1 \ldots N\}\} \tag{4}$$

The formulation of the average detector is given by:

$$X_{\text{avg}} = \frac{1}{N} \sum_{n=0}^{N-1} X[n] \tag{5}$$

The formulation of the rms detector is given by:

$$X_{\text{rms}} = \sqrt{\frac{1}{N} \sum_{n=0}^{N-1} X^2[n]} \tag{6}$$

2.5 Analog Quasipeak Detector

In Fig. 3 the circuit of the analog quasipeak detector is shown. The quasipeak detector was originally described to perform a weighting of transient signals according to the disturbance of AM radio reception [9]. The demodulated signal $u(t)$ of the conventional EMI-receiver charges the capacitor C by the resistor R_1 as long as $u(t)$ is above $u_2(t)$. This is a typical RC-charging with the time constant τ_c. If the input signal $u(t)$ is lower than $u_2(t)$, the capacitor C is discharged via resistor R_2. A critically damped meter with the time constant τ_m is used to display the amplitude. A buffer is necessary to avoid any repercussions. The maximum of the shown value is taken as quasipeak value. The values of time constants and IF bandwidths for CISPR 16-1 [10] are shown in Table 1.

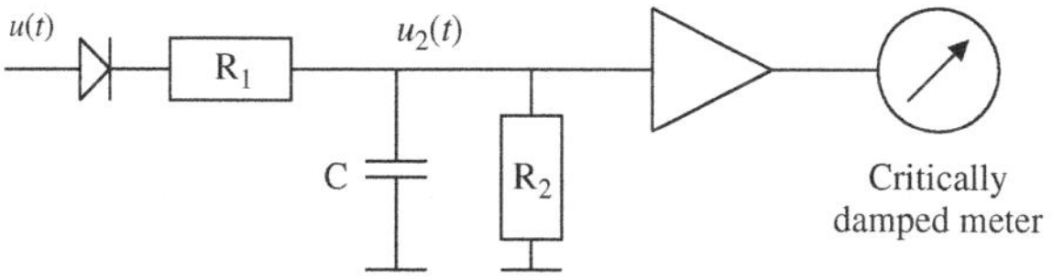

Fig. 3 Analog quasipeak detektor

Table 1 CISPR specifications for quasipeak detector

	Frequency range		
	Band A 9–150 kHz	Band B 0,15–30 MHz	Band C/D 30–1000 MHz
IF 6dB Bandwidth	0,2 kHz	9 kHz	120 kHz
τ_c	45 ms	1 ms	1 ms
τ_d	500 ms	160 ms	550 ms
τ_m	160 ms	160 ms	100 ms

2.6 Digital Quasipeak Detector

In order to integrate a quasipeak detector into the TDEMI measurement system a digital model is developed. The modelling is performed via Infinite Impulse Response (IIR) – filters. The signal $u(t)$ is provided at a single discrete frequency bin from the calculated spectrogram.

For modelling the charging and discharging processes digital IIR1-filters are used. Figure 4 shows the signal flow of an IIR1-filter. For calculating the filter coefficients, the following transfer function is used:

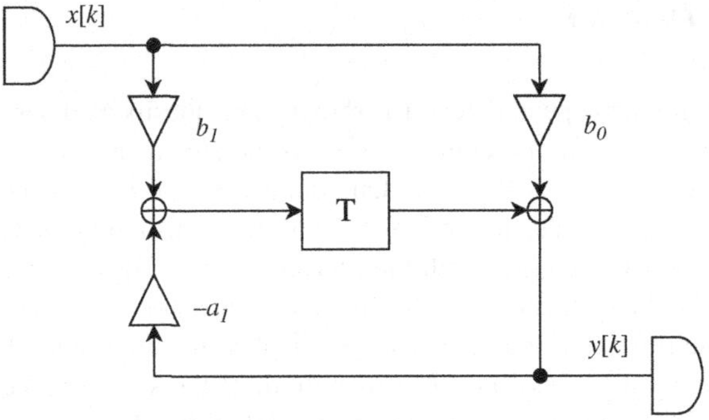

Fig. 4 Signal flow of a IIR1-filter

$$H(s) = \frac{1}{1 + \tau s} \tag{7}$$

Also the bilinear transformation for characterizing the RC-system is used. The bilinear transformation is defined as follows:

$$s = \frac{2}{\Delta t} \frac{z-1}{z+1}, \tag{8}$$

where s is the frequency-domain variable, z is the discrete time-domain variable and Δt is the time-domain discretisation. In order to get better numeric precision not the normal topology of a IIR2-system, but a cascading of two identical IIR1-filters for modelling the critically damped meter were used. The maximum of the output signal is taken as quasipeak value. Charging and discharging is decided by a comparator, while in discharging mode the coefficient b_0 is set to zero.

3 Dynamic of a Time-Domain EMI Measurement System with One ADC

In this section we discuss the dynamic range that can be achieved with a time-domain EMI measurement system when a single signal interval is digitized. We only consider quantization noise that is added to the signal during analog-to-digital conversion. We have to consider the signal-to-noise ratio of a single spectrum calculated by the short-time fast Fourier transformation [11]. By this way we can describe the dynamic parameters that can be compared with conventional EMI receivers.

For a general signal the SNR is given by [8]:

$$\mathrm{SNR} = \frac{P_{\mathrm{Signal}}}{P_{\mathrm{Noise}}} \tag{9}$$

while the power of the signal is given by (10):

$$P_{\text{Signal}} = \sigma_s^2 = \left\langle (x(t) - \langle x(t) \rangle)^2 \right\rangle \tag{10}$$

The power of the quantization noise is given by (11)

$$P_{\text{Noise}} = \sigma_e^2 = \frac{\triangle^2}{12} \tag{11}$$

with $\triangle$ as the quantization step of the analog-to-digital-converter (ADC).

Spectral estimation is performed using the Discrete Fourier Transform (DFT). The DFT is used to calculate the amplitude spectrum of a signal sample of length N. In addition DFT is based upon the assumption that this sample is continued periodically in time. So the time record $x[n]$ needs to be multiplied with a window $w[n]$ of length N to avoid spectral leakage:

$$z[n] = x[n]w[n] \tag{12}$$

For the window function $w[n]$ a Gaussian window is used. In order to get at each discrete frequency value the same filter response as the IF-filter of the conventional EMI receiver the window needs to be adapted to the IF-filter regarding impulse-bandwidth and equivalent noise-bandwidth. The DFT being applied to each of the data blocks and is defined as follows [12]:

$$Z[k] = \sum_{n=0}^{N-1} z[n]\, e^{-j\frac{2\pi kn}{N}} \qquad k = 1 \ldots N \tag{13}$$

The digitized signal $\hat{x}[n]$ is the result of the superposition of the original signal $x[n]$ and the quantization noise $e[n]$. Using (12) we obtain:

$$z[n] = x[n]\, w[n] = \hat{x}[n]\, w[n] + e[n]\, w[n], \tag{14}$$

where $e[n]$ is described as a statistical signal with constant probability density function in the interval $[-\frac{\triangle}{2}; +\frac{\triangle}{2}]$ according to a midriser quantization [8]. Applying (12), (9) and (10) we obtain a SNR for each calculated spectrum as shown in [13]:

$$\text{SNR} = \frac{\frac{1}{N}\sum_{n=1}^{N}\left(\hat{x}[n]w[n] - \frac{1}{N}\sum_{k=1}^{N}\hat{x}[k]w[k]\right)^2}{\frac{1}{N}\sum_{n=1}^{N}\left(e[n]w[n] - \frac{1}{N}\sum_{k=1}^{N}e[k]w[k]\right)^2} \tag{15}$$

For $e[n] \ll x[n]$ we can replace $\hat{x}[n]$ by $x[n]$ and obtain with (15):

$$\text{SNR} \approx \frac{\frac{1}{N}\sum_{n=1}^{N}\left(x[n]w[n] - \frac{1}{N}\sum_{k=1}^{N}x[k]w[k]\right)^2}{\frac{1}{N}\sum_{n=1}^{N}\left(e[n]w[n] - \frac{1}{N}\sum_{k=1}^{N}e[k]w[k]\right)^2} \tag{16}$$

4 Dynamic Range of a Multiresolution Time-Domain EMI Measurement System

In this section the dynamic performance of the multiresolution time-domain EMI measurement system consisting of several ADCs is investigated. Several ADCs are used in parallel to digitize the analog signal. In the following an algorithm is described that extracts a floating point value from the discrete values obtained from all ADCs.

4.1 Floating Point A/D Conversion

In Fig. 5 the intervals that result of a configuration of 3 ADCs is depicted which was taken from [13]. Each analog value is simultaneously digitized by all ADCs. The lower and upper boundaries of the amplitude interval i are donated with a a_i^l and a_i^h respectively. If the value is within the range $[a_1^l, a_1^h]$ the digitized value is extracted from the first ADC. If the value is between $[a_2^l, a_1^l[\cup]a_1^h, a_2^h]$ the value is taken from the second ADC, and so on. Each extracted value is corrected by the known gain of the ADC. The result is a signal that is digitized as a floating point with a higher dynamic range.

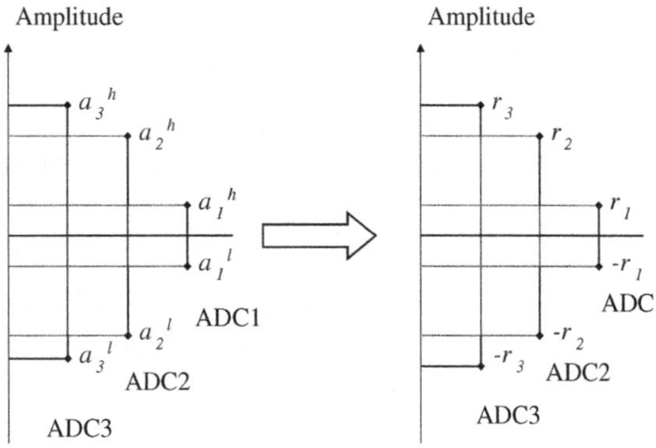

Fig. 5 Intervals for a Configuration of Three ADCS. From [13]

4.2 Quantization Noise

A further possibility to enhance to maximize the SNR and SFDR is to set up the intervals show in Fig. 5 for a maximum SNR.

This can be achieved by adding controlled PIN-attenuators between the amplifiers and the ADCs. In the following the relationship between the selected amplitude scales for a specific input signal and the SNR are described.

In a practical measurement system by a first measurement with one ADC the absolute frequency of the digitized values $H[k]$ can be obtained. By the known absolute frequency of digitized values the optimum parameters are determined. Several measurements can be done to refine the setup.

4.2.1 Optimum Amplitude Intervals

A sample that is digitized within the amplitude range $[a_1^l, a_1^h]$ shows a quantization error that is determined by the quantization interval d_1 of the first ADC. If the sample is within the amplitude range $[a_2^l, a_1^l] \cup [a_1^h, a_2^h]$ the quantization noise is determined by the quantization step d_2 of the second ADC, and so on. We consider the quantization noise as random error according a midriser quantization and obtain the quantization noise for each ADC according to (11). The absolute frequency of digitized values $H[k]$ can be used to estimate the random noise that is added by a system of several ADCs. We assume that the total noise is a summation of the quantization noise of all ADCs. The Quantization Noise $P_{N,I}$ of a system consisting of I ADCs is described as

$$P_{N,I} = \frac{1}{12N}(d_1^2 \sum_{k=a_1^l}^{a_1^h} H[k] + \sum_{i=2}^{i=I}(d_i^2 \alpha_i) \tag{17}$$

with

$$\alpha_i = \left(\sum_{k=a_i^l}^{a_{i-1}^h} H[k] + \sum_{k=a_{i-1}^h}^{a_i^h} H[k] \right). \tag{18}$$

With the used PIN-Attenuators we can only change the amplification. Any DC-offset is eliminated by a high-pass filter. We can assume that $H(k) \approx H(-k)$. We get from (17) and (18)

$$P_{N,I}(r_1, \ldots r_{I-1}) \approx \frac{2}{12N} \sum_{i=1}^{I} \left(d_i^2 \sum_{k=r_{i-1}}^{r_i} H(k) \right) \tag{19}$$

with

$$r_0 = 0. \tag{20}$$

The quantization step d_i of the ADC i can be calculated as follows:

$$d_i = 2^{-B_i} r_i \tag{21}$$

B_i is the number of bits of the ADC i. In order to maximize the SNR the quantization noise has to be minimized. By a numerical algorithm the parameters $r_1, .., r_{I-1}$ are obtained where $P_{N,I}$ shows the absolute minimum.

An example of the parameter obtained for a configuration of 3 ADCs appears in Fig. 6. The ADC 3 digitizes the full amplitude scale, while the ADC 2 digitizes the range between the values -26 and $+26$. The interval between -2 and $+2$ is digitized by ADC 1.

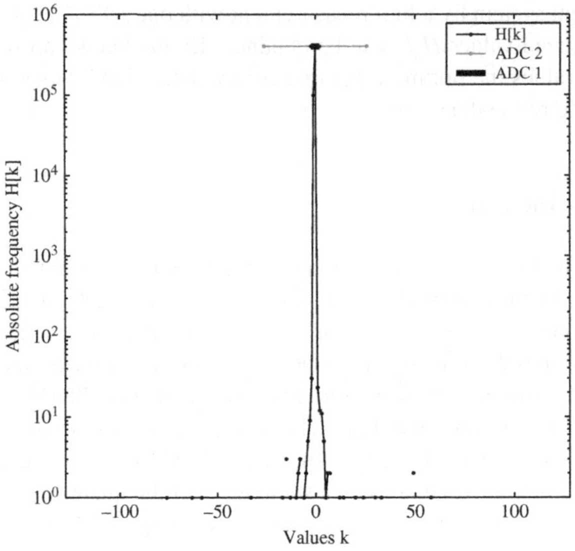

Fig. 6 Optimum amplitude intervals

4.3 Thermal Noise

The received noise power of a system is determined by

$$P_N = kT_0B, \tag{22}$$

where k is the is the Boltzmann constant and $T_0 = 290\,\mathrm{K}$. B is the bandwidth of the receiving system. The noise figure of a system is defined by:

$$F = \frac{SNR_{\text{output}}}{SNR_{\text{input}}} \tag{23}$$

The noise power of a system at the output which exhibits the noise figure F is determined by:

$$P_N = FkT_0B. \tag{24}$$

In order to calculate the spectral noise density at the input of each ADC we calculate the noise figure of each channel according to (25).

The cascaded noise figure of the low-noise limiter circuit and the amplifiers is calculated according to:

$$F = F_1 + \frac{F2-1}{G1} + \frac{F3-1}{G1G2} + \cdots + \frac{F_N-1}{\prod_{k=1}^{N-1} G_k}. \tag{25}$$

For the most sensitive channel a noise figure of 7.6 dB has calculated.

The average noise floor of the TDEMI measurement system can be determined from (24), where B is the equivalent noise bandwidth of the modelled IF-Filter. The equivalent noise bandwidth can be determined according to:

$$B = \int_{-\infty}^{+\infty} H(f)^2 df, \tag{26}$$

where $H(f)$ is the transfer function of the modelled IF-Filter.

The TDEMI measurement system shows a noise figure of about 7.6 dB. With a modelled IF-bandwidth of 120 kHz a equivalent noise bandwidth of 90.3 kHz is obtained. Thus for the TDEMI measurement system an average noise floor of about $-10\,\mathrm{dB}\mu\mathrm{V}$ is calculated. For a 9 kHz IF-filter an average noise floor of about $-21\,\mathrm{dB}\mu\mathrm{V}$ is obtained.

5 Digital Signal Processing

The calculation of a spectrogram is equivalent to a bank of parallel digital filters. By this way a set of several thousand intermediate frequency filters can be simulated digitally. The short-time fast Fourier transform and the Gaussian window function are used to model the behavior of the intermediate frequency filter of a conventional EMI receiver.

5.1 Digital Downconversion

The block diagram of the digital-down-conversion unit is shown in Fig. 7. By a sine-cosine lockup table and the digital mixer one of the eight sub-bands is moved towards its baseband. In order to reduce the sampling frequency and to fulfill the Nyquist criterion a polyphase decimation filter is implemented for the inphase and quadrature channel. The output sampling frequency is 250 MHz, while the bandwidth is 125 MHz.

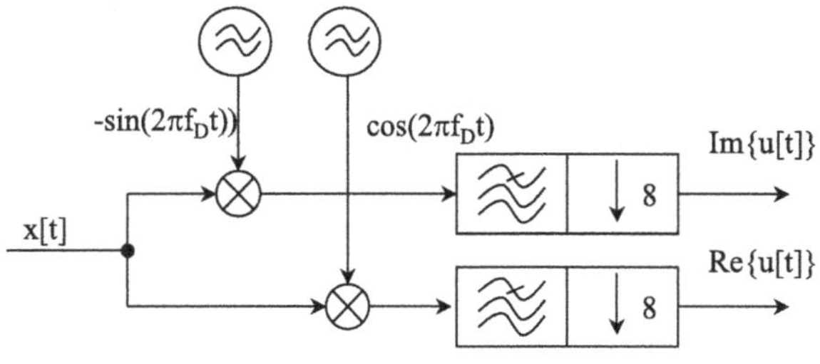

Fig. 7 Digital Downconversion

5.2 Short-Time Fast Fourier Transform

In Fig. 8 the block diagram of the implemented short-time fast Fourier transform is shown. By the use of decimation in time [8] and the symmetry of the Gaussian window the showed block diagram can be derived. In order to improve the frequency resolution a numerical oscillator is implemented, that allows us to move the signal frequency domain in small frequency steps of typically 200 Hz. By this way a higher frequency resolution is achieved by performing several measurements.

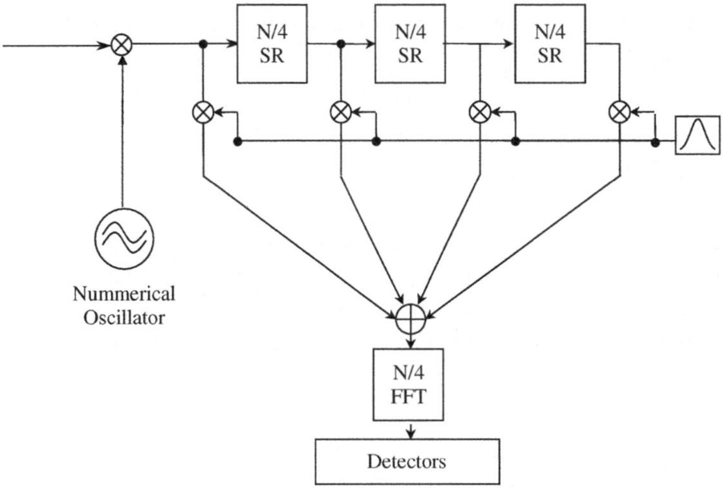

Fig. 8 Short time fast Fourier transform

6 Hardware Implementation

6.1 Broadband Analog-to-Digital Converter System

The emitted signal of a device under test is digitized by an analog-to-digital converter unit with a bandwidth of 1 GHz, shown in Fig. 9. The realtime signal processing, like reconstruction if the emitted signal and STFFT, is performed on two field programmable gate arrays and the data is transferred to a personal computer via a connection to an Universal Serial Bus (USB) Controller. The analog-to-digital converter unit uses several parallel ADCs with different sensitivities to perform a floating-point analog-to-digital conversion with a high dynamic range. By this way the measurement time has been reduced by a factor of up to 8000 in comparison to conventional measurement systems.

Fig. 9 Analog-to-Digital Converter DSP Unit

6.2 Asymmetric Low Noise Power Splitter

The power of the signal, received from the antenna and bandlimited by a lowpass, is divided by an asymmetric power splitter into three channels each with different insertion loss respectively gain. The first channel exhibits an amplification of 24.5 dB to digitize low signal amplitudes up to 30 mV. The channel uses a series of low noise low reflection limiter circuits and low noise amplifiers. In order to suppress aliasing effects, the thermal noise of the broadband amplifiers has to be attenuated. Otherwise noise at frequencies above 1 GHz could be down converted to the detecting frequency range, due to the broad frequency range the system is designed for. Such aliasing of the thermal noise caused by inter-modulation in consequence of the nonlinearities of the ADCs leads to a degradation of the noise performance of the system. Thus a slight attenuation of 1.4 dB is performed above 1 GHz, which is shown in Fig. 12. Up to 1.5 GHz the transmission coefficient of channel one decreases an attenuation of 3 dB. The second channel provides an attenuation of 2dB to signals range up to 630 mV and a frequency response of –1.7 dB in the operating range up to 1 GHz. Channel three can theoretically be used to digitize signal up to 9 V by providing an attenuation of 25.5 dB. The frequency response of channel number three add up to –1 dB at 1.5 GHz. Additionally all channels features limiter diodes to protect the input ports of the ADCs in respect of out-of-range voltages. The hybrid circuit is shown in Fig. 10.

The transmission of the power splitter has been investigated. The result is shown in Fig. 11.

The noise figure of the sensitive channel one has been characterized by an HP8970B noise figure meter and is depicted in Fig. 12. The noise figure in the range of 100 MHz–1 GHz is about 7.5 dB and shows an increasing level towards

Fig. 10 Low noise power splitter

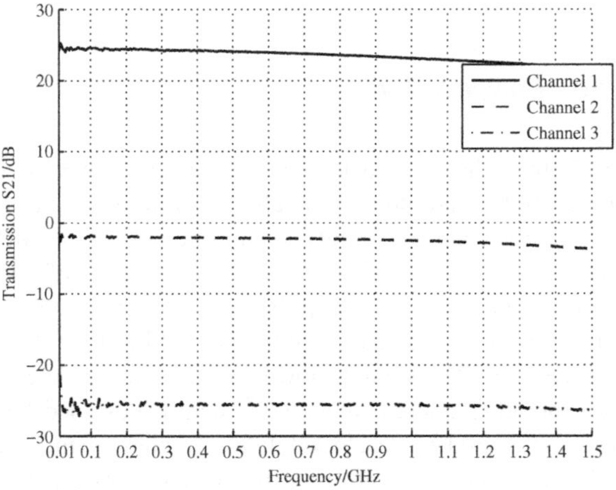

Fig. 11 Transmission of Powersplitter

lower frequencies because of degradation by shot noise and 1/f noise. Calculating the cascaded noise figure for channel one leads to the result of 7.6 dB.

6.3 Measurement Procedure

6.3.1 EMI Receivers

Measurements of emssions are traditionally performed by measurement systems operating in frequency domain. During the dwell time the input signal is observed at

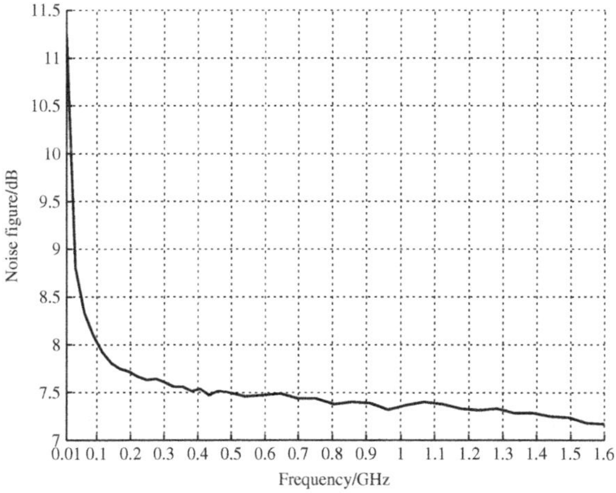

Fig. 12 Noise Figure of Channel 1

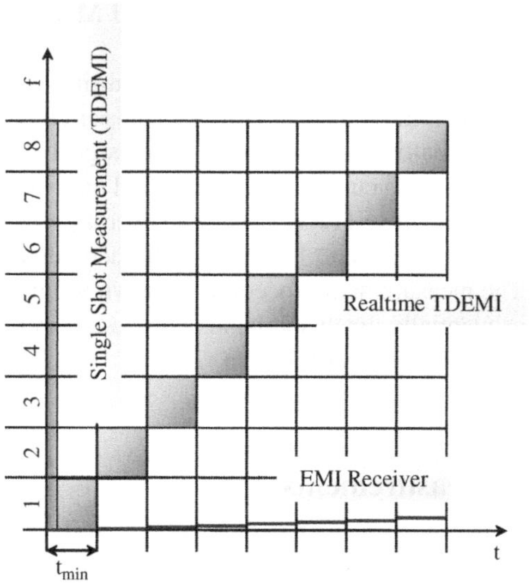

Fig. 13 Measurement Procedures for Single Shot TDEMI, EMI receiver and Realtime TDEMI Measurement System

a single frequency and weightend with a detector e.g. the quasipeak detector. During the complete dwell time the signal is processed via the analog and digital stages of an EMI receiver in a continuous way. Thus for the complete characterization of a device under test at a number of several thousand frequencies the measurement has to be repeated. For long dwell times which may be required by the time constants of the detector mode or due the discontinous disturbance of the device under test, the complete characterization of the device under test takes several hours. The stepped measurement procedure is shown in Fig. 13.

6.3.2 Single Shot Measurement in Time-Domain

Time-domain EMI Measurement Systems capture for the record time t_{rec} the EMI signal and store it. The signal is processed by the fast Fourier transform. By a Gaussian windows function the IF-Filter is modelled. By the short time fast Fourier transform the recorded signal can be processed and a spectrogram generated. By a digital simulation of the detector modes the spectra under the peak, average, quasipeak and rms detector mode could be calculated. However the record time t_{rec} is limited by the memory depth of the available high-speed memory. Today several hundred μ s can be recorded but the detector modes like the quasipeak detector require a dwell time of about 1.5 s. The procedure of such a measurement system is shown in Fig. 13. By such a system the measurement time can be reduced by up to 5 orders of magnitude. The drawback of such a system is the limitation by the depth of the used high speed memory. A continuous disturbance analysis as requested by CISPR 14 cannot be achieved by such as system.

6.3.3 Realtime Time-Domain EMI Measurement System

In order to solve this problem a real-time time-domain EMI measurement system has been implemented [14]. The measurement system performs the signal processing in realtime without the digitized signal. The digitized signal is processed continuously. With current available digital hardware the it is not economically possible to calculate the STFFT in realtime over a frequency range from DC–1 GHz. Thus the frequency range to 1 GHz is subdivided into 8 bands that are processed sequentially. The measurement procedure is shown in Fig. 13. All eight sub-bands are processed sequentially. Each sub-band is processed in real-time by the real-time time-domain EMI Measurement System.

7 Measurements

An emission measurement of a laptop has been carried out in the quasipeak detector mode. The dwell time has been 2 s. In Fig. 14 a subregion of the measured frequency range 30 MHz–1 GHz is shown. The deviation is below 0.5 dB.

An emission measurement has been carried out in average detector mode. The dwell time has been 4 s. The bin width is 30 kHz. The result is shown in Fig. 15. The maximum difference between the measurement performed with the TDEMI measurement system and the results obtained by the EMI receiver ESCS30 is below 0.2 dB.

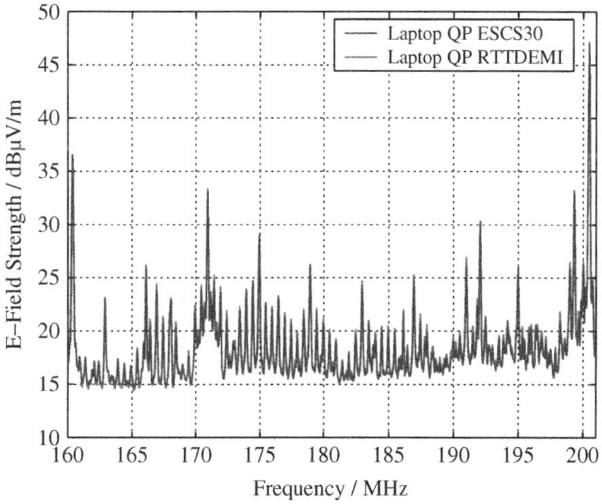

Fig. 14 Emission of a Laptop in Quasipeak Detector mode

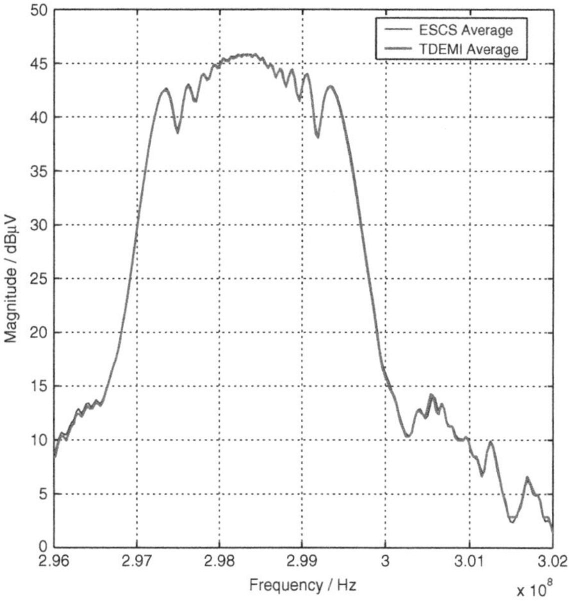

Fig. 15 Emission of a desktop PC, average detector mode

References

1. C. R. Barhydt, "Radio noise meter and its application," in *General Electric Rev.* Vol. 36, pp. 201–205, 1933.
2. Rohde&Schwarz, *Ultra Broadband Antenna HL562 ULTRALOG*. Data sheet, 2001.

3. S. Braun and P. Russer, "A Low-Noise Multiresolution High-Dynamic Ultra-Broad-Band Time-Domain EMI Measurement System," *IEEE Transactions on Microwave Theory and Techniques*, Vol. 53, pp. 3354–3363, Nov 2005.
4. F. Krug, T. Hermann, and P. Russer, "Signal Processing Strategies with the TDEMI Measurement System," in *2003 IEEE Instrumentation and Measurement Technology Conference Proceeding*, May 20–22, Vail, USA, pp. 832–837, 2003.
5. C. Keller and K. Feser, "Fast Emission Measurement In Time Domain," in *14th International Zurich Symposium On Electromagnetic Compatibility, Zurich, Switzerland, 20.–22.2.2001*, no. 70K7, 2001.
6. S. Braun, F. Krug, and P. Russer, "A Novel Automatic Digital Quasi-Peak Detector for a Time Domain Measurement System," in *2004 IEEE International Symposium On Electromagnetic Compatibility Digest*, August 9–14, Santa Clara, USA, Vol. 3, pp. 919–924, Aug. 2004.
7. S. Braun, M. Aidam, and P. Russer, "Development of a Multiresolution Time-Domain EMI Measurement System that Fulfils CISPR 16-1," in *2005 IEEE International Symposium On Electromagnetic Compatibility*, Chicago, USA, pp. 388–393, 2005.
8. A. V. Oppenheim and R. W. Schafer, *Discrete–Time Signal Processing*. ISBN 0-13-214107-8, Prentice-Hall, 1999.
9. C. M. Burill, "An evaluation of radio noise meter performance in terms of listening experience," in *Institute of Radio Engineers*, Vol. 30, no 10, pp. 473–478, 1942.
10. CISPR 16-1-1, *Specification for radio disturbance and immunity measuring apparatus and methods Part 1-1: Radio disturbance and immunity measuring apparatus Measuring apparatus*. International Electrotechnical Commission, 2006.
11. L. Cohen, "Time-Frequency Distributions – A Review," in *Proceeding of the IEEE*, Vol. 77, no. 7, pp. 941–981, 1989.
12. S. L. Marple, *Digital Spectral Analysis with Applications*. ISBN 0-8493-7892-3, Prentice-Hall, 1987.
13. S. Braun, A. Alt, and P. Russer, "A Novel Multiresolution High-dynamic Ultra-broadband Time-domain EMI Measurement System," in *IEEE MTT-S 2005 International Microwave Symposium*, Long Beach, USA, pp. 1303–1306, 2005.
14. S. Braun, M. Al-Qedra, and P. Russer, "A novel Realtime Time-domain EMI Measurement System based on Field Programmable Gate Arrays," in *17th International Zurich Symposium on Electromagnetic Compatibility, Digest*, (Singapore), pp. 501–504, Feb. 2006.

Space Mapping Optimization and Modeling of Microwave Devices with MEFiSTo

Slawomir Koziel and John W. Bandler

Abstract We review recent developments in space mapping techniques for optimization and modeling of microwave devices. We present a general formulation of space mapping optimization algorithms as well as the standard and enhanced space mapping modeling methodologies. Application examples of optimization and modeling of microwave structures with the time-domain electromagnetic simulator MEFiSTo as a fine model evaluator are provided.

Keywords Computer-aided design (CAD) · EM modeling · space mapping · surrogate modeling · MEFiSTo · time-domain simulation · electromagnetic simulation

1 Introduction

Space mapping [1, 2, 3, 4, 5] is a novel but already recognized methodology that allows efficient optimization of expensive or *fine* models by means of the iterative optimization and updating of so called *coarse* models which are less accurate but cheaper to evaluate. Provided that the misalignment between the fine and coarse models is not significant, space-mapping-based algorithms typically provide excellent results after only a few evaluations of the fine model. A similar idea is shared by other surrogate-model-based methods [6, 7, 8, 9, 10, 11, 12], however, many of them do not use a physics-based coarse model: the surrogate model is created by direct approximation of the available fine model data.

Slawomir Koziel
McMaster University, Hamilton, ON, Canada L8S 4K1,
e-mail: koziels@mcmaster.ca
Now with Reykjavik University, Reykjavik, Iceland, e-mail: koziel@ru.is

John W. Bandler
McMaster University, Hamilton, ON, Canada L8S 4K1, e-mail: bandler@mcmaster.ca

P. Russer, U. Siart (eds.), *Time Domain Methods in Electrodynamics*,

Space mapping is widely used in the optimization of microwave devices [1, 2, 3], [13, 14, 15, 16], where fine models are often based on full-wave electromagnetic simulations, whereas coarse models may be physically based circuit models. Recently, space mapping techniques have been applied to design problems in a growing number of areas (e.g., [17, 18, 19]). A number of papers cover different aspects of space mapping, including the development of new algorithms ([2, 3, 20]), space-mapping modeling and neuro-space mapping modeling [21, 22, 23, 24, 25, 26, 27, 28, 29, 30, 31, 32], theoretical foundations ([20, 33, 34]), etc.

There are two, in some sense complementary, aspects of space mapping: optimization and modeling. In optimization, a space mapping surrogate model is created and updated iteratively with the main focus put on its local behavior in the neighborhood of the current solution. In modeling, a space mapping surrogate model is created upfront based on a certain amount of data consisting of fine model evaluations at base points suitably distributed throughout the region of interest. In this case the model should be valid in the whole region of interest without being biased to any particular location. Leaving alone the algorithmic details of the iterative update of the solution and of the surrogate in space mapping optimization, both optimization and modeling implement similar methods of creating the surrogate model based on the coarse model and available fine model data.

The most characteristic feature of space mapping is that the optimization or modeling process is based on the computationally cheap coarse model. It is critical that the coarse model is a physics-based model, i.e., it is a model that describes the same physical phenomena as the fine model. This ensures that the general properties of the coarse model are the same as the general properties of the fine model. Consequently, only a small amount of fine model data should be necessary to correct the coarse model through space mapping in order to get a good quasi-global match between the fine model and the space mapping surrogate. Typically, the fine model is implemented using an electromagnetic simulator such as HFSS, Sonnet **em** or MEFiSTo. The coarse model may use the same simulator with a coarser mesh or be a circuit equivalent of the microwave device under consideration.

In this paper we review space mapping optimization and modeling techniques and provide illustrative application examples. As a fine model evaluator we use the time-domain electromagnetic simulator MEFiSTo [35]. Due to its parameterization and batch processing capabilities, MEFiSTo can be easily linked into an automated space mapping optimization loop.

2 Space Mapping Optimization

Let $\mathbf{R}_f$ and $\mathbf{R}_c$ denote the response vectors of the fine and coarse model of the device of interest, respectively. In the microwave area, the model response is typically the evaluation of some characteristics of the device, e.g., the scattering parameter $|S_{21}|$, at a given set of frequencies. Both models depend on a number of parameters called the design variables. The vector of design variables will be denoted by $\mathbf{x}$. Our goal is to find an optimal solution (design) of the fine model with respect to given specifications, which can be formally written as

$$\mathbf{x}_f^* = \arg\min_{\mathbf{x}} U(\mathbf{R}_f(\mathbf{x})) \tag{1}$$

where U is a given objective function, e.g., minimax [33].

According to space mapping methodology, instead of directly solving (1), which may be prohibitive because of the high computational cost of evaluating the fine model, we consider an iterative algorithm that generates a sequence of points $\mathbf{x}^{(i)}$, $i = 0, 1, 2, \ldots$, and a family of surrogate models $\mathbf{R}_s^{(i)}$ so that the next point is an optimal solution of the current surrogate model, i.e.,

$$\mathbf{x}^{(i+1)} = \arg\min_{\mathbf{x}} U(\mathbf{R}_s^{(i)}(\mathbf{x})) \tag{2}$$

while the surrogate $\mathbf{R}_s^{(i)}$ is a suitable enhancement of the coarse model defined as

$$\mathbf{R}_s^{(i)}(\mathbf{x}) = \overline{\mathbf{R}}_s(\mathbf{x}, \mathbf{p}^{(i)}) \tag{3}$$

with

$$\mathbf{p}^{(i)} = \arg\min_{\mathbf{p}} \sum_{k=0}^{i} w_{i.k} \left\| \mathbf{R}_f(\mathbf{x}^{(k)}) - \overline{\mathbf{R}}_s(\mathbf{x}^{(k)}) \right\| \tag{4}$$

$\overline{\mathbf{R}}_s$ in (3) and (4) is a generic space mapping surrogate model, i.e., the coarse model composed with suitable space mapping transformations; $\mathbf{p}^{(i)}$ is a vector of model parameters obtained using the so-called parameter extraction process (4); $w_{i.k}$ are weighting factors. There is a number of space mapping models that are utilized in practice (see, e.g., [2, 3, 4, 5, 20]). The most popular one is undoubtedly the input space mapping [1], in which the generic space mapping surrogate model takes the form $\overline{\mathbf{R}}_s(\mathbf{x}, \mathbf{p}) = \overline{\mathbf{R}}_s(\mathbf{x}, \mathbf{B}, \mathbf{c}) = \mathbf{R}_c(\mathbf{B} \cdot \mathbf{x} + \mathbf{c})$. The starting point for space mapping optimization is typically the optimal solution of the coarse model.

The space mapping optimization algorithm flow can be described as follows.

Step 1	Set $i = 0$;
Step 2	Evaluate the fine model to find $\mathbf{R}_f(\mathbf{x}^{(i)})$;
Step 3	Obtain the surrogate model $\mathbf{R}_s^{(i)}$ using (3) and (4);
Step 4	Given $\mathbf{x}^{(i)}$ and $\mathbf{R}_s^{(i)}$, obtain $\mathbf{x}^{(i+1)}$ using (2);
Step 5	If the termination condition is not satisfied go to Step 2; else terminate the algorithm.

Usually, the algorithm is terminated through convergence or by exceeding the user-defined maximum number of iterations or maximum number of fine model evaluations. In order to improve the convergence properties, space mapping algorithms are often enhanced by trust region methods [36, 37].

Consider the six-section H plane waveguide filter [38] shown in Fig. 1. The fine model is simulated using MEFiSTo [35] in a 2D mode. The Matlab coarse model (Fig. 2) has lumped inductances and dispersive transmission line sections.

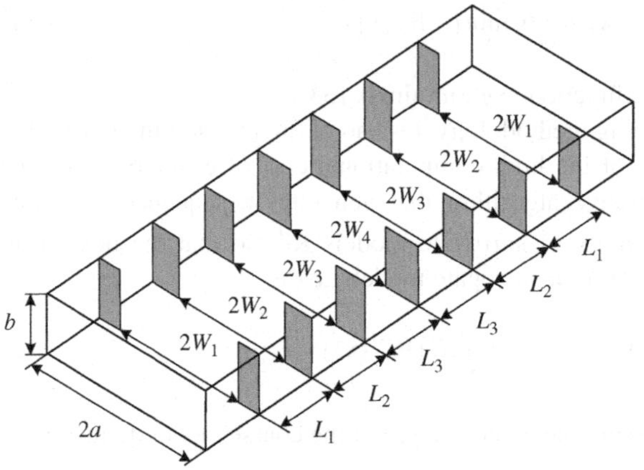

Fig. 1 The six-section H-plane waveguide filter: the 3D view [38]

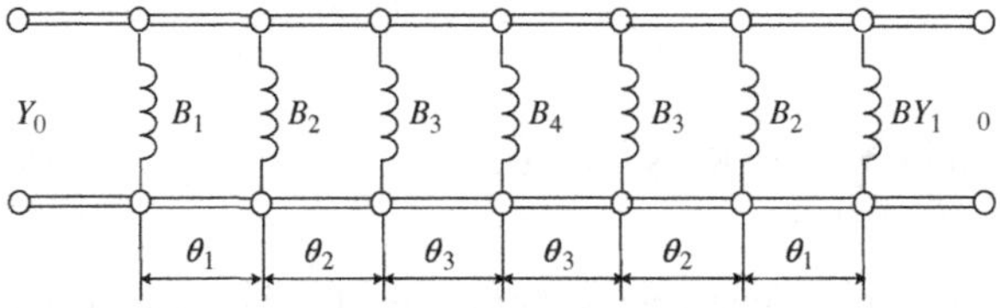

Fig. 2 The six-section H-plane waveguide filter: the equivalent empirical circuit model [38]

We simplify formulas due to Marcuvitz for the inductive susceptances corresponding to the H-plane septa. Design parameters are $\mathbf{x} = [L_1\ L_2\ L_3\ W_1\ W_2\ W_3\ W_4]^T$. The design specifications are $|S_{11}| \leq 0.16$ for $5.4\,\text{GHz} \leq \omega \leq 9.0\,\text{GHz}$, $|S_{11}| \geq 0.85$ for $5.0\,\text{GHz} \leq \omega \leq 5.2\,\text{GHz}$, and $|S_{11}| \geq 0.5$ for $9.5\,\text{GHz} \leq \omega \leq 10.0\,\text{GHz}$. The starting point is $\mathbf{x}^{(0)} = [16.49\ 16.71\ 17.16\ 12.83\ 11.78\ 11.23\ 11.09]^T$ mm (coarse model optimal solution).

For this problem we use input space mapping and output space mapping [20]. In particular, the space mapping surrogate model is defined as $\mathbf{R}_s^{(i)}(\mathbf{x}) = \mathbf{R}_c(\mathbf{x}+\mathbf{c}^{(i)}) + \mathbf{d}^{(i)}$, where vector $\mathbf{c}^{(i)}$ is found using parameter extraction, after which $\mathbf{d}^{(i)}$ is the residual vector evaluated by $\mathbf{d}^{(i)} = \mathbf{R}_f(\mathbf{x}^{(i)}) - \mathbf{R}_c(\mathbf{x}^{(i)} + \mathbf{c}^{(i)})$. Figure 3 shows the fine and coarse model responses at the starting point $\mathbf{x}^{(0)}$. The fine model specification error at $\mathbf{x}^{(0)}$ is +0.184. The solution obtained after 11 space mapping iterations (20 fine model evaluations) is $\mathbf{x}^* = [16.11\ 16.23\ 16.51\ 12.77\ 11.82\ 11.42\ 11.11]^T$ and the corresponding specification is −0.027. Figure 4 shows the fine model response at $\mathbf{x}^*$. Figure 5 shows the specification error versus iteration index.

The performance of the space mapping algorithm may be improved by employing frequency scaling of the coarse model. It follows from Fig. 3 that most of the misalignment between the fine and coarse model at $\mathbf{x}^{(0)}$ can be removed by proper scaling of the coarse model response with respect to frequency. We use a polynomial scaling of the form $\omega \leftarrow f_0 + f_1\omega + f_2\omega^2$. Coefficients f_0, f_1

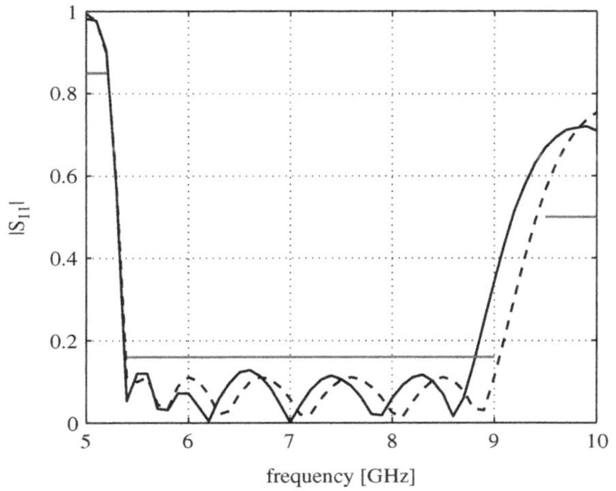

Fig. 3 H-plane filter: fine (*solid line*) and coarse (*dashed line*) model responses at the coarse model optimal solution $\mathbf{x}^{(0)} = [16.49\ 16.71\ 17.16\ 12.83\ 11.78\ 11.23\ 11.09]^T$ mm

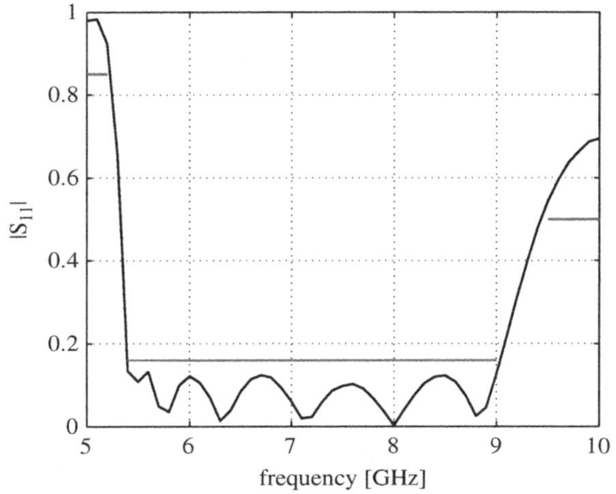

Fig. 4 H-plane filter: fine model response at the solution found using the space mapping optimization algorithm $\mathbf{x}^* = [16.11\ 16.23\ 16.51\ 12.77\ 11.82\ 11.42\ 11.11]^T$ mm

and f_2 are obtained and encoded in the coarse model, and the scaled coarse model is created and used instead of the original one in the space mapping optimization process. The optimal solution of the frequency-scaled coarse model $\mathbf{x}_s^{(0)} = [16.08\ 16.17\ 16.62\ 12.81\ 11.83\ 11.29\ 11.16]^T$ is a starting point for the space mapping optimization process. Figure 6 shows the fine and frequency-scaled coarse model response at $\mathbf{x}_s^{(0)}$. The fine model specification error at $\mathbf{x}^{(0)}$ is -0.011, i.e., the fine model satisfies the design specifications at the starting point. The solution obtained after 14 space mapping iterations (22 fine model evaluations) is

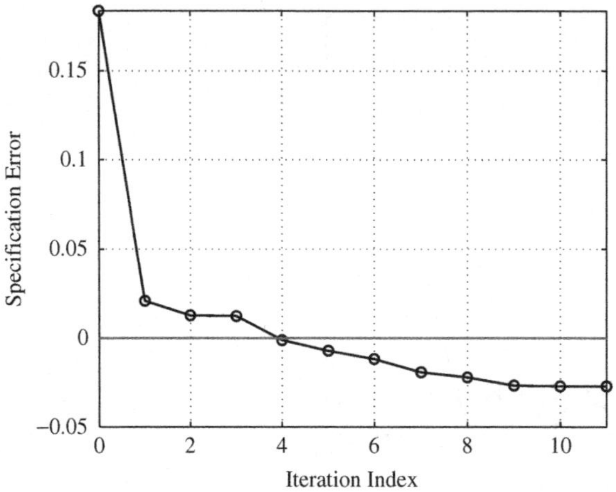

Fig. 5 H-plane filter: specification error versus iteration number for the space mapping optimization algorithm with the original coarse model

$\mathbf{x}_s^* = [16.14\ 16.17\ 16.50\ 12.78\ 11.92\ 11.50\ 11.12]^T$ and the corresponding specification is −0.037. Figure 7 shows the fine model response at $\mathbf{x}_s^*$. Figure 8 shows the specification error versus iteration number.

It is seen that the quality of the solution found with the frequency-scaled coarse model is better than the quality of the solution found with the original coarse model, while the computational cost is similar in both cases.

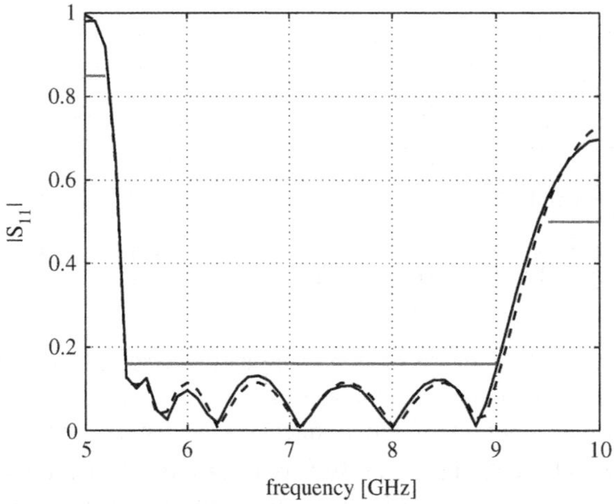

Fig. 6 H-plane filter: fine (*solid line*) and frequency-scaled coarse (*dashed line*) model responses at the frequency-scaled coarse model optimal solution $\mathbf{x}_s^{(0)} = [16.08\ 16.17\ 16.62\ 12.81\ 11.83\ 11.29\ 11.16]^T$ mm

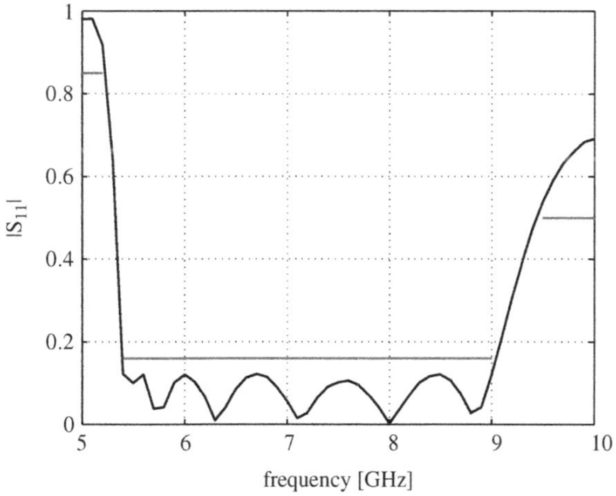

Fig. 7 H-plane filter: fine model response at the solution found using the space mapping optimization algorithm with the frequency-scaled coarse model $\mathbf{x}_s^* = [16.14\ 16.17\ 16.50\ 12.78\ 11.92\ 11.50\ 11.12]^T$ mm

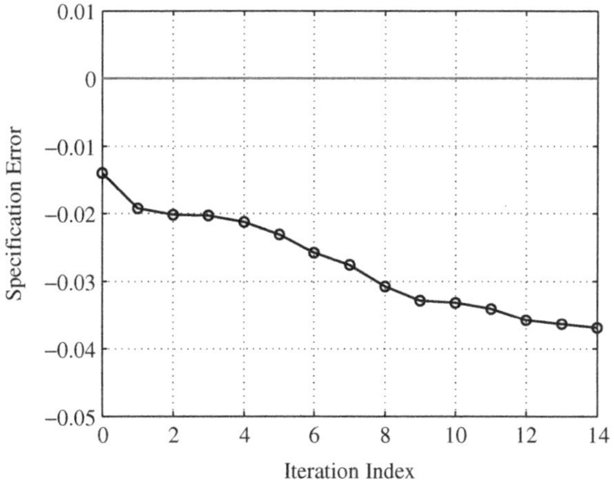

Fig. 8 H-plane filter: specification error versus iteration number for the space mapping optimization algorithm with the frequency-scaled coarse model

For comparison purposes we also performed direct optimization of the fine model using a Matlab minimax optimizer [39]. The solution obtained after 311 fine model evaluations resulted in the specification error equal to −0.030.

Table 1 summarizes the results for the H-plane filter obtained through space mapping algorithms and direct optimization. The computational cost of the space mapping algorithm is similar in both cases (with the original and frequency-scaled coarse model), although the frequency-scaled coarse model allows us to obtain a

Table 1 H-plane filter: comparison of optimization results for space mapping and direct optimization

Optimization technique	Final specification error	Number of MEFiSTo evaluations
Space mapping (original coarse model)	−0.027	20
Space mapping (frequency-scaled coarse model)	−0.037	22
Direct optimization (Matlab minimax optimization routine)	−0.030	311

better value of the final specification error. The computational cost of the space mapping is substantially lower than the cost of direct optimization, which is typical for this methodology.

3 Space Mapping Modeling

Let X_R denote the region of interest in which we want enhanced matching between the surrogate and the fine model. Typically, we assume that X_R is an n dimensional interval in $\Re^n$ with center at reference point $\mathbf{x}_0$ and size δ; n is the number of design variables. Suppose we have the base set $X_B = \{\mathbf{x}^1, \mathbf{x}^2, \ldots, \mathbf{x}^N\} \subset X_R$, where N is the number of base points, such that the fine model response is known at all points $\mathbf{x}^j$, $j = 1, 2, \ldots, N$. In general, we do not assume any particular location of these base points. As before, let $\overline{\mathbf{R}}_s(\mathbf{x}, \mathbf{p})$ be a generic space mapping surrogate model, i.e., the coarse model composed with suitable space mapping transformations. Here, $\mathbf{x}$ is a vector of design variables, and $\mathbf{p}$ is a vector of model parameters.

The standard space mapping modeling technique (SM-Standard) [22] assumes that the surrogate model is created by performing parameter extraction on the base points as follows:

$$\mathbf{R}_s(\mathbf{x}) = \overline{\mathbf{R}}_s(\mathbf{x}, \overline{\mathbf{p}}) \tag{5}$$

where

$$\overline{\mathbf{p}} = \arg\min_{\mathbf{p}} \sum_{\mathbf{y} \in X_B} ||\mathbf{R}_f(\mathbf{y}) - \overline{\mathbf{R}}_s(\mathbf{y}, \mathbf{p})|| \tag{6}$$

Note that the definition of $\mathbf{R}_s$ is practically the same as the definition (3), (4). The only difference is the location of the base points. In modeling, the base points are distributed using a suitable design of experiments technique, e.g., some kind of factorial design (typically a star distribution [22]) or uniform sampling.

The standard space mapping surrogate model is very simple and fast, because once the space mapping parameters are established, model evaluation cost is roughly the same as the evaluation cost of the coarse model which is assumed to be much cheaper than the fine model. Also, because the coarse model is assumed to be physics-based, good modeling accuracy can be obtained with relatively few base points. The limitation of the standard space mapping modeling is that it may not be able to provide accuracy sufficient for some applications. Moreover, including more base points has little effect on the model's accuracy because of the finite number of model parameters, and also because the parameters are extracted in one shot for the whole region of interest.

The above limitations can be overcome using space mapping modeling with variable weight coefficients (SM-VWC) [25, 26]. This methodology uses the same generic surrogate model (5), however, the model parameters are obtained in the following parameter extraction process:

$$\overline{\mathbf{p}} = \arg\min_{\mathbf{p}} \sum_{\mathbf{y}\in X_B} w(\mathbf{x},\mathbf{y})||\mathbf{R}_f(\mathbf{y}) - \overline{\mathbf{R}}_s(\mathbf{y},\mathbf{p})|| \tag{7}$$

The weighting coefficients w in (7) are functions of $\mathbf{x}$. They are calculated as $w(\mathbf{x},\mathbf{y}) = \exp(-||\mathbf{x}-\mathbf{y}||^2/(C\cdot\gamma^2))$, $k = 1,2,\ldots,N$, where $\mathbf{x}$ is the evaluation point, $\gamma = \gamma(\delta,N) = 2||\delta||_1/(nN^{1/n})$ is a characteristic distance depending on the size of the region of interest and the number of base points [25]. Constant $C > 0$ determines how fast the weighting coefficients decrease with increase of base-point distance from $\mathbf{x}$. Paper [25] contains a discussion on the implementation details of the method. In can be shown that this space mapping modeling methodology with variable weight coefficients can assure any required accuracy, provided that the base set is sufficiently dense, even if the number of model parameters is fixed [26]. The disadvantage of the method is the significant increase in the evaluation time, which is due to a separate parameter extraction process required for each evaluation of the surrogate model. This limits potential applications of the method.

A recent development in space mapping modeling is the combination of space mapping with radial basis functions (SM-RBF) [27]. In particular, on top of the standard space mapping surrogate, a correction term is added that interpolates the difference between the fine model $\mathbf{R}_f$ and the standard space mapping surrogate $\mathbf{R}_s$ through radial basis functions [40], [41]. Let $\mathbf{R}_f(\mathbf{x}) = [\mathbf{R}_{f.1}(\mathbf{x})\ldots\mathbf{R}_{f.m}(\mathbf{x})]^T$ and $\mathbf{R}_s(\mathbf{x}) = [\mathbf{R}_{s.1}(\mathbf{x})\ldots\mathbf{R}_{s.m}(\mathbf{x})]^T$. The radial basis function model $\tilde{\mathbf{R}}_s$ is defined as

$$\tilde{\mathbf{R}}_s(\mathbf{x}) = \begin{bmatrix} \sum_{j=1}^{N} \lambda_{1.j}\phi(||\mathbf{x}-\mathbf{x}^j||/\gamma) \\ \cdots \\ \sum_{j=1}^{N} \lambda_{m.j}\phi(||\mathbf{x}-\mathbf{x}^j||/\gamma) \end{bmatrix} \tag{8}$$

where $||\cdot||$ denotes the Euclidean norm. The parameters $\lambda_{k.j}$ are calculated so that they satisfy an interpolation requirement $\Phi\lambda_k = \mathbf{F}_k$, $k = 1,2,\ldots,m$, where $\lambda_k = [\lambda_{k.1}\lambda_{k.2}\ldots\lambda_{k.N}]^T$, $\mathbf{F}_k = [R_{f.k}(\mathbf{x}^1) - R_{s.k}(\mathbf{x}^1)\ldots R_{f.k}(\mathbf{x}^N) - R_{s.k}(\mathbf{x}^N)]^T$, and Φ is an $N\times N$ matrix with elements $\Phi_{ij} = \phi(||\mathbf{x}^i - \mathbf{x}^j||/\gamma)$. As before, γ is a characteristic distance of the base set and is used as a normalization factor. In this paper we use a Gaussian basis function defined as $\phi(r) = e^{-cr^2}$, $r \geq 0$, $c > 0$. Other choices of basis functions can be found in the literature [40].

The combined surrogate model $\breve{\mathbf{R}}_s$ is defined as

$$\breve{\mathbf{R}}_s(\mathbf{x}) = \mathbf{R}_s(\mathbf{x}) + \tilde{\mathbf{R}}_s(\mathbf{x}) \tag{9}$$

where $\mathbf{R}_s$ is a standard space mapping surrogate model. Once coefficients λ are found, evaluation of (8) is fast, which means that the evaluation cost of model (9) is not significantly larger than the evaluation cost of the standard space mapping surrogate model. This is in contrast with the modeling technique [25], requiring a separate parameter extraction for each evaluation of the surrogate model, which involves a number of coarse model evaluations (typically hundreds or even thousands).

Let us consider the single-resonator filter [42] shown in Fig. 9. The design parameters are resonator length d and septum width W. The rectangular waveguide width and length are fixed. The frequency range is $3.0\,\text{GHz} \le \omega \le 5.0\,\text{GHz}$ with a step of 0.04 GHz. The fine model employs MEFiSTo [35] in a 2D mode with $1\text{mm} \times 1\text{mm}$ mesh size utilizing the rubber cell feature to perform interpolation. The coarse Matlab model consists of lumped inductances and dispersive transmission line sections connected by circuit theory. The inductive susceptances, simplified formulas due to Marcuvitz, are used to model the resonator septa. The reference point is $\mathbf{x}_0 = [35\ 15]^T$ mm and the region of interest size is $\delta = [6\ 6]^T$.

For our test problem, we performed a number of experiments using the following surrogate models: SM-Standard, SM-VWC, and SM-RBF. For the standard model we use input, output and frequency space mapping [22]. Table 2 shows the details of the base sets used in our experiments. The base sets have growing numbers of points (and decreasing values of characteristic distances γ) in order to examine the dependence of the modeling error on the amount of fine model data used to create the model.

Accuracy was tested using 30 test points randomly distributed in the region of interest. The error measure used was the l_2 norm of the difference between the fine model response and the corresponding surrogate model response.

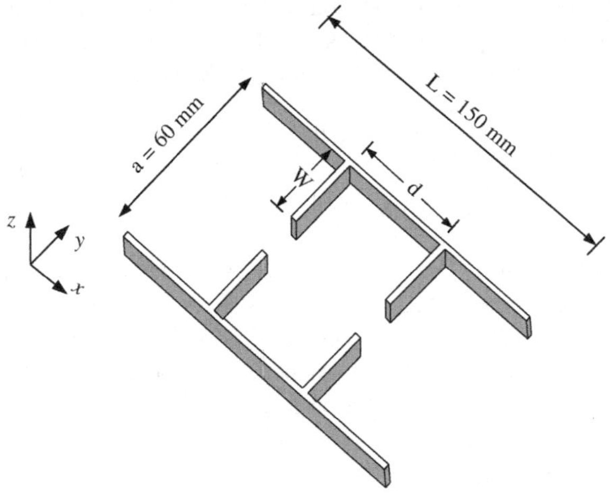

Fig. 9 Topology of the single-resonator filter [42]

Table 2 Base set data for the single resonator filter problem

Base Set	Base set description	Number of base points	γ
X_{B1}	Uniform mesh of density 3	9	4.00
X_{B2}	Uniform mesh of density 5	25	2.40
X_{B3}	Uniform mesh of density 7	49	1.71

Table 3 shows numerical results (error statistics) for the models with the various base sets considered. Figures 10 and 11 show error plots (the modulus of the difference between the fine model and the corresponding surrogate model response versus frequency) for the SM-Standard and SM-RBF, respectively, with base set X_{B3}. Figure 12 shows the dependence of the average modeling error on the characteristic distance γ. Table 4 presents a qualitative comparison of the surrogate model evaluation cost for all the considered methods.

Table 3 Modeling results for single resonator filter problem. Verification for 30 random test points

Model	Base set	Average error	Maximum error	Standard deviation
SM-Standard		0.0656	0.1229	0.0248
SM-VWC	X_{B1}	0.0476	0.1015	0.0294
SM-RBF		0.0545	0.1187	0.0338
SM-Standard		0.0534	0.0850	0.0178
SM-VWC	X_{B2}	0.0274	0.0754	0.0189
SM-RBF		0.0306	0.1143	0.0273
SM-Standard		0.0546	0.0977	0.0214
SM-VWC	X_{B3}	0.0278	0.0772	0.0234
SM-RBF		0.0239	0.0713	0.0241

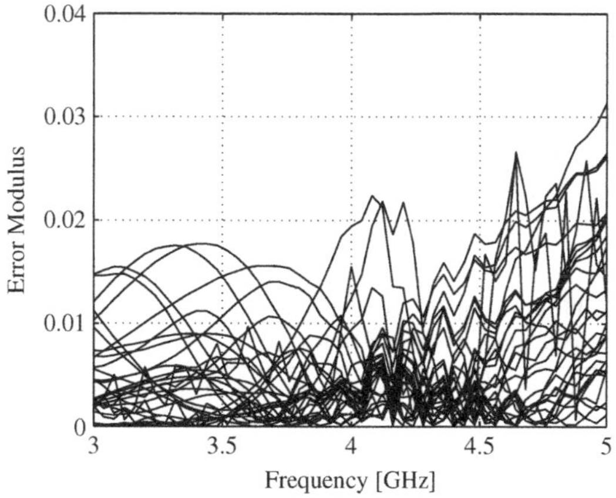

Fig. 10 Single resonator filter: error plots for the SM-Standard surrogate model with base set X_{B3} (30 test points)

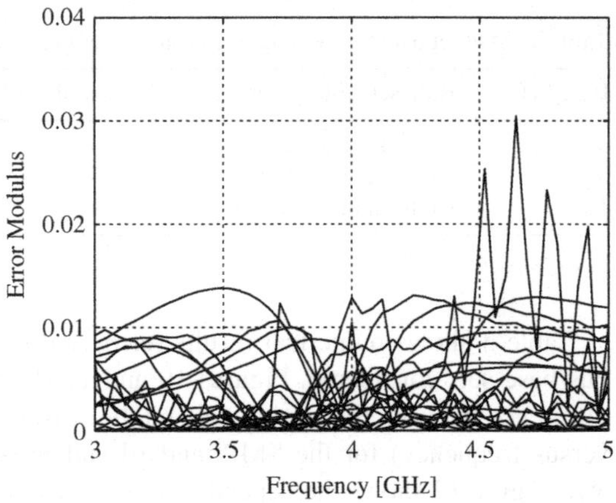

Fig. 11 Single resonator filter: error plots for the SM-RBF surrogate model with base set X_{B3} (30 test points)

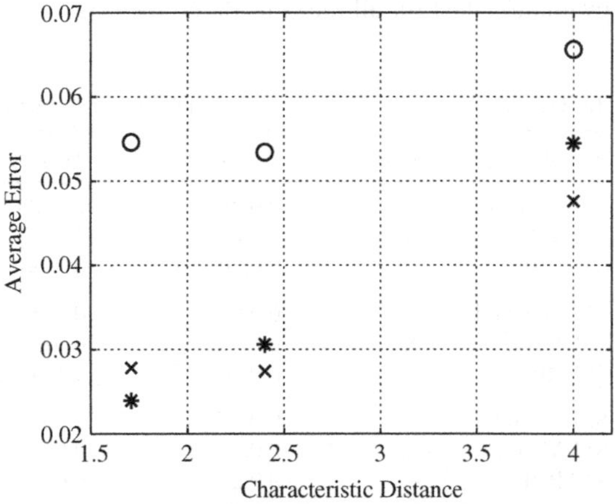

Fig. 12 Single resonator filter: average modeling error versus characteristic distance γ. Data for SM-Standard ($\circ$), SM-VWC ($\times$), and SM-RBF ($*$)

The results show that both SM-VWC and SM-RBF outperform the standard SM model and have the property of improving the modeling accuracy with decreasing value of the characteristic distance of the base set. As expected [26], the modeling accuracy of the SM-Standard method is almost independent of the density of the base set. Although for the single resonator filter the performance of SM-VWC is almost the same as the performance of SM-RBF, other experiments, e.g., [27],

Table 4 Qualitative comparison of computational cost of evaluating the surrogate model for space mapping modeling techniques

Surrogate model	Main sources of computational cost	Relative evaluation cost
SM-Standard	Coarse model evaluation	Similar to coarse model
SM-VWC	Parameter extraction	Much higher than coarse model
SM-RBF	Coarse model evaluation	Similar to coarse model

show that SM-RBF is superior to SM-VWC for most test cases. Moreover, the computational cost of the SM-RBF model is significantly lower than the cost of the SM-VWC model and almost the same as for the standard SM model. The reason is that the computational cost of SM-RBF and SM Standard is virtually the same as the cost of the coarse model (after their parameters are established). The evaluation cost of the SM-VWC model is much higher because of the fact that each evaluation requires a separate parameter extraction, which typically amounts to hundreds of coarse model evaluations.

4 Conclusions

A review of space mapping optimization and modeling techniques is presented. Application examples with the time-domain simulator MEFiSTo as a fine model evaluation engine are demonstrated.

Acknowledgements The authors thank Dr. W.J.R. Hoefer, Faustus Scientific Corporation, Victoria, BC, for useful discussions and making the MEFiSTo software available.

References

1. Bandler J.W., Biernacki R.M., Chen S.H., Grobelny P.A., Hemmers R.H. (1994) Space mapping technique for electromagnetic optimization. IEEE Trans. Microwave Theory Tech. 42: 536–544
2. Bandler J.W., Cheng Q.S., Nikolova N.K., Ismail M.A. (2004) Implicit space mapping optimization exploiting preassigned parameters. IEEE Trans. Microwave Theory Tech. 52: 378–385
3. Bandler J.W., Cheng Q.S., Gebre-Mariam D.H., Madsen K., Pedersen F., Søndergaard J. (2003) EM-based surrogate modeling and design exploiting implicit, frequency and output space mappings. IEEE MTT-S IMS Digest. 1003–1006
4. Bandler J.W., Cheng Q.S., Dakroury S.A., Mohamed A.S., Bakr M.H., Madsen K., Søndergaard J. (2004) Space mapping: the state of the art. IEEE Trans. Microwave Theory Tech. 52: 337–361
5. Echeverria D., Hemker P.W. (2005) Space mapping and defect correction. CMAM The International Mathematical J. Comput. Methods Appl. Math. 5: 107–136
6. Alexandrov N.M., Lewis R.M. (2001) An overview of first-order model management for engineering optimization. Optimization Eng. 2: 413–430
7. Booker A.J., Dennis J.E., Frank P.D., Serafini D.B., Torczon V., Trosset M.W. (1999) A rigorous framework for optimization of expensive functions by surrogates. Structural Optimization. 17: 1–13
8. Dennis J.E., Torczon V. (1997) Managing approximation models in optimization. Multidisciplinary Design Optimization, N.M. Alexandrov and M.Y. Hussaini, eds., SIAM: Philadelphia, USA, 330–374
9. Leary S.J., Bhaskar A., Keane A.J. (2003) A knowledge-based approach to response surface modeling in multifidelity optimization. Global Optimization. 26: 297–319

10. Gano S.E., Renaud J.E., Sanders B. (2004) Variable fidelity optimization using a kriging based scaling function. Proc. 10th AIAA/ISSMO Multidisciplinary Analysis and Optimization Conf., Albany, New York
11. Simpson T.W., Peplinski J., Koch P.N., Allen J.K. (2001) Metamodels for computer-based engineering design: survey and recommendations. Engineering with Computers. 17: 129–150
12. Queipo N.V., Haftka R.T., Shyy W., Goel T., Vaidynathan R., Tucker P.K. (2005) Surrogate based analysis and optimization. Progress in Aerospace Sciences. 41: 1–28
13. Ismail M.A., Smith D., Panariello A., Wang Y., Yu M. (2004) EM-based design of large-scale dielectric-resonator filters and multiplexers by space mapping. IEEE Trans. Microwave Theory Tech. 52: 386–392
14. Wu K.L., Zhao Y.J., Wang J., Cheng M.K.K. (2004) An effective dynamic coarse model for optimization design of LTCC RF circuits with aggressive space mapping. IEEE Trans. Microwave Theory Tech. 52: 393–402
15. Amari S., LeDrew C., Menzel W. (2006) Space-mapping optimization of planar coupled-resonator microwave filters. IEEE Trans. Microwave Theory Tech. 54: 2153–2159
16. Dorica M., Giannacopoulos D.D. (2006) Response surface space mapping for electromagnetic optimization. IEEE Trans. Magn. 42: 1123–1126
17. Leary S.J., Bhaskar A., Keane A.J. (2001) A constraint mapping approach to the structural optimization of an expensive model using surrogates. Optimization Eng. 2: 385–398
18. Redhe M., Nilsson L. (2002) Using space mapping and surrogate models to optimize vehicle crashworthiness design. 9th AIAA/ISSMO Multidisciplinary Analysis and Optimization Symp., Atlanta, GA, Sept. 2002, Paper AIAA-2002-5536
19. Choi H.S., Kim D.H., Park I.H., Hahn S.Y. (2001) A new design technique of magnetic systems using space mapping algorithm. IEEE Trans. Magn. 37: 3627–3630
20. Koziel S., Bandler J.W., Madsen K. (2006) A space mapping framework for engineering optimization: theory and implementation. IEEE Trans. Microwave Theory Tech. 54: 3721–3730
21. Bandler J.W., Georgieva N., Ismail M.A., Rayas-Sánchez J.E., Zhang Q.J. (2001) A generalized space mapping tableau approach to device modeling. IEEE Trans. Microwave Theory Tech. 49: 67–79
22. Koziel S., Bandler J.W., Mohamed A.S., Madsen K. (2005) Enhanced surrogate models for statistical design exploiting space mapping technology. IEEE MTT S IMS Digest, Long Beach, CA: 1609–1612
23. Cheng Q.S., Koziel S., Bandler J.W. (2006) Simplified space mapping approach to enhancement of microwave device models. Int. J. RF and Microwave Computer-Aided Eng. 16: 518–535
24. Rautio J.C. (2004) A space mapped model of thick, tightly coupled conductors for planar electromagnetic analysis. IEEE Microwave Magazine. 5: 62–72
25. Koziel S., Bandler J.W. (2006) Space-mapping-based modeling utilizing parameter extraction with variable weight coefficients and a data base. IEEE MTT-S Int. Microwave Symp. Dig., San Francisco, CA, 1763–1766
26. Koziel S., Bandler J.W., Madsen K. (2006) Theoretical justification of space-mapping-based modeling utilizing a data base and on-demand parameter extraction. IEEE Trans. Microwave Theory Tech. 54: 4316–4322
27. Koziel S., Bandler J.W. (2007) Microwave device modeling using space-mapping and radial basis functions. IEEE MTT-S Int. Microwave Symp. Dig., Honolulu, Hawaii. 799–802
28. Zhang L., Xu J.J., Yagoub M., Ding R.T., Zhang Q.J. (2003) Neuro-space mapping technique for nonlinear device modeling and large signal simulation. IEEE MTT-S Int. Microwave Symp. Dig., Philadelphia, PA, 173–176
29. Devabhaktuni V.K., Chattaraj B., Yagoub M.C.E., Zhang Q.J. (2003) Advanced microwave modeling framework exploiting automatic model generation, knowledge neural networks, and space mapping. IEEE Trans. Microwave Theory Tech. 51: 1822–1833
30. Rayas-Sánchez J.E. (2004) EM-based optimization of microwave circuits using artificial neural networks: the state-of-the-art. IEEE Trans. Microwave Theory Tech. 52: 420–435

31. Rayas-Sánchez J.E., Lara-Rojo F., Martinez-Guerrero E. (2005) A linear inverse space-mapping (LISM) algorithm to design linear and nonlinear RF and microwave circuits. IEEE Trans. Microwave Theory Tech. 53: 960–968
32. Zhang L., Xu J., Yagoub M.C.E., Ding R., Zhang Q.J. (2005) Efficient analytical formulation and sensitivity analysis of neuro-space mapping for nonlinear microwave device modeling. IEEE Trans. Microwave Theory Tech. 53: 2752–2767
33. Koziel S., Bandler J.W., Madsen K. (2005) Towards a rigorous formulation of the space mapping technique for engineering design. Proc. Int. Symp. Circuits, Syst., ISCAS, Kobe, Japan, 5605–5608
34. Madsen K., Søndergaard J. (2004) Convergence of hybrid space mapping algorithms. Optimization Eng. 5: 145–156
35. MEFiSTo-3D Pro, version 4.0, Faustus Scientific Corporation, 1256 Beach Drive, Victoria, BC, V8S 2N3, Canada, 2006
36. Conn A.R., Gould N.I.M., Toint P.L. (2000) Trust Region Methods, MPS-SIAM Series on Optimization
37. Koziel S., Bandler J.W., Madsen K. (2006) Space-mapping based interpolation for engineering optimization. IEEE Trans. Microwave Theory and Tech. 54: 2410–2421
38. Bakr M.H., Bandler J.W., Georgieva N., Madsen K. (1999) A hybrid aggressive space mapping algorithm for EM optimization. IEEE Trans. Microwave Theory Tech. 47: 2440–2449
39. Matlab, ver. 7.1, The MathWorks Inc., Natick, MA, 2005
40. Powell M.J.D. (1987) Radial basis functions for multivariate interpolation: a review. In Algorithms for Approximation, J.C. Mason and M.G. Cox, Eds., Clarendon Press, Oxford
41. Buhmann M.D., Ablowitz M.J. (2003) Radial Basis Functions: Theory and Implementations. Cambridge University
42. Bakr M.H., So P.P.M., Hoefer W.J.R. (2002) The generation of optimal microwave topologies using time-domain field synthesis. IEEE Trans. Microwave Theory Tech. 50: 2537–2544

Index

The manufacturer's authorised representative in the EU is Springer
Nature Customer Service Centre GmbH, Europaplatz 3, 69115 Heidelberg,
Germany. If you have any concerns regarding our products, please
contact ProductSafety@springernature.com

Printed and bound by CPI Group (UK) Ltd, Croydon, CR0 4YY
30/09/2026
02225599-0019